Tribute to Emil Wolf

Science and Engineering Legacy of Physical Optics

Tribute to Emil Wolf

Science and Engineering Legacy of Physical Optics

Tomasz P. Jannson

Editor

SPIE PRESS

Bellingham, Washington USA

Library of Congress Cataloging-in-Publication Data

Tribute to Emil Wolf: science and engineering legacy of physical optics / [edited by] T.P. Jannson.
   p. cm.
   Includes bibliographical references and index.
   ISBN 0-8194-5441-9
   1. Optics. 2. Coherence (Optics). 3. Wolf, Emil—Contributions in optics. I. Wolf, Emil.
   II. Jannson, Tomasz. III. Title.

QC355.3.T75 2004
535—dc22   2004018387

Published by

SPIE—The International Society for Optical Engineering
P.O. Box 10
Bellingham, Washington 98227-0010 USA
Phone: +1 360 676 3290
Fax: +1 360 647 1445
Email: spie@spie.org
Web: http://spie.org

SPIE Press gratefully acknowledges the generous contributions of the Army Research Lab, Army
Research Office, Air Force Research Lab, Duke University, Physical Optics Corporation,
Waveband Corporation, and Tomasz and Joanna Jannson for the publishing of this book.

CONTENTS

Chapter 19. Backward Thinking: Holography and the Inverse Problem / 401
H. John Caulfield and Michael A. Fiddy

Chapter 20. Several Controversial Topics in Contemporary Optics: Dispersive Pulse Dynamics and the Question of Superluminal Pulse Velocities / 421
Kurt E. Oughstun

LIST OF CONTRIBUTORS

Girish S. Agarwal
Physical Research Lab
Amedabad, India

Henk F. Arnoldus
Mississippi State University
Mississippi State, MS USA

Christian Brosseau
Univ. de Bretagne Occidentale
Brest Cedex, France

P. Scott Carney
University of Illinois
Urbana, IL USA

H. John Caulfield
Alabama A&M Research
Institute
Normal, AL USA

Mikael Ciftan
U.S. Army Research Office
Research Triangle Park, NC
USA

Anthony J. Devaney
Northeastern University
Boston, MA USA

Aristide C. Dogariu
Univ of Central Florida
Orlando, FL USA

Michael Fiddy
Univ of North Carolina
Charlotte, NC USA

David G. Fischer
NASA Glenn Research Center
Cleveland, OH USA

John T. Foley
Mississippi State University
Mississippi State, MS USA

Ari T. Friberg
Royal Institute of Technology
Kista, Sweden

Asher A. Friesem
Weizmann Institute of Science
Rehovot, Israel

Zu-Han Gu
Surface Optics Corporation
San Diego, CA USA

Pengyi Guo
Northeastern University
Boston, MA USA

Tomasz P. Jannson
Physical Optics Corporation
Torrance, California USA

Norbert Lauinger
Corrysy-Datron GmbH
Wetzlar, Germany

Tamara A. Leskova
University of California
Irvine, CA USA

Renat R. Letfullin
Mississippi State University
Mississippi State, MS USA

Vladilen S. Letokhov
Institute of Spectroscopy
Troitsk, Russia

Robert G. Littlejohn
University of California
Berkeley, CA USA

Adolf W. Lohmann
Nürnberg University,
Erlangen, Germany

Vladimir A. Manasson
WaveBand Corp
Torrance, CA USA

Alexei A. Maradudin
University of California
Irvine, CA USA

Kurt E. Oughstun
University of Vermont
Burlington, VT USA

Avi Pe'er
Weizmann Institute of Science,
Rehovot, Israel

Jan Peřina
Palacký University
Olomouc, Czech Republic

Lev S. Sadovnik
WaveBand Corp
Torrance, CA USA

K.A. Snail
Naval Research Laboratory
Washington, DC USA

Yupin Sun
Light Prescriptions Innovators,
LLC
Irvine, CA USA

Valerian I. Tatarskii
NOAA
Boulder, CO USA

Brian J. Thompson
University of Rochester
Pittsford, NY USA

Jari Turunen
University of Joensuu
Joensuu, Finland

Taco D. Visser
Free University
Amsterdam, The Netherlands

Roland Winston
University of California
Merced, CA USA

Emil Wolf
University of Rochester
Rochester, NY USA

Frank Wyrowski
University of Jena
Jena, Germany

PREFACE

In his classic text: "*Principles of Optics*," written with Nobel laureate Max Born, Emil Wolf laid the foundations of contemporary physical optics. By frequency of citation, that book is one of the three most popular physics books. In the first edition, published in 1959, Emil Wolf described the almost unknown concept of *spatial coherence* before lasers were introduced. He was also the first to document in a book a new concept: Gabor's *holography*. The basic idea of publishing a modern book on physical optics came from Max Born, but the fact that the closely related concepts of spatial coherence and holography appeared so early in textbook form had a formidable impact on science and physical optics engineering. At present we can identify at least 250 companies and corporate divisions in the English-language zone alone (U.S.A., Great Britain, Australia, and Canada), the origins of which are easily traced to modern physical optics in general, and to the book *Principles of Optics* in particular. Moreover, several multibillion dollar industries can also be traced to this legacy, including liquid crystal and LED displays and screens, screens for direct-projection and rear-projection TV, and many other advanced illumination systems, sensors, and nonimaging optical devices.

This SPIE Press book pays tribute to Emil Wolf (see Fig. 1) for his pioneering contributions to the science and engineering of physical optics. His close friend, collaborator, and well-known authority on diffractive optics, Professor Brian Thompson, refers to Alexander Pope's "An Essay on Man" in characterizing Emil Wolf's contributions. With the ever growing impact of Prof. Wolf's fundamental ideas on the nature of light, Pope's famous epitaph for Newton comes to mind:

> Nature and Nature's laws lay hid in Night:
> God said, *Let NEWTON be*! And all was Light.

Even though his position that spatial coherence is critical to physical optics was to some degree opposed by Max Born, it has, over the years, become a very pow-

Figure 1 Emil Wolf and his wife Marlies at the SPIE Annual Meeting, August 2003, San Diego, California.

erful concept in many areas of physical optics, some of which are presented herein. These include diffraction optics, statistical optics, polarization of light, electromagnetic theory of optical coherence, microscopic theory of spatial coherence, physical radiometry (radiance), physical optics modeling of millimeter wave antennas, coherent optical microscopy, color vision, and Wolf's wavelength shift. Professor Jan Peřina reviews optics in the Czech Republic (then Czechloslovakia, where Prof. Wolf was born). Others address coherence-based light scattering, new aspects of the Sommerfeld half-plane problem as well as Young's experiment, comparison between Doppler and Wolf's shifts, phase and information, wave-optical engineering, and holography and the inverse and scattering problems. Also discussed here are controversial topics in contemporary optics, total-internal-reflection tomography, coherence-mode analysis, nano-optics, special problems in coherence, and Newton-Goethe controversy.

All the chapters of this book are presented by major experts in the field (see Fig. 2), many of them closely connected to Emil Wolf's University of Rochester School of Optics. The ideas they express are their own, subject only to peer review. The papers are part science and part memoir, but all are suffused with love and admiration of Emil Wolf, and for his contributions to science and engineering.

Figure 2(a) Some attendees at SPIE AM 100 Conference, Tribute to Emil Wolf, Engineering Legacy of Physical Optics.

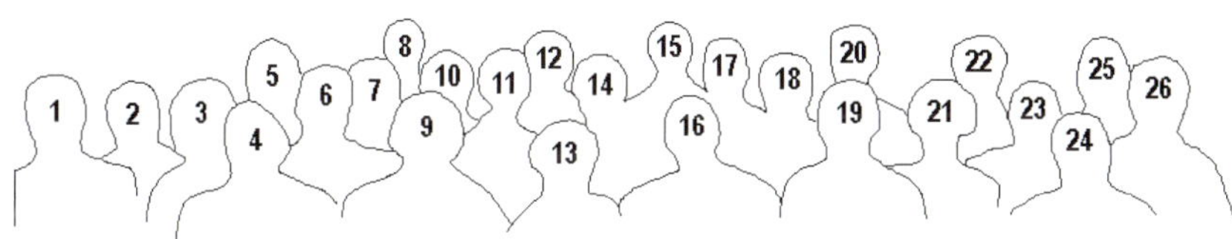

Figure 2(b) Silhouettes of attendees from Fig. 2(a): (1) Frank Wyrowski, Friedrick-Schiller, Univ. of Jena, Germany; (2) Gajendra Savant, Physical Optics Corp., Torrance, CA; (3) Sharon Peet, Physical Optics Corp.; (4) John Foley, Mississippi State Univ.; (5) Ari Friberg, Royal Inst. of Technology, Sweden; (6) Aristide Dogariu, CREOL, Univ. of Florida; (7) Chrysostomos L. Nikias, Univ. of Southern Calif.; (8) Nitin Savant, Physical Optics Corp.; (9) Tomasz Jannson, Physical Optics Corp.; (10) Joseph Kunc, Univ. of Southern Calif.; (11) Kristina M. Johnson, Duke Univ.; (12) Kurt Oughstun, Univ. of Vermont; (13) Marlies Wolf, Emil's wife; (14) James Bilbro, SPIE 2004 President, and NASA Marshall Space Center; (15) David Fischer, NASA Glenn Research Ctr.; (16) Emil Wolf, Univ. of Rochester; (17) Riccardo Borghi, Univ. di Roma Tre, Italy; (18) Christian Brosseau, Univ. de Bretagne Occidentale, France; (19) Taco Visser, Free Univ. Netherlands; (20) Petr Smid, Palacky Univ., The Czech Republic; (21) Anya van der Meulen-Visser, Taco's wife; (22) Pavel Horwath, Palacky Univ., The Czech Republic; (23) Mikael Ciftan, Army Res. Center; (24) G.S. Agarwal, Physical Research Lab., India; (25) Mark Bennahmias, Physical Optics Corp.; (26) Zu-Han Gu, Surface Optics, San Diego, CA.

This book is based on the authors' presentations at SPIE Conference AM100: Tribute to Emil Wolf: Engineering Legacy of Physical Optics, T.P. Jannson, Chair, at the SPIE Annual Meeting in August 2003 in San Diego, California. Most chapters in this book are extended versions of those conference presentations.

Tomasz P. Jannson
James C. Wyant
James W. Bilbro
September, 2004

ACKNOWLEDGMENT

This book is mostly a product of the SPIE Conference: Tribute to Emil Wolf, August 2003, San Diego, California. The idea for this conference came from James W. Bilbro, NASA Marshall Space Flight Center, and 2004 SPIE President; Dennis H. Goldstein, SPIE Fellow, Air Force Research Lab.; James C. Wyant, Director, Optical Science Center/University of Arizona; and H. John Caulfield, SPIE Fellow, Fisk University, and the organizer of the two previous SPIE Conferences—also tributes to pioneers in optics (Adolf W. Lohmann; Yuri N. Denisyuk and Emmett N. Leith).

However, for whole logistics, as well as all organizational issues, we are all in debt to the SPIE and the Conference Organization Committee members who provided, through their parent organizations, a significant financial contribution to support both the conference and this book. The committee members include James W. Bilbro, H. John Caulfield, Mikael Ciftan, Program Manager Army Research Office; Dennis Goldstein, Kristina Johnson, Dean, Pratt School of Engineering, Duke Univ.; Joseph Kunc, Professor, Univ. of Southern California; Joseph Mait, SPIE Fellow, Army Research Lab.; Chrysostomos L. Nikias, Dean, School of Engineering, Univ. of Southern California; John M. Pellegrino, Director, Army Research Lab.; Lev Sadovnik, CEO, WaveBand Corp.; Todd D. Steiner, Program Manager, Air Force Research Lab.; and James C. Wyant.

Tomasz Jannson
September 2004

𝔇𝔲𝔨𝔢 𝔘𝔫𝔦𝔟𝔢𝔯𝔰𝔦𝔱𝔶

𝔈𝔡𝔪𝔲𝔫𝔡 𝔗. 𝔓𝔯𝔞𝔱𝔱, 𝔍𝔯. 𝔖𝔠𝔥𝔬𝔬𝔩 𝔬𝔣 𝔈𝔫𝔤𝔦𝔫𝔢𝔢𝔯𝔦𝔫𝔤

DURHAM, NORTH CAROLINA 27708-0271

OFFICE OF THE DEAN   TELEPHONE (919) 660-5386
BOX 90271   FAX (919) 668-0656

August 2004

Dear Emil,

It was an honor and a pleasure to be part of the Society of Photo-Instrumentation Engineers Annual Meeting Tribute to Emil Wolf: Engineering Legacy of Physical Optics, recognizing your seminal contributions to this field. Your energy, enthusiasm, and passion for optics and life have inspired generations of students, including myself and fellow graduate students at Stanford, University of Colorado, and now Duke University.

I recall the first time we met. It was in Cuernavaca, Mexico where many of the luminaries in the field gathered in February of 1981 to enjoy lively discussions on all aspects of physical optics, as well as the good weather. There were only a handful of students at the meeting, and we were a bit intimidated by the stellar participants. Always a teacher, coach and mentor, you and Marlies took us under your wings and spent most of the week with us. That serendipitous meeting started a lifelong friendship that means the world to me.

Our tradition of swimming at 7:00 a.m. before optics meetings started in the freezing waters in Mexico. We continued to swim in Tucson at the 1982 OSA meeting, where you and Marlies met my mother, establishing another friendship

Marlies Wolf, Kristina Johnson, and Emil Wolf at SPIE's Annual Meeting 2003. (Courtesy of Valerian Tatarskii, copyright 2003.)

xxiii

Kristina Johnson and Emil Wolf, 2003.

that lasted until her death in 1999. We swam in New Orleans (1983), Rochester (with Bailey's Irish Cream, 1985), and in San Diego (2003).

You showed me how to build community within the academy. When I became an assistant professor of electrical engineering at the University of Colorado, you invited me to give a talk at Rochester on May 2, 1990. You and Marlies hosted a dinner party for me and other participants of the Cuernavaca meeting, including Profs. Nicholas George and Brian Thompson. We had a fabulous evening of interesting conversation, good food, and great friends. And, of course, we swam.

Emil, you taught me about mutual coherence theory, the "Wolf Shift," and how to be a leader in the field. And, like my advisor, Joseph Goodman at Stanford, you taught me that great men of optics are simply great men.

I look forward to your visit to Duke University this fall and to your lecture in our Fitzpatrick Center distinguished speaker series. We will again enjoy eating and thinking. Bring your swimsuit.

With my love, respect, and admiration,

Kristina M. Johnson
Professor and Dean

Tribute to Emil Wolf

Science and Engineering
Legacy of Physical Optics

CHAPTER 1

GUIDE, PHILOSOPHER, AND FRIEND

Brian J. Thompson

Emil Wolf, "Thou wert [and art] my guide, philosopher, and friend."[*]

1.1 Introduction

Having good mentors during the graduate student years and in the early-career professional experience provides a major stimulus to a productive life as a scholar and teacher. Our international university doctoral and postdoctoral programs are at their best when they fully integrate research, educational, and teaching opportunities so that intellectual development and scholarship go hand-in-hand with clear insight, exposition, and formal teaching.

In my own case I was blessed with three major mentors. The first was Prof. Henry Lipson, Chairman of the Physics Department in the Faculty of Technology of the University of Manchester [UMIST—University of Manchester Institute of Science and Technology, as it is now called]. I had been an undergraduate in Lipson's department, graduating in 1955. Before I entered this program I had served for two years in the British Army in the Royal Electrical and Mechanical Engineers, where I had the opportunity to learn as a technician about radar systems and predictors (i.e., early single-function electronic computers).

Immediately upon graduation with my bachelor's degree, I entered the doctoral program at UMIST and continued to have Henry Lipson as a mentor and acquired my second mentor—Dr. Charles Taylor, who was my thesis advisor. My third mentor was Dr. Emil Wolf, who came into my life in late 1955.

---

[*] Alexander Pope, "An Essay on Man," Epistle iv 1.390

1.2 Manchester 1955–1959

Both Lipson and Taylor were x-ray crystallographers [1–3] with a deep interest in developing optical analog to x-ray diffraction and the analog computing opportunity that this optical approach provided. This matched my own interest in optical science and its applications. Thus, I had access to the optical diffractometer, the relatively new device that had been developed starting in 1949 [4–7].

The optical system of the diffractometer is shown in Fig. 1. The source was a high-pressure mercury arc operated with one of the hot-spots in the arc imaged onto a pinhole at S_1. The light emitted by this effective secondary incoherent

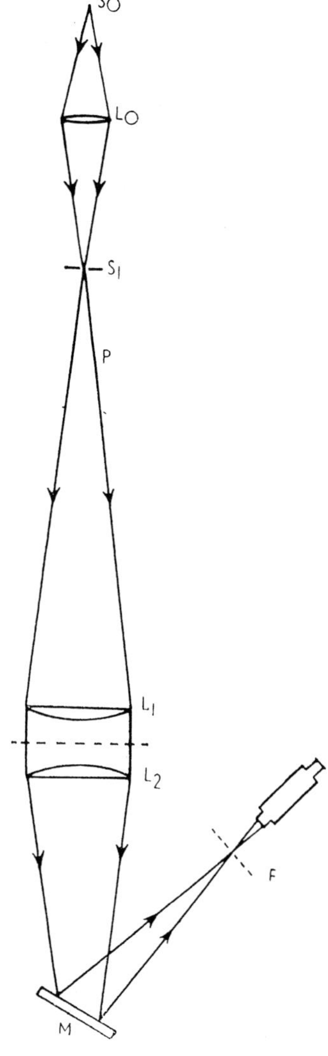

Figure 1 Schematic diagram of the optical diffractometer.

source was collimated by lens L_1, a 5-foot focal length telescope lens. The diffracting object was placed between lens L_1 and L_2 (a matching 5-foot focal length telescope objective lens) and the diffraction pattern was formed in the focal plane of L_2, where it was observed, recorded, and measured. Figure 2 shows an old photograph of the lower half of the diffractometer that was mounted on a vertical I-beam—I-beams and H-beams became quite popular later on for long optical benches.

My initial tasks were to fully characterize this instrument and its performance and solve a number of specific problems of optical and mechanical alignment, focusing, resolution, and coherence control. Clearly it was very important to think about this basically simple system in terms of its expected performance as a generator of diffraction patterns of planar two-dimensional binary objects (i.e., a diffracting mask containing circular holes laid out in various specific geometries). For example, Fig. 3 shows in (a) a representation of a projection of a molecule of hexamethylbenzene in which each atom in the molecule is represented by a circular hole; (b) shows the diffraction pattern of (a). It is relatively easy to recognize the symmetry relationships and see the reciprocal diffraction structure of the original benzene ring. Quantitative positional information is readily available. Figure 4 shows a much more complicated arrangement of holes representing part of the projected structure of deoxyribose nucleic acid and its corresponding diffraction pattern.

Making a quantitative analysis of the performance of the optical diffractometer and interpreting the relatively complicated and detailed diffraction patterns required a significant knowledge of the coherence properties of the illumination of the diffracting mask. Thus, I studied the papers of Zernike, van Cittert, H.H. Hopkins, and most particularly the writings of Emil Wolf. I also reread the work of Michelson on the stellar interferometer. I could not believe my good fortune in finding that Dr. Wolf was a resident in Manchester as a Research Fellow in the Theoretical Physics Department at Owens College of the University of Manchester, having arrived there the previous year (1954). Thus, my third major mentor, Emil Wolf, became a significant influence in my life and my work starting in 1955. Our relationship and friendship have continued over the intervening years and I am pleased to say exists today (I should also note that my professional and social relationship with Henry Lipson and Charles Taylor continued throughout their lives).

Emil Wolf was born in Prague, Czechoslovakia, almost exactly 10 years before I was born in Glossop, England. I believe he came to England as a teenager and then entered Bristol University in 1941, receiving his B.Sc. degree in 1945 and his Ph.D. in 1948; these degrees were in mathematics and physics. He then went to Cambridge University as a postdoctoral fellow (1948–1951), next moved to Edinburgh University and spent a period of several years as a University lecturer and

Figure 2 Photograph of the lower half of the diffractometer.

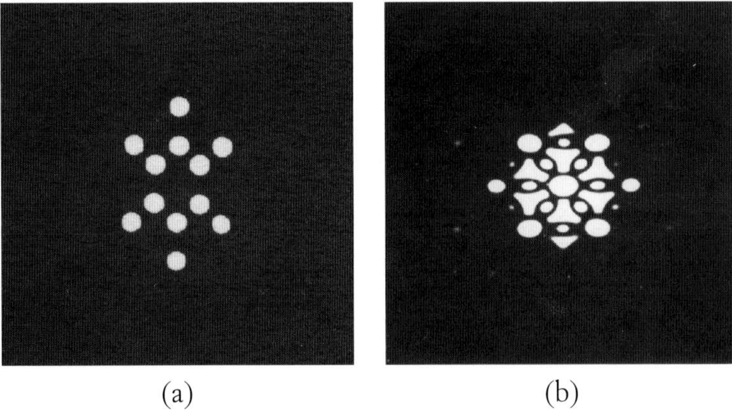

(a)   (b)

Figure 3 (a) A representation of a projection of a molecule of hexamethylbenzene;
(b) the diffraction pattern of (a) (from [8]).

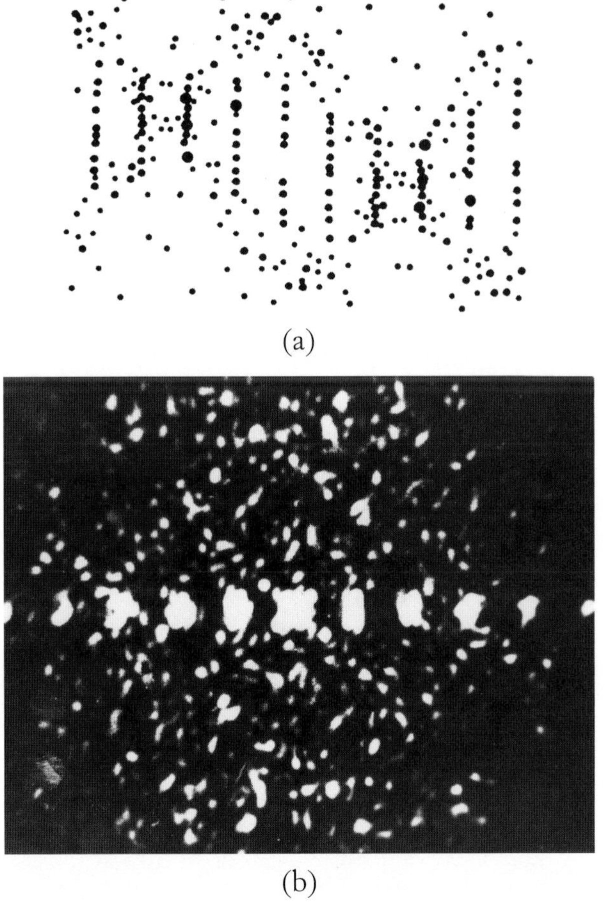

(a)

(b)

Figure 4 (a) A representation of part of the projected structure of deoxyribose nucleic
acid; and (b) its corresponding diffraction pattern (from [8]).

assistant to Prof. Max Born, and hence to Manchester as a Research Fellow from 1954 to 1958. I should also note that he received a D.Sc. degree from Edinburgh in 1955. (Coincidentally, my wife began her undergraduate studies at St. Matthias College of Bristol University the very year that Emil left Bristol.)

1.2.1 Two-beam interference

When I met Emil, he was already a very distinguished scholar with a portfolio of some 25 major publications and he was working hard on the book *Principles of Optics*. Our major collaboration was our work on two-beam interference with partially coherent light. I still have in my own archives some of the original experimental results. One of these sets of results obtained with the optical diffractometer is shown in Fig. 5. In this particular set of results the diameter of the incoherent

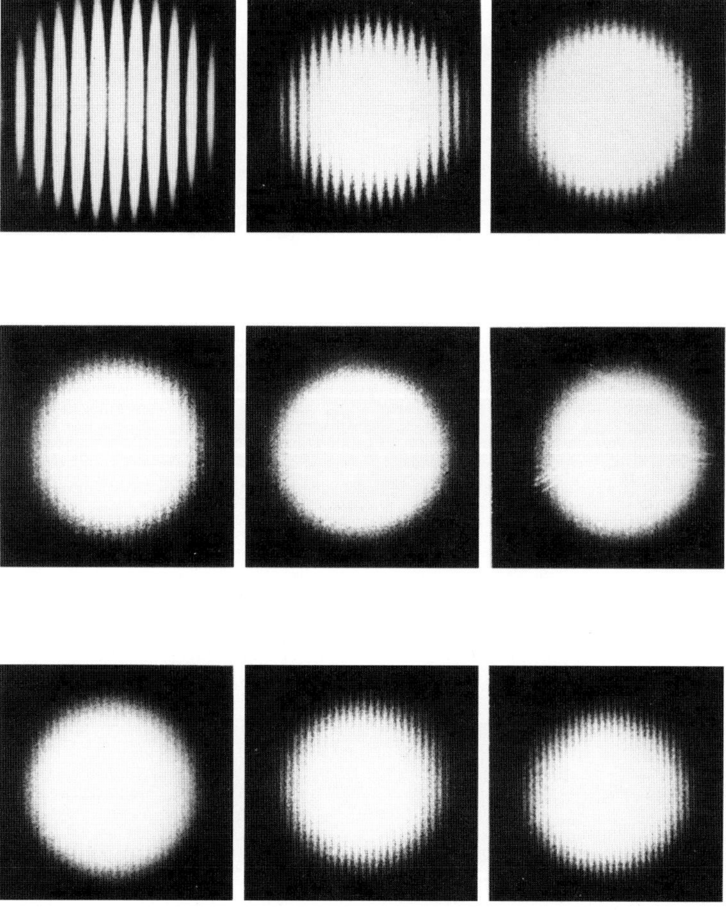

Figure 5 Two-beam interference figures with partially coherent light. Parameters are listed in the text (from [8]).

source was 49 μm and the wavelength was 5941 Å (mercury green). Two points in the optical field were selected by two small circular apertures 0.14 cm in diameter, with varying separations from 1.0 cm to 5 cm covering two zeros of the degree of spatial coherence. A related set of results were the ones published in our joint paper entitled "Two-beam interference with partially coherent light" that was submitted in December of 1956 and published in October of 1957 (see [9]). In this paper we correlated the results with the theoretical predictions. Figure 6 shows two of these historical results that were reproduced as part of six illustrations in *Principles of Optics* published in 1959. We are pleased to note that they have been used in many texts, review papers, and articles. These examples in Fig. 6 have the important parameters listed; note the phase change in the spatial coherence in the right-hand result, which provides a central minimum in the fringe field, as opposed to a central maximum. In a follow-up study [10], I left the fringe spacing constant

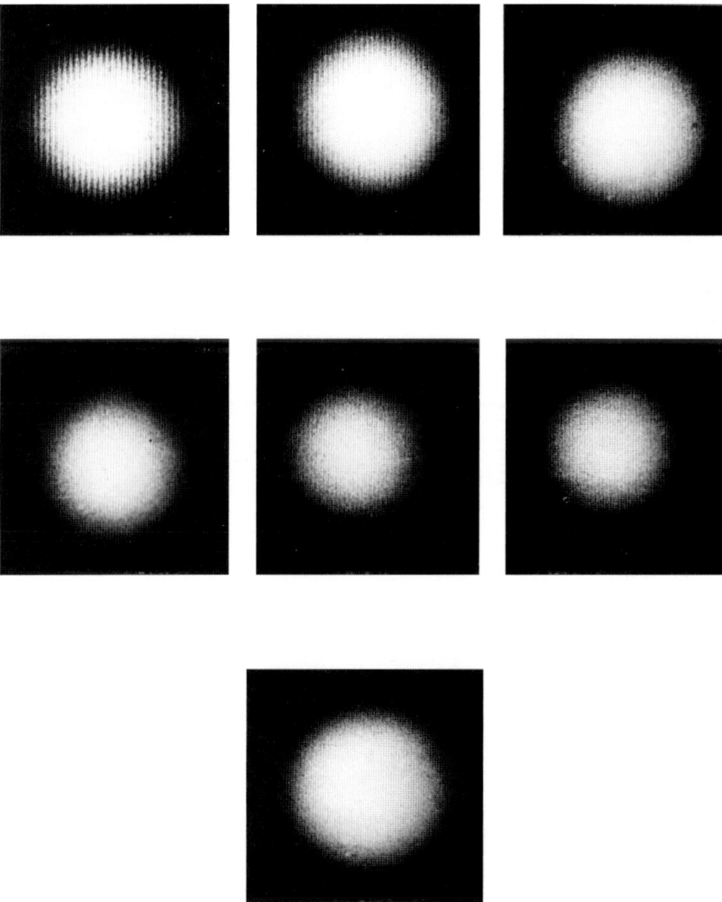

Figure 5 (Continued).

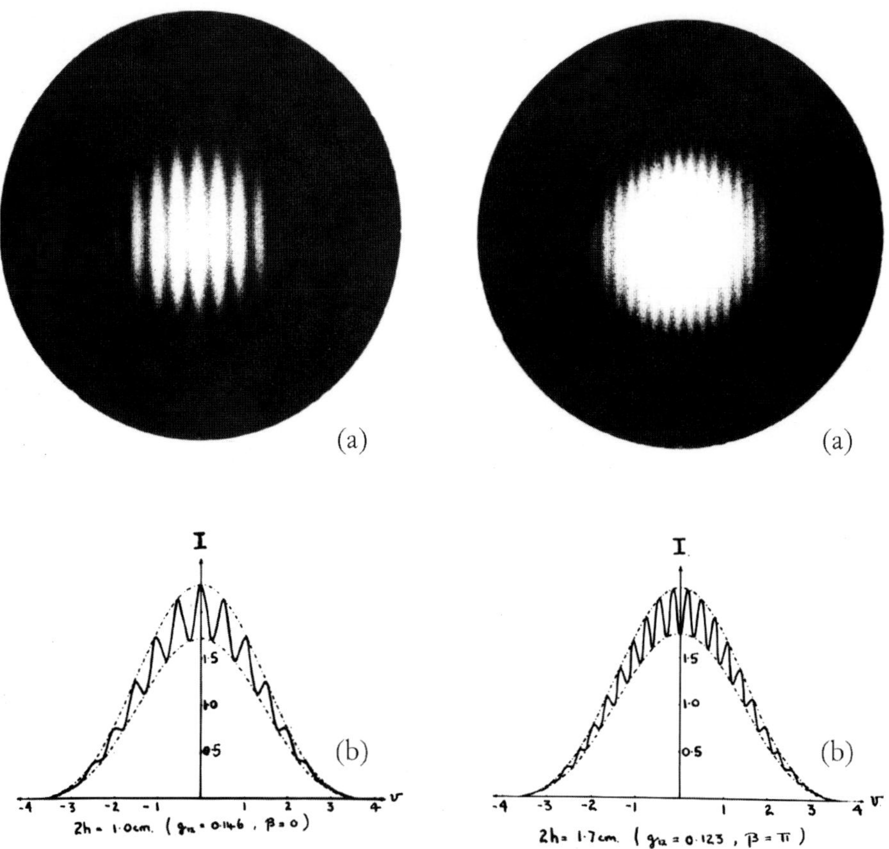

Figure 6 Two specific examples from the published set of six illustrations; example 2h is the separation of the two pinholes, g_{12} is the modulus of the degree of spatial coherence, and β is the phase of the degree of coherence [8,9].

and changed the spatial coherence by changing the size of the incoherent source, thus producing a very nice quantitative illustration of the phase change (Fig. 7).

There is an interesting anecdote associated with our joint paper. It appeared that there was a discrepancy between the theoretical results and the experimental results. Of course, my assumption was that there was a mistake in the calculation, but Emil's take was that there was an error in the experiment! I went back and checked all the parameters and indeed found a small error in the measurement of the diameter of one of the pinholes (an important lesson to make sure of all measurements at all times).

1.2.2 Coherence control

During this same time period a great deal of progress was made on dealing with other issues associated with the diffractometer, including alignment, practical rules

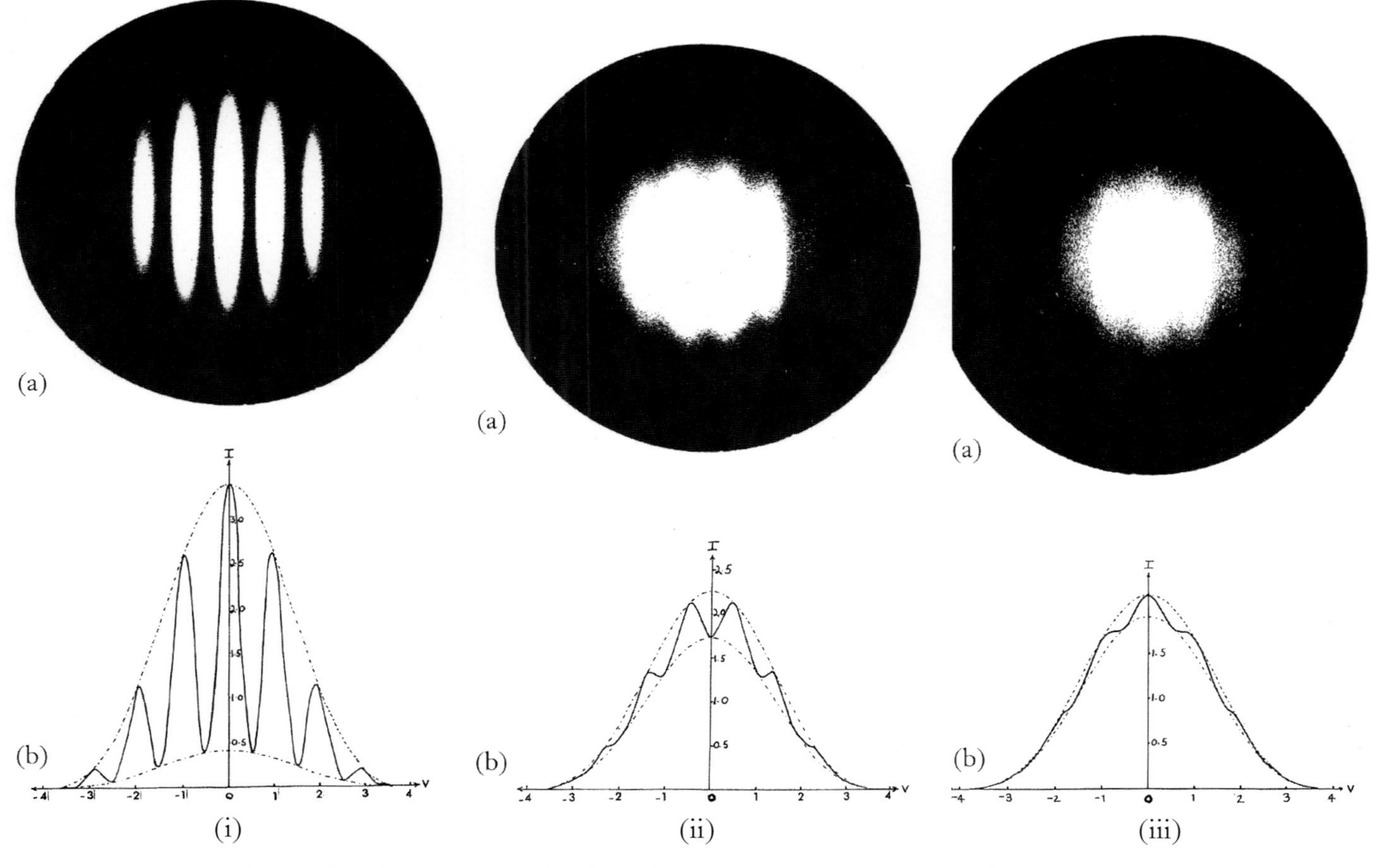

Figure 7 An illustration of the phase change in the degree of spatial coherence [8,10].

for controlling the illumination and its spatial coherence tailored to the particular problem being studied, and improvements in the recording techniques [11].

A significant number of other related techniques were developed using control of the spatial coherence. One of these was to change the secondary source to be spatially coherent itself by stopping down the aperture of the lens that imaged the arc onto the pinhole so that the pinhole was smaller than the Airy disc produced in that plane by the aperture of the imaging lens. The resultant illumination of the diffraction plane was spatially coherent but with an amplitude taper. Another control technique is illustrated in Fig. 8. We wished to achieve a bright display of a diffraction pattern for live display (here we use the example of the projection of the molecule of hexamethylbenzene). A mask was made of an array of these representations, shown in (a). If the illumination is caused to be spatially coherent over an individual molecule, but is effectively incoherent from molecule to molecule, then no interference terms are produced and a bright diffraction pattern of the single molecule (b) appears some 56 times brighter than that produced by 2a single molecule alone, as illustrated in (c).

1.2.3 The Brussels Universal and International Exhibition, 1958

Prof. H. Lipson, Dr. C.A. Taylor, and I were invited by Sir W. Lawrence Bragg, F.R.S., Director, Davy Faraday Research Laboratory, to contribute an explanatory exhibit for the International Science Hall at the Brussels Exhibition. The overall theme was the sequence: atom, crystal, molecule, and cell. Our component was an *Introduction to Diffraction* [Catalogue and Handbook for International Science Hall (1958), Section 0-15], the task was to devise and implement a display suitable for public viewing to illustrate diffraction of x rays by crystals. It was clear that an optical analog display was the way to go. Figure 9 shows a mock-up of part of our display consisting of two elements; first (on the right-hand side), large-scale enlargements of optical diffraction patterns of various structures representing projections of real molecular crystal structures; and second (on the left-hand side), a direct viewing of a quite bright aerial image of some of these patterns when looking through a "window." The "window" was in fact a large two-dimensional array of the structure whose diffraction pattern was to be viewed. The basic scheme was similar to the earlier discussion of Fig. 8, i.e., coherence control with the addition of some irregularity of position of each "molecule." We devised a step-and-repeat camera to produce these windows on film with clear transparent circular "holes" on a dark background. First a small irregular array of the molecule was produced with each molecule in the same orientation; this unit was then repeated in a more-or-less regular two-dimensional array over a sheet of film that became the window. Behind the window a bright, but small source of light was reflected by a polished stainless steel ball to produce a demagnified image of the original source. The observer looked through the window, and in the plane of the source an image of a relatively

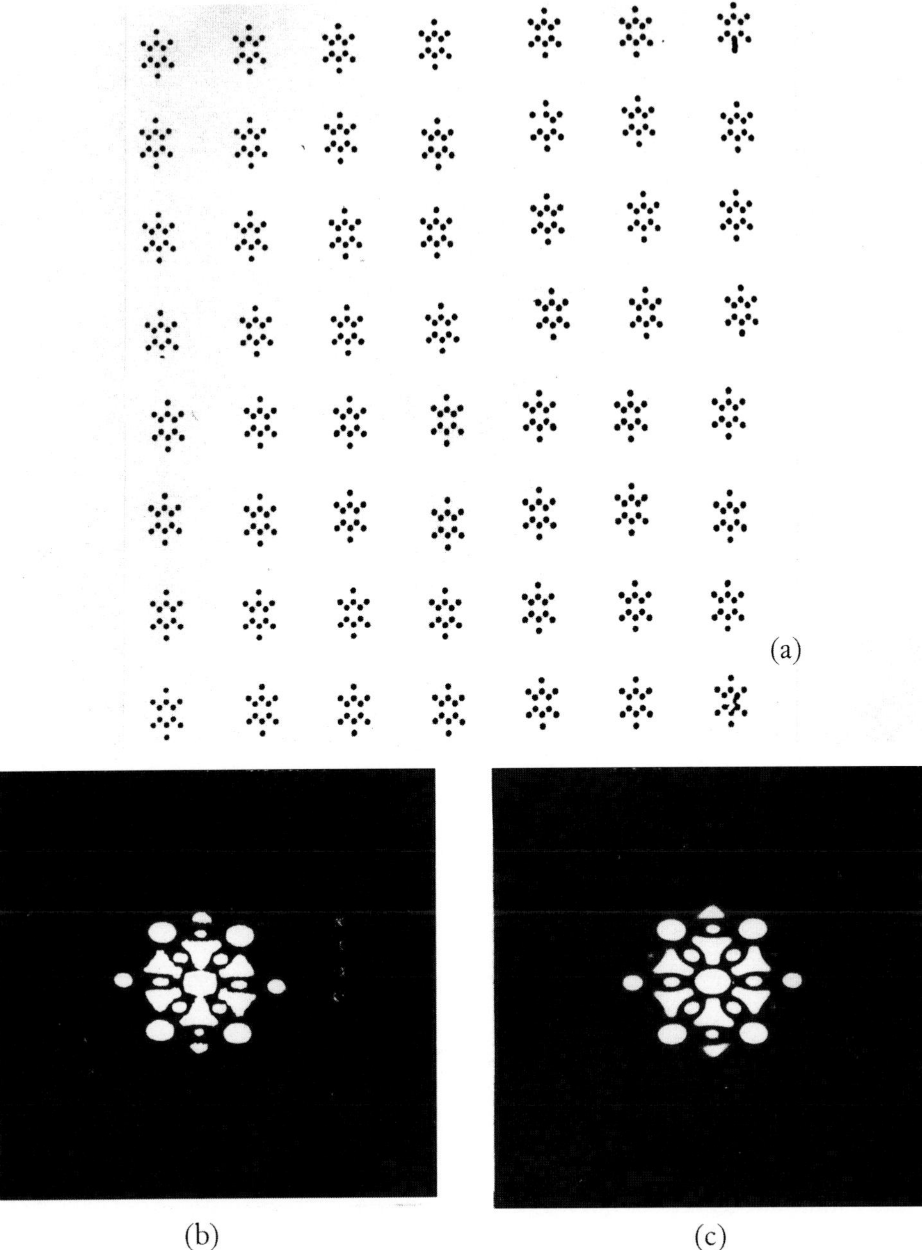

Figure 8 Control of the spatial coherence: (a) a diffracting mask with a two-dimensional array of a representation of the projection of a molecule of hexamethylbenzene; (b) the composite diffraction pattern of this array with the coherence controlled to be coherent over each molecule but approximately incoherent between molecules; (c) for comparison, the pattern of a single molecule alone. Note: (b) would be visually 56 times brighter than (c) [8].

Figure 9 "Mock-up" of the display for the International Science Hall at the Brussels Exhibition (see text for detail).

bright diffraction pattern was seen. The final product looked very professional, and Charles Taylor and I delivered it to Sir Lawrence Bragg at The Royal Institution in London for integration and shipment to Brussels. I next saw it on site in the International Science Hall. Yes, that is a young Brian Thompson looking through one of the "windows" at our collective handiwork in Fig. 9!

At it turned out, Emil Wolf and his doctoral student George Parrent (of Beran and Parrent fame, 1964 [12]) were going to Liège, Belgium, to the International Symposium on Radio Wave Propagation and then on to Brussels. Emil's paper was entitled "Some aspects of rigorous scalar treatment of electromagnetic waves" [13]. So we all went together. Figure 10 shows early morning photographs of the Wolfs and the Parrents and the author on board the cross-channel ferry. Note Emil Wolf's ever constant pipe!

1.2.4 Books

Fourier Transforms and X-ray Diffraction, *by H. Lipson and C.A. Taylor*. During the same period of time, Lipson and Taylor published the small but elegant book on Fourier transforms and x-ray diffraction [1]. I was very pleased to contribute an appendix showing the optical transforms of a hypothetical molecule, two centrosymmetrically related units (equivalent to the real part of the transform of the

(a)

(b)

Figure 10 "On the way to Brussels." (a) Left to right: The Parrents and the Wolfs on board the early morning cross-channel ferry; (b) Brian Thompson photographed by George Parrent.

single molecule transform, two adjacent unit cells, four adjacent unit cells, and finally the optical transform of many unit cells. Real and imaginary parts as well as phase and modulus of the transforms were calculated by Keith A. Morley for comparison with the optical transforms. I am pleased to have a dedicated copy of this book in my library.

Principles of Optics, a.k.a. *"Born and Wolf."* It is a great tribute to the authors of this monumental book that it is usually referred to as "Born and Wolf" [14] rather than by its title! Over 40 years since its first publication in 1959, it is still going strong and in its seventh edition. I was very pleased to have an association with this volume: the Thompson and Wolf results discussed above became a two-page spread, and we also contributed a number of other illustrations on diffraction. In addition, I had the opportunity to proofread a number of the chapters. Emil's dedicated efforts and his scholarship are much to be admired—I believe he checked every reference in the original. I am very pleased to have a dedicated copy of the

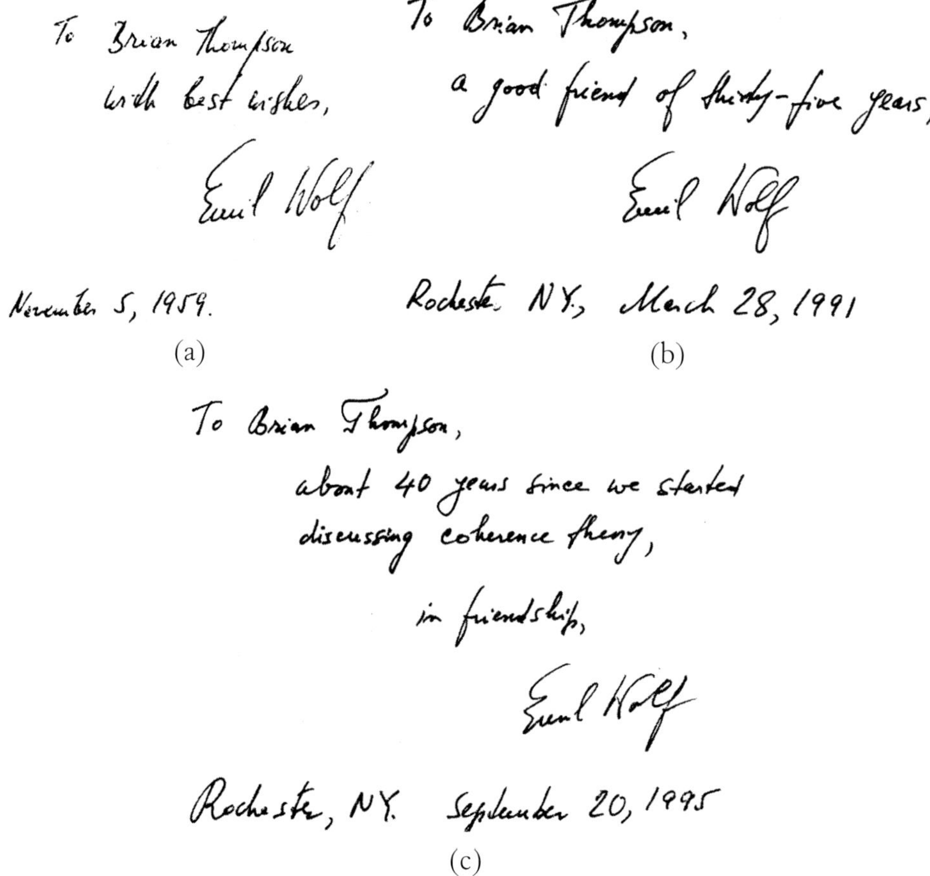

Figure 11 (a) Dedication in the first edition of *Principles of Optics*, 1959, (b) Dedication in the sixth edition, 1991, (c) Dedication in *Optical Coherence and Quantum Optics*, 1995.

first edition of this book along with subsequent editions, as well as other volumes produced by Emil (see Fig. 11).

1.2.5 The three-dimensional structure of two-dimensional diffraction

In my work I needed to find a technique for "focusing" the diffractometer, or what some people call the diffraction-image. More precisely, we needed to be able to locate the Fourier transform plane as accurately as possible. This plane is, of course, the image plane of the illuminating source, but getting the best focus for that image was not generally good enough because of the effective depth-of-focus. Wolf's work on "the light distribution near focus is an error-free diffraction image" [15,16] gave me an idea since the minimums along the axis were more sharply defined than the central maximum. Thus, if we locate the first minimum on either side of the required focal plane, then the plane we need is then halfway between [11]. It works!

I confess I could not resist engaging in a full experimental investigation of the diffraction region for a variety of aperture functions, circular, annular, and rectangular. Calculating the equivalent two-dimensional intensity distribution was a real chore, involving me in many weeks of work with tables of Lommel functions; however, the results were very worthwhile [17]. It was Emil Wolf who helped me get acquainted with Lommel functions. (For circular and annular apertures see [18].)

Figure 12 shows one of these results for a square aperture; (a) shows the photographic record of the two-dimensional intensity distribution in the plane of the second axial zero; (b) is an equal-intensity (isophote) plot for the same example [8,19].

1.2.6 Fourier synthesis—spatial filtering

A final activity during this period was some very early work on optical Fourier synthesis (or what was called spatial filtering and the more general optical image and/or data processing). The optical diffractometer was modified by adding a second lens system to form an image of the original mask by retransforming the diffraction pattern, then limiting the contents of the diffraction pattern that contributed to the image (i.e., spatial frequency filtering). One issue that was evaluated was the effects of so-called series termination errors in x-ray diffraction, i.e., how is the final calculated crystal structure affected by the limitation of the collection of the diffracted x rays over a wide angle. Figure 13 shows an example of these results [2,8].

To conclude this section (and to jump forward in time for a moment relative to the optical analogs to x-ray diffraction), it is worth noting that Lipson and

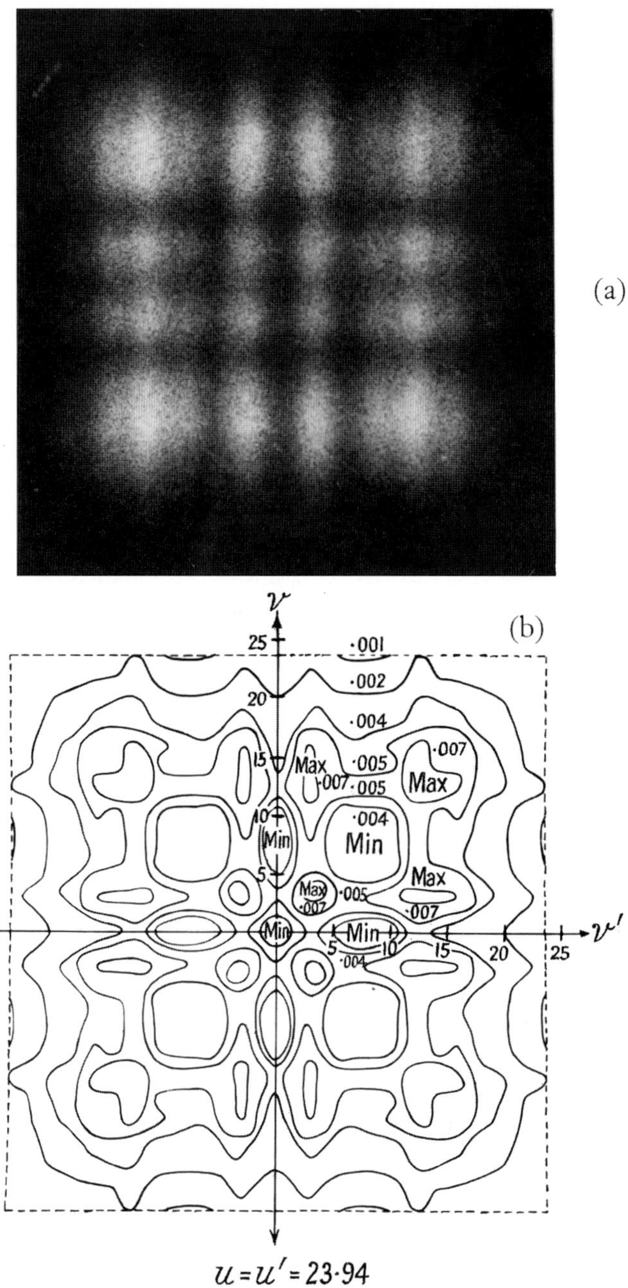

Figure 12 Diffraction pattern near focus for a square aperture at the location of the first axial minimum. (a) Photograph of the two-dimensional intensity distribution in a plane perpendicular to the optical axis; (b) calculated isophote diagram corresponding to (a) [8,19].

Taylor contributed a long article entitled "X-ray crystal-structure determination as a branch of physical optics" to Volume V of *Progress in Optics*, that excellent series started by Emil Wolf in 1961 (and still going strong with Volume 46 appearing this year and many to follow). In return I was asked to contribute two chapters to a book entitled *Optical Transforms*, edited by H. Lipson in 1972. These chapters were titled "Coherence requirements" and "Optical data processing" [20]. Finally, all these reciprocal insights came together in a review article "Optical transforms and coherent processing systems—with insights from crystallography" [21].

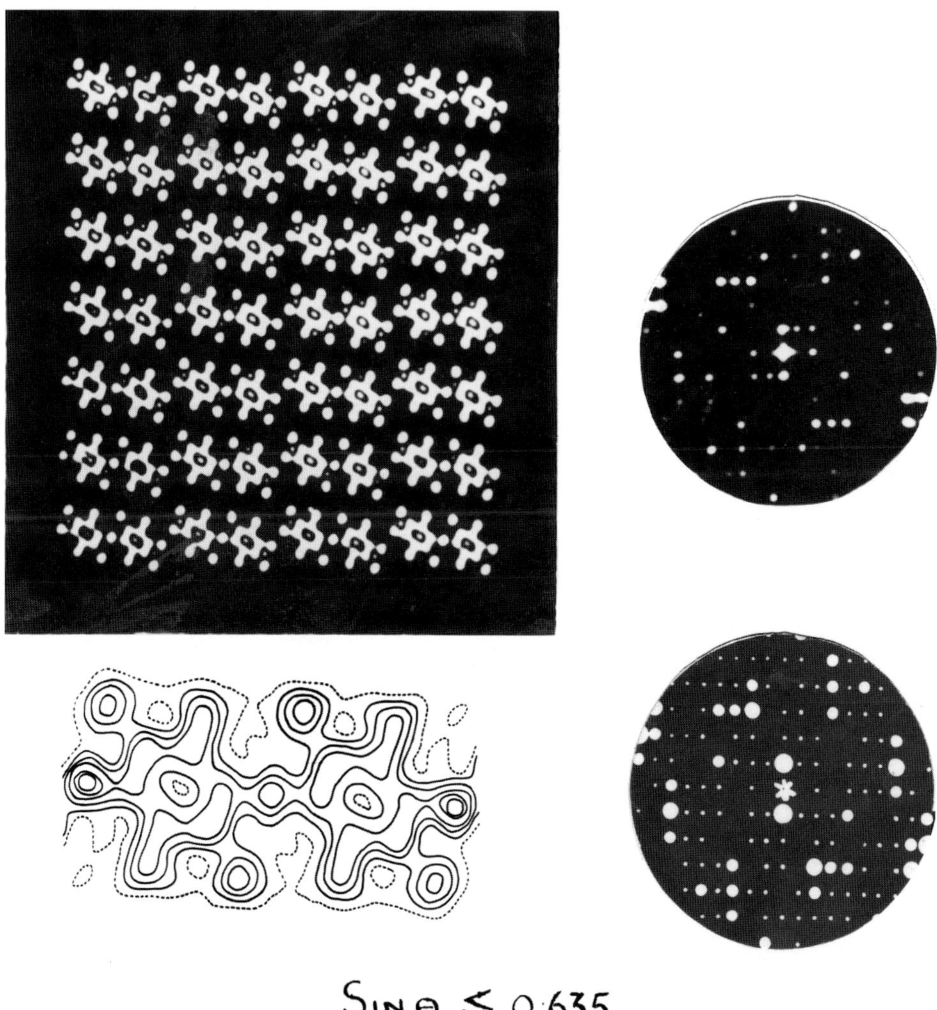

Figure 13 An example of an early spatial filtering experiment to illustrate series termination errors in x-ray diffraction by an optical analogue.

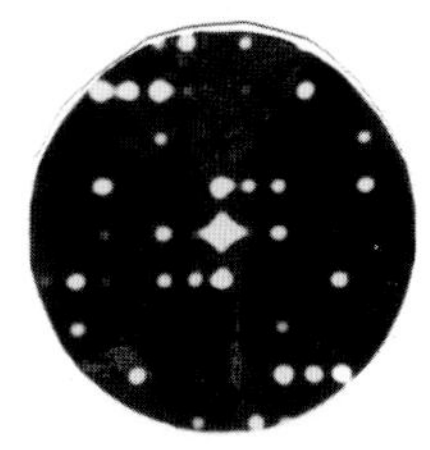

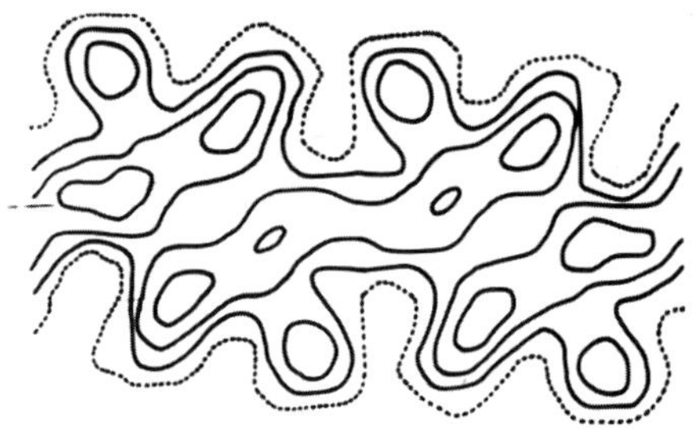

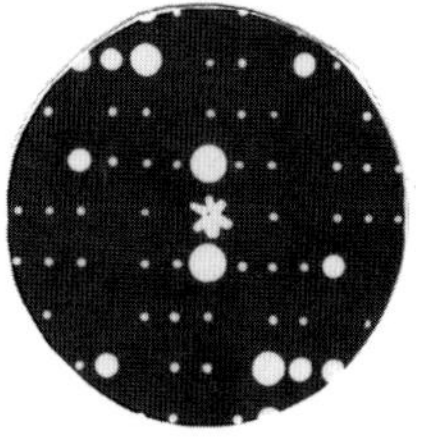

Figure 13 (Continued).

1.3 1958–1968 Various Locations

Emil Wolf left Manchester after spending some time in New York City at the Courant Institute. He arrived in Rochester, New York, in 1959 to join The Institute of Optics at the University of Rochester; and what a distinguished career he has had in Rochester. My own travels took me to Leeds University as Lecturer in Physics in 1959, where I was involved with soft x-ray spectroscopy of metals and alloys. One of my main colleagues was Dr. Colin Curry, who had written some very fine texts, including one that I admired on "Wave optics interference and diffraction" [22]. Then in 1963 I moved from England to the U.S. and into Emil Wolf's sphere of influence. With a little persuasion from Emil and George

Parrent, I joined Technical Operations, Inc. (Tech/Ops), in Burlington, Massachusetts, where George Parrent had started a fine optics research and development group. Physical optics was our main pursuit, and we had many years of exiting and productive activity, including the development of far-field holography and its application to dynamic particle size measurement (initially fog), coherent image formation and its applications, multiple beam interference with partially coherent light, storing color images in black and white film with Fourier readout (Peter Mueller gets most of the credit for that development), and some detailed study of diffraction with partially coherent light. The years 1963–1964 saw the start of a series of articles that George and I prepared for the Journal of SPIE under the general title *Physical Optics Notebook* [23]. Then late in 1969 it became a hardbound book with the 16 articles that had been published between 1963 and 1967. After several reprintings, a new and very much expanded version was prepared and published in 1989 [24]. We had great team at Tech/Ops that was certainly influenced by the time Parrent and I spent with Wolf. Our contacts with Emil were frequent and at various meeting locations around the world. In September 1964, both Emil and I attended the International Commission for Optics meeting in Tokyo and Ky-

Figure 14 Emil Wolf in Sydney 1964.

Figure 15 "At the Intensity Interferometer" in Australia, Hanbury-Brown in the center with hands outstretched is seen talking with part of our group. Toraldo de Francia is on the immediate right (photograph taken in 1964).

oto, Japan, after a satellite meeting in Sydney, Australia, in August. As part of the Sydney meeting we went out to the site of the intensity interferometer. Figure 15 shows a group of people at that site, with Hanbury-Brown talking as we toured.

At the end of my "tour of duty" at Tech/Ops in Massachusetts, I spent a year at Tech/Ops West and Beckman & Whitley in California worrying about ultra-high-speed photography, optical scanners, and electro-optic modulators. Then, at Emil's request, I prepared a review article for *Progress in Optics* on "Image formation with partially coherent light" [25]. By the time the article was published I had accepted the position as Professor of Optics and Director of the Institute of Optics and assumed those duties in September of 1968. Thus, I was reunited with Emil Wolf on the same campus—the University of Rochester.

1.4 The University of Rochester 1968 –

I fully intended to limit my remarks in this paper to the earlier years 1955–1968, but I couldn't resist a few comments about the 35 years in Rochester since 1968. Others, of course, will talk more fully about Emil during that extended period of productive scholarship on such topics as inverse scattering, evanescent waves, radiometric models, focused fields, quantum optics, partial coherence in the space frequency domain, red shifts and blue shifts (Wolf shift), and many, many more too numerous to mention here. As for me, I was working on holography, optical processing (including Knox-Thompson algorithm), apodization of coherent imaging and beam propagation systems, hybrid processing, two-step phase microscopy, and the Lau effect, amongst several others.

1.4.1 Books and editing

Over the 35-year period *Principles of Optics* continued to be a major source book, and its seventh edition (much expanded) came out in 1999. In addition the volume entitled *Optical Coherence and Quantum Optics* written by Mandel and Wolf was published in 1995. Finally, a fine volume *Selected Works of Emil Wolf with Commentary* appeared in 2001 [26]. This book will make a great companion to the volume in which this current paper appears. For my own part, the much revised version of the original *Physical Optics Notebook* retitled *The New Physical Optics Notebook—Tutorial in Fourier Optics*, with an expanded list of authors that includes George Reynolds and John DeVelis, was published by SPIE's Optical Engineering Press [24].

In the world of editing Emil's founding of, and continued editing of, *Progress in Optics* (46 volumes and counting [27]) has been of immense value to our community. Together with Len Mandel, Wolf [31] had put together a volume entitled *Selected Papers on Coherence and Fluctuations of Light* [28]. Late in 1984 SPIE founded the *Milestone Series of Selected Papers* on topics in optical science and engineering [21]. As the series editor, I was pleased that we were able to reprint Mandel's and Wolf's volume of selected papers listed above (see *Milestone Volume 19*).

Emil Wolf together with Robert Hopkins had started the "Rochester Conferences on Coherence," later to be called the "Rochester Conferences on Coherence and Quantum Optics." Emil Wolf was to be coeditor of five of the proceedings of these conferences (3rd–7th, 1973–1996). (See Mandel and Wolf 1973–1984 [29–31].)

Attempting in vain to keep up with Emil's productivity, I was eager to accept the invitation to be the series editor of Marcel Dekker's new book series on optical engineering, which recently produced its 85th volume. Additionally, The International Society for Optical Engineering's main referenced journal is called *Optical Engineering* [32] and has published many papers that are a part of Emil's legacy in physical optics. My involvement in the early days of the journal, and then as its editor from 1990–1997, was a very rewarding experience.

1.4.2 Miscellaneous highlights

Michelson Interferometry and Film, March 12, 1980

The above title was that of a symposium presented by The Institute of Optics to commemorate the Eastman Kodak Centennial. The highlight of this program was the presentation by Dorothy M. Livingston, Albert Michelson's youngest daughter, based on her book *The Master of Light: A Biography of Albert A. Michelson* [33]. Figure 16 shows three happy people together on this occasion.

Kingslake Seminar, March 1985

Figure 17 shows two images from this important event as we celebrated the contributions of both Rudolph and Hilda Kingslake—both very good friends and colleagues of the Wolfs and the Thompsons. The two images are of Emil and myself each with Lem Hyde, a former director of The Institute, whom I succeeded in office.

Wolf-Mandel Day, October 24, 1987

We celebrated Wolf's and Mandel's separate and joint contributions at a day-long event and Symposium on the Coherence, Propagation, and Fluctuations of Light. Figure 18 shows these two worthy scholars on that occasion and I will let you write the caption; it clearly needs a balloon to record Emil's words!

Farewell Party Goer, March 28, 1995

Emil has always been fully engaged in the total life of the optics community. He is seen here (Fig. 19) with Duncan Moore (also a former director of The Institute

Figure 16 Emil Wolf, Dorothy Livingston, and Brian Thompson on March 12, 1980, at the Symposium "Michelson Interferometry and Film," The Institute of Optics, University of Rochester.

Figure 17 Emil Wolf and the author, each with Lem Hyde on the occasion of the Kingslake Seminar, March 1988 (courtesy The Kingslake Archives, University of Rochester).

Figure 18 Emil Wolf and Len Mandel at the celebration of Wolf-Mandel Day, October 24, 1987 (courtesy The Kingslake Archives, University of Rochester).

Figure 19 Emil and Marlies Wolf with Duncan Moore at a reception at the University of Rochester (courtesy The Kingslake Archives, University of Rochester).

Figure 20 Emil Wolf chatting with Boris Stoicheff at a reception for Optical Society of America directors and guests at the historic Patrick Barry House.

of Optics) and Mrs. Wolf; Marlies was his constant companion at all these events and they are attending a farewell reception for Dean Bruce Arden on this occasion.

Optical Society of America Annual Meeting, Rochester, N.Y., October 20, 1996

We had a very pleasant reception at the Provost residence, the historic Patrick Barry House, for the OSA directors and other guests. Here (Fig. 20) is Emil (the party goer) talking with old friend, Boris Stoicheff (University of Toronto). Emil had spent a sabbatical at the University of Toronto in 1974–1975.

1.5 Conclusions

I remember being asked many years ago by one of my Ph.D. students "who is my grandfather?" I was puzzled at first by this question—if he didn't know who his grandfather was, how would I know! He explained that since I was his thesis advisor I was his "academic father," and thus my thesis advisor would be his "grandfather" [it crossed my mind at the time that maybe he thought that my academic ancestry had more value for him for name recognition than his current thesis advisor's name! But perish the thought!] I was, however, able to give him great satisfaction: Professor Charles Taylor, F.R.S., was my advisor; in turn, Taylor's advisor was Professor Henry Lipson, F.R.S., whose advisor was Sir Lawrence Bragg, F.R.S., and Nobel Laureate (this last honor shared with his father Sir William Bragg, F.R.S.—the father was the son's academic advisor. The frosting on the cake came when I informed the student that he could count Prof. Emil Wolf as his academic great uncle.

There is one paper that sums up Emil's approach to theoretical work and that is really the title of his *Nuovo Cimento* paper, "Optics in terms of observable quantities" [34].

Emil, we celebrate your contributions to the scholarship and teaching that you have provided for us all. Thank you.

Acknowledgments

I would like to thank Andrew Thompson for excellent work in resurrecting some of the early images in this paper from my old slides and old prints. I also acknowledge the University of Rochester River Campus Library, Rare Book Section, for permission to use a number of images from the Kingslake Archives. Finally, thanks to Harriet Cartwright for formatting and preparing the final version of this manuscript in the "stipulated form."

References

1. H. Lipson and C.A. Taylor, *Fourier Transforms and X-ray Diffraction*, G. Bell & Sons Ltd. (1958).

2. H. Lipson and B.J. Thompson, "Fourier synthesis of crystal structures: its physical significance," *Bull. Natl. Inst. of Sci. India* **14**, 80–90 (1959). Symposium on Crystal Physics, Aug. 25–26 (1958). See also: Thompson, B.J. [8].

3. H. Lipson and C.A. Taylor, "X-ray crystal-structure determination as branch of physical optics," *Progress in Optics*, Vol. 5, E. Wolf, Ed., 289–350, North-Holland, Amsterdam (1969).

4. A.W. Hanson, H. Lipson, and C.A. Taylor, "The application of the principles of physical optics to crystal structure determination," *Proc. Roy. Soc. A* **218**, 371–384 (1953).

5. W. Hughes and C.A. Taylor, "Apparatus used in the development of optical-diffraction methods for the solution of problems in x-ray diffraction," *J. Sci. Instrum.* **30**, 105–110 (1953).

6. P.R. Pinnock and C.A. Taylor, "The determination of the signs of structure factors by optical methods," *Acta. Cryst.* **8**(11), 687–691 (1955).

7. C.A. Taylor, R.M. Hinde, and H. Lipson, "Optical methods in x-ray analysis, I (Imperfect Structures)," *Acta. Cryst.* **4**(3), 261–266 (1951).

8. B.J. Thompson, "The study of partial coherence and diffraction phenomena," Ph.D. thesis, University of Manchester, Manchester, England (1959).

9. B.J. Thompson and E. Wolf, "Two-beam interference with partially coherent light," *J. Opt. Soc. Amer.* **47**(10), 895–902 (1957).

10. B.J. Thompson, "Illustration of the phase change in two-beam interference with partially coherent light," *J. Opt. Soc. Amer.* **48**(2), 95–97 (1958).

11. C.A. Taylor and B.J. Thompson, "Some improvements in the operation of the optical diffractometer," *J. Sci. Instrum.* **34**, 439–447 (1957).

12. M.J. Beran and G.B. Parrent, *Theory of Partial Coherence*, Prentice Hall, NJ (1964).

13. E. Wolf, "Some aspects of rigorous scalar treatment of electromagnetic wave propagation," *Proc. International Symposium on Radio Wave Propagation* (Liège, Belgium), 119–125, Academic Press, PLACE (1960).

14. M. Born and E. Wolf, *Principles of Optics*, Pergamon Press, Oxford (1959).

15. E.H. Linfoot and E. Wolf, "Phase distribution near focus in an aberration-free diffraction image," *Proc. Phys. Soc. B* (London) **69**, 823 (1956).

16. E. Wolf, "Light distribution near focus in an error-free diffraction image," *Proc. Roy. Soc. A* **204**, 533–548 (1951).

17. C.A. Taylor and B.J. Thompson, "A critical method of focusing on optical diffractometer," *J. Sci. Instrum.* **35**, 294–297 (1958).

18. C.A. Taylor and B.J. Thompson, "Attempt to investigate experimentally the intensity distribution near the focus in the error-free diffraction patterns of circular and annular apertures," *J. Opt. Soc. Amer.* **48**, 844–850 (1958).

19. B.J. Thompson, "The three-dimensional intensity distribution near the focus of wave diffracted by slit and rectangular apertures," *Proc. Roy. Soc.* **73**, 905–915 (1959).

20. B.J. Thompson, "Coherence Requirement" and "Optical Data Processing," in *Optical Transforms*, H. Lipson, Ed., 27–69, 267–298, Academic Press, PLACE (1972).

21. B.J. Thompson, "Optical transforms and coherent processing systems—with insights from crystallography," Chapter 2 in *Topics in Applied Physics* 23, D. Casasent, Ed., 17–52, Springer-Verlag, Berlin (1978).

22. C. Curry, *Wave Optics, Interference and Diffraction*, Edward Arnold, London (1957).

23. G.B. Parrent and B.J. Thompson, *Physical Optics Notebook*, SPIE Press, Bellingham, WA (1969).

24. G.O. Reynolds, J.B. DeVelis, G.B. Parrent, and B.J. Thompson, *The New Physical Optics Notebook—Tutorials in Fourier Optics*, SPIE Press, Bellingham, WA (1989).

25. B.J. Thompson, "Image formation with partially coherent light," *Progress in Optics*, Vol. 7, E. Wolf, Ed., 169–230, North-Holland, Amsterdam (1969).

26. E. Wolf, *Selected Works of Emil Wolf with Commentary*, World Scientific Publ. Co., Singapore (2001).

27. E. Wolf, Ed., *Progress in Optics*, Vols. 1–44, North-Holland, Amsterdam (1961–2003).

28. L. Mandel and E. Wolf, Eds., *Selected Papers on Coherence and Fluctuations of Light*, Vol. 1 (1850–1960); Vol. 2 (1961–1966), Dover Publications, New York (1970).

29. J. Eberly, L. Mandel, and E. Wolf, Eds., *Proceedings of the Rochester Conference on Coherence and Quantum Optics*, 6th and 7th, Plenum Press, New York (1990 and 1996).

30. L. Mandel and E. Wolf, Eds., *Proceedings of the Rochester Conference on Coherence and Quantum Optics*, 3rd, 4th, and 5th, Plenum Press, New York (1973, 1978, and 1984).

31. L. Mandel and E. Wolf, *Optical Coherence and Quantum Optics*, Cambridge University Press, Cambridge and New York (1995).

32. "Optical Engineering" SPIE—The International Society for Optical Engineering Vol. 1. Various editors including D. Sinclair, J. DeVelis, J. Gaskill, B.J. Thompson and D. O'Shea.

33. D.M. Livingston, *The Master of Light*, Scribners, New York (1973).

34. E. Wolf, "Optics in terms of observable quantities," *Nuovo Cimento* **12**, 884–888 (1954).

Chapter 2

Recollections of Max Born

Emil Wolf

At the request of the organizers of the SPIE International Symposium on Optical Science and Technology, which was held in San Diego, CA, on 3–8 August 2003, I gave an after-dinner speech at the Symposium Banquet. I spoke about my collaboration with Max Born, a half century earlier. The talk followed closely an article that was originally published in *Optics News* **9**, 10–16 (1983) and is reprinted below.

The editor of this volume, Dr. Tomasz Jannson, asked me to add some remarks about the early days of holography and coherence that might be of special interest to the reader. The brief remarks that follow were written in response to this request.

Max Born knew well the inventor of holography, Dennis Gabor, and because of it, we learned about Gabor's invention long before the great discovery became generally known and appreciated. In fact, *Principles of Optics* was, I believe, the first book in which the principles of holography were explained. Gabor was very pleased that our book presented an account of his invention, as will become evident on reading the article that follows.

The subject of coherence was, at the time of my collaboration with Max Born, in its infancy. I became aware of it when I was working on the chapter concerning interference for our book. The theory of interference, as described in optics textbooks of that time, dealt mainly with monochromatic waves, not with wave-fields that randomly fluctuate. These more complicated waves, which, in general, are partially coherent, can be adequately described only in statistical terms. While attempting to develop in our book a more satisfactory treatment of interference by using elementary statistical concepts, I was able to introduce a more realistic treatment of interference. It was a very fortunate coincidence that only a year after our book was published, the first lasers were developed, which triggered great interest in questions concerning the coherence of light.

More about these two topics is briefly mentioned in the pages that follow.

2.1 Introduction

The invitation to address this meeting has given me the rare opportunity to set aside my customary activities and try to recall a period of my life several decades ago when I had the great fortune of being able to collaborate with Max Born. As the title of my talk suggests, this will be a rather personal account, but I will do my best to present a true image of a scientist who has contributed in a decisive way to modern physics in general and to optics in particular; it will also present glimpses of a man who, under a somewhat brusque exterior, was a very humane and kind person and who in the words of Bertrand Russell was brilliant, humble, and completely without fear in public utterances.

The early part of my story is closely interwoven with another great scientist, Dennis Gabor, through whose friendship I became acquainted with Born.

I completed my graduate studies in 1948 at Bristol University. My Ph.D. thesis supervisor was E. H. Linfoot, who at just about that time was appointed Assistant Director of the Cambridge University Observatory. He offered me, and I accepted, a position as his assistant in Cambridge. During the next two years while I worked in Cambridge I frequently traveled to London to attend the meetings of the Optical Group of the British Physical Society. They were usually held at Imperial College and were often attended by Gabor, whose office was in the same

Figure 1 Max Born at his desk, ca 1950. (Credit: AIP Niels Bohr Library.)

complex of buildings. From time to time I presented short papers at these meetings. At the end of some of the meetings Gabor would invite me to his office for a chat. He would comment on the talks, make suggestions regarding my work, and speak about his own researches. Gabor liked young people, and he always offered encouragement to them. He knew Born from Germany, and he had great admiration for him.

Through Gabor I learned in 1950 that Born was thinking of preparing a new book on optics, somewhat along the lines of his earlier German book *Optik*, published in 1933, but modernized to include accounts of the more important developments that had taken place in the nearly 20 years that had gone by since then. At that time Born was the Tait Professor of Natural Philosophy at the University of Edinburgh, a post he had held since 1936, and in 1950 he was 67 years old, close to his retirement. He wanted to find some scientists who specialized in modern optics and who would be willing to collaborate with him in this project. Born approached Gabor for advice, and at first it was planned that the book would be written jointly by him, Gabor, and H. H. Hopkins. The book was to include a few contributed sections on some specialized topics, and Gabor invited me to write a section on diffraction theory of aberrations, a topic I was particularly interested in at that time. Later it turned out that Hopkins felt he could not devote adequate time to the project, and in October of 1950, Gabor, with Born's agreement, wrote

Figure 2 Dennis Gabor.

to Linfoot and me asking if either of us, or both, would be willing to take Hopkins' place. After some lengthy negotiations it was agreed that Born, Gabor, and I would co-author the book.

2.2 The Start of Collaboration

I was, of course, delighted with this opportunity, but there was the problem of my finding the necessary time to work on this project while holding a full-time appointment with Linfoot at Cambridge. I mentioned this to Gabor, and I told him that if there were any possibility of obtaining an appointment with Born, which would allow me to spend most of my time working on the book, I would gladly leave Cambridge and go to Edinburgh.

Gabor took up the matter with Born, who was interested. Toward the end of November 1950, Gabor wrote me that Born would be in London a few days later and that he (Gabor) was arranging for the three of us to meet the following weekend. It was agreed that I would come to Gabor's office at Imperial College on the following Saturday morning, December 2, 1950, and that we would then go to his home in South Kensington, within walking distance of Imperial College. Born was to come directly to Gabor's home from his London hotel, and the three of us and Mrs. Gabor would have lunch there.

I arrived at Gabor's office just before lunch, and I have a vivid recollection of that meeting. There was a long staircase leading to the entrance hall of the building. As we were walking down the staircase, Gabor suddenly became somewhat apprehensive. He knew that our luncheon meeting might lead to an appointment for me with Born, and he said to me, "Wolf, if you let me down, I will never forgive you. Do you know who Born's last assistant was? Heisenberg!" This statement was not accurate. Born had other assistants after Heisenberg, but the remark shows how nervous Gabor was on that particular occasion. Fortunately, all turned out well, and Gabor certainly seemed in later years well satisfied with the consequences of our luncheon with Born.

During that meeting Born asked me a few questions, mainly about my scientific interests, and before the lunch was over he invited me to become his assistant in Edinburgh, subject to the approval of Edinburgh University. It seemed to me remarkable that Born should have made up his mind so quickly, without asking for even a single letter of reference, especially since I had published only a few papers by that time and was quite unknown to the scientific community.

Later, when I got to know Born better, I realized that his quick decision was very much in line with one trait of his personality; he greatly trusted the judgment of his friends. Since Gabor recommended me, Born considered further inquiries about me to be superfluous. Unfortunately, as I also learned later, Born's implicit trust in people whom he considered to be his friends was occasionally misplaced and sometimes created problems for him.

A few days after our meeting I received a telegram from Born inviting me to a formal interview at Edinburgh University. The interview took place about two weeks later, and the next day Born wrote me saying that the committee which interviewed me recommended my appointment as his private assistant, beginning January 23, 1951. I resigned my post in Cambridge and took up the new appointment. Later I learned that committee approval was not really needed because my salary was to be paid from an industrial grant that was entirely at Born's disposal. However, on this occasion Born was careful, because some time earlier he had had on his staff Klaus Fuchs, who turned out to be a spy for the Russian, and Born got rather bad publicity from that.

Now, the name Fuchs means fox in German, and before inviting me to Edinburgh, Born apparently wrote to Sir Edward Appleton, the Principal of Edinburgh University at that time, saying that he felt the decision about this particular appointment should not be made by him alone; since he would like to appoint a Wolf after a Fox!

2.3 Arrival at Edinburgh

I arrived in Edinburgh toward the end of January 1951, eager to start on our project. Born's Department of Applied Mathematics was located in the basement of an old building on Drummond Street. I was surprised by the small size of the department. Physically it consisted of Born's office; an adjacent large room for

Figure 3 The building on Drummond Street in Edinburgh that housed Max Born's Department of Applied Mathematics.

all of his scientific collaborators, about five at that time; a small office for Mrs. Chester, his secretary; two rooms for the two permanent members of his academic staff, Robert Schlapp, a senior lecturer, and Andrew Nisbet, a lecturer; and one lecture room. The rest of the building was occupied by experimental physicists under the direction of Professor Norman Feather. In earlier days the building housed a hospital, in which Lord Lister, a famous surgeon known particularly for his work on antiseptics, also worked.

In spite of his advanced age Born was very active and, as throughout all his adult life, a prolific writer. He had a definite work routine. After coming to his office he would dictate to his secretary answers to the letters that arrived in large numbers almost daily. Afterward he would go to the adjacent room where all his collaborators were seated around a large U-shaped table. He would start at one end of it, stop opposite each person in turn, and ask the same question: "What have you done since yesterday?" After listening to the answer he would discuss the particular research activity and make suggestions. Not everyone, however, was happy with this procedure. I remember a physicist in this group who became visibly nervous each day as Born approached to ask his usual question, and one day he told me that he found the strain too much and that he would leave as soon as he could find another position. He indeed did so a few months later. At first I too was not entirely comfortable with Born's question, since obviously when one is doing research and writing there are sometimes periods of low productivity. One day when Born stood opposite me at the U-shaped table and asked, "Wolf, what have you done since yesterday?" I said simply, "Nothing!" Born seemed a bit startled, but he did not complain and just moved on to the next person, asking the same kind of question again.

Born was always direct in expressing his views and feelings, but he did not mind if others did the same, as this small incident indicates. There will be more examples of this later.

2.4 Work at Edinburgh

We started working on the optics book as soon as I came to Edinburgh. It was understood right from the beginning that Born's main contribution would consist of making material available from his German *Optik*, but he was to take part in the planning of the new book, make suggestions, and provide general advice. Most of the actual writing was to be done by Gabor and me and a few contributors. However, like Hopkins earlier on, Gabor soon found it difficult to devote the necessary time to the project, and it was mutually agreed that he would not be a co-author after all, but would just contribute a section on electron optics. So in the end it became my task to do most of the actual writing. Fortunately I was rather young then, and so I had the energy needed for what turned out to be a very large project. I was

Figure 4 Max Born as *Privatdozent* in Göttingen. (Reproduced from *Hilbert* by Constance Reid, 1970.)

in fact 40 years younger than Born. This large age gap is undoubtedly responsible for a question I am sometimes asked, whether I am a son of the Emil Wolf who co-authored *Principles of Optics* with Max Born!

Although I did most of the writing, Born read the manuscript and made suggestions for improvements. We signed a contract with the publishers about a year after I came to Edinburgh, and we hoped to complete the manuscript by the time Born was to retire, one-and-a-half years later. However, we were much too optimistic. The writing of the book took about eight years altogether.

Throughout his life Born was a quick, prolific writer, publishing more than 300 scientific papers, about 31 books (not counting different editions and translations), apart from numerous articles on nonscientific topics.[*] In spite of my relative youth I could not compete with the speed with which Born wrote, even at his advanced age, and it soon became clear to me that he was not too pleased with my slow progress.

One day when I started writing an Appendix on Calculus of Variations, Born said that the best treatment of that subject he knew of was in his notes of lectures given by the great mathematician David Hilbert in Göttingen in the early years of

---

[*] A bibliography of Born's scientific publications is given in "Max Born," by N. Kemmer and R. Schlapp in Biographical Memoirs of the Royal Society, 17, London: the Royal Society, 1971, pp. 17–52. Born's autobiography [1] was published posthumously, first in German in 1975 and is, therefore, not included in that bibliography.

Figure 5 Max Born in the 1920s.

this century. Born suggested that he dictate the Appendix to me, following in the main Hilbert's presentation, and that we acknowledge this in the preface to our book. So we started in this way. After each dictating session I was to rewrite the notes and give them to Born the next day for his comments. But we did not get very far this way. After about two dictating sessions Born said he could prepare the whole Appendix himself much faster without my help, which he then did. It is essentially in this version, written by Born, that the Appendix on Calculus of Variations appears in our book.

2.5 Born's Revered Teacher

David Hilbert, whose presentation Born closely followed, was one of Born's great heroes. To physicists Hilbert is mainly known in connection with the concept of the Hilbert space and as co-author of the classic text *Methods of Mathematical Physics*, referred to generally as "Courant-Hilbert." But Hilbert contributed in a fundamental way to many branches of mathematics and was generally considered to have been the greatest mathematician of his time. Born became acquainted with Hilbert soon after coming to Göttingen in 1905, later becoming Hilbert's private assistant. In one of his later writings Born refers to Hilbert as his "revered teacher and friend," and in a biography of Hilbert by Constance Reid [2], Born is quoted as

Figure 6 David Hilbert, 1912.

saying that his job with Hilbert was to him "precious beyond description because it enabled [him] to see and talk to him every day."

Born had an encyclopedic knowledge of physics and whatever problem one brought to him, he was able to offer a useful insight or suggest a pertinent reference. He also knew personally all the leading physicists of his time and would often recall interesting stories about them.

Optics in those days—remember we are talking about optics in pre-laser days—was not a subject that most physicists would consider exciting; in fact, relatively little advanced optics was taught at universities in those days. The fashion then was nuclear physics, particle physics, high energy physics, and solid state physics. Born was quite different in this respect from most of his colleagues. To him all physics was important, and rather than distinguish between "fashionable" and "unfashionable" physics he would only distinguish between good and bad physics research.

Born was equally broad-minded about the techniques used by physicists in their research. For example, when we were writing a section on certain mathematical methods needed to evaluate the performance of optical systems, we found that although the results given in a basic paper on this subject were correct, the derivation contained serious flaws. I was rather indignant about this, but Born just said something like, "In pioneering work everything is allowed, as long as one gets the right answer. Real justification can come later."

One of the earliest occasions when many physics students encounter Born's name comes when they start studying quantum theory of scattering. Here they soon learn about the *Born approximation*. This term also occurs in many of the papers on potential scattering that have been published in the more than half a century that has gone by since Born wrote a basic paper on this subject. Yet Born was rather irritated when the Born approximation was mentioned. He once said to me, "I developed in that paper the whole perturbation expansion for the scattered field, valid to all orders, yet I am only given credit for the first term in that series!"

2.6 Resistance to New Discoveries

It was not always easy for Born's collaborators to convince him quickly of new discoveries. Let me illustrate this by an example from my own experience. In the early 1950s I became very interested in problems of partial coherence. One day I found a result in this area of optics that seemed to me remarkable. I phoned Born from my home one morning, told him I had an exciting new result, and asked him for an appointment to discuss it. We arranged to have lunch together that day.

When I came to his office just before lunch, Born wanted to know straight away what the excitement was all about. I told him I had found that not only an optical field, but also its coherence properties, characterized by an appropriate correlation function (now known as the mutual coherence function), are propagated in the form of waves. Born looked at me rather skeptically, put his arm on my shoulder and said, "Wolf, you have always been such a sensible fellow, but now you have become completely crazy!" Actually after a few days he accepted my result, and I suspect he then no longer doubted my sanity.

This incident illustrates a fact well known to Born's collaborators—that Born had a certain resistance to accept new results obtained by others. Nonetheless, he continued thinking about them, and if they were correct he would eventually apologize for doubting them in the first place.

This trait of Born's personality is very well described by the Polish physicist Leopold Infeld, who collaborated with Born in Cambridge in the 1930s. I will quote shortly some very perceptive observations Infeld made about Born in his biography [3]; but before doing so I would like to mention a small incident relating to this book.

One day I browsed through a bookstore in Edinburgh and found a used copy of Infeld's book. I was astonished to note that the book had Born's signature on its first page. I purchased it and asked Born the next day whether he knew the book.

He said, "Yes, I had a copy of it and there is a funny description of me in it; but I lent it to someone and it was never returned. I cannot remember whom I lent it to." The book I had purchased was obviously Born's missing copy, so I gave it to him, much to his delight.

In the book Infeld describes some of his experiences in Cambridge. He started working with Dirac but found him rather uncommunicative. Later Infeld attended some of Born's lectures. During one of them Born gave an account of some results that he had recently obtained. Infeld could not understand one of Born's arguments. He borrowed his notes so that he could study the argument more closely later. Let me now quote from Infeld's biography ([3], p. 208, et seq.):

> On the evening of the day I received the paper the point suddenly became clear to me. I knew that the mass of the electron was wrongly evaluated in Born's paper and I knew how to find the right value. My whole argument seemed simple and convincing to me. I could hardly wait to tell it to Born, sure that he would see my point immediately. The next day I went to him after his lecture and said: "I read your paper; the mass of the electron is wrong." Born's face looked even more tense than usual. He said: "This is very interesting. Show me why." Two of his audience were still present in the lecture room. I took a piece of chalk and wrote a relativistic formula for the mass density. Born interrupted me angrily: "This problem has nothing to do with relativity theory. I don't like such a formal approach. I find nothing wrong with the way I introduced the mass." Then he turned toward the two students who were listening to our stormy discussion. "What do you think of my derivation?" They nodded their heads in full approval. I put down the piece of chalk and did not even try to defend my point. Born felt a little uneasy. Leaving the lecture room, he said, "I shall think it over."

Infeld then goes on to say:

> I was annoyed at Born's behavior as well as at my own and was, for one afternoon, disgusted with Cambridge. I thought: "Here I met two great physicists. One of them does not talk. I could as easily read his papers in Poland as here. The other talks, but he is rude." The next day I went again to Born's lecture. He stood at the door before the lecture room. When I passed him he said to me. "I am waiting for you. You were quite right. We will talk it over after the lecture. You must not mind my being rude. Everyone who has worked with me knows it. I have a resistance against accepting something from outside. I get angry and swear but always accept it after a time if it is right."
>
> Our collaboration had begun with a quarrel, but a day later complete peace and understanding had been restored.

A little further on in his biography, Infeld speaks about Born again, and this is what he says:

> I marveled at the way in which he managed his heavy correspondence, answering letters with incredible dispatch, at the same time looking through scientific papers. His tremendous collection of reprints was well ordered; even the reprints from cranks and lunatics were kept, under the heading "Idiots." Born functioned like an entire institution, combining vivid imagination with splendid organization he worked quickly and in a restless mood. As in the case of nearly all scientists, not only the result was important but the fact that he had achieved it.

Infeld later continues:

> There was something childish and attractive in Born's eagerness to go ahead quickly, in his restlessness and his moods, which changed suddenly from high enthusiasm to deep depression. Sometimes when I would come with a new idea he would say rudely, "I think it is rubbish," but he never minded if I applied the same phrase to some of his ideas. But the great, the celebrated Born was as happy and as pleased as a young student at words of praise and encouragement. In his enthusiastic attitude, in the vividness of his mind, the impulsiveness with which he grasped and rejected ideas, lay his great charm.

I regard these remarks of Infeld as a true and very perceptive description of Born's mode of work and of Born's personality.

2.7 Kind and Compassionate

In spite of Born's occasional irritation and impatience, he was a person who cared deeply for the well-being of his fellow scientists and collaborators. His wife, Hedwig Born, was likewise a person with deep compassion for others. She too was a remarkable and gifted person, Mrs. Born published a number of books, especially poetry, and around 1938 became a Quaker, remaining active in the Quaker movement for the rest of her life.

Figure 7 Mrs. Hedwig Born, 1961.

I would like to give just one example from my own experience, which illustrates Born's concern for others. A few months after I began working with Born, I was getting married. In those days it was difficult to rent an apartment in Edinburgh. One day during the time when we were searching for a home I received a letter from Mrs. Born, who was then with Professor Born on a visit to Germany. She said that they had heard about our problem and were very concerned that we might have to postpone getting married if we did not find somewhere to live. She then offered to help us, suggesting that we share with them their small house in Edinburgh. In the end we found an apartment elsewhere; but this small episode is an indication of the warmth of their personalities and of their willingness to make a personal sacrifice to help, when help was needed.

"In an Age of mediocrity and moral pygmies, the lives of Albert Einstein and Max Born shine with an intense beauty. Something of this is reflected in their correspondence, and the world is the richer for its publication." [†]

Bertrand Russell

I mentioned earlier, that one of Born's great heroes was the mathematician David Hilbert. But there was another, even greater hero in Born's life: Albert Einstein, with whom he and also Mrs. Born maintained close personal friendships for almost half a century. Unfortunately, after Einstein left Europe for America in

Figure 8 The house of Max and Hedwig Born in Edinburgh, at 84 Grange Loan.

[†] From Bertrand Russell's Foreword to *The Born-Einstein Letters*, Ref. [4].

Figure 9 Albert Einstein in the 1920s. (Credit: AIP Niels Bohr Library.)

1932 they did not see each other again, but they carried on extensive correspondence until Einstein's death in 1955. The letters they exchanged were published in 1971, together with Born's commentary, and the volume [4] is a precious contribution to the history of physics and of the times in which they lived.

There is an episode I would like to relate briefly in connection with Born's friendship with Einstein. In the early 1950s, when Sir Edmund Whittaker was preparing the second volume of his classic work *A History of the Theories of Aether and Electricity*, he sent Born the manuscript of a section dealing with the special theory of relativity. Whittaker's treatment placed a heavy emphasis on the work of Poincaré and Lorentz and dismissed Einstein's contribution as being of rather minor significance. Born, who himself wrote a book on the theory of relativity, was most unhappy with Whittaker's manuscript and sent him a long report in which he analyzed in detail the various contributions, indicating why he considered Einstein's contribution to be much more fundamental.

Figure 10 Sir Edmund Whittaker. (Reproduced by the courtesy of the University of Edinburgh.)

However, Born did not succeed in changing Whittaker's opinion.[‡] In September of 1953, around the time Whittaker's book was published, Born wrote to Einstein about this. Let me quote from Born's letter [4, p. 197]: "Many people may now think (even if you do not) that I played a rather ugly role in this business. After all it is common knowledge that you and I do not see eye to eye over the question of determinism."

Einstein was not concerned. This is what he said in his reply to Born [4, p. 199]: "Don't lose any sleep over your friend's book .... If he manages to convince others, that is their own affair. I myself have certainly found satisfaction in my efforts...." and then Einstein added, "After all, I do not need to read the thing."

Born retired that year, in 1953. The accompanying photograph shows Born with the members of his department at the time of his retirement.

[‡] Born's opinion on this question rather than Whittaker's is generally accepted. See, for example, D. Martin's biographical note about E. T. Whittaker in *Dictionary of Scientific Biography*, C. C. Gillespie, editor-in-chief (Charles Scribner's Sons, New York, 1976), Vol. XIV, p. 317; or A. Pais: *Subtle Is the Lord, The Science and the Life of Albert Einstein* (Clarendon Press, Oxford, and Oxford University Press, New York, 1982), p. 168.

Figure 11 Members of Max Born's department at the time of his retirement (1953) from the Tait Chair of Natural Philosophy at the University of Edinburgh. Standing (from left to right) E. Wolf, D. J. Hooton, A. Nisbet. Sitting: Mrs Chester (secretary), M. Born, R. Schlapp.

"Everybody who had contact with him remembers him not only as a brilliant scientist but also as a man of human warmth and greatness." [§]

2.8 Life in Retirement

Soon afterward the Born's left Edinburgh and settled in Bad Pyrmont, a spa in West Germany, not far from Göttingen, where they built a small house. When they left Edinburgh our book was far from finished. We corresponded about it, and I visited Born in his new home several times. Born was hoping that he and Mrs. Born would be able to lead a more quiet life in Bad Pyrmont, but he told me on one of my visits that this proved difficult to achieve. For example, soon after they settled, in Bad Pyrmont, Born was invited to address a meeting of a West German physical society. He declined the invitation, saying that he was too old to travel. He received a reply stating that in view of this the meeting would be moved to Bad Pyrmont!

[§] From an introduction by Victor F. Weisskopf to an article by Max Born entitled "Man and the Atom," published by the Society for Social Responsibility of Science (Pamphlet #4) Southhampton, Pa., and the American Friends Service Committee, Philadelphia, Pa.

Figure 12 Max Born in Bad Pyrmont feeding pigeons.

In 1954, the year after his retirement, Born was awarded the Nobel Prize. He was, of course, delighted, but I am quite sure he felt, as many others did, that this great recognition had come somewhat late. The Nobel Prize was awarded to him for contributions that he made almost 30 years earlier. However, as his son Gustav later noted in a postscript to Born's memoirs [1, p. 296], it came at the right time to add weight to his main retirement occupation, which was to educate thinking people in Germany and elsewhere in the social, economic, and political consequences of science and also of the dangers of nuclear weapons and re-armament.

In 1957 I was a Visiting Scientist at the Courant Institute of New York University, still working on our book. One day I received a letter from Born asking me why the book was not yet finished. I replied that practically the whole manuscript was completed, except for a chapter on partial coherence on which I was still working. Born wrote back almost at once, saying something like, "Who apart from you is interested in partial coherence? Leave that chapter out and send the rest of the

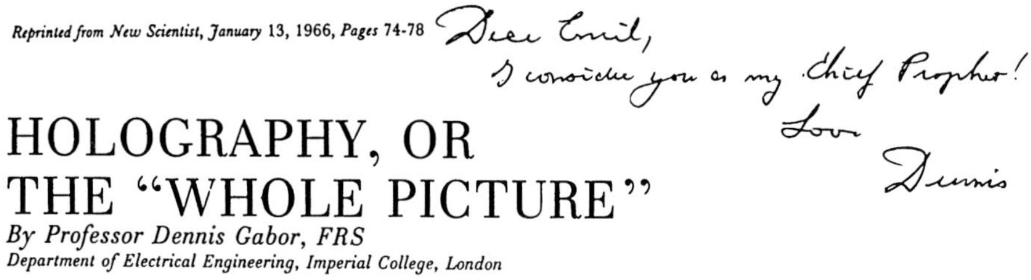

Reprinted from New Scientist, January 13, 1966, Pages 74-78

HOLOGRAPHY, OR
THE "WHOLE PICTURE"

By Professor Dennis Gabor, FRS
Department of Electrical Engineering, Imperial College, London

Figure 13 A dedication from Dennis Gabor.

manuscript to the printers." Actually I completed that chapter shortly afterward and it was included in the book.

It so happened that within about two years after the publication of our book (in 1959) the laser was invented and optical physicists and engineers then became greatly interested in questions of coherence. Our book was the first that dealt in depth with this subject, and Born was then as pleased as I was that the chapter was included.

Our book was also one of the first textbooks containing an account of holography. Gabor was very happy about it. Later, when holography became popular and useful, he sent me a reprint of one of his papers with a charming dedication (Fig. 13).

As I approach the end of my reminiscences about Max Born, I would like to say that I hope my talk conveyed to you the warmth and the affection with which he remains in my memory, not only as a great scientist, but also as a kind and remarkable human being. My feelings about our collaboration are well described by exactly the same words that Born used when he spoke about his association with David Hilbert, quoted earlier, namely that my appointment with him was precious to me beyond description, because it enabled me to see and to talk to him every day.

2.9 Olivia

Before ending I would like to show you a few pictures taken in Bad Pyrmont during Born's retirement and also to mention one more episode. One shows Professor and Mrs. Born with one of their daughters, Irene. Some years ago I learned that Irene is the mother of a lady who has achieved fame comparable to that of Max Born himself, but in an entirely different field. I am speaking of the pop singer Olivia Newton-John. Shortly after I learned that Olivia Newton-John was Max Born's granddaughter I was on a sabbatical leave at the University of Toronto. Olivia was scheduled to give a concert in Toronto while I was there. I wrote to her, told her I had collaborated with her grandfather in the writing of a book, and asked her whether we could meet. I received a charming reply in which she invited me to

Figure 14 Hedwig Born and Max Born, with their daughter Irene Newton-John in Bad Pyrmont, 1957. (Credit: AIP Niels Bohr Library.)

Figure 15 Max Born in front of his library at his home in Bad Pyrmont.

Figure 16 Max Born with two of his grandsons, Max and Sebastian (children of Gustav) in Bad Pyrmont. (Credit: AIP Niels Bohr Library.)

Figure 17 *Left*: Olivia Newton-John, granddaughter of Max Born. *Right*: Max Born. (Credit: Lotte Meitner-Graf.)

meet her after the concert. We met then and talked mainly about her grandparents. Before I left Olivia gave me two autographed photos of herself. Let me add that to some of my students I am known not so much as the co-author of *Principles of Optics* but rather as the person who knows Olivia Newton-John and who has a picture of her hanging in his office signed "To Emil, Love, Olivia."

I cannot bring you the voice of Max Born, but I will end my presentation with one of the songs that made Olivia famous. (The lectures on which this article is based concluded with an excerpt from the song "If You Love Me Let Me Know.")

Acknowledgments

In preparing this article for publication I received assistance with obtaining some of the photographs, determining the approximate dates when they were taken and with checking some of the references. I am particularly obliged to G. V. R. Born (London University), R. M. Sillitto and S. D. Fletcher (University of Edinburgh), L. H. Caren (University of Rochester), and D. Dublin (American Institute of Physics) for their help.

References

1. M. Born, *My Life*, Taylor and Francis, London, and Charles Scribner's Sons, New York, p. 296 (1978).
2. C. Reid, *Hilbert*, Springer-Verlag, New York, Heidelberg, Berlin, p. 103 (1970).
3. L. Infeld, *Quest: The Evolution of a Scientist*, Doubleday, Doran and Co., New York (1941).
4. M. Born, *The Born–Einstein Letters*, Walker and Company, New York (1971).

CHAPTER 3

WHAT POLARIZATION OF LIGHT IS: THE CONTRIBUTION OF EMIL WOLF

Christian Brosseau

3.1 Introduction and Scope

Whenever I teach my polarization optics course, one of the central messages I try to get across early is that polarization of light abounds with dichotomies. This has been known for more than three centuries, and there is no question that it has generated a considerable amount of excitement among researchers in the last decades. It is an aspect of the visual world detected by insects and many vertebrates other than mammals but is hidden from us, its origins rooted deep in statistical physics and electromagnetism. Its applications involve areas as diverse as photonics, information technology, and biology, yet its understanding is still incomplete. Before starting to consider the details of the theory of polarized light, I would like to draw the reader's attention to a brief consideration of the historical background to illustrate that Emil Wolf is a most influential and contemporary theoretical physicist in the development of polarization optics.

3.1.1 Pulling the strands of Emil Wolf's contributions to polarization optics

Emil Wolf is a living legend in the field of physical optics. Born in Prague, Czechoslovakia, Emil Wolf began research on the behavior and physics of light under the auspices of Prof. Linfoot at the University of Bristol, U.K. After holding several research positions at Cambridge and Edinburgh, Prof. Wolf moved to the United States (University of Rochester) in 1959, where he was soon making classic contributions to the theories of coherence and polarization of light. Not only has Wolf's productivity continued unabated, his work has been a turning point in the history of modern optics.

Wolf's ideas on partial coherence and partial polarization were first published in his 1954 paper "Optics in terms of observable quantities," [1] and later discussed in his 1959 magnum opus *Principles of Optics* [2] coauthored with Nobel laureate Max Born, which is among, perhaps, five of the most famous books ever written on optics. A generation of students (including my own when I was an undergraduate) has learned the basics of optics thanks in no small part to courses based on *Principles of Optics*. Through seven editions it has established an enviable record for high-quality presentation, with the author showing a remarkable ability to make both basic concepts and cutting-edge research topics accessible to readers [3]. I remember that this book was my first exposure to the amazing facts of optics, and it also taught me some remarkable mathematics that I could actually see for myself made sense. Wolf has also written standard works on a large variety of topics ranging from medical imaging to astronomy, and a pioneering textbook [4] on the coherence of light coauthored with the late Leonard Mandel, which is today the undisputed bible of the subject. His prolific publications have influenced all aspects of the discipline and are actively discussed in academic literature (e.g., correlation-induced shift is now identified with the adjective "Wolf"), as well as in engineering fields (e.g., diffraction tomography).

It is a daunting task to integrate the many facets of the extraordinary career of Emil Wolf into a unified whole. Rather than trying to do that, I focus here on his work on coherence and polarization, which were early influences on my interests in optics. Wolf's growing influence on the statistical description of polarized light was recognized as long ago as 1954 [2], when he introduced a precise measure of the correlations between the fluctuating field variables at two space-time points. The idea of correlations represents a milestone in the history of polarization optics and has been highly successful. Still, it was Wolf who gave us the alphabet from which the field of coherence and polarization optics was written. We celebrate his work and hope to live up to it in some small way.

3.1.2 Structure of the review

The remainder of this introduction presents an overview of the salient historical and experimental facts and qualitatively describes the ideas and issues that have been shown to be important for understanding the phenomenon of polarization. In Sect. 3.2.1, it will be shown how the polarization and coherence concepts call for a statistical method that can handle the second-order description of the fluctuations of the electric field vector of light. A number of questions related to scalar invariants are considered. Section 3.2.1 describes the statistical method in just the right amount of detail for the reader to appreciate its use in polarization theory. The application of the general concepts to the problem of light scattering is then given in

Sect. 3.2.3. Motivated by the problem of multiple scattering of polarized light by a spatially random medium composed of uncorrelated and noninteracting spherical dielectric particles, the task of numerically establishing the size and polarization dependence of the characteristic depolarization length is undertaken. To cope with numerical difficulties encountered, a Monte Carlo technique is developed, which allows us to study the statistical behavior of the wave propagation. With its help, strong numerical evidence is found, suggesting that the the size of the spherical particles and the optical depth play an important role on entropy production. It is also shown that one of the most remarkable aspects of this problem where no energy exchange between radiation and scatterer takes place is that the stationary state corresponds to both the state of minimum production of radiation entropy and to the state of maximum entropy. Section 3.2.3 presents opinions on the current state of the field as well as the areas of activity with the brightest outlook for future work. The final section contains concluding remarks.

3.1.3 Historical overview

Polarization effects have historically captured the interest of physicists and it is natural to look at the background research on this subject to see the progressive development of ideas and concepts in their historical context.[*] A number of authors have discussed the establishment of the facts of the past, and the importance and unique value of archival research in this connection is in evidence throughout all these works [5]; Refs. [6–8] list a few of my favorites. The history of polarized light is a long one and exciting applications for polarization of electromagnetic waves continue to be discovered. While I am unable in this brief review to discuss

---

[*] At this point it may be worthwhile to pose a general question: Does one need to know anything about the history of polarized light to appreciate the subject matter? In fact, there seems to be a recent trend in textbooks to include snippets of history and biography of individual scientists. This is certainly a harmless way to add human interest to what might otherwise seem to be "dry" physics, but may not by itself make the subject matter more understandable. It is much easier to convey the facts of Wolf's life than to explain to undergraduates what he accomplished physically. I note that aside from the human interest involved in biographical studies, there may be some intellectual value in retracing the way optical ideas have developed. The development is often messy, however. Occasionally good ideas emerge prematurely in obscure places and are forgotten for a time, only to be rediscovered independently. Sometimes the original motivation for an investigation looks a bit eccentric to later generations, as in the case of Paul Soleillet's approach to what we know as Mueller polarization matrices. But, in the end, one is often just curious to know where the currently accepted ideas came from. Whatever one's view may be on the role of the history of optics in teaching or research, probably most people will agree that it is more challenging to deal with the twentieth century than with the immediately preceding centuries. Optics tends to be hierarchical, making it difficult to appreciate later work without a substantial foundation in earlier work. Now that optics is developing more rapidly and in more places by more people, tracing the development of an idea does not become easier.

it in any depth, I shall mention three major breakthroughs to highlight recent work in polarization optics. These are just a few of the high points; the list could be extended much further [9].[*]

In the first, the evidence that light can be polarized was gleaned in 1669 by Erasmus Bartholinus. By carefully studying crystals of Iceland spar (i.e., calcite), he discovered double refraction. The optical properties of crystals gave the *starting impetus* of the discipline of polarization optics. After Bartholinus's discovery, Huygen's goal of research was to interpret the double-refraction phenomenon from his conception of his spherical lightwave (i.e., envelope construction), and he observed that each of the two beams arising from the double-refraction phenomenon can be extinguished by passing through a second calcite crystal that is rotated about the direction of the beam. His investigations also showed that the two beams have different polarization directions. In 1808, Etienne-Louis Malus, a military French engineer, discovered the polarization of natural light by reflection while experimenting with a crystal of Iceland spar and light reflected by the windows of the Palais du Luxembourg in Paris. By extensive experimentation, he showed by purely geometric reasoning how to express the intensity of light emerging from a polarizing crystal when the light it receives is linearly polarized along a direction making a specific angle with its axis, i.e., Malus' law. A major advance in the understanding of light polarization was made by Augustin Jean Fresnel. In 1823, he derived on the basis of the elastic theory of ether[†] his famous reflection and transmission formulas for a plane wave that is incident on a static and plane interface between two dielectric isotropic media. To Fresnel credit must be assigned for discovering the modern concept of polarization and stimulating the efforts that put the wave theory of light on a firm foundation. In the years 1812–1815 came the important milestone by the French physicist Dominique François Arago at the Paris Observatory. On the theoretical side, his principal contribution was the discovery of the interference laws published in a joint paper with Fresnel, which played a key role in the demonstration of the transverse nature of lightwaves propagating in free space. Another major advance to the field came by Sir Georges Gabriel Stokes.[‡] Stokes

---

[*] A note on referencing policy: computerized literature searches citing "polarized light" as key words find thousands of articles that are scattered through the literatures of different subareas of physics, including optics, astrophysics, biology, and materials science. Although I have tried to identify original key papers whenever possible, our references put more emphasis on recent works from which earlier papers can be found. Obviously the choice of these is highly subjective and indeed arbitrary, and I hope that the many authors whose papers I have failed to reference will not attribute this to malice.

[†] The concept of an ethereal medium, filling space, was formulated by Descartes two centuries before.

[‡] The name of Stokes, a contemporary of Maxwell, has become well known to generations of international scientists, mathematicians, opticists, and engineers, through its association with various physical laws and mathematical formulas. In standard textbooks of physics, mathematics, and

introduced four measurable quantities that now bear the name of Stokes parameters for describing properties of polarized light. Before the entry of the seminal electromagnetic theory of Maxwell, the gauntlet was thus thrown, and a search for the correct physical mechanism began. During this time the predominant belief in the field was that the transverse wave theory of light could provide an understanding of the major optical phenomena discovered at the the time of Maxwell's treatise: propagation, polarization, diffraction, and interference. However, there was little agreement as to the details of how this worked, and in spite of many difficulties, the mechanical theory of the elastic ether persisted.

In the next and second step based on the theory of electromagnetism, polarization properties are closely connected to the electric field vector distribution. The quest for a better understanding of the macroscopic world in terms of underlying fundamental microscopic laws has informed the history of science and natural philosophy. The best theories are the ones that have settled, either by virtue of their actual genesis or more commonly through their subsequent evolution, at the right level of generality. They must be sufficiently general to encompass problems of broad interest and generality, but not so super general as to allow for an expanse of phenomena not amenable to any sort of reasonable taxonomy. There is of course a litany of subjects of questionable merit that fail to satisfy one of these criteria (and sometimes both), but in these terms it is difficult to imagine one that meets them more spectacularly than Maxwell's theory of electromagnetism. Modern science has provided an admirable powerful theory, and mathematical tool as well, to address this issue in a sensible and productive way, the gift of Maxwellian Electrodynamics.[§]

---

engineering, we find Stokes law, Stokes theorem, and the Navier–Stokes equations, in addition to the Stokes parameters. His major advance was in the wave theory of light. He was by then well established at the University of Cambridge (where he spent all of his working life occupying the Lucasian Chair of mathematics from 1849 until his death in 1903), examining mathematically the properties of the ether, which he treated as an incompressible elastic medium. This enabled him to obtain major results on the mathematical theory of diffraction, which he confirmed by experiment, and on fluorescence, which led him into the field of spectrum analysis. His last major paper on light was his study of the dynamical theory of double refraction, presented in 1862. As a special comment, it is interesting to quote to the reader the leading article of *The Times*, which appeared two days after his death: "It is sometimes supposed—and instances in point may sometimes be adduced—that minds conversant with the higher mathematics are unfit to deal with the ordinary affairs of life. Sir George Gabriel Stokes was a living proof that if the mathematician is only big enough, his intellect will handle practical questions so easily and as well as mathematical formulas." See also Lord Kelvin, Obituary of Sir G. G. Stokes, *Nature* **67**, 337 (1903), and Lord Rayleigh, Obituary of Sir G. G. Stokes, *Proc. Roy. Soc.* **75**, 199 (1905).

[§] Among the enduring legacies of nineteenth-century science, James Clerk Maxwell's equations of electrodynamics have long held a preferential place in the hearts of physicists. One of today's more outspoken physicists, Steven Weinberg, has argued that the equations constitute a noncontingent fact, without which contemporary physics would be unimaginable.

A chief contributor to the fundamental aspects of early electromagnetic theory was the British physicist Michael Faraday. In 1831, he discovered electromagnetic induction, subsequently explained para- and diamagnetism and interpreted them through his field theory. When Faraday postulated the physical laws of electromagnetism, he had in mind a mechanical picture, deduced from geometric reasoning, using the two concepts of lines of force traversing all space and actions-at-a-distance exerted between the particles in a medium. In 1864 came another turning point: the Scottish physicist James Clerk Maxwell completed a six-page memoir entitled *A Dynamical Theory of the Electromagnetic Field*, in which he developed the mathematical theory required for the description of how electromagnetic waves propagate [10]. His eponymous equations summarize the fundamental relations between electricity and magnetism and became the cornerstone on which generations of scientists have based their theoretical studies. Maxwell put Faraday's concepts into the elegant mathematical form of four differential equations, and one of his major innovations was to introduce the notion of displacement current. Indeed, were it not for the displacement current, it would not be possible to deduce from Maxwell's equations that electromagnetic waves have the property of light. His electromagnetic theory was confirmed by Heinrich Hertz's discovery of electromagnetic waves,[*] which in turn led to remarkable advances in physics, astronomy, and technology.[†] According to Einstein, Michael Faraday,[‡] along with

---

[*] In the years 1887 and 1888, Heinrich Rudolf Hertz, a German physicist at the Technical University in Karlsruhe, produced and detected electric waves in air, demonstrating the application of the concepts of the electromagnetism theory to the microwave and radio regions of the spectrum. In this respect, it is also worth noting the close intertwining of the theory and experiment in nineteenth-century electrodynamics, which is distinguished by the fact that all leading theorists were active in the laboratory. Today, Hertz is remembered for the unit of frequency named after him.

[†] In the 19th century, the value of Maxwell's work was appreciated by experts working on similar problems, but in the scientific community as a whole his achievements were less famous than those of Kelvin or Helmholtz. For example, it is amusing to note that when Albert Einstein studied at the Swiss Federal Polytechnical School (ETH), Maxwell's electromagnetic theory was not covered in any of the courses there so he had to study it on his own.

[‡] Writing in a special issue of the London *Times* in 1931 to celebrate the centenary of Michael Faraday's discovery of electromagentic induction, Lord Rutherford said "The more we study the work of Faraday, with the perspective of time, the more we are impressed by his unrivalled genius as an experimenter and a natural philosopher. When we consider the magnitude and extent of his discoveries and their influence on the progress of science and of industry, there is no honor too great to pay to the memory of Michael Faraday—one of the greatest scientific discoverers of all time." In the physical sciences, apart from inventing the dynamo and the transformer, Faraday established the identity of electricity from various sources as well as investigating the discharges of electricity through gases, electrostatics, electrodeposition, and discovering the magneto-optical effect, which was the first proof that light had a magnetic component. But it is not just his extraordinary experimental skills and intellectual power that has made Faraday so fascinating to all succeeding generations; he also possessed intuition, insight and moral perfection. Aldous Huxley

James Clerk Maxwell, was responsible for the greatest change in the axiomatic basis of physics since Newton [11].

The publication of Poincaré's great treatise *Théorie Mathématique de la Lumière* in 1892 broke a new intellectual ground.[§] In his book Poincaré introduced the Poincaré sphere and the complex plane representations to specify the state of polarization. Poincaré's heavy use of geometry, including many unfamiliar propositions, makes his analysis nearly impenetrable to the contemporary reader. But it was Poincaré, more than any other, who truly saw the physical implications of geometry in polarization optics. Using a stereographic projection, he mapped each point on the plane into a sphere whose points are in one-to-one correspondence with all the possible states of polarization of a light beam. One of the conveniences of the Poincaré sphere is that it provides an intuitively geometric view of the transformation of a polarized light when it interacts with optical devices in terms of rotations of states.

Near the end of the nineteenth century, John William Strutt, the third baron Rayleigh (more familiar as Lord Rayleigh), published many fascinating articles in optics. One of his major contributions came in 1871, when he derived the polarization at 90-deg law, the inverse fourth-power law for the intensity of light scattered by particles whose size is much smaller than the wavelength of the light and explained that the degree of polarization of the scattered light depends on the angle of scattering from the elastic-solid theory of the "luminiferous ether."

The third step in the development of a detailed, predictive understanding of polarized light, which took place between 1905 and 1954, stands out as one of the

---

reflected on this "conquering man of genius" in 1925 and wrote: "If I could be born again and choose what I should be in my next existence, I should desire to be a man of science .... But even if I could be Shakespeare, I think I would still choose to be Faraday. True, the posthumous glory of Shakespeare is greater than that of Faraday.... Posthumous fame brings nobody much satisfaction this side of the grave."

[§] Mathematician of the first rank, Jules Henri Poincaré is one of France's greatest scientific genius of the nineteenth century, the range of his interests and achievements being hard to conceive. He is a fruitful subject for historical enquiry, as he left behind a large archival trail. With his polymathic interest, he has attracted much attention, e.g., H. Gispert, "La France mathématique: La Société Mathématique de France (1870–1914)," *Cahiers d'Histoire et de Philosophie des Sciences* **34**, 11 (1991). Poincaré made many contributions to mathematics and to other sciences, including celestial mechanics, fluid mechanics, the special theory of relativity, and optics, to cite but a few; he is often described as the last universalist in mathematics. As a Poincaré aficionado, I strongly believe that the questions posed and the techniques developed to answer them are thoroughly modern. Poincaré wrote in *Mathematical Definitions in Education*, Georges Carré, Paris (1904): "It is by logic we prove, it is by intuition that we invent." The breadth of his research led to him being the only member of the French Academy of Sciences in every one of the five sections (geometry, mechanics, physics, geography, and navigation) of the Academy. Note that Poincaré's family produced other men of great distinction during his lifetime, e.g., Raymond Poincaré, who was prime minister of France several times and president of the French Republic during World War I.

most exciting advances in theoretical optics of the last half of the last century. It is interesting to remember that, at the turn of the century, theorists believed that optics was a mature field and nobody could believe that optics would return from the physics rearguard to the forefront. The idea that polarized light is a "static" geometric concept has persisted since the time of Bartholinus. One interesting thing is that it seems that most scientists considered that the world is deterministic—in the sense of Isaac Newton. Yet one of the thrusts of twentieth-century physics has been that the world is not. The Heisenberg uncertainty principle, statistical mechanics, and many other parts of the modern theory give us substantive reason to think that certain forms of physical information are unknowable. In particular, statistical physics is one of the pillars of modern physics, explaining the macroscopic world on the basis of the dynamics of its microscopic components. It was only in the twentieth century that polarization evolved from a geometric concept into a statistical concept, the analysis of radiation fluctuations being powerfully addressed.

The deepest and most revolutionary insights arose from statistical physics. These are at two levels. First is the very basic fact that probability is central to modern physics. The electric field of light is then described as a random process. The introduction of probability into the fundamental nature of the physical world by Maxwell and Boltzmann provided part of the foundation for the so-called "probabilistic revolution"[*] that affected all areas of science between 1840 and 1940. At this point, it is worth noting that James Clerk Maxwell allowed probabilistic physics to bring him to the verge of mysticism: "It is the peculiar function of physical science to lead us to the confines of the incomprehensible, and to bid us behold and receive it in faith, till such time as the mystery shall open." At that time, the use of statistics as a mathematical tool of all the sciences provoked passionate disputes between philosophers and physicists. The second level of insight is that autocorrelation and cross-correlation functions between field variables at two space-time points were introduced to describe correlations of random processes in electromagnetic fields. An influential result was the introduction of the density matrix formalism by John von Neumann, which has much to do with the coherency matrix formalism pioneered by Norbert Wiener in 1930.[†]

---

[*] The interested reader may consult *The Probabilistic Revolution*, edited by Lorenz Krüger, Lorraine J. Daston, Michael Heidelberger, Gerd Gigenrenzer, and Mary S. Morgan, MIT, Cambridge, MA (1987).

[†] What is seldom appreciated is that Wiener belongs to that small group of theoretical physicists who shaped modern coherence theory. Although Wiener gave real grounds for the concept of coherence, his ideas did not gain general recognition. To quote Levinson in *Selected Papers of Norman Levinson*, N. Levinson, J.A. Nohel, and D.H. Sattinger, Eds., Vol. 1, p. 13, Birkhäuser, Boston, MA (1998), "Most of Wiener's important work was inspired by physics or engineering and in this sense he was very much an applied mathematician. He formulated his theories in the framework of rigorous mathematics and as consequence his impact on engineering was very much delayed."

This brings us to the modern concept of polarization of light. Observe that the statistical nature of quantum mechanics is different from that of classical physics, as it invokes variables with values that are not merely unknown but unknowable. Theorists quickly appreciated the significance of this result, and by the middle of the twentieth century the development of this field was coming to a close. An important aspect in interpreting the statistical properties of light has been pointed out by Emil Wolf. In 1954, Wolf launched the idea that "... correlation functions of optical fields, not the fields themselves, provide a description of optical phenomena in terms of observable quantities" [2]. The foundations of the modern polarization theory were laid. The starting point for most calculations having a bearing on optical coherence theory is the Wolf coherency matrix. Wolf also introduced more general three-dimensional tensors for dealing with nonplane waves. The statistical description of the properties of light provided a key impetus for a new generation of high-precision experiments, e.g., the Hanbury-Brown and Twiss experiment. With the year 1954 came another big step in optics, when Charles Townes and his coworkers realized the first maser. This laid eventually the ground work for development of the laser, but this is another story [12]. Thus in the early 1960s, the conceptual basis of the modern theory of light polarization was thoroughly formulated.

Other notable contributors in the story of polarized light are Paul Soleillet, who is now largely forgotten but was eventually the discoverer of what we know as the Mueller polarization matrix; Francis Perrin, whose meticulous work in the study of the symmetry property of scattering by particles was very important; Robert Clark Jones,[§] who invented the Jones calculus in polarization optics; Hans Mueller, who described the effect of nonimage forming optical systems and scattering media in terms of the Mueller formalism; Edwin H. Land,[†] who invented the sheet polarizers; van de Hulst, who explained the polarization characteristics of the glory; S. Pancharatnam, who introduced the concept of spectral functions to deal with the description of the polarization properties of a polychromatic beam and eventually was first to introduce the concept of geometric phase[‡] in his study of the interfer-

---

[§] Prof. Russell Chipman kindly informed readers that Robert Clark Jones is still living in the Boston area.

[†] Edwin H. Land received more than 500 patents related to different areas of research, including polarization, photography, and human color vision. The interested reader is referred to the special issue of *Optics & Photonics News* **5**, 9 (1994), dedicated to the memory of Land in recognition of his pioneering contributions to science and technology.

[‡] The geometric phase concerns the phase change of a light beam whose polarization state is made to trace out a cycle on the Poincaré sphere. Its quantal counterpart was discovered by Michael Berry, who proved the existence of geometrical phases in cyclic adiabatic evolutions. It is remarkable that when Pancharatnam discovered this important effect, he was only 22 years of age. The interested reader may consult M.V. Berry, *Current Science* **67**, 220 (1994), and *Geometric Phases in*

ence of light in distinct states of polarization; and Richard Barakat,[*] who added the concept of spectral coherency matrix from a different point of view than the one introduced by Wiener.

The discovery of the laser brought optics back to prominence. There followed the discovery of of nonlinear optics, coherent optics, and quantum optics. Although there are still open questions about the details of the quantum description of polarization, considerable progress has been made. Polarization optics today has reached maturity; certainly the future of polarization optics will be as exciting and fruitful as the past has been.

3.2 Basic Principles and Some Applications

The above brief account suffices to set the stage for the main subject. However, one of the most difficult parts of learning polarization optics is to get a feel for how abstract formalism can be applied to actual phenomena in the laboratory. This section has three basic objectives. The first of these is to clarify operationally what is really meant by polarized light. This section does not attempt to be a comprehensive review of polarized light and of its interaction with optical systems. Rather, it is intended to be tutorial in nature, and the intended reader is a graduate student about to embark on research, either experimental or theoretical, in this area. The goal therefore has been to set out as clearly as possible a set of concepts basic to the understanding of light polarization, and to discuss how they relate to existing experiments. Related to this, the second objective is to lay out the physics of multiple scattering of polarized light (in the visible range) in disordered arrays of dielectric scatterers. The discussion is limited to the particle size and state of polarization dependence of the depolarization lengths. The third and final objective is to explain in direct physical terms why the concept of the scalar invariant is important to describe the second-order statistics of the electromagnetic field.

3.2.1 Polarized light: a statistical optics approach

What is the nature of polarized light? This simple question is, in fact, hard to answer from either the theoretical or observational point of view. However, it is of the utmost importance if we really want to detect and predict the consequences of

---

Physics, edited by A. Shapere and F. Wilczek, World Scientific, Singapore (1989), for historical comments on the development of geometric phases in polarization optics.

[*] Barakat was a hero of mine of whom I had read many pioneering papers. When I came to Harvard University in the beginning of 1989 as a postdoctoral fellow, I experienced the privilege of enjoying such collaboration. One lesson I understood from Dick's way of working is that it is more useful to learn from one's peers than from one's teachers. I have already given elsewhere, i.e., C. Brosseau, *J. Opt. A: Pure Appl. Opt.* **2**, R9–R15 (2000) an account of Barakat's contributions to polarization optics.

polarized light interacting with an optical system or a scattering medium. Before we jump into the physics of polarized light, it will help in getting started if we have in common a bit of vocabulary indicating what the different concepts will mean here.

The objective of this section is not a review of the theoretical exploration of the coherency (density) matrix, which has been done extensively in the past [4,7,8]. Instead the reader is presented with resources adequate to generate a basic familiarity with the principles and language of density matrix theory. I have discussed statistical optics concepts and their implications to polarization optics at length elsewhere [5], although there is still more to say. For technical details of many of the topics discussed here, the reader is referred to [1,4,5]. The usual development is in terms of the coherency matrix Φ, but in order to deal with dimensionless forms of the Stokes parameters we have found it convenient to employ the density matrix approach, i.e.,

$$\mathbf{D} \equiv \frac{\Phi}{\mathrm{tr}(\Phi)},$$

where tr denotes the trace of a matrix.

It has gradually become clear, building on pioneering contributions of Falkoff and MacDonald [13], Fano [14], and Wolf [4,7], that the density matrix formalism has a broad range of applications in the theory of partial coherence of optical fields. Because of the close analogy that exists between the theory of partial coherence and the theory of partial polarization, one might expect that the density matrix is also a workhorse in polarization theory. In that theory, diagonal (respectively off-diagonal) elements of the density matrix elements are interpreted as autocorrelation (respectively cross-correlation) functions between the random components of the analytic signal representation of the electric vector at a particular point in space. It is worth noting that this approach is restricted to second-order statistics of the fluctuating field: the polarization density matrix $\mathbf{D}$ is a 2×2 matrix (hereafter, noted $\mathbf{D}_2$) and is a complete description of a Gaussian distributed plane wavefield. For a non-Gaussian optical field, higher-order statistics of the electromagnetic field is required; however, the second-order approximation may still be a good one provided corrections due to higher-order correlations are small. The theoretical discussions up to now have mainly dealt with light in the form of plane waves. This is due, in part, to the fact that the polarization states of a plane wave can be described by means of the Poincaré sphere representation, which has a simple topology, and in part, to the fact that the assumption of plane wave leads to results adequate for most practical applications. Here we wish to calculate the entropy of a partially polarized wavefield, not necessarily plane. To this end, we will consider polarization density matrices $\mathbf{D}_\mathbf{N}$ for arbitrary N.

3.2.1.1 Lie group expansion of the density matrix and Stokes parameters

Consider a narrowband optical field that can be represented by an ensemble of realizations, which we shall assume to be statistically stationary, at least in the wide sense. Each realization of the fluctuating electric field vector is represented by a complex analytic signal. The Stokes parameters, defined as the covariances of the analytic signal components, are the observables of the field vector at optical frequencies. The key mathematical idea utilized here is that the available information on the wavefield is the density matrix that describes the second-order stochastics of the electric field components at a given point in space. So far no assumption has been made about the statistics that governs the light fluctuations: we limit our description of the statistical properties of the underlying radiation fields to second order.

By definition, $\mathbf{D_N}$ is nonnegative definite and Hermitian: $\mathbf{D_N}$ can be diagonalized by a unitary transformation and its N eigenvalues are real and nonnegative. On the basis of this description, we introduce the normalized Stokes parameters $\Theta_j^{(N)}$, which are defined by the scalar coefficients in the expansion of $\mathbf{D_N}$ in terms of the N^2 Hermitian, trace orthogonal and linearly independent $\mathbf{O}_j^{(N)}$ matrices. An important point to appreciate here is that these real parameters form a quorum of observables that completely specify the state of polarization of the optical field. Let us formally introduce the $N \times N$ polarization density matrix as a linear combination of N^2 independent Hermitian matrices:

$$\mathbf{D_N} = \frac{1}{N} \sum_{j=0}^{N^2-1} \Theta_j^{(N)} \mathbf{O}_j^{(N)}. \tag{1}$$

Equation (1) is known as the special unitary group $SU(N)$ expansion of the polarization density matrix $\mathbf{D_N}$. It is convenient to work with a normalized version of the Stokes parameters $\Theta_j^{(N)}$, since they take a dimensionless form. The expectation value of a physical observable, characterizing the light at any point, described by the density matrix $\mathbf{D_N}$ is given by

$$\left\langle \mathbf{O}_j^{(N)} \right\rangle = \mathrm{tr}\left(\mathbf{O}_j^{(N)} \mathbf{D_N} \right) = \Theta_j^{(N)}, \tag{2}$$

where the angular brackets in the left-hand side denote the average taken over the statistical ensemble representing the fluctuating field. Two important points should be stressed. First, it is usually convenient that one $\mathbf{O}_j^{(N)}$ be the unit $N \times N$ matrix, indicated as $\mathbf{O}_0^{(N)}$. The second important feature is the trace relations, namely the normalization condition

$$\mathrm{tr}\left(\mathbf{O}_j^{(N)}\right) = N\delta_{j0} \tag{3a}$$

and the orthogonality condition

$$\mathrm{tr}\left(\mathbf{O}_j^{(N)}\mathbf{O}_k^{(N)}\right) = 2\delta_{jk}. \tag{3b}$$

For $N = 2$, $\mathbf{O}_j^{(2)}$ are the Pauli matrices. Other sets include the Gell-Mann matrices for $N = 3$ and the Dirac matrices for $N = 4$. At this point it is also worth commenting on several important properties of $\mathbf{D_N}$. Scalar functions of $\mathbf{D_N}$ that are invariant, with respect to all transformations, $\mathbf{D_N} \to \mathbf{U}\mathbf{D_N}\mathbf{U}^{-1}$, where $\mathbf{U}$ is a general unitary matrix, are the scalar invariants of $\mathbf{D_N}$. Simple examples are the trace and the determinant of $\mathbf{D_N}$. The Cayley–Hamilton theorem implies that these basic invariant quantities can be obtained by direct evaluation of the traces of powers of $\mathbf{D_N}$. The physical significance of these scalar invariants as measures of the degree of polarization of the optical field is discussed in the next sections.

3.2.1.2 Entropy of a partially polarized light

We consider an optical field in the form of plane waves propagating in some direction characterized by the unit vector $\mathbf{e}_3$. The transverse field is resolved into two orthogonal components along the directions characterized by the unit vectors $\mathbf{e}_1$ and $\mathbf{e}_2$ corresponding to orthogonal linear polarizations. Note that all matrix quantities will be defined in this linear polarization basis. At this point, we recall that we employ normalized Stokes parameters $\langle \sigma_j \rangle = S_j/S_0$, where $S_j, j = 0, 1, 2, 3$ is the notation for the usual Stokes parameters having the physical dimensions of intensity (or irradiance). The physical interpretation of these parameters is as follows: S_0 is the total intensity, S_1 describes the excess of linearly horizontal polarized light over linearly vertical polarized light, S_2 specifies the excess intensity of 45 deg linearly polarized light over −45 deg linearly polarized light, and S_3 is the excess of right over left circularly polarized light [1,5].

Our starting point is the von Neumann entropy S of the radiation field in the impure (mixed) state represented by the density matrix $\mathbf{D}_2$, which is defined according the usual dimensionless version appearing in quantum statistical mechanics as

$$S = -\mathrm{tr}\left[\mathbf{D}_2\ln(\mathbf{D}_2)\right], \tag{4}$$

where the density matrix reads, in the linear polarization basis, as

$$\mathbf{D}_2 = \frac{1}{2}\begin{bmatrix} 1 + \langle \sigma_1 \rangle & \langle \sigma_2 \rangle - i\langle \sigma_3 \rangle \\ \langle \sigma_2 \rangle + i\langle \sigma_3 \rangle & 1 - \langle \sigma_1 \rangle \end{bmatrix}. \tag{5}$$

The von Neumann measure is a quantitative measure of the amount of information that would be gained by switching from a mixed state to a pure state. This orthogonal decomposition is the most economical representation of $\mathbf{D}_2$ (in the sense of entropy minimization) [5]. Several properties of S are of importance for us. First, it is worth observing that Eq. (5) is basis independent (i.e., the entropy remains invariant under a similarity transformation of the density matrix $\mathbf{D}_2 \rightarrow \mathbf{R}\mathbf{D}_2\mathbf{R}^{-1}$). This is expected since polarization properties must be unaffected by the particular choice of basis. Second, we note that the mapping of $\mathbf{D}_2 \rightarrow S(\mathbf{D}_2)$ is concave: the entropy of a mixed state is greater than the constituent entropies weighted as in the mixing. Consequently, taking linear combinations $\sum_j p_j \mathbf{D}_{2j}$ of density matrices $\mathbf{D}_{2j}$ with real positive coefficients $0 \leq p_j \leq 1$ summing to unity $\sum_j p_j = 1$, we have

$$\sum_j p_j S(\mathbf{D}_{2j}) \leq S(\mathbf{D}_2) \leq \sum_j p_j S(\mathbf{D}_{2j}) - \sum_j p_j \ln(p_j). \tag{6}$$

Equation (6) is an optimal inequality in the sense that equality holds on the left if all $\mathbf{D}_{2j}$ are equal, and on the right if all $\mathbf{D}_{2j}$ have disjoint support. Let us remark in passing that both matrices $\mathbf{D}_2$ and $\ln(\mathbf{D}_2)$ are diagonalized by the same unitary transformation; this comes from the fact that $\mathbf{D}_2$ commutes with $\ln(\mathbf{D}_2)$.

Given these background remarks, we now sketch two simple approaches for deriving the degree of polarization dependence of the radiation entropy. Actually, the methods complement each other in that the first approach involves an eigenvalue problem, whereas the second approach involves a geometric property of the set of polarization states. To begin with, we express Eq. (4) differently in the representation in which $\mathbf{D}_2$ is diagonal. Denoting the eigenvalues of $\mathbf{D}_2$ by λ_j, the entropy can be expressed, via Eq. (4), as

$$S = -\sum_j \lambda_j \ln(\lambda_j). \tag{7}$$

Consequently the problem shifts to the analytic evaluation of these eigenvalues. In this case, computation of these eigenvalues via Eq. (5) gives

$$\lambda_j = \frac{1 \pm P}{2}, \tag{8}$$

where we have set

$$P = \left(\sum_{j=1}^{3} \langle \sigma_j \rangle^2 \right)^{\frac{1}{2}},$$

which is the degree of polarization of the plane wave. Now by substituting Eq. (8) into Eq. (7), the result takes a relatively simple form, namely,

$$S(P) = -\ln[s(P)], \tag{9}$$

with

$$s(P) = \frac{1}{2}(1-P)^{\frac{1-P}{2}}(1+P)^{\frac{1+P}{2}}. \tag{10}$$

Thus a closed-form equation has been arrived at: it is a measure of purity of the states of polarization. Equation (9) demonstrates that one needs only a single parameter, P, to characterize the entropy of a radiation field in the form of plane waves. The entropy varies between 0 and $\ln(2)$ inclusive and is displayed in Fig. 1 as a function of the scalar invariant. From Fig. 1 we observe that the curve undergoes a monotone decrease as the degree of polarization is increased. It does not differentiate pure states, i.e., irrespective of the form of the pure state of polarization considered, linear or circular, we have $S = 0$. These symbols denote respectively linear horizontal, linear vertical, linear $+45$ deg, linear -45 deg, right-handed circular and left-handed circular states of polarization. Minimum entropy states define the pure states and are located at the surface of the so-called Poincaré sphere. Another interesting limit is obtained by considering the maximum entropy state corresponding to the completely unpolarized state, located at the center of the ball Σ_1^3. In all other cases such $0 < S < \ln(2)$, mixed states are the points inside the ball Σ_1^3 [5]. As pointed out earlier, this property reflects the isotropy of the Poincaré sphere Σ_1^2 (Fig. 2).

An alternative approach to the computation of the entropy is via the convexity property of the states of polarization (Fig. 3). This property is of a topological nature and may be simply visualized through the Poincaré sphere representation

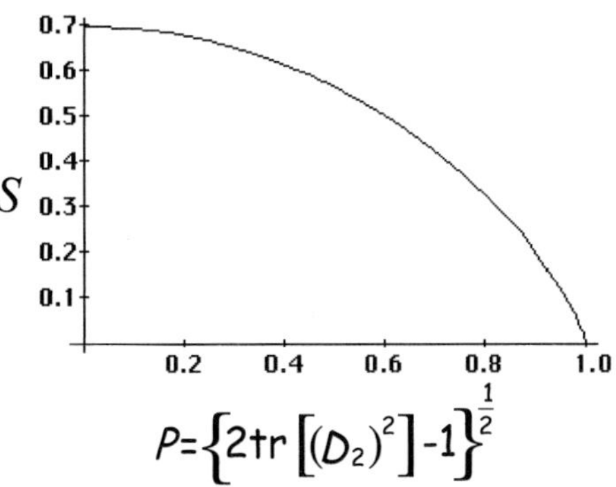

$$P = \left\{2\mathrm{tr}\left[(D_2)^2\right]-1\right\}^{\frac{1}{2}}$$

Figure 1 Plot of the entropy S as function of the degree of polarization P, i.e., scalar invariant $\mathrm{tr}(\mathbf{D}_2^2)$, of a plane wave.

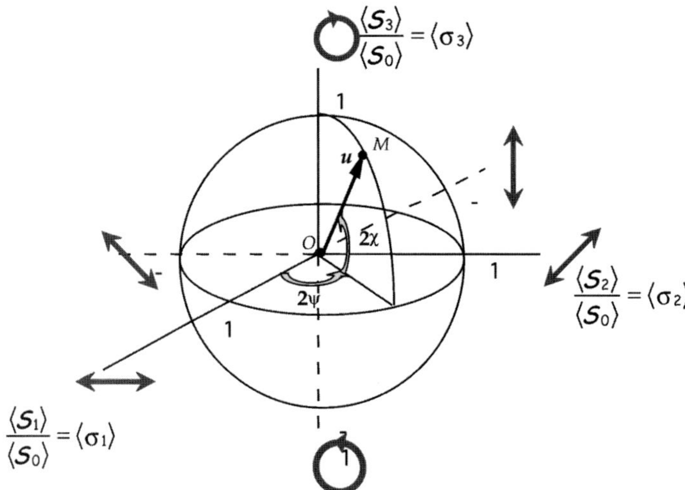

Figure 2 The Poincaré sphere Σ_1^2 is the unit sphere surrounding the origin of the Cartesian coordinate orthonormal basis $(\mathbf{e}_1, \mathbf{e}_2)$. The normalized Stokes parameters $[\langle\sigma_1\rangle, \langle\sigma_2\rangle, \langle\sigma_3\rangle]^T$ constitute the components of the Poincaré vector $\mathbf{u}$ that represents the state of polarization of an arbitrary pure state of polarization ($|\mathbf{u}| = 1$). The longitude 2ψ and latitude 2χ of point M are related to the azimuth and the ellipticity angles of the polarization ellipse of the wave. Each point on Σ_1^2 corresponds to a unique state of polarization. The north pole N ($[0, 0, 1]^T$) represents left circularly polarized light. The south pole S ($[0, 0, -1]^T$) represents right circularly polarized light. Points on the equator ($2\chi = 0$) represent linearly polarized light. Elliptical polarization states lie between the poles and equator. The positive directions of the angle 2ψ and 2χ are defined according the adopted sign convention.

embedded in the three-dimensional Stokes space. For instance, the Krein-Millman theorem, which states that a compact convex set is completely determined by its extreme points, implies that every mixed state can be written as a nonunique convex combination of pure states. States of polarization specified by $\mathbf{D}_2$ form a convex set with pure states being the extremal points of the set. Here the convex set of states possesses two strata of dimensions 2, Σ_1^2, and 3, Σ_1^3, respectively [5]. We can also state a useful theorem that will be employed in the subsequent analysis: any mixed state can be uniquely decomposed into a purely polarized part and an unpolarized part. The relative weight of each component is determined by the degree of polarization P. Thus the density matrix of any mixed state is given by a convex sum

$$\mathbf{D}_2 = P\mathbf{D}_{2p} + (1 - P)\mathbf{D}_{2u}, \tag{11}$$

with $\mathbf{D}_{2u} = 1/2\sigma_0$. Once this is done, it becomes straightforward to derive the entropy. Upon substituting Eq. (11) into Eq. (4), we explicitly obtain Eq. (9).

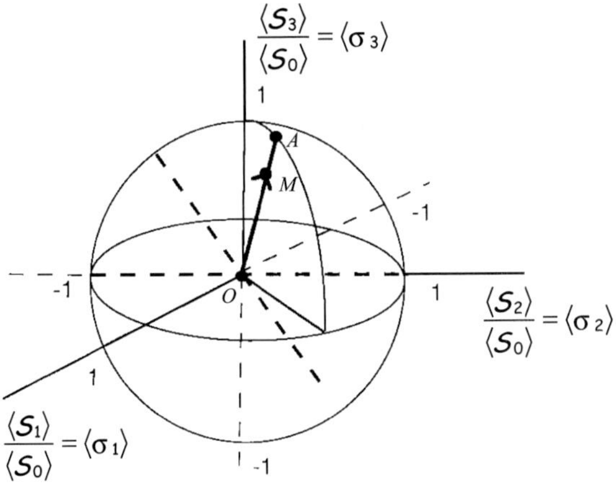

Figure 3 Schematic illustration of the convexity property of the set of polarization states on the unit ball Σ_1^3. Pure states correspond to surface points (e.g., A) and mixed states to interior points (e.g., M). The partially polarized state, described by point M, is represented by the vector **OM** whose length is the degree of polarization P.

3.2.1.3 Thermodynamics of a radiation field

Before we can proceed any further, we are naturally led to inquire whether a more general thermodynamic treatment of partial polarization exists. The purpose of this section is to present such treatment. We first wish to introduce the analogy of the two-level description of a partially polarized wave with a one-dimensional Ising spin system in contact with a heat bath. The Hamiltonian assigned to a particular configuration of spins is $-J \sum_{\langle ij \rangle} \sigma_i \sigma_j$, with each site $\mathbf{r_i}$ having a spin $\sigma_i = \pm 1$. The expression $\langle ij \rangle$ refers to nearest neighbors i and j sites and J stands for the spin-spin coupling. After some calculations, it can be shown that the entropy per spin of such a one-dimensional system may be expressed as

$$\frac{S(y)}{Nk} = \ln\left[2\cosh(y)\right] - y\tanh(y), \tag{12}$$

where $y = J\beta$, $\beta = 1/(kT)$ is the inverse temperature and k is Boltzmann's constant [5,16]. By a straightforward calculation, Eq. (12) can be verified to be identically equal to Eq. (9) if one makes use of the following expression:

$$\frac{1}{\tau} = \frac{1}{2}\left[\ln\left(\frac{1+P}{1-P}\right)\right], \tag{13}$$

and sets $\tau = kT/\beta$, which defines an effective polarization temperature. It is to be emphasized, in connection with Eq. (13), that it should not be confused with the radiance temperature obtained using Planck's spectral law. In Fig. 3 we depict the behavior of P as a function of P. We first observe that τ is a monotonic decreasing function of P. It also clearly shows that $\tau \sim P^{-1}$ for very small values of P and we observe a dramatic change as P approaches 1 from below. At this point two comments are in order. Just as in the study of the Ising system, one can determine the thermodynamic functions for the partially polarized radiation field. Using the thermodynamics of the canonical ensemble the partition function is given by

$$Z = 2(1 - P^2)^{-\frac{1}{2}}. \tag{14}$$

Equation (14) can be used to find other thermodynamic quantities such as the free energy and the internal energy [5,16]. Similarly, as one defines in thermodynamics the equilibrium temperature such as $1/T = \partial S/\partial U$, we may prove that the polarization temperature verifies $1/\tau = \partial S/\partial U = -\partial S/\partial P$. The "specific heat" can be found by a simple expression:

$$C = \frac{1 - P^2}{\tau^2}. \tag{15}$$

For later purposes it will also be useful to consider a correlation length by analogy with the Ising model. We assume a finite correlation length, for $S(\mathbf{r})$, where as usual is defined by the following expression

$$\langle S(r_i)S(r_j)\rangle \sim \exp\left(-\frac{|r_i - r_j|}{\xi}\right). \tag{16}$$

Upon making the substitution, we found that the correlation length behaves as $[\ln(P)]^{-1}$. At small P, we therefore expect the correlations to decay to zero. We will use this expession explicitly in Sect. 3.2.2.2.

While interest has been focused here on the plane wave solution ($N = 2$), it is interesting to note that Brosseau [5] has recently discussed the general case and showed that the entropy depends $N - 1$ scalar invariants and not on the full N invariants. In this author's opinion, there is no clear a priori justification to study, for partial polarization purpose, high dimensions ($N > 3$) of the $SU(N)$ parametrization. However, calculations by Brandenberger et al. [17], have shown that the entropy of stochastic fields with many degrees of freedom has interesting applications in general relativity and cosmology; thus an a posteriori justification for its interest can be claimed.

To summarize, this section contains two important features that motivate our formal considerations below. First, we introduced the concept of scalar invariant. This is undoubtedly the most remarkable property of the density (coherency) matrix description of the electromagnetic field. Conceptually, this is the connection between the geometric definition of polarized light and its algebraic representation in terms of second-order statistical moments. Second, we found that these scalar invariants completely define the entropy of the stochastic radiation field. Thus, we anticipate that depolarization is connected to a process of entropy production.

3.2.2 Multiple scattering of polarized light by spherical diffusers

Apart from these interesting problems, a long-term goal of our investigations is the explicit consideration of polarization effects in multiple scattering of light. Multiple scattering of an initially pure state of polarization from a slab composed of randomly distributed scatterers gives rise to observable phenomena that cannot be explained by single-scattering arguments. Scattering and absorption by particles is traditionally handled using Rayleigh or Mie theories [2].

We now come to the important problem of characterizing the entropy produced by a depolarizing radiation/matter interaction. Despite intensive studies of this problem, progress in this area has been slow. There have been some efforts to determine the entropy transformation by scattering in specific cases, but no rigorous calculation of the entropy production has been presented in the context of multiple scattering of light by randomly positioned ensembles of particles.

In recent years, new physical effects have been identified in elastic multiple scattering of light by ensembles of particles, e.g., significant backscattering enhancement observed in the form of a well-defined narrow peak in the angular distribution of the far-field intensity of the incoherent component of the scattered light at scattering angles near 180 deg [18]. In this section, quantitative expressions are derived for the degree of polarization when incident light in the form of pure states is incident on a spatially random optical medium. The purpose of this section is to analyze in detail the behavior of entropy production during the irreversible evolution of the state of polarization. More specifically, we investigate the consequences of multiple scattering of light by a dense random collection of dielectric spheres on entropy production.

It should be noted, at the outset, that this situation differs from those considered by Enk and Nienhuis [19], and Eu and Mao [20]. The former were mostly interested in investigating the connection between entropy production and kinetic effects of light on atoms or molecules, such as laser cooling and macroscopic flows in gases. The latter introduced a set of semiclassical Boltzmann equations to describe the interaction of a nonequilibrium photon (ideal) gas with matter. What we are principally concerned with is the relation of the depolarization of an incident pure state of polarization to a process of entropy production and the determination of its characteristic length scale. Most of the theoretical work on the transport properties of multiply scattered light has used two characteristic length scales: the elastic mean-free path ℓ, and the transport mean-free path ℓ^{*}, which is defined as the length over which momentum transfer becomes uncorrelated. Using a Monte

Carlo simulation, we discovered recently that another length occurs in the problem, the depolarization length ξ_i (the index $i = L, C$ refers to linear and circular incident pure states of polarization) [21].

There have been two different approaches to the important problem of irreversibility of light scattering. The first approach was initiated by Chandrasekhar [22] and Rozenberg [23] many years ago on the basis of the Boltzmann equation for an isotropic scattering medium and followed by Callies [24] and by Gudkov [25]. The second approach was initiated by Jones [26]. In the first approach, a kinetic theory for treating the optical transport by a phenomenological radiative transfer approach has been introduced. In these theories, use of a statistical description for the covariances of the field by the Bethe-Salpeter integral equation characterizes irreversibility in the multiple scattering process. It was also shown theoretically, not long ago, that the Bethe-Salpeter equation under the ladder approximation of uncorrelated discrete scatterers results in the usual vector radiative transfer equation [22,27–30]. In the second approach, the author had a particular objective: he wished to understand what is different between reversible (e.g., specular reflection of a plane wave incident on a plane surface between two homogeneous isotropic media) and irreversible (e.g., wave scattering by an incoherent array) manipulations of waves. As an additional comment, we note from these earlier studies that the entropy production criterion in the context of multiple scattering of waves by a disordered dielectric medium is a topic that has not been explicitly investigated.

Propagation and scattering of electromagnetic waves in an inhomogeneous medium depends critically on the ratio between the wavelength and the scale lengths of the inhomogeneities. We emphasize at the outset that wave transport through a medium with randomly positioned scatterers can be characterized by a set of significant scale lengths. The first scale is the thickness d of the optical medium. The second scale is the elastic mean-free path $\ell = 1/(\phi\sigma)$, i.e., the average length the wave travels before it suffers an elastic collision. Here ϕ is the concentration of scatterers and σ is a scattering cross section. The third scale worth considering is the transport mean-free path ℓ^*, which is defined as the average distance over which momentum transfer becomes uncorrelated; i.e., the wave propagates a distance of the order of ℓ^* before it forgets completely about its initial direction of propagation. The fourth scale is the wavelength λ. The fifth scale is the size of the scatterers, namely a. Note that in the general case of a polydisperse system, the size, shape, and refractive index distributions of the scatterers are characterized by a distribution of lengths. We will find it useful to introduce a dimensionless size parameter qa, where a is the radius of the particles and q is the wavenumber of the wave. Waves with short wavelengths see a smoothly varying medium, while long waves essentially do not feel the inhomogeneities. We would like to outline here why the sixth scale, i.e., the length ξ of the path over which a polarized wave becomes depolarized, emerges from this analysis and how it depends on whether it is initially linearly or circularly polarized, of the size of the particles, and of the anisotropy of the diffusers that scatter light.

If $d \leq \ell^*$, the inhomogeneities in the medium give rise to only weak elastic scattering. Consequently most of the incident wave is unscattered upon emerging from the sample. The single scattering events dominate; thus, the scattered wave can be conveniently described by the Born approximation in standard scattering theory. At the opposite multiple scattering regime $d \gg \ell^*$ (but still in the weak-localized regime $\lambda \ll \ell^*$), a wave scatters many times before emerging from the sample. The unscattered component of the transmitted wave is exponentially attenuated, by a factor $\exp(-d/\ell^*)$. The typical number of scattering events by a particular wave propagating across the sample is $\sim (d/\ell^*)^2 \gg 1$.

The remainder of this section is organized as follows: we specialize our discussion to the case where the radius of the particles is much smaller than the wavelength of light in the supporting medium (Rayleigh regime), then the results established earlier for the Rayleigh regime in terms of the Mueller scattering matrix are extended to treat the Mie regime.

3.2.2.1 Rayleigh scattering

In this section we want to study the propagation of an incident pure state of polarization in a medium with randomly positioned particles such that multiple scattering effects cannot be neglected. We present a self-contained review of the physics behind the scattering of electromagnetic radiation from a pointlike suspension. The derivation here focuses on the problem of determining the full Stokes vector for a multiply scattered wave.

Let us suppose a quasi-monochromatic, of mean frequency ν_0, plane wave is incident onto the left side of a scattering three-dimensional random medium that occupies a finite volume Ω in free space, as displayed schematically in Fig. 4. The output wave intensity pattern will be a complicated speckle pattern. To describe the scattering of a polarized lightwave we use a right-handed Cartesian coordinate system, referred to as the laboratory reference frame. Figure 4 shows a possible path of a wave entering normally to the system. A useful physical picture for the propagation of the wave as it enters the sample is one in which a wave undergoes a random walk. Each trajectory is composed of straight-line segments and sudden interruptions that randomly change the wave's propagation direction. The average length of each random typical step is the mean-free path ℓ. In the weakly scattering regime, i.e., $\lambda \ll \ell^*$, the wave intensity satisfies the classical diffusion equation. Then for distances that are much larger than ℓ^*, beyond which the direction of light propagation is randomized, light transport can be regarded as a diffusion process with diffusion constant $D = 1/3v\ell^*$, where v is the transport velocity, i.e., the speed of light in the medium. For pointlike scatterers, v is equal to the phase

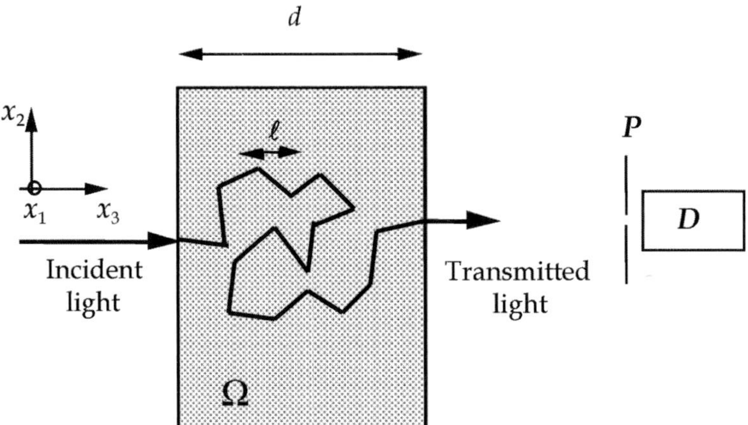

Figure 4 Schematic diagram and notation relating to the scattering geometry. An incident field, in the form of plane waves is scattered by a nonabsorbing medium that occupies a finite volume Ω and of thickness d, consisting of uncorrelated spherical pointlike particles (Rayleigh scattering). Typical scattering path executing zigzag random walk through the medium (propagating "channel"). The mean-free path ℓ is the typical step size.

velocity, which is approximately equal to the velocity of light divided by the index of refraction. It is worth mentioning that, unlike the transport mean-free path ℓ^* that is obtained experimentally from steady-state measurements, the diffusion constant D is obtained from dynamical measurements. The density of scatterers must be small enough to allow the weak scattering approximation to be valid.

Besides being nonabsorbing, the scattering medium is assumed to be time-invariant, nonmagnetic, spatially nondispersive, and such that the spatial fluctuations of its dielectric susceptibility $u_{ij}(\mathbf{r})$ tensor are statistically homogeneous and stationary in space (at least in the wide sense). The incident and scattered beams are normal to the surfaces of the scattering medium and the coordinates system lies parallel to the slab faces. Typical realization of such medium would be a collection of discrete pointlike, optically inactive scattering centers whose size is very small compared to the wavelength of the scattered radiation (i.e., $qa \ll 1$). This approximation permits the small sphere to be treated as a dipolar oscillator with its polarizability determined by the optical constants of the particle.

We also assume that the temporal fluctuations of the scatterers are sufficiently slow relative to the period of the field oscillations so that the scattering medium behaves as if it is essentially time-invariant (i.e., adiabatic approximation). The usual boundary conditions require continuity of the magnetic field $\mathbf{H}$ and tangential electric field at every discontinuity surface. From the above assumptions, we may characterize the dielectric susceptibility of the three-dimensional medium by

$u_{ij}(\mathbf{r}) = u(\mathbf{r})\delta_{ij}$, of zero mean and white-noise correlation function:

$$\langle u(\mathbf{r}_1)u(\mathbf{r}_2)\rangle = u\delta(\mathbf{r}_1 - \mathbf{r}_2) \text{ when } r_1 \in \Omega \text{ and } r_2 \in \Omega$$
$$= 0 \text{ otherwise,} \qquad (17)$$

where u is a constant that is a measure of the scattering potential. Let us add a further condition: we will consider only weak disorders such that the elastic mean-free path $\ell = 6\pi/(uq_0^4)$ is much larger than the wavelength of the radiation (i.e., $q_0\ell \gg 1$), q_0 is the free space wavenumber associated with the frequency ν_0; consequently, the wavefield propagation may be described by a classical diffusion process. Finally, we assume that the fluctuations of the medium and the fluctuations of the incident field are statistically independent. These restrictions do not present severe difficulties to experimental practice.

All information about an elastic scattering process is contained in the 16-element Mueller matrix. A number of restrictions are placed at the outset on the form of the **M**-matrix depending on the symmetry and reciprocity requirements. On the one hand, the Mueller matrix should show Perrin symmetry that holds for elastic scattering from isotropic suspensions of particles; i.e., in that case **M** is diagonal for normal incidence. On the other hand, the optical medium is nondissipative. Upon introduction of these symmetries, the general form of the Mueller matrix **M** can be written [5] as

$$\mathbf{M} = \begin{bmatrix} 1 & 0 & 0 & 0 \\ 0 & m_{11} & 0 & 0 \\ 0 & 0 & m_{11} & 0 \\ 0 & 0 & 0 & m_{33} \end{bmatrix}. \qquad (18)$$

Now we analytically evaluate the elements m_{ii}, $i = 1,2,3$ in Eq. (18) by an argument of maximum entropy. We proceed as follows. The entropy production per scattering reads as

$$\Delta S(n) = S[P(n+1)] - S[P(n)] = h_s(n), \qquad (19)$$

where $P(n)$ denotes the degree of polarization after $n+1$ scattering events. Here $S(P)$ is given by Eq. (9) and the subscript s indicates that h_s depends on the particular state of polarization. With the help of Eq. (19), the total entropy production after $n+2$ scatterings reduces to

$$\Delta S = S[P(n+1)] - S[P(0)] = \sum_{j=0}^{n} h_s(j) = \ln\left\{\frac{s[P(0)]}{s[P(n+1)]}\right\}. \qquad (20)$$

The function $h_s(x)$ is taken to be a monotonically decreasing function from $h_s(0)$ down to zero; this is to be expected from the theory of irreversible thermodynamics. We are seeking a candidate function $h_s(x)$ in the metric space L^2 and satisfies the condition $h_s''(x) > 0$, where the prime indicates differentiation with respect to x. At this point it is necessary to postulate the functional form of $h_s(x)$: we have chosen to work out the function $h_s(x) = \psi\exp(-\chi x)$, which meets the above requirements.

Next, we consider an incident pure state of polarization that is linearly polarized $[P_{n+1} = m_{11}(n)]$. From Eq. (20), we arrive at the relation

$$s(m_{11}) = \exp\left[-\sum_{j=0}^{n} h_s(j)\right] = \exp\left\{-\psi\left[\frac{1-\exp(-\chi n)}{1-\exp(-\chi)}\right]\right\}. \tag{21}$$

This is equivalent to saying that

$$m_{11}(n) = s^{-1}\left(A\sum_{j=0}^{n-1} B^j\right), \tag{22}$$

where we have set for notational convenience $A = \exp(-\psi)$ and $B = \exp(-\chi)$. We call attention to the important fact that B can be written from Eq. (6.4) as $B = 1 - S[P(1)]/\ln(2)$; consequently, B is fully determined by double scattering. Moreover, when maximum entropy is achieved (i.e., in the limit $n \to \infty$), we require that $A^{1/(1-B)} = 1/2$; i.e., in the limit $n \to \infty$. Putting everything together, we get the final expression for m_{11}:

$$m_{11}(n) = s^{-1}\left[2^{(B^n-1)}\right]. \tag{23}$$

Physically this procedure allows the successive orders of iteration to be expressed in terms of the sole parameter B. It is worth noting that the same formula will apply for a pure state which is circularly polarized [i.e., $P_{n+1} = m_{33}(n)$] with a change in the value of B. Note that this method is quite general and may be used for more involved Mueller matrices, but it does not tell us what kind of trial functions $h_s(x)$ are to be used. This maximum entropy argument can incorporate any function $h_s(x)$ that satisfies the physical constraints.

Now that we have the above result, we can compare with the exact result. The problem shifts to the explicit calculation of the dependence of the degree of polarization of light in function of the number of scattering events when the medium fulfills the assumptions stated above. Indeed, one can evaluate exactly the Mueller matrix elements $m_{ii}(n)$ by the Bethe-Salpeter equation handled in the latter approximation. We emphasize that this derivation is valid over distances greater than the mean-free path. The problem of evaluating the coherency matrix reduces to a matrix eigenvalue problem. We leave the details of formulas to the references [5,21]. Next we consider a pure state of polarization and arbitrary degree of spatial coherence, of unit intensity incident normally on the slab-shaped medium; its Stokes vector writes as follows:

$$
\mathbf{S}_i =
\begin{bmatrix}
\langle |E_1|^2 \rangle + \langle |E_2|^2 \rangle = 1 \\
\langle \sigma_1 \rangle = \langle |E_1|^2 \rangle - \langle |E_2|^2 \rangle \\
\langle \sigma_2 \rangle = \langle E_1^* E_2 + E_1 E_2^* \rangle \\
\langle \sigma_3 \rangle = i \langle E_1^* E_2 - E_1 E_2^* \rangle
\end{bmatrix}.
\tag{24}
$$

In the limit of weak scattering, the linear response of the scattering medium is determined by the ensemble averaged covariance satisfying the Bethe-Salpeter equation [5,21]. Following this approach, we obtain the expression for the output Stokes vector:

$$
\mathbf{S}_o =
\begin{bmatrix}
1 \\
\langle \sigma_1 \rangle \, m_{11}(n) \\
\langle \sigma_2 \rangle \, m_{11}(n) \\
\langle \sigma_3 \rangle \, m_{33}(n)
\end{bmatrix},
\tag{25}
$$

where the subscripts 1 and 2 label components with respect to the Cartesian coordinate system chosen, $m_{11}(n) = 3(7/10)^n/[2 + (7/10)^n]$, and $m_{33}(n) = 3(1/2)^n/[2 + (7/10)^n]$, $n + 1$ being the number of scattering events. It is readily verified from Eq. (25) that the Mueller matrix of the scattering medium has the kind of symmetry we expect from Eq. (18). Having found the form of the Stokes vector $\mathbf{S}_o$, we might naturally inquire as to what form the output degree of polarization has. From Eq. (25), P_o takes the form

$$
P_o = m_{11}(n) \left[\langle \sigma_1 \rangle^2 + \langle \sigma_2 \rangle^2 + \langle \sigma_3 \rangle^2 \left(\frac{5}{7} \right)^{2n} \right]^{\frac{1}{2}},
\tag{26}
$$

which involves three independent parameters.

Several comments may be in order. This equation is in accordance with the fact that single scattering (i.e., $n = 0$) by pointlike particles having spherical symmetry preserves the state and degree of polarization. For concreteness, it is worthwhile to specialize Eq. (26) to some special cases of interest. For instance, an input linear

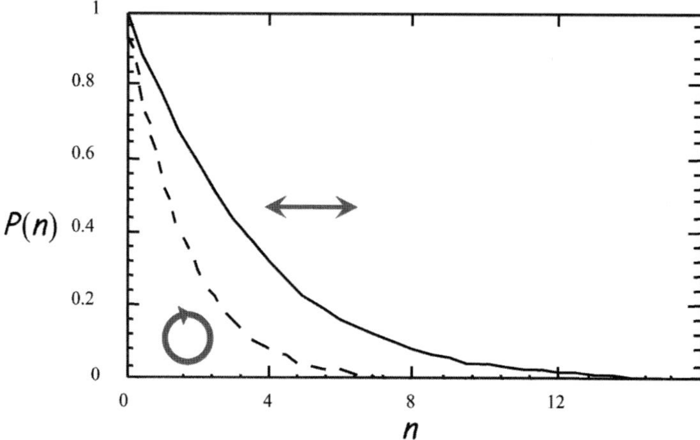

Figure 5 Degree of polarization of scattered light as a function of the number of scatterings n (Eq. [28]) for an input pure state of linear parallel polarization (solid line), right circular polarization (dashed line).

polarization state $(\mathbf{E} = \mathbf{e}_1)$ has for output Stokes vector

$$\begin{bmatrix} 1 \\ m_{11}(n) \\ 0 \\ 0 \end{bmatrix},$$

degree of polarization $P_o = m_{11}(n)$, which is a monotonically decreasing function of the number of scatterings (see Fig. 5). Similarly, for an input right-handed circular polarization state $[\mathbf{E} = (1/\sqrt{2})(\mathbf{e}_1 - i\mathbf{e}_2)]$, one gets

$$\begin{bmatrix} 1 \\ 0 \\ 0 \\ m_{33}(n) \end{bmatrix},$$

$P_o = m_{33}(n)$. Then, the two functions m_{11} and m_{33} have clear physical meanings. A curve showing the behavior of $m_{33}(n)$ is also shown in Fig. 5. The process of depolarization cannot be assimilated to an isotropic contraction of the Poincaré sphere but induces a symmetry breaking, i.e., the symmetry of $SO(3)$ is broken. Figure 6 illustrates the asymmetric depolarization, as n is increased, through the change of symmetry of the surface $S_2^{P(n)}$ passing from a sphere $(n = 0)$ to a prolate ellipsoid $(n > 0)$.

To further discuss the physical significance of these results, we have also plotted the variation of the entropy production $\Delta S(n) = S[P(n + 1)] - S[P(n)]$ with the

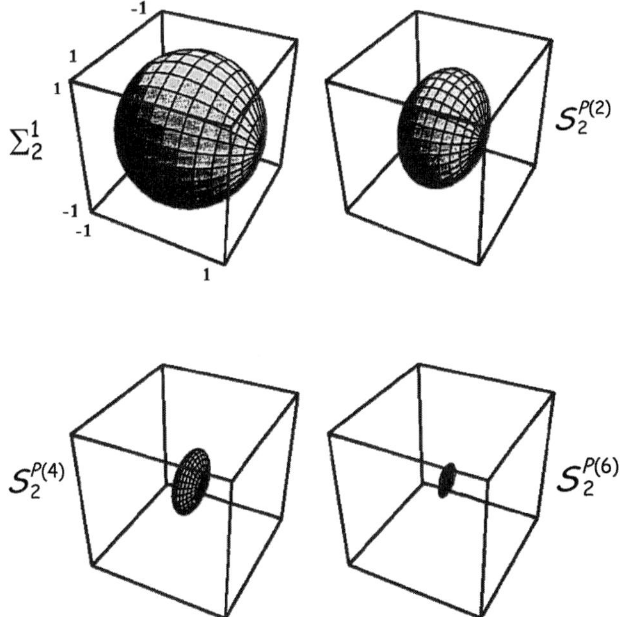

Figure 6 Parametric plots of the surface $S_2^{P(n)}$ for different values of the number of scattering events n. Rayleigh scattering by pointlike dielectric spheres: (a) $n = 0$, (b) $n = 2$, (c) $n = 4$, and (d) $n = 6$.

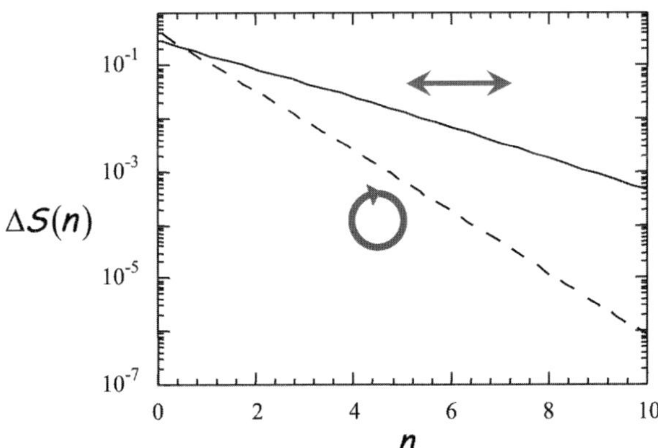

Figure 7 Entropy production $\Delta S(n) = S[P(n + 1)] - S[P(n)]$ as a function of the number of scattering events. Same symbols as in Fig. 5.

number of scattering events (Fig. 7). As can be seen, $\Delta S(n)$ is well represented by an exponential decay $\Delta S(n) = a\exp(-bn)$, with a and b depending on the particular state of polarization. For large values of n (say, $n \geq 10$), the entropy of radiation is unaffected by further scattering; it defines the steady state of maximum entropy $[S(P = 0) = \ln(2)]$ attainable by multiple scattering. In closing, it is worth observing that in the limit of large number of scattering events, Eq. (18) approaches the Mueller matrix of an ideal depolarizer [5]

$$\lim_{n \to \infty} \mathbf{M} = \begin{bmatrix} 1 & 0 & 0 & 0 \\ 0 & 0 & 0 & 0 \\ 0 & 0 & 0 & 0 \\ 0 & 0 & 0 & 0 \end{bmatrix}. \tag{27}$$

3.2.2.2 Mie scattering

Thus far the analysis in the previous subsection has been confined to situations in which the size of the scatterers is very small relative to the wavelength of the probing radiation, i.e., Rayleigh region. We now focus attention on a natural extension of the theory to cover the interesting situation of scattering by objects whose size is of the order of the wavelength or larger. A computational Monte Carlo algorithm is used to perform a simulation of the complete Stokes parameters for multiply scattered radiation in an inhomogeneous system composed of uncorrelated spherical particles. These simulations indicate that the amount of depolarization generated from multiple scattering depends on such factors as the size and shape distribution and index of refraction. The primary aim of this subsection is to present a theory for predicting the effect of particle size on the quantity of our primary interest, i.e., the degree of polarization, which is then compared with Monte Carlo simulation studies. These numerical results are compared to measurements on suspensions of polystyrene latex spheres in water.

The method of Monte Carlo simulation is well known in the context of statistical mechanics and condensed-matter physics; for a recent review see Lewis and Miller [31]. The Monte Carlo modeling technique provides a way of finding solutions to multiple scattering effects by tracing histories, i.e., sequences of events, that statistically occur to waves propagating through an optical medium. The following is a develoment of such a method. Here we are particularly concerned about the size parameter and optical depth dependences of the characteristic lengths of depolarization. A recent review article on Monte Carlo results that is complementary to the present subsection was presented by Bruscaglioni and colleagues [32]. The mathematical and statistical assumptions inherent in this procedure are well known in the literature and are not covered in depth in this subsection.

In this numerical experiment, one generates a realization of the random medium and calculates the resulting wavefield. We have used the SLAB Monte Carlo simulation code to analyze the depolarization behavior of a wave propagating through a slab of finite thickness and composed of uncorrelated spherical particles [21,33]. This simulation technique was developed to study the three-dimensional random walklike multiply scattering process of the wave propagation. In the SLAB code, the three-dimensional paths for waves are followed from one scattering to the next as the wave propagates into the medium. Each scattering is assumed to be elastic and is described by the standard Mie theory. A single numerical simulation consists in launching some number of waves, at a source, along a specified axis. Referring to Fig. 4, a typical scattering path consists of a series of linear translations of random length (of average value equal to the mean-free path ℓ), each of which is followed by a change of flight direction. Selection of the new flight direction is made by generating a random number from a scattering distribution function. The numerical implementation of this algorithm was checked through comparison with the Rayleigh regime, for which exact analytic results are known. The theoretical details pertinent to the testing of the Monte Carlo code are reviewed in Martinez [33].

We still assume in our subsequent discussion that the situation of weak scattering limit and absorption can be regarded as negligible. Consider a quasi-monochromatic plane wavefield that is incident normally along the x_3 axis upon a plane-parallel slab, of finite thickness d $(d \gg \ell)$ and of infinite extent in the x_1, x_2 directions, composed of uncorrelated spherical particles of radius a (Fig. 4). We begin by computing the degree of polarization of the light transmitted by the scattering medium for incident linearly (P_L) and circularly (P_C) polarized light. To do this, one must evaluate the different contributions of light following many different paths. Take a particular sequence of scattering events. The number of steps in this path of length s is $n = s/\ell$. The number of scattering paths of length s is simply given by the Green's function $G(n, d)$ of the diffusion equation. The degrees of polarization are given by a proper weighting $G(n, d)$ of scattering paths of length s. The resulting expressions are

$$P_i = \frac{\displaystyle\int_0^\infty f_i(n)G(n,d)\,dn}{\displaystyle\int_0^\infty G(n,d)\,dn}, \tag{28}$$

where we have adopted the notation $i = L$ for linear and C for circular states of polarization. In the multiple scattering regime, the functions f express the dependences of the output degrees of polarization for a number of scatterings equal to $n + 1$. In the large n limit, the f's simply reduce to

$$f_L(n) \cong \frac{3}{2}\exp\left(-n\frac{\ell}{\zeta_L}\right)$$

and

$$f_C(n) \cong \frac{3}{2}\exp\left(-n\frac{\ell}{\zeta_C}\right)$$

[21]. The ζ's define characteristic lengths of depolarization for a path of $n + 1$

scatterings: $\zeta_L = \ell/\ln(10/7)$ and $\zeta_C = \ell/\ln(2)$. From the numerical values of the ζ's we find that $\zeta_L \cong 2\zeta_C$. Upon performing the integration in (28), we obtain

$$P_i = \frac{d}{\ell} \frac{\sinh\left(\ell/\xi_i\right)}{\sinh\left(d/\xi_i\right)}, \tag{29}$$

where $\xi_i = (\zeta_i\ell/3)^{1/2}$, with $i = L$, and C defining the characteristic lengths of depolarization for the slab geometry for linear and circular states of polarization, respectively. Since $d \gg \xi_i$, the degree of polarization of the transmitted light in the far field can be approximated by

$$P_i \cong \frac{2d}{\ell}\sinh\left(\frac{\ell}{\xi_i}\right)\exp\left(-\frac{d}{\xi_i}\right). \tag{30}$$

Thus we see that the characteristic length of depolarization for incident linearly polarized light is greater (by a factor of $\sqrt{2}$) than the corresponding length for incident circularly polarized light. This analysis should apply equally well for large spheres provided that ℓ is changed into the transport mean-free path $\ell^* = 1/(\phi\sigma^*) = \ell/[1-\langle\cos(\theta)\rangle]$, where $\langle\cos(\theta)\rangle$ is the mean cosine of the scattering angle θ, and that the appropriate size dependence of the f's is inserted therein. Here the transport scattering cross section for each scatterer is defined in the usual way as $\sigma^* = \int_\Omega \sigma(\theta)[1-\cos(\theta)]d\Omega$.

We now move on to the numerical results. Through the use of the above numerical algorithm, a set of different simulations was performed to investigate the effects of the particle size and medium thickness. The input parameters are the relative refractive index, $m = n_S/n_M = 1.20$, where n_S and n_M are the refractive index of the spheres ($n_S = 1.59$ for polystyrene) and of the surrounding medium ($n_M = 1.33$ for water), the size parameter qa and $k\ell^* = 1000$. These parameters were chosen for the purpose of comparison with experimental data. The experiments were carried out at room temperature, using a setup similar to that described in Bicout and Brosseau [21], which contains all relevant details. A semiconductor laser emitting at 0.67 µm was used as the source beam. The beam was normally incident on one side of the sample (3-mm thickness) and the scattered light transmitted through the back wall of the sample cell was detected within a solid angle of collection of 2 deg. The scattering medium consists of various concentrations of polystyrene spheres (PolySciences, Inc.), with mean diameters of 0.22, 0.48, and 1.05 µm, and which were mixed into filtered distilled water and serve as the scattering centers.

Results of simulations that were carried out are presented in Figs. 8–12. For a starting point, we refer to Eq. (30) and find, in the limit $qa \ll 1$, that $\xi_L = 0.967\ell$

and $\xi_C = 0.684\ell$. The wave becomes depolarized over a distance that is of the order of the mean-free path. Now examine Fig. 8, which shows the degrees of polarization for linearly and circularly polarized light as a function of d/ℓ for $qa = 1.19$ (i.e., intermediate Rayleigh-Mie region) and $qa = 6.43$ (Mie region) in a semilogarithmic plot. The curves of Fig. 8 exhibit linear behavior in this plot. It is interesting to observe that the effect of the input polarization state in these regions is markedly different from that in the Rayleigh region. It is important to appreciate that for $qa \sim 1$, the slopes of these plots do not depend on the incident state of polarization. This is in contrast with the region $qa > 1$, for which these slopes (ℓ/ξ_i) now depend strongly on polarization: the slope being greater for linearly polarized light than for circularly polarized light. Having considered the numerics, we now proceed to compare with data from measurements on suspensions of polystyrene-latex spheres. Variations of the degrees of polarization for three values of the size parameter, viz., $qa = 1.23$, 2.69, and 5.89 are shown in Fig. 9. It is remarkable that the experiment gives an exponential decay over several decades. The behavior of these data is consistent with the simulation results of Fig. 9. Figure 10 shows the characteristic lengths of depolarization for incident linearly, ξ_L/ℓ, and circularly, ξ_C/ℓ, polarized light as a function of the dimensionless size parameter qa. In the case of large particles when compared to the wavelength, a linearly or circularly polarized wave becomes depolarized over a distance that is significantly greater than the mean-free path. Both the experimental data and simulations show that the ξ are nearly equal in the region $qa \sim 1$, as can be seen in the inset of Fig. 10, where we have plotted the variation of the

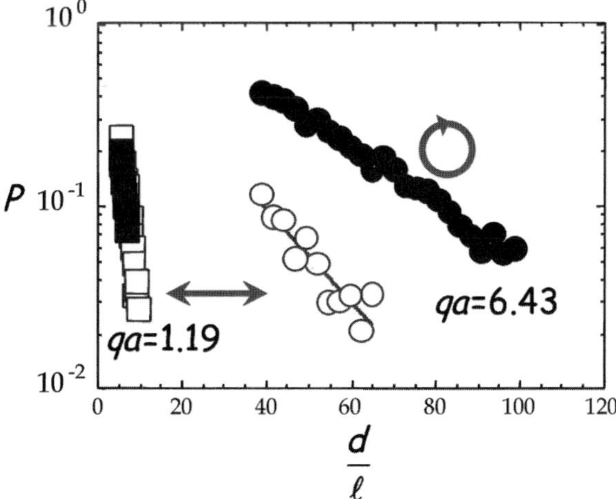

Figure 8 Semilogarithmic plot of the degrees of polarization for linearly polarized (open symbols) and circularly polarized (filled symbols) light as a function of the optical depth d/ℓ. Squares are for the intermediate region $qa = 1.19$ and circles correspond to the Mie region $qa = 6.43$. The lines are exponential fits to the data.

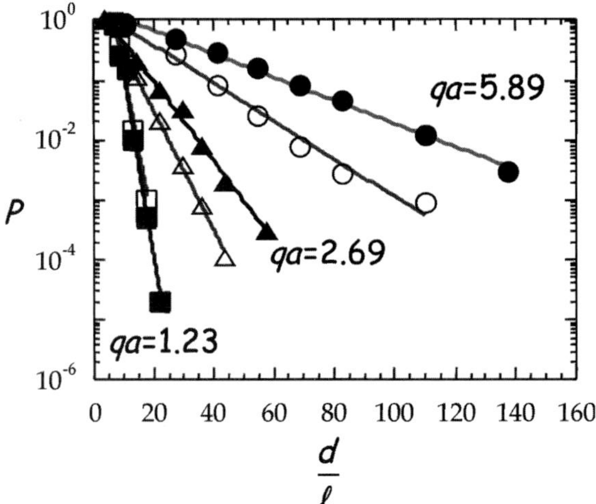

Figure 9 Same as Fig. 8. Experimental data correspond to measurements on suspensions of polystyrene-latex spheres in water (0.22 μm (circle), 0.48 μm (triangle), and 1.05 μm (square)).

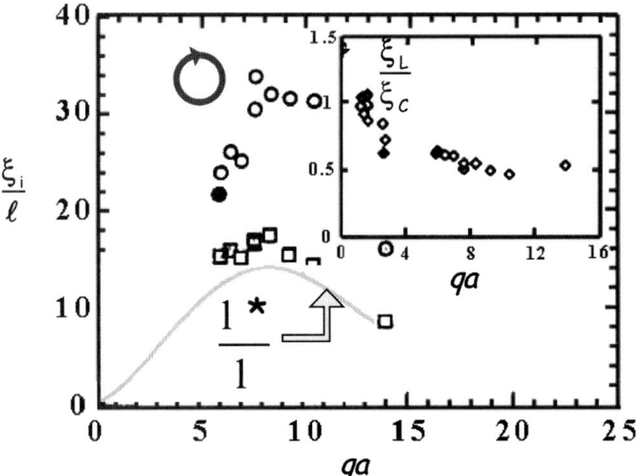

Figure 10 The characteristic lengths of depolarization, ξ_L/ℓ, for linearly polarized (square) and, ξ_C/ℓ, for circularly polarized (circle) light as a function of the dimensionless size parameter qa, as calculated by Monte Carlo simulation. Filled symbols indicate experimental data corresponding to measurements on suspensions of polystyrene-latex spheres in water. The inset shows the ratio of the characteristic lengths ξ_L/ξ_C (open diamonds) as a function of qa. Experimental data (full diamonds) correspond to measurements on suspensions of polystyrene-latex spheres in water. One sees that there is a good agreement with experiment. The solid curve shows the variation of ℓ^*/ℓ as a function of qa.

length ratio ξ_L/ξ_C with the size parameter qa. For the Rayleigh region, this ratio can be computed exactly using Eq. (30) and is equal to $\sqrt{2}$. In the range of sizes investigated, this ratio is a decreasing function of the parameter qa. As can be seen, the numerical calculations are in good agreement with experimental data.

The discussion would be incomplete without considering some additional related developments concerning entropy production

$$\Delta S = S\!\left(qa, \frac{d + \ell^*}{\ell^*}\right) - S\!\left(qa, \frac{d}{\ell^*}\right).$$

The effect of optical thickness is illustrated in Fig. 11 for an incident circularly polarized wave. Since identical behavior is observed for incident linearly polarized waves, we concentrate on the circularly polarized case only. It is clear from looking at this figure that the entropy production falls off exponentially with the optical thickness for $d/\ell^* \gg 1$. The dependence of the entropy production on qa, hence on the degree of forwardness of the scattering is depicted in Fig. 12. This figure shows a comparison between the size dependence of the entropy production for fixed values of d/ℓ^*. Notice that a minimum of entropy production at $qa \sim 1$, corresponding to a maximum of entropy, separates a domain of decreasing entropy production for small particles from a domain of increasing entropy production for large particles. As before, this behavior is interpreted as arising from the anisotropic property of the scattering. Furthermore, if we compare the three curves in Fig. 11, it is interesting to observe that ΔS is significantly smaller when d/ℓ^*

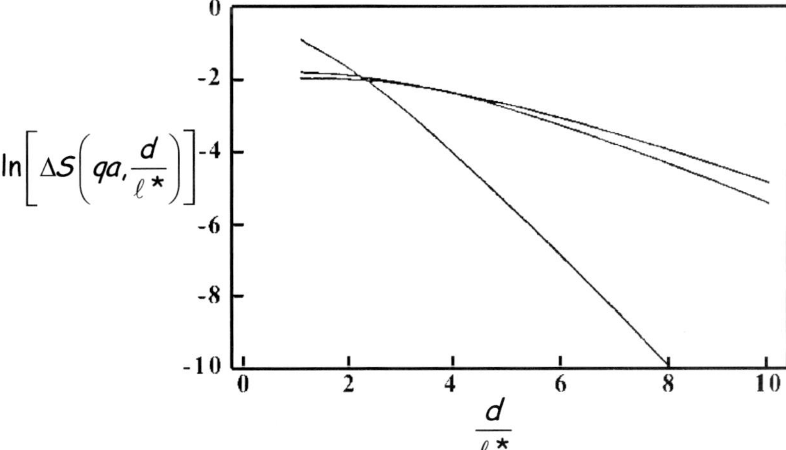

Figure 11 Optical-thickness dependence of entropy production $\Delta S = S[qa, (d + \ell^*)/\ell^*] - S(qa, d/\ell^*)$ plotted as $\ln(\Delta S)$ vs. d/ℓ^* for a circularly polarized incident light beam and fixed values of the size parameters qa. The values of qa from the top are 3.5, 5, and 7.

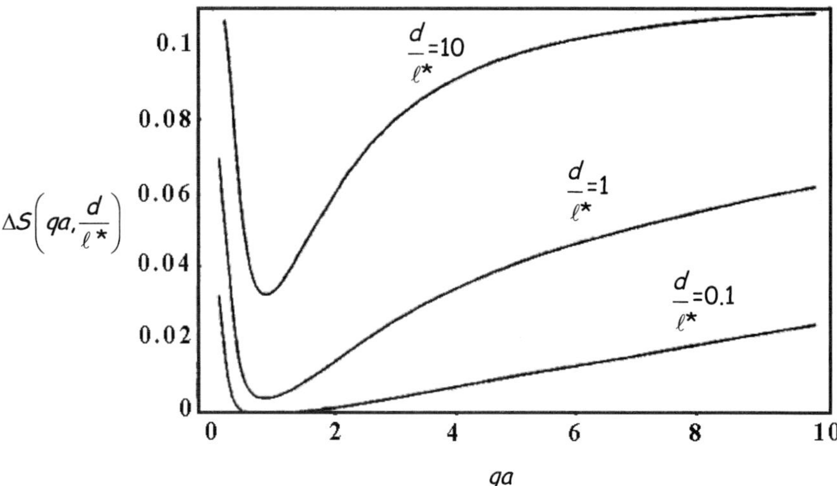

Figure 12 Entropy production as a function of size parameter qa for a circularly polarized incident light beam and fixed values of the optical thickness d/ℓ^*. The values of d/ℓ^* from the top are 10, 1, and 0.1.

increases. We refer to Refs. [5] and [21] for a discussion of the physical reasons for why the depolarization process actually implies that the system should tend to produce entropy according an exponential law.

The difference between the two types of behavior corresponding to $qa \ll 1$ and $qa > 1$ stems from the anisotropic property of the scattering. In the Mie regime, the scattering is predominantly in the forward direction, while in the Rayleigh region, forward and backward scattered directions are treated on an equal footing. It is also appropriate to recall that the direction of linear polarization is not affected by a single scattering of light from a particle in the backward direction, regardless of particle size. It should be noted that the variations of ξ_L/ℓ and ℓ^*/ℓ as function of qa are close to each other (cf. solid line in Fig. 10). This supports the idea that the mechanism of depolarization for an incident linearly polarized state originates from the randomization of the direction of the wave. On the other hand, backscattering acts as an optical mirror (right $\leftrightarrow$ left) for circularly polarized states, i.e., helicity flip. The helicity being preserved over large distances compared to ℓ^* for large scattering particles explains why the characteristic length for incident circularly polarized light is greater than the corresponding length for incident linearly polarized light. Two distinct mechanisms contribute to the depolarization of circularly polarized light: the randomization of the direction and the randomization of the helicity. However, it is difficult to infer from the preceding arguments what the contribution of these two mechanisms to the total length ξ_C will actually be.

Two comments may be in order here. Since in most experiments absorption is weak, it has been neglected. However we would like to stress that the effect of absorption may be an important issue in many cases. Moreover, it is now well established that when scatterers of a finite size are considered, Mie resonance effects (not considered here) should be taken into account. In the resonance region of particle size parameters, the Rayleigh and geometric-optics approximations are inapplicable and numerical methods for characterizing scattering must be based on directly solving Maxwell's equation.

3.2.3 New applications

Over the past decade polarization and coherence notions have continued to find new applications in their traditional areas of use, such as optical vortex generation and singularities in paraxial vector fields. Recently, several groups [34–37] have pursued a morphological approach of polarization patterns in the transverse plane and showed that these patterns contain different types of singular points of polarization that are characterized by conserved topological numbers (singularities). Singularities are directions in anisotropic media in which two or three waves have coincident phase velocities [38,39]. Berry [34] showed that in complex vector waves, there are two sorts of polarization singularity. On the one hand, the polarization is purely circular on lines in space or points in the plane (C singularities). On the other hand, the polarization is purely linear on lines in space for general vector fields, and surfaces in space or lines in the plane for transverse fields (L singularities).

Another interesting example among the many uses of the polarization parameterization is the current research focused on the development of optical fiber transmission systems. Since the invention of the Er-doped optical amplifier a decade ago, commercial optical fiber communication systems have increased in transmission capacity by almost four orders of magnitude. This increase has been made possible by the advent of wavelength division multiplexing (WDM), in which a large number of wavelength channels are used simultaneously to transmit information. Given the complexity and cost of modern-day optical communication systems, accurate design modeling of optical fiber transmission has become absolutely necessary. One of the key problems to be modeled is the polarization mode dispersion. The combination of polarization-dependent loss, polarization-dependent gain, and polarization mode dispersion can lead to fading in long-distance undersea systems. This effect can be counteracted by polarization-scrambling the signal, but polarization-dependent loss can lead to repolarization.

Physics and its techniques have played a significant role in biology for decades. X-ray crystallography and nuclear magnetic resonance are essential tools for structural biologists. Medicine is also beginning to discover the benefits of polarized light. This is not surprising since of the four basic forces in nature, only the electromagnetic interaction is effective on a scale comparable to that of biological organisms. One of the most promising applications lies in the field of biomedical diagnosis of tissues. Optical materials found in biology provide numerous starting places for biomimetic design or direct use of biomolecular materials. There is considerable recent interest in using polarized characteristics as contrast parameters to investigate turbid biological media. This originates from the fact that most mammalian tissues are weakly absorbing in the 600–1300-nm wavelength range; however, the scattering properties of tisues at these wavelengths are significant, meaning that the near-IR light is subjected to extensive scattering. An excellent review by Barbour et al. [40] provides a comprehensive view of the experimental studies devoted to the use of polarized light in biomedical applications. The hope is that polarized light will be useful to isolate ballistic photons from the diffuse background and thus enhance the spatial resolution in optical tomographic methods. A number of experimental studies have appeared showing that relevant information can be obtained from measurements of the spatially dependent response of the medium to a polarized point source [41–43]. Scientists at the forefront of this research have shown how polarized light is already being used to diagnose skin diseases by the in vivo and in vitro characterizations of the biological tissues [44,45]. In a similar way, the polarization information encoded in the diffusely scattered light is potentially useful for body glucose monitoring [46,47] and microstructural bioimaging [41,43]. Vitkin and coworkers [48] have clearly shown that the dependence of the degree of polarization on pigmentation of tissues, and on optical properties in general, must be fully accounted for if the polarization information is to be used for quantitative analysis of biological tissues. Most recent experiments merely add to the list of problems yet to be understood.

As already noted at the beginning of this chapter, polarization is an important aspect of the visual world that is hidden to us. However, polarization is detected by most insects and by many vertebrates other than mammals. Sweeney et al. [49] have recently shown how polarized light is used by several species of butterflies, which are physiologically sensitive to polarization for mate recognition, or detection of any object in a terrestrial environment.

3.3 Additional Remarks and Future Directions

In this chapter the focus has been on some basic aspects of polarization optics, summarizing the optics community's current understanding of polarization and the various formalisms that describe how polarized radiation interacts with matter. It is fascinating to try to write about this. We have known ever since Stokes and Poincaré that polarization is specific to the vector nature of the electromagnetic field. But it is only through decades of trying to measure the correlations between the fluctuating field variables that we have come up with a formulation of polarized light in terms of statistical physics concepts. The usefulness of the entropic formalism for studying an interaction of radiation with matter has been noted before, but for the first time we have exploited the information contained in the polarization of the radiation. Thus, unlike previous workers who have dealt with a scalar description of light, our equations make use of the coherency (density) matrix formalism. A careful analysis of the key concepts shows that the low-dimensional unitary groups, $SU(N)$, can be applied to a useful parametrization of the density matrix. On the basis of this work, its use can be expected to characterize non-image-forming optical devices, scattering media, and anisotropic media. These results depend on the assumption that the optical field is narrowband, but the results are independent of the spectral distribution. Various formulas were then quoted and physical arguments presented, to motivate the use of density matrix in partial polarization theory. It should perhaps be emphasized once again that the theoretical arguments presented in this paper are limited to second-order statistics of the radiation field.

There do, of course, remain a number of points to be investigated regarding the fundamental issue of the entropic description of non-Gaussian electromagnetic fields, which require a higher than second order analysis. The solution of the fourth-order moment problem is of particular importance since it contains information about the intensity fluctuations and their correlations. For the case of non-Gaussian distributed radiation field, the fourth-order entropy is expected to differ from the second-order entropy version. At a still more general level, we conclude that, although the treatment discussed in this paper was developed to calculate the entropy of a stochastic wavefield in classical terms exclusively, there is no corresponding quantum approach of the entropic properties of a radiation field based on the use of Stokes operators. In classical statistical mechanics, dynamics and thermodynamics are separable. In writing down simple nondynamical models such as the Ising model, we typically take advantage of this simplicity. An interesting problem that deserves further work is the extension of this analysis dealing with the noncommutation of the Stokes operators [5,50].

Space limitations for this article preclude a fuller description of the possible applications for which polarization concepts may be useful. The physical evaluation of polarization properties of light is an area of prolific work at present in both the scientific and technical communities, with reasonably complete surveys of the literature probably being possible only through computer search methods. However, the brief descriptions above are a good indication of the broad range of these possibilities, and the reader is referred to the works mentioned above, and the works cited in the references, for further information.

I end with several remarks about open questions. The list of future challenges, both for theorists and experimentalists, is long. Perhaps it is appropriate to close by enumerating some of them: (1) despite many years of intensive investigation, we still do not fully understand the details of the propagation of a polarized wave in multiply scattering media, e.g., to understand the physical origin of the non-Gaussian distribution of the electric field when a small number of scatterers is present in the scattering volume; (2) lots of people, myself included, expect that physics will have increasingly important things to say about biological processes. In particular, much work remains to be done to establish quantitative applications of polarized light diagnostics in biomedicine, e.g., to quantify the concentration of optically active (chiral) metabolites such as glucose in optically turbide tissues; (3) the propagation of polarized light in randomly varying birefringent optical fibers, i.e., to determine how randomly varying dispersion affects the polarization state on the Poincaré's sphere is important for the development of optical networks; (4) defining new polarization descriptors for an arbitrary (nonplane) electromagnetic wave is one of the major research directions in the contexts of nonparallel light, optical near fields, and nano-optics. Fortunately, as with any healthy science, the theory is now being driven by a large amount of new measurements and observations. Far from being an obscure phenomenon, polarization has great significance throughout science and technology and there is every indication more excitement is yet to come.

All of the applications mentioned in the previous paragraphs rely entirely on what we have learned about the theoretical description of polarized light. This excursion into the domain of polarization optics brings to mind that the early great contribution made by Emil Wolf came by way of statistical physics through the concept of coherency matrix, i.e., correlations between the fluctuating field variables at two space-time points. This concept has initiated a flurry of further work and suggests many further directions of research to keep practitioners of polarization optics occupied well into the new century. Most scientists only dream of contributing to a paradigm shift—Wolf personally initiated one. Most focus their research efforts in one narrow area of specialization—Wolf's work ranged over the breadth of optics. Most influence the training of only a small cohort of students—Wolf's contributions to education reach hundreds of thousands, at all levels of sophistication. Wolf's creations are a constant presence in today's optics.[*] His abstract, elegant style is now the standard language of statistical optics.[†] I hope this account of Wolf's contribution to polarization optics will serve to remind the present generation of optical physicists and engineers that Wolf is one of those on

---

[*] It is worth observing that Wolf recently presented [51] a unifield theory of coherence and polarization of random electromagnetic fields, which clearly brings out the intimate relationship that exists between these two phenomena and makes it possible to predict the changes in the state of polarization of a partially coherent electromagnetic beam on propagation.

[†] In this day of high-tech gadgets growing ever smarter and more powerful, Wolf's breakthrough in polarization optics came not with the help of complex computer programs, or online resources. Instead, Wolf relied on two ancient tools: pen and paper.

whose shoulders they stand. Clearly, numerous challenges remain, but the rate of advance in polarization science over the past decade promises that this area will deliver exciting developments in the early twenty-first century. It is a great time to be involved in polarization research. We who follow Wolf must rededicate ourselves to a strict adherence to his standards of excellence both in research and teaching.

Acknowledgments

I express my gratitude to the late Dick Barakat, with whom I had very fruitful collaborations and discussions over the years. The year I spent with Dick at Harvard University was extraordinarily influential in shaping my life as a scientist. I thank also all past and present collaborators with whom I had stimulating discussions over the years. Special thanks go to Prof. Michael Berry for an enlightening seminar on polarization singularities, which he delivered at the 2002 Third Stokes Summer Workshop held at Skreen, County Sligo, Ireland, and insightful discussions with him on diabolical points. Conversations with Dominique Bicout at the Joseph Fourier University, Grenoble, France, during various phases of some of the work I have reviewed in this article were essential to the development of the ideas presented in the paper. I also want to thank Philippe Elies at the Université de Bretagne Occidentale for very helpful suggestions during the preparation of this manuscript.

References

1. E. Wolf, *Nuovo Cimento* **12**, 884 (1954). See also E. Wolf, *Proc. Roy. Soc. A* (London) **225**, 96 (1954).
2. M. Born and E. Wolf, *Principles of Optics*, 7th ed., Cambridge Univ. Press, New York (1999).
3. The recent Esther Hoffman Beller (OSA) award was presented to Emil Wolf for his "numerous outstanding contributions as an educator, but especially, for the influence of his books, which have been educating optical scientists and engineers for more than forty years." Wolf's texts are not easy. But his writing is elegant, compact, and logically precise. There are no gaps in the arguments. Even professionals discover fresh perspectives. In this sense there is a positive similarity with Landau and Lifshitz's textbooks, which I routinely recommend at every level.
4. L. Mandel and E. Wolf, *Optical Coherence and Quantum Optics*, Cambridge University Press, New York (1995).
5. C. Brosseau, *Polarized Light: A Statistical Optics Approach*, Wiley, New York (1998). My lecture notes for the graduate students became a book in which I tried to give a comprehensive review of Polarization Optics-related literature

up to 1998, and from which I have taught polarized light at Fourier University, Grenoble, France, and at the Université de Bretagne Occidentale, Brest, until about 2002. To our knowledge it contains the largest body of cited work that appears in one place for that topic up to that time.

6. G.P. Konnen, *Polarized Light in Nature*, Cambridge Univ. Press, Cambridge (1985). See also W. A. Shurcliff and S. S. Ballard, *Polarized Light*, van Nostrand, Princeton, N.J. (1964).

7. L. Mandel and E. Wolf, *Rev. Mod. Phys.* **37**, 231 (1965). This 1965 paper on coherence properties of optical fields was designated as a Citation Classic in 1980, and in 1988 was listed as one of the 100 most cited articles published in the *Reviews of Modern Physics* since 1955.

8. E. Wolf, *Selected Works of Emil Wolf*, World Scientific, New York (2001). This voluminous book is devoted to Wolf's various intellectual pursuits. It presents most of the important papers of Emil Wolf published over a half a century, concerning electromagnetic theory, physical optics, and the phenomenon of partial coherence.

9. Contrary to the development of science, which is unlike the smooth and monotonically progressive path, the history of science tends to be streamlined, ignoring numerous false trails and blind alleys on the road to modern knowledge.

10. E. Whittaker, *A History of the Theories of Aether and Electricity*, Vols. 1 and 2, Harper & Row, New York (1960).

11. J.Z. Buchwald, *The Rise of the Wave Theory of Light*, Univ. of Chicago Press, Chicago (1989).

12. M. Bertolotti, *Masers and Lasers: A Historical Approach*, Adam Hilger, Bristol (1983).

13. D.L. Falkoff and J.E. MacDonald, *J. Opt. Soc. Am.* **41**, 861 (1951).

14. U. Fano, *Rev. Mod. Phys.* **29**, 74 (1957).

15. C. Brosseau, *C. R. Acad. Sci. Paris, Série B* **310** 191 (1990). See also C. Brosseau, *Optik* **88**, 109 (1990).

16. C. Brosseau and D. Bicout, *Phys. Rev. E* **50**, 4997 (1994).

17. R. Brandenberger, Y. Mukhanov, and T. Prokopec, *Phys. Rev. Lett.* **69**, 3606 (1992).

18. For a review of some of these contributions, see, for instance, F.C. Mackintosh and Sajeev John, *Phys. Rev. B* **40**, 2383 (1989); M.P. van Albada and A. Lagendjik, *Phys. Rev. Lett.* **55**, 2692 (1985); P.E. Wolf and G. Maret, *Phys. Rev. Lett.* **55**, 2696 (1989); and Y. Kuga and A. Ishimaru, *J. Opt. Soc. Am. A* **1**, 831 (1984).

19. S.J. van Enk and G. Nienhuis, *Phys. Rev. A* **46**, 1438 (1992).

20. B. Chan Eu and K. Mao, *Physica A* **180**, 65 (1992).

21. D. Bicout and C. Brosseau, *J. Phys. I France* **2**, 2047 (1992). See also D. Bicout, C. Brosseau, A.S. Martinez, and J.M. Schmitt, *Phys. Rev. E* **49**, 1767 (1994).

22. S. Chandrasekhar, *Radiative Transfer*, Dover, New York (1960).

23. G.V. Rozenberg, *Usp. Fiz. Nauk* **56**, 77 (1955).

24. U. Callies, *Beitr. Phys. Atmosph.* **62**, 212 (1989). See also U. Callies and F. Herbert, *J. Appl. Math. Phys.* **39**, 242 (1988).

25. N.D. Gudkov, *Opt. Spectros. (USSR)* **68**, 130 (1990).

26. R. Clark Jones, *J. Opt. Soc. Am.* **53**, 1314 (1963).

27. A. Ishimaru, *Wave Propagation and Scattering in Random Media*, Academic Press, New York (1978).

28. A.A. Golunbentzev, *Zh. Eksp. Teor. Fiz.* **86**, 47 (1984) [*Sov. Phys.-JETP* **59**, 26 (1984)].

29. Yu.A. Kravtsov, S.M. Rytov, and V.I. Tatarskii, *Sov. Phys. Uspekhi* **18**, 118 (1975).

30. B. Wen, L. Tsang, D.P. Winenbrenner, and A. Ishimaru, *IEEE Trans. Geosci. Remote Sens.* **28**, 46 (1990).

31. E.E. Lewis and W.F. Miller, *Computational Methods of Neutron Transport*, Wiley, New York (1984).

32. P. Bruscaglioni, G. Zaccanti, and Q. Wei, *Appl. Opt.* **32**, 6142 (1993). See also P. Bruscaglioni and G. Zaccanti, *Multiple Scattering in Dense Media*, M. Nieto-Vesperinas and J.C. Dainty, Eds., Elsevier, New York (1990).

33. A.S. Martinez, Ph.D. diss. Fourier University, Grenoble, France (1993).

34. M.V. Berry, *J. Mod. Opt.* **45**, 1845 (1998).

35. J.F. Nye, *Natural Focusing and Fine Structure of Light*, Institute of Physics Publ., Bristol (1999).

36. M.R. Dennis, *Opt. Commun.* **213**, 201 (2002).

37. I. Freund, *Opt. Commun.* **201**, 251 (2002).

38. V.I. Alshits, A.V. Sarychev, and A.L. Shuvalov, *Sov. Phys. JETP* **62**, 531 (1985).

39. A.L. Shuvalov, *Proc. Roy. Soc. London A* **454**, 2911 (1998).

40. B.A. Barbour, H.B. Barnes, C.P. Lewis, P.E.A. Lindquist, M.W. Jones, and M. Nohon, *Opt. Phot. News* **44**, 44 (1988).

41. J.M. Schmitt, A.H. Gandjbakhche, and R.F. Bonner, *Appl. Opt.* **31**, 6535 (1992).

42. R.R. Anderson, *Arch. Dermatol.* **127**, 1000 (1991).

43. S.L. Jacques, L.H. Wang, D.V. Stephens, and M. Ostermeyer, in *Lasers in Surgery: Advanced Characterization, Therapeutics, and Systems VI*, R.R. Anderson, Ed., 199, *SPIE Proc.*, Vol. 2671, Bellingham, WA (1996).

44. Yu. Ushenko, *Opt. Spectros.* **93**, 321 (2002).

45. D.A. Zimnyakov, Yu.P. Sinichkin, I.V. Kiseleva, and D.N. Agafonov, *Opt. Spectros.* **92**, 765 (2002).

46. D.J. Mailand and J.T. Walsh, *Lasers Surg. Med.* **20**, 310 (1997).

47. V. Sankaran and J.T. Walsh, *Photochem. Photobiol.* **68**, 846 (1998).

48. I.A. Viktin and R.C.N. Studinski, *Opt. Commun.* **190**, 37 (2001).

49. A. Sweeney, C. Jiggins, and S. Johnsen, *Nature* **423**, 31 (2003).

50. J.M. Jauch and F. Rohrlich, *The Theory of Photons and Electrons*, Addison-Wesley, Reading, PA (1955). See also P. Roman, *Advanced Quantum Theory*, Addison-Wesley, New York (1965).

51. E. Wolf, *Phys. Lett. A* **312**, 263 (2003). [That changes may occur in the state of polarization of a light beam as it propagates has only recently attracted attention, i.e., D.F.V. James, *J. Opt. Soc. Am. A* **11**, 1641 (1994), and F. Gori, M. Santarsiero, S. Vacalvi, R. Borghi, and G. Gvattari, *Pure Appl. Opt.* **7**, 941 (1998).]

Emil Wolf's Influence

For almost six decades, the work of Emil Wolf has gained considerable attention and has become the basic reference for researchers and students in optics. As a matter of fact, Emil Wolf is a living legend in the field of physical optics. I first came across the work of Emil Wolf when I was an undergraduate student fascinated by the mystery of polarization phenomena. In my undergraduate years, I learned that light passing through a birefringent waveplate has a different phase velocity depending on the direction of its electric field vector to special axes of the crystal, i.e., the ordinary and extraordinary waves with orthogonal polarization states. I found a comprehensive treatment of this effect in his 1959 *magnum opus Principles of Optics* co-authored with Nobel laureate Max Born, which is among the maybe five most famous books ever written on optics. As I write this, I am remembering that it was my first exposure to the amazing facts of optics. A generation of students have learned the basics of optics thanks in no small part to courses based on *Principles of Optics*. It also taught to me some remarkable mathematics that I could actually see for myself made sense. His prolific publications have influenced all aspects of the discipline and are actively discussed in the academic literature, e.g., correlation-induced shift now identified with the adjective "Wolf," as well as in engineering fields, e.g., diffraction tomography. It is a daunting task of integrating the many facets of the extraordinary career of Emil Wolf into a unified whole. Wolf's growing influence on the statistical description of polarized light was recognized as long ago as 1954, when he introduced a precise measure of the correlations between the fluctuating field variables at two space-time points. The idea of correlations represents a landmark in the history of polarization optics and has been highly successful. Still, it was Wolf who gave us the alphabet from which the field of coherence and polarization optics was written.

Christian Brosseau received his Ph.D. in physics from Fourier University, Grenoble, France, in 1989. After holding a postdoctoral fellow position at Harvard University, he returned to Fourier University to become a Research Associate. He became an associate professor, and professor at the University of Brest, France, in 1994 and 1997, respectively. He led the wave-matter interaction, modeling, and simulation group in the physics department, and supervises Ph.D. students and postdoctoral research associates. Dr. Brosseau has published more than 80 refereed journal articles on a wide variety of theoretical and experimental topics and presented over 70 papers at conferences. He has also written the book *Fundamentals of Polarized Light: A Statistical Approach*, published in 1998. His research interests include polarization optics, all aspects of computational electromagnetics, nanomagnetism, and electromagnetic wave propagation in complex media.

CHAPTER 4

ELECTROMAGNETIC THEORY OF OPTICAL COHERENCE

Ari T. Friberg

4.1 Introduction

Almost half a century has passed since the polarization of light beams and the theory of optical coherence, in classical and quantized forms, were formulated in a systematic manner. In a classic paper [1] published in 1955, Emil Wolf introduced the two-point space-time correlation function, now known as the mutual function, and showed that in free space this function obeys two wave equations (see Fig. 1). This demonstrated the fundamental phenomenon that not only the field but also the spatial coherence propagates in the form of waves. In another pioneering work [2], Wolf analyzed the state of polarization of a light beam in terms of its "coherency matrix" and the now well-familiar Stokes parameters. Using the properties of the 2×2 coherence matrix, the degree of polarization could be introduced in an unambiguous manner. The formal theory of space-time coherence of arbitrary stationary electromagnetic fields was put forward in twin papers in 1960 by Roman and Wolf [3,4]. In these works the four general 3×3 correlation tensors (electric, magnetic, and mixed coherence matrices) were introduced and their properties were analyzed. This research, which took place before or around the time the first lasers were produced, has become the cornerstone of most of the subsequent studies on polarization and electromagnetic coherence. The quantum theory of coherence was formulated soon afterwards [5].

Entirely new physical insights were subsequently gained through the formulation of optical coherence phenomena in the space-frequency domain, in terms of the cross-spectral density tensors (matrices) [6]. This is natural in many circumstances since, for example, the material response (e.g., refractive index) in optics is frequency dependent. A novel but rather subtle quantity, the spectral degree of coherence, which is a measure of the spatial coherence of a statistically stationary

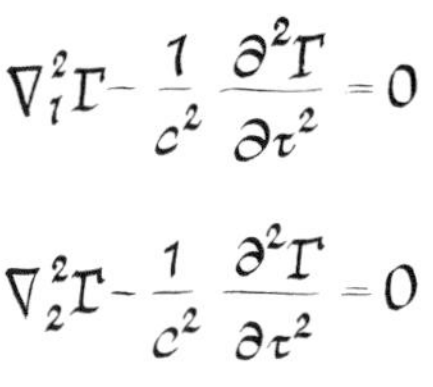

$$\nabla_1^2 \Gamma - \frac{1}{c^2}\frac{\partial^2 \Gamma}{\partial \tau^2} = 0$$

$$\nabla_2^2 \Gamma - \frac{1}{c^2}\frac{\partial^2 \Gamma}{\partial \tau^2} = 0$$

Figure 1 Two wave equations for mutual coherence function Γ, in Wolf's own handwriting. [From the cover of special issue on physical optics and coherence theory in honor of Prof. Emil Wolf's 75th birthday, *J. Eur. Opt. Soc. A: Pure Appl. Opt.* **7**, September (1998). Copyright (1998) by Institute of Physics Publishing.]

field at a given frequency, was introduced [7]. The space-frequency theory of optical coherence was extensively employed by Wolf already in the 1970s in studies of radiative energy transfer [8] and generalized radiometry [9]. Many of these subjects are reviewed in the comprehensive textbook by Mandel and Wolf [10].

The electromagnetic description of light, with its inherent polarization and vectorial coherence properties, has quite recently attracted increased attention in many areas of modern optical science and engineering such as diffractive and microoptics, near-field physics and spectroscopy, and nanophotonics. The electromagnetic correlations of light are altered on propagation and scattering, resulting in corresponding changes in the spectrum [11], the (spatial and temporal) coherence, and the polarization state. In general, all these effects are described by a unified theory of coherence and polarization of random electromagnetic fields, which lately has become a topic of intensive research.

4.2 Fundamental Scalar Results

To begin, briefly recall from the scalar theory of optical coherence some key concepts and results that have a direct bearing on the subsequent electromagnetic analysis. In coherence theory the fluctuating field is represented by a complex analytic signal $U(\mathbf{r},t)$, where $\mathbf{r}$ denotes position and t time. The real and imaginary parts of $U(\mathbf{r},t)$ form a Hilbert transform pair [10]. Assuming the field is stationary in time and ergodic, the mutual coherence function is defined as

$$\Gamma(\mathbf{r}_1,\mathbf{r}_2,\tau) = \langle U^*(\mathbf{r}_1,t)U(\mathbf{r}_2,t+\tau)\rangle, \tag{1}$$

where the asterisk denotes a complex conjugate and the angular brackets indicate time or ensemble averaging. In the space-frequency representation, the central quantity is the cross-spectral density function, which is obtained from Eq. (1) via the (generalized) Wiener-Khintchine theorem [10]

$$W(\mathbf{r}_1,\mathbf{r}_2,\omega) = \frac{1}{2\pi}\int_{-\infty}^{\infty} \Gamma(\mathbf{r}_1,\mathbf{r}_2,\tau)\exp(i\omega\tau)\,d\tau. \tag{2}$$

The cross-spectral density $W(\mathbf{r}_1, \mathbf{r}_2, \omega)$, which is zero for $\omega < 0$, is applicable irrespective of the spectral bandwidth of the radiation.

Correlation functions are subject to stringent mathematical conditions. The cross-spectral density $W(\mathbf{r}_1, \mathbf{r}_2, \omega)$ is Hermitian and nonnegative definite [10]. In usual circumstances it is square-integrable with respect to $\mathbf{r}_1$ and $\mathbf{r}_2$ in the domain of observation D. Hence, $W(\mathbf{r}_1, \mathbf{r}_2, \omega)$ may be regarded as a Hermitian, nonnegative-definite Hilbert-Schmidt kernel, and by Mercer's theorem it admits a uniformly and absolutely convergent expansion [12]

$$W(\mathbf{r}_1, \mathbf{r}_2, \omega) = \sum_n \lambda_n(\omega)\phi_n^*(\mathbf{r}_1, \omega)\phi_n(\mathbf{r}_2, \omega), \tag{3}$$

where $\lambda_n(\omega)$ are the eigenvalues and $\phi_n(\mathbf{r}, \omega)$ the eigenfunctions of the Fredholm integral equation

$$\int_D W(\mathbf{r}_1, \mathbf{r}_2, \omega)\phi_n(\mathbf{r}_1, \omega)\, \mathrm{d}^3 r_1 = \lambda_n(\omega)\phi_n(\mathbf{r}_2, \omega); \tag{4}$$

the domain D may also be other than three-dimensional (3D). The eigenvalues $\lambda_n(\omega)$ are real and nonnegative, and the eigenfunctions $\phi_n(\mathbf{r}, \omega)$ are orthonormal (or can be chosen so in a degenerate case) within D.

The expansion in Eq. (3) is known as the coherent-mode decomposition [10] of the cross-spectral density function $W(\mathbf{r}_1, \mathbf{r}_2, \omega)$. It represents $W(\mathbf{r}_1, \mathbf{r}_2, \omega)$ as an incoherent superposition of elementary modes that are spatially fully coherent at frequency ω. The spectral degree of (spatial) coherence, bounded in magnitude between 0 and 1, is defined as [7]

$$\mu(\mathbf{r}_1, \mathbf{r}_2, \omega) = \frac{W(\mathbf{r}_1, \mathbf{r}_2, \omega)}{[S(\mathbf{r}_1, \omega)S(\mathbf{r}_2, \omega)]^{1/2}}, \tag{5}$$

where

$$S(\mathbf{r}, \omega) = W(\mathbf{r}, \mathbf{r}, \omega) \tag{6}$$

is the spectrum (or spectral density) of the field. The complete spatial coherence of the modes follows directly from the fact that each term in Eq. (3) factors in the two spatial variables. A distinction must be made between the traditional complex degree of coherence [10] $\gamma(\mathbf{r}_1, \mathbf{r}_2, \tau)$, which is a normalized form of Eq. (1), and the spectral degree of coherence, Eq. (5); they are usually quite different [13]. It is remarkable, and perhaps even somewhat counterintuitive, that a stationary field analyzed at a given frequency ω is, as a rule, spatially partially coherent.

By making use of the coherent-mode representation, Wolf also demonstrated that the cross-spectral density may be expressed as a correlation over an ensemble of monochromatic functions, i.e.,

$$W(\mathbf{r}_1, \mathbf{r}_2, \omega) = \langle \tilde{U}^*(\mathbf{r}_1, \omega) \tilde{U}(\mathbf{r}_2, \omega) \rangle, \tag{7}$$

where the brackets now stand for an ensemble average [12]. The functions $\tilde{U}(\mathbf{r}, \omega)$ are not the Fourier transforms of field realizations $U(\mathbf{r},t)$ [10]. This is a fundamental result that justifies the analysis of stationary random scalar wavefields frequency by frequency. The electromagnetic analog of the coherent-mode decomposition will be addressed in the subsequent sections.

4.3 Electric Cross-Spectral Density Matrix

For arbitrary fluctuating electromagnetic fields, the correlation properties are fully characterized by the four coupled coherence tensors [10]. As is common in optics, only the electric field $\mathbf{E}(\mathbf{r},t)$ will be considered. In full analogy with Eqs. (1) and (2), the elements of the electric space-time coherence matrix and cross-spectral density matrix are [3,10]

$$\mathcal{E}_{jk}(\mathbf{r}_1, \mathbf{r}_2, \tau) = \left\langle E_j^*(\mathbf{r}_1,t) E_k(\mathbf{r}_2, t + \tau) \right\rangle \tag{8}$$

and

$$\mathcal{W}_{jk}(\mathbf{r}_1, \mathbf{r}_2, \omega) = \frac{1}{2\pi} \int_{-\infty}^{\infty} \mathcal{E}_{jk}(\mathbf{r}_1, \mathbf{r}_2, \tau) \exp(i\omega\tau)\, d\tau, \tag{9}$$

where $(j,k) = (x,y,z)$ denote the Cartesian coordinates. In the space-frequency representation the central object, in matrix form, then is explicitly:

$$\mathcal{W}(\mathbf{r}_1, \mathbf{r}_2, \omega) = \begin{bmatrix} W_{xx}(\mathbf{r}_1,\mathbf{r}_2,\omega) & W_{xy}(\mathbf{r}_1,\mathbf{r}_2,\omega) & W_{xz}(\mathbf{r}_1,\mathbf{r}_2,\omega) \\ W_{yx}(\mathbf{r}_1,\mathbf{r}_2,\omega) & W_{yy}(\mathbf{r}_1,\mathbf{r}_2,\omega) & W_{yz}(\mathbf{r}_1,\mathbf{r}_2,\omega) \\ W_{zx}(\mathbf{r}_1,\mathbf{r}_2,\omega) & W_{zy}(\mathbf{r}_1,\mathbf{r}_2,\omega) & W_{zz}(\mathbf{r}_1,\mathbf{r}_2,\omega) \end{bmatrix}. \tag{10}$$

The electric cross-spectral density matrix in Eq. (10) contains all the information about the electric-field correlations. It is well defined and applicable irrespective of whether the $\mathbf{E}$-field has one component (plane polarized), two components (transversely polarized), or three components, as in optical near fields. Within the paraxial approximation, a related 2×2 quantity is sometimes called the beam coherence-polarization (BCP) matrix [14].

The correlation matrix $\mathcal{W}(\mathbf{r}_1, \mathbf{r}_2, \omega)$ is a mathematically subtle quantity. It is not a Hermitian matrix as written in Eq. (10), but it is Hermitian if the points $\mathbf{r}_1$ and $\mathbf{r}_2$ are also interchanged, i.e.,

$$\mathcal{W}^{\dagger}(\mathbf{r}_2, \mathbf{r}_1, \omega) = \mathcal{W}(\mathbf{r}_1, \mathbf{r}_2, \omega), \tag{11}$$

where the superscript † means Hermitian conjugate. Thus the "diagonal element" $\mathcal{W}(\mathbf{r}, \mathbf{r}, \omega)$ is a 3×3 Hermitian matrix. Furthermore, the diagonal elements of Eq. (10), such as $\mathcal{W}_{xx}(\mathbf{r}_1, \mathbf{r}_2, \omega)$, are nonnegative definite scalar quantities, since they are autocorrelation functions, while the off-diagonal elements, for example $\mathcal{W}_{xy}(\mathbf{r}_1, \mathbf{r}_2, \omega)$, are not subject to the nonnegative definiteness requirement, since they are cross-correlation functions. In general, it can be shown that $\mathcal{W}(\mathbf{r}_1, \mathbf{r}_2, \omega)$ satisfies the matrix-form nonnegative definiteness condition [15]

$$\iint_D \mathbf{g}^{\dagger}(\mathbf{r}_1) \cdot \mathcal{W}(\mathbf{r}_1, \mathbf{r}_2, \omega) \cdot \mathbf{g}(\mathbf{r}_2) \, \mathrm{d}^3 r_1 \mathrm{d}^3 r_2 \geq 0, \tag{12}$$

where $\mathbf{g}(\mathbf{r})$ is an arbitrary (well-behaved) vector-valued function. Additionally, the cross-spectral density $\mathcal{W}(\mathbf{r}_1, \mathbf{r}_2, \omega)$, and its magnetic and mixed counterparts, satisfy a number of propagation laws and similar relations [3,4,10], because the electromagnetic field obeys Maxwell's equations.

In decreasing complexity, the cross-spectral density matrix $\mathcal{W}(\mathbf{r}_1, \mathbf{r}_2, \omega)$ in Eq. (10) provides the following information about the field:

$$\text{General } \mathcal{W}(\mathbf{r}_1, \mathbf{r}_2, \omega) \;\Rightarrow\; \text{Electromagnetic coherence}$$

$$\text{Diagonal element } \mathcal{W}(\mathbf{r}, \mathbf{r}, \omega) \;\Rightarrow\; \text{Polarization state}$$

$$\text{tr } \mathcal{W}(\mathbf{r}, \mathbf{r}, \omega) \;\Rightarrow\; \text{Spectrum}$$

Here tr denotes the trace of the matrix. Next, these topics are discussed separately, starting from the simplest one.

4.4 Spectral Changes

Inspired by the coherence-induced spectral changes [11], the spatial coherence properties and spectra of electromagnetic fields emanated from semi-infinite sources in thermal equilibrium have recently been examined [16,17]. The half-space boundary introduces a surface that breaks the overall symmetry and may introduce various types of surface excitations. The current-density correlations within the source are given by the fluctuation-dissipation theorem of statistical physics and the ensuing electric-field coherence properties, as described by Eq. (10), are then found with the help of the appropriate Green dyadic.

Several new and unexpected results have been obtained. The half-space source supports surface waves (polaritons), which are confined electromagnetic modes near the boundary. Surface plasmons are collective electron-density waves that correspond to poles in p-polarized wave transmission, while surface phonons are lattice vibrations in polar material. For example, silicon carbide (SiC) supports a surface phonon at wavelength $\lambda = 11.36$ μm. As a result, the spectrum

$$S(\mathbf{r}, \omega) = \text{tr}\, \boldsymbol{\mathcal{W}}(\mathbf{r}, \mathbf{r}, \omega) \tag{13}$$

exhibits marked changes on propagation. The spectra of thermal emission from a room-temperature sample of SiC at different heights above the surface are shown in Fig. 2. The mean wavelength of radiation is about $\lambda \approx 10$ μm. In the near field [Fig. 2(c)] the spectrum is almost monochromatic, while in the far field [Fig. 2(a)] it has evolved into a broadband distribution containing a distinctive nonradiation region where the near-field peak was located. The peak corresponds to surface phonons, whose effect disappears at distances comparable to a wavelength.

The main observation is that the spectrum of light in the near field is quite different from what one might expect on the basis of spectroscopic far-field mea-

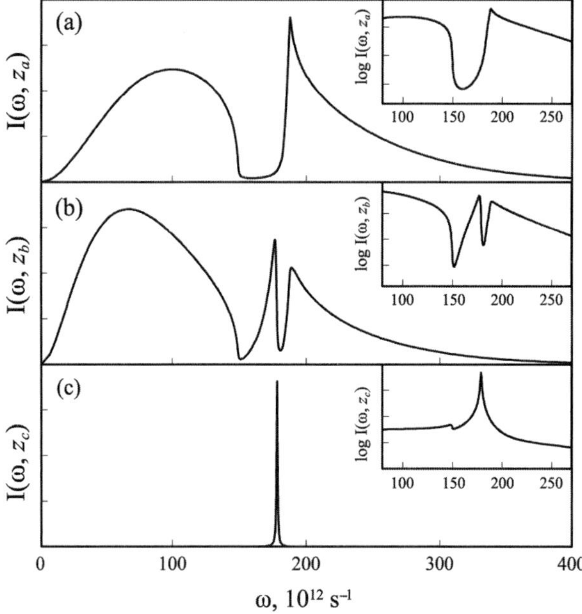

Figure 2 Spectra of thermal emission of a semi-infinite SiC sample at $T = 300$ K at heights: (a) $z_a = 1000$ μm, (b) $z_b = 2$ μm, and (c) $z_c = 0.1$ μm. The insets show the spectra on a semilog scale in the range of strong contribution from evanescent surface modes. [Reprinted with permission from A.V. Shchegrov, K. Joulain, R. Carminati, and J.J. Greffet, "Near-field spectral effects, due to electromagnetic surface excitations," *Phys. Rev. Lett.* **85**, 1548–1551 (2000). Copyright (2000) by the American Physical Society.]

surements. These results may have implications in surface physics and near-field optics, such as particle manipulation by tweezers and spanners, where the field in which the object is immersed obviously is of particular importance.

4.5 Degree of Polarization

The traditional polarization analysis applies to fields that have planar wavefronts, such as uniform optical beams or wide-angle far fields. The electric field vector is transverse to the direction of propagation, and the 2D coherence (or polarization) matrix $\mathbf{J}_2$ contains all necessary information. As Wolf [2] has shown, the coherence matrix can be decomposed in the 2×2 unit matrix and the three Pauli spin matrices [generators of SU(2) group]. The expansion coefficients are the Stokes parameters, which all have clear physical meanings (Poincaré sphere). In particular, the degree of polarization P_2 of the beam can, equivalently, be expressed using the coherence matrix $\mathbf{J}_2$ or the four Stokes parameters [10].

In general cases, such as near fields or tightly focused beams, the electric field contains all three components (E_x, E_y, E_z). The coherence matrix describing the partial polarization of the field, at point $\mathbf{r}$ and frequency ω, then is

$$\mathbf{J}_3(\mathbf{r}, \omega) = \left[J_{jk}(\mathbf{r}, \omega) \right] = \left[\mathcal{W}_{jk}(\mathbf{r}, \mathbf{r}, \omega) \right], \tag{14}$$

where the diagonal elements $J_{jj}(\mathbf{r}, \omega)$ ($j = x, y, z$) are the "intensities" of the components while the off-diagonal elements $J_{jk}(\mathbf{r}, \omega)$ ($j \neq k$) characterize their correlations. One may introduce a coefficient

$$\mu_{jk}(\mathbf{r}, \omega) = \frac{J_{jk}(\mathbf{r}, \omega)}{[J_{jj}(\mathbf{r}, \omega) J_{kk}(\mathbf{r}, \omega)]^{1/2}}, \tag{15}$$

with absolute value between 0 and 1. Since $\mathbf{J}_3$ is a Hermitian and nonnegative definite matrix, its eigenvalues are real and nonnegative. Using the mathematical properties of $\mathbf{J}_3$, as in the 2D case, a degree of polarization P_3 may be introduced that is applicable for any random electromagnetic fields [18,19]. The matrix $\mathbf{J}_3$ does not, however, admit a resolution as a sum of two matrices that correspond to fully polarized and unpolarized 3D fields. As I show later, P_3 is related to a quantity that was proposed as a general measure of the second-order spectral electromagnetic correlations.

The coherence matrix $\mathbf{J}_3(\mathbf{r}, \omega)$ is expanded in terms of proper basis matrices, chosen to be the Gell-Mann matrices, as [18]

$$\mathbf{J}_3(\mathbf{r}, \omega) = \frac{1}{3} \sum_{j=0}^{8} \Lambda_j(\mathbf{r}, \omega) \lambda_j. \tag{16}$$

Unit matrix λ_0 and the eight Gell-Mann matrices λ_j $(j = 1, \ldots, 8)$, generators of $SU(3)$ symmetry group, are Hermitian, trace-orthogonal, and linearly independent 3×3 matrices. The nine expansion coefficients Λ_j $(j = 0, \ldots, 8)$ can be shown to have well-defined physical interpretations similar to those of the four Stokes parameters in the beam formalism, and therefore these coefficients are regarded as generalized Stokes parameters [18].

In analogy with P_2 for beams, the degree of polarization P_3 of an arbitrary random 3D electromagnetic field is defined by the formula ($\mathbf{r}$ and ω suppressed)

$$P_3^2 = \frac{3}{2}\left[\frac{\mathrm{tr}(\mathbf{J}_3^2)}{\mathrm{tr}^2(\mathbf{J}_3)} - \frac{1}{3}\right] = \frac{1}{3}\frac{\sum_{j=1}^{8}\Lambda_j^2}{\Lambda_0^2}. \tag{17}$$

Evidently P_3 is invariant in unitary transformations, so it does not depend on the orientation of the coordinate system. It can further be shown that the degree of polarization is bounded such that $0 \leq P_3 \leq 1$, the limits representing the extremes of unpolarized and fully polarized fields.

Additional physical insight is gained by expressing P_3 explicitly in terms of the correlation coefficients $|\mu_{jk}|$ of the electric-field components as [18]

$$P_3^2 \geq \frac{|\mu_{xy}|^2 J_{xx}J_{yy} + |\mu_{xz}|^2 J_{xx}J_{zz} + |\mu_{yz}|^2 J_{yy}J_{zz}}{J_{xx}J_{yy} + J_{xx}J_{zz} + J_{yy}J_{zz}}. \tag{18}$$

This result shows that P_3^2 is always greater than or equal to the averaged squared correlations of the electric-field components weighted by the corresponding intensities. The left-hand side of Eq. (18) does not depend on the coordinate orientations, but the right-hand side does. The equality holds for a system in which the diagonal elements (electric-field component intensities) are the same. In such a situation P_3^2 reduces to an average of the squared correlations. These conclusions are in agreement with the analogous 2D results. Hence, the degree of polarization P_3 is a measure of the correlations that, at any point in space, exist among the three orthogonal $\mathbf{E}$-field components.

In is important to stress that the 3D formalism, in general, gives for the degree of polarization different values than the conventional beam method. For example, while a fully unpolarized plane wave in the 2D analysis has a zero degree of polarization ($P_2 = 0$), in the 3D treatment its degree of polarization P_3 must be nonzero since the electric field is confined to a transverse plane ($E_z = 0$). In fact, for plane waves P_3 is restricted to values $0.5 \leq P_3 \leq 1$. Intuitively the differences can be understood by considering Fig. 3. In the upper row, an unpolarized planar wave ($J_{xx} = J_{yy}$, no correlations) traverses a polarizer. The usual beam formalism gives $P_2 = 0$ and $P_2 = 1$ for the wave before and after the polarizer, respectively. In the lower row, an unpolarized 3D electromagnetic field ($J_{xx} = J_{yy} = J_{zz}$,

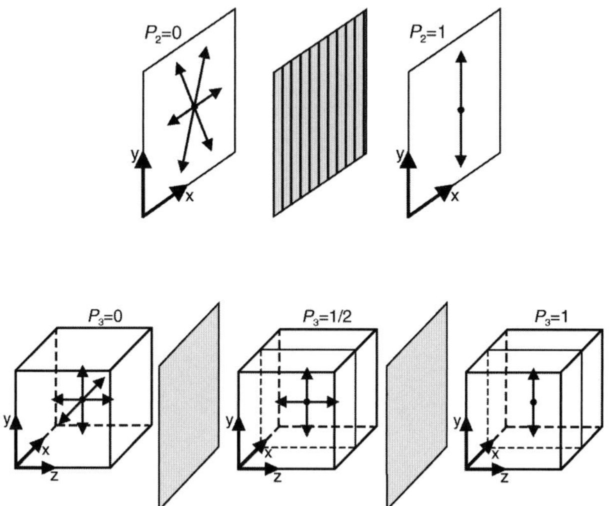

Figure 3 Illustration of the differences between the traditional 2D (top row) and the general 3D (bottom row) coherence-matrix formalisms in the treatment of the degree of polarization of an electromagnetic wavefield.

no correlations between any components) is passed through two devices that cut off the electric-field components in orthogonal directions. For the initial field, which cannot be described using the beam method, the 3D formalism gives zero for P_3. On passing through the first device the field becomes partially polarized ($J_{xx} = 0, J_{yy} = J_{zz}$, no correlations between y and z components) and consequently $P_3 = 0.5$. After the second device, the field is fully polarized ($P_3 = 1$), since the oscillations now take place only in a single direction.

4.5.1 Degree of polarization in near fields

The general 3D formalism of partial polarization has been applied to the electromagnetic near fields emitted by semi-infinite thermal half-space sources [19]. The methods for obtaining the cross-spectral density matrix $\mathcal{W}(\mathbf{r}_1, \mathbf{r}_2, \omega)$ in Eq. (10), as well as the polarization-degree calculations and the results, are explained in more detail elsewhere [16,20].

In Fig. 4 the behavior of the degree of polarization P_3, introduced in Eq. (17) for an arbitrary point $\mathbf{r}$ and frequency ω, is illustrated as a function of distance z from the boundary of a source consisting of gold (Au) and SiC (both at two wavelengths), and glass. Always $P_3(0) = 0.25$, but the far-field value depends on source material. At $\lambda = 620$ nm, gold exhibits a surface-plasmon resonance, and since plasmon waves are strongly polarized, the near field attains a high peak value for $P_3(z)$. Similar behavior is noted in the near field of SiC source at $\lambda = 11.36$ μm, corresponding to a surface phonon. However, tuning the wavelength off

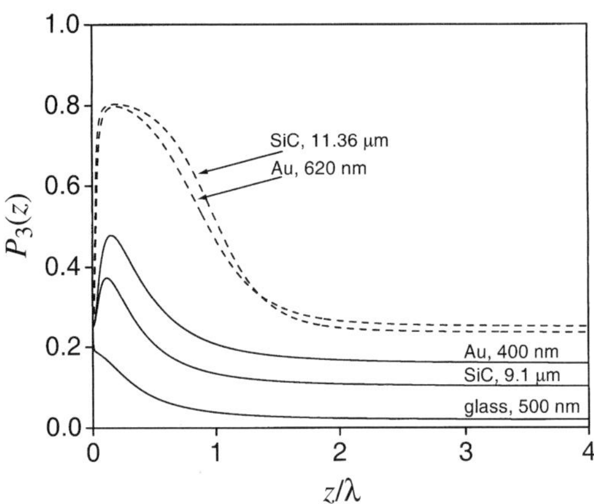

Figure 4 Degree of polarization $P_3(z)$ in the optical near fields, at certain wavelengths, emitted by thermal half-space sources consisting of glass, gold (Au), and silicon carbide (SiC). At large distances, $P_3(z)$ does not tend to zero.

resonance, the near-field degree of polarization is greatly reduced, as is evidenced by the curves for gold at $\lambda = 400$ nm and SiC at $\lambda = 9.1$ μm. Further, it is seen that $P_3(z)$ for a thermal glass sample decays monotonically as the distance z from the source increases. Glass does not support surface waves, and the decay is directly related to the loss of the evanescent modes.

4.6 Coherence of Electromagnetic Fields

Conventional wisdom in optical coherence theory says that spatial correlations in electromagnetic fields necessarily extend, at least, over distances on the order of the wavelength [9,21]. This assertion has recently been tested by rigorous electromagnetic calculations pertaining to fields emitted by thermal half-space sources [22]. Some key results are presented in Fig. 5. If the source consists of slightly lossy glass, the field correlations, such as the component $W_{xx}(\mathbf{r}_1, \mathbf{r}_2, \omega)$, very close to the boundary indeed behave as $\sin(k\rho)/k\rho$, where $k = \omega/c = 2\pi/\lambda$ is the wave-number and $\rho = |\mathbf{r}_1 - \mathbf{r}_2|$, characteristic of blackbody radiation [10]. However, if the medium is tungsten (W), which shows strong absorption at $\lambda = 500$ nm, spatial correlations close to the source's surface are very short, only on the order of 0.06λ, corresponding to about the skin depth of tungsten. On propagation to larger distances, the field coherence assumes the blackbody form. The extremely narrow, quite unusual, spatial correlations in the immediate vicinity of a tungsten source are explained by absorption, and thereby by decorrelation of the field, over a skin-depth distance within the medium [22].

The coherence behaves quite differently at wavelengths that correspond to surface-polariton resonances. For example, at $\lambda = 620$ nm, both silver (Ag) and

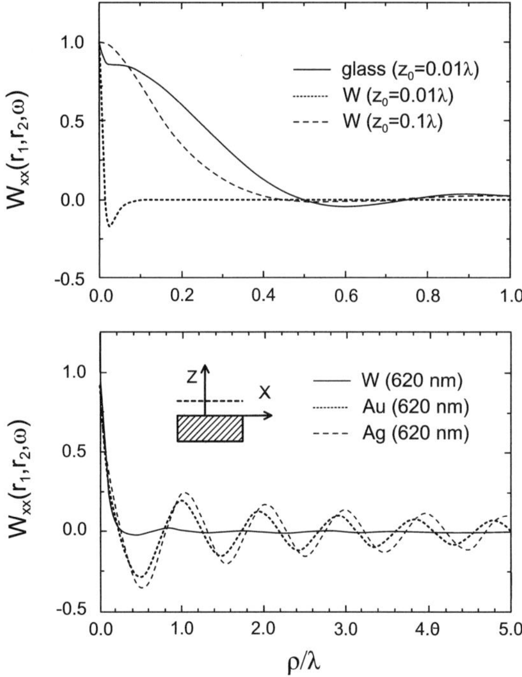

Figure 5 Normalized transverse spatial coherence at a fixed height z_0 above a semi-infinite thermal half-space source. Top: lossy glass and tungsten (at two z_0), at $\lambda = 500$ nm. Bottom: tungsten, gold, and silver at $\lambda = 620$ nm, at height $z_0 = 0.05\lambda$. [Reprinted from R. Carminati and J.J. Greffet, "Near-field effects in spatial coherence of thermal sources," *Phys. Rev. Lett.* **82**, 1660–1663 (1999). Copyright (1999) by the American Physical Society.]

gold (Au) support surface plasmons, while tungsten does not. The surface waves are manifested in field correlations, e.g., component $\mathcal{W}_{xx}(\mathbf{r}_1, \mathbf{r}_2, \omega)$, that persist in an oscillatory manner over tens of wavelengths (bottom part of Fig. 5). The surface phonon of SiC at $\lambda = 11.36$ μm leads to similar long-range correlation features [22]. These results illustrate that surface polaritons, which are associated with highly polarized waves decaying exponentially above the surface, are able to transport electromagnetic coherence near the boundary over large distances. Grating couplers etched on the source can convert surface-polariton fields into usual propagating waves [23]. This leads to a directional emission of almost coherent radiation at certain wavelengths from thermal sources.

4.6.1 Young's interference experiment

Since early analyses [24] employing the usual space-time representation of electromagnetic coherence, Young's interference experiment with vector fields has quite recently attracted renewed interest, in the space-frequency domain [25]. The clas-

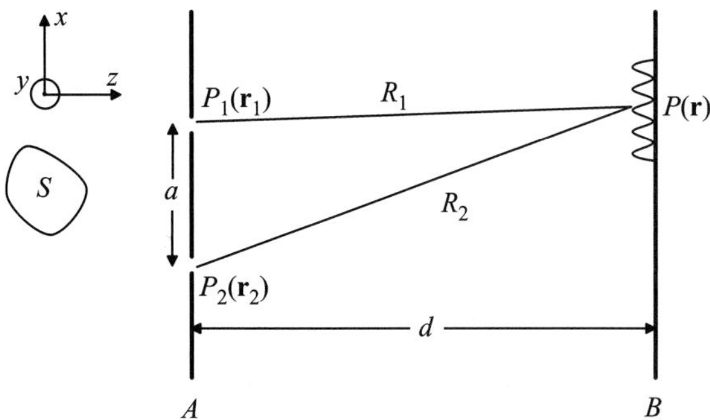

Figure 6 Illustration of the geometry and notation related to Young's interference arrangement. Source S gives rise to a random electromagnetic field that is incident on the two-pinhole setup.

sic arrangement is schematically illustrated in Fig. 6. A fluctuating electric field with two components (E_x, E_y) impinges on two pinholes, located at $P_1(\mathbf{r}_1)$ and $P_2(\mathbf{r}_2)$ in plane A, and the interference fringes are observed some distance away on screen B. The spectrum at point $P(\mathbf{r})$, as defined by Eq. (13), is given by the spectral interference law [7,10,25]

$$S(\mathbf{r}, \omega) = 2S^{(1)}(\mathbf{r}, \omega)\left[1 + |\xi(\mathbf{r}_1, \mathbf{r}_2, \omega)|\cos\alpha(\mathbf{r}_1, \mathbf{r}_2, \omega) + \delta\right], \qquad (19)$$

where $S^{(1)}(\mathbf{r}, \omega)$ is the spectrum of light reaching $P(\mathbf{r})$ through a single opening (spectra from both holes are assumed equal). Moreover [24,25],

$$\xi(\mathbf{r}_1, \mathbf{r}_2, \omega) = \frac{\mathrm{tr}\, \mathcal{W}(\mathbf{r}_1, \mathbf{r}_2, \omega)}{[S(\mathbf{r}_1, \omega)S(\mathbf{r}_2, \omega)]^{1/2}}, \qquad (20)$$

where $\mathcal{W}(\mathbf{r}_1, \mathbf{r}_2, \omega)$ is the 2×2 cross-spectral density matrix, $\alpha(\mathbf{r}_1, \mathbf{r}_2, \omega)$ is the phase (argument) of $\xi(\mathbf{r}_1, \mathbf{r}_2, \omega)$, and $\delta = (R_2 - R_1)/c$. Hence, the visibility of the fringes is directly obtained as $\mathcal{V}(\mathbf{r}, \omega) = |\xi(\mathbf{r}_1, \mathbf{r}_2, \omega)|$, while $\alpha(\mathbf{r}_1, \mathbf{r}_2, \omega)$ determines the fringe location.

If the electric field across the pinholes in plane A has only one component (or if the polarization state is uniform), the situation corresponds to usual scalar-wave optics (a 1D case) and Eq. (20) reduces to Eq. (5) for the spectral degree of coherence $\mu(\mathbf{r}_1, \mathbf{r}_2, \omega)$. Hence by analogy to scalar coherence theory, and following Zernike's classic work [26], the quantity $\xi(\mathbf{r}_1, \mathbf{r}_2, \omega)$ in Eq. (20) has been called [24,25] the spectral degree of coherence of the electric field at points $P_1(\mathbf{r}_1)$ and $P_2(\mathbf{r}_2)$. Clearly the interference term in Eq. (19) disappears if $\xi(\mathbf{r}_1, \mathbf{r}_2, \omega) = 0$ and fringes of maximum contrast are produced when $|\xi(\mathbf{r}_1, \mathbf{r}_2, \omega)| = 1$.

The trace-quantity in Eq. (20), which only involves the diagonal elements of $\mathcal{W}(\mathbf{r}_1, \mathbf{r}_2, \omega)$, has been used as a characteristic measure of spectral coherence in random electromagnetic fields [21,27], and some of its consequences have been analyzed [28,29]. Electromagnetic fringe formation depends not only on the coherence of the electric fields at the pinholes but also on their states of polarization, since orthogonal components do not interfere. Hence, for example, a wave plate that rotates the incident-wave polarization across one of the openings will alter the visibility, consequently changing $\xi(\mathbf{r}_1, \mathbf{r}_2, \omega)$. Further, Eq. (20) is not invariant in transformation to curvilinear coordinates frequently used in far fields.

4.6.2 Degree of electromagnetic correlations

To arrive at a more stringent quantity free from these drawbacks, one must also include the off-diagonal elements of $\mathcal{W}(\mathbf{r}_1, \mathbf{r}_2, \omega)$. To this end, let me first consider intensity interferometry for beams of Gaussian statistics (Hanbury-Brown and Twiss experiment in space-frequency domain). The fourth-order field correlations of the incident radiation are observed. For a Gaussian scalar wave $U(\mathbf{r}, \omega)$, the ensemble average of intensity fluctuation $\Delta S(\mathbf{r}, \omega) = U^*(\mathbf{r}, \omega)U(\mathbf{r}, \omega) - S(\mathbf{r}, \omega)$, at points $P_1(\mathbf{r}_1)$ and $P_2(\mathbf{r}_2)$, is given by [10]

$$\langle \Delta S(\mathbf{r}_1, \omega)\Delta S(\mathbf{r}_2, \omega)\rangle = S(\mathbf{r}_1, \omega)S(\mathbf{r}_2, \omega)|\mu(\mathbf{r}_1, \mathbf{r}_2, \omega)|^2, \tag{21}$$

where $\mu(\mathbf{r}_1, \mathbf{r}_2, \omega)$ and $S(\mathbf{r}, \omega)$ are the coherence degree and spectrum defined by Eqs. (5) and (6), and the Gaussian moment theorem was used. This shows that intensity interferometry yields the degree of coherence, and the fringe visibility, for Gaussian scalar waves. What if the field is treated electromagnetically? For Gaussian vector beams (a 2D case) $\mathbf{E}(\mathbf{r}, \omega) = [E_x(\mathbf{r}, \omega), E_y(\mathbf{r}, \omega)]$, the two-point average of intensity fluctuation $\Delta S(\mathbf{r}, \omega) = \mathbf{E}^*(\mathbf{r}, \omega) \cdot \mathbf{E}(\mathbf{r}, \omega) - S(\mathbf{r}, \omega)$ takes on the form [10]

$$\langle \Delta S(\mathbf{r}_1, \omega)\Delta S(\mathbf{r}_2, \omega)\rangle = \sum_{jk} |\mathcal{W}_{jk}(\mathbf{r}_1, \mathbf{r}_2, \omega)|^2, \tag{22}$$

where $[\mathcal{W}_{jk}(\mathbf{r}_1, \mathbf{r}_2, \omega)]$ is a 2×2 analog of Eq. (10) and $S(\mathbf{r}, \omega)$ now is given by Eq. (13). The normalized correlation of intensity fluctuations for Gaussian vector beams then becomes

$$|\mu_G(\mathbf{r}_1, \mathbf{r}_2, \omega)|^2 = \frac{\sum_{jk} |\mathcal{W}_{jk}(\mathbf{r}_1, \mathbf{r}_2, \omega)|^2}{\sum_j \mathcal{W}_{jj}(\mathbf{r}_1, \mathbf{r}_1, \omega) \sum_k \mathcal{W}_{kk}(\mathbf{r}_2, \mathbf{r}_2, \omega)}, \tag{23}$$

which, in view of Eq. (21), may be identified as the square of the degree of spatial coherence. While the derivation of Eq. (23) is specific to Gaussian beams, one may relax this condition and define the spectral degree of correlations (or coherence [29]) for an arbitrary electromagnetic field as

$$\zeta^2(\mathbf{r}_1,\mathbf{r}_2,\omega) = \frac{\mathrm{tr}\left[\boldsymbol{\mathcal{W}}(\mathbf{r}_1,\mathbf{r}_2,\omega)\boldsymbol{\mathcal{W}}^\dagger(\mathbf{r}_1,\mathbf{r}_2,\omega)\right]}{\mathrm{tr}\,\boldsymbol{\mathcal{W}}(\mathbf{r}_1,\mathbf{r}_1,\omega)\,\mathrm{tr}\,\boldsymbol{\mathcal{W}}(\mathbf{r}_2,\mathbf{r}_2,\omega)}, \tag{24}$$

where Eq. (11) was used. A more compact form for Eq. (24) is

$$\zeta^2(\mathbf{r}_1,\mathbf{r}_2,\omega) = \frac{\|\boldsymbol{\mathcal{W}}(\mathbf{r}_1,\mathbf{r}_2,\omega)\|_{\mathrm{F}}^2}{S(\mathbf{r}_1,\omega)\,S(\mathbf{r}_2,\omega)}, \tag{25}$$

with $\|\cdot\|_{\mathrm{F}}$ denoting the Frobenius (or Euclidean) norm of a matrix.

This expression for the spectral degree of correlations is equally valid for 1D (scalar), 2D (beams), and 3D electromagnetic fields. The quantity $\zeta(\mathbf{r}_1,\mathbf{r}_2,\omega)$ in Eq. (25) is always real and is normalized such that [29]

$$0 \le \zeta(\mathbf{r}_1,\mathbf{r}_2,\omega) \le 1. \tag{26}$$

With scalar waves, $\zeta(\mathbf{r}_1,\mathbf{r}_2,\omega)$ reduces to the absolute value of the usual spectral degree of coherence $|\mu(\mathbf{r}_1,\mathbf{r}_2,\omega)|$. For beam fields, the elements of $\boldsymbol{\mathcal{W}}(\mathbf{r}_1,\mathbf{r}_2,\omega)$, and therefore $\zeta(\mathbf{r}_1,\mathbf{r}_2,\omega)$, are measurable by use of polarizers [29], but in the general case, such as for electromagnetic near fields, more elaborate techniques must be employed.

One of the most important consequences [30] of Eq. (25) is that the cross-spectral density $\boldsymbol{\mathcal{W}}(\mathbf{r}_1,\mathbf{r}_2,\omega)$ factors in its two spatial variables if, and only if, $\zeta(\mathbf{r}_1,\mathbf{r}_2,\omega) = 1$. Such a factorization property is a fundamental characteristic of a completely coherent field [5,10]. Analogous results pertaining to fully coherent electromagnetic fields in the space-time domain have also been examined [31].

4.6.3 Coherence and polarization

Equation (14) shows that the equal-point ($\mathbf{r}_1 = \mathbf{r}_2$) electromagnetic coherence is closely related to the degree of polarization. From the general definition of Eq. (24) $\zeta(\mathbf{r}_1,\mathbf{r}_2,\omega)$ it follows, after some algebra, that the degrees of self-correlations are:

$$\text{1D:} \quad \zeta(\mathbf{r},\mathbf{r},\omega) = 1 \qquad\qquad\qquad \text{(scalar)},$$

$$\text{2D:} \quad \zeta(\mathbf{r},\mathbf{r},\omega) = \left[\left(P_2^2(\mathbf{r},\omega) + 1\right)/2\right]^{1/2} \quad \text{(EM beam)},$$

$$\text{3D:} \quad \zeta(\mathbf{r},\mathbf{r},\omega) = \left[\left(2P_3^2(\mathbf{r},\omega) + 1\right)/3\right]^{1/2} \quad \text{(EM field)},$$

where P_3 is given by Eq. (17) and P_2 is the 2D degree of polarization [2,10].

It is of interest to note that the equal-point degree of correlations equals unity if, and only if, the field is completely polarized (for scalar waves this is always the case naturally). An electromagnetic field cannot be fully correlated (coherent) at a point if its orthogonal components are only partially correlated, manifested in partial polarization. Likewise, for multicomponent fields $\zeta(\mathbf{r},\mathbf{r},\omega)$ never actually reaches zero; it has a lower bound, $1/\sqrt{2}$ for 2D beams and $1/\sqrt{3}$ in the general case. The origin of the nonzero lowest attainable value is that even for unpolarized fields each Cartesian component at any point (fully) correlates with itself. These results indicate that electromagnetic coherence is fundamentally different from the familiar scalar-wave coherence.

4.6.4 Coherent-mode representation

Wolf's coherent-mode decomposition of scalar fields, given by Eq. (3), has led to a number of important applications in radiation and propagation, scattering, and inverse problems. Extensions of this key result to partially coherent and partially polarized paraxial [32] and full vector fields [15] were introduced very recently. In view of Eqs. (11) and (12), the electric cross-spectral density matrix $\mathbf{W}(\mathbf{r}_1,\mathbf{r}_2,\omega)$ may be regarded as a Hermitian, nonnegative definite Hilbert-Schmidt kernel, and so it can be expanded as [15]

$$\mathbf{W}(\mathbf{r}_1,\mathbf{r}_2,\omega) = \sum_n \lambda_n(\omega)\boldsymbol{\phi}_n^\dagger(\mathbf{r}_1,\omega)\boldsymbol{\phi}_n(\mathbf{r}_2,\omega), \qquad (27)$$

where $\lambda_n(\omega)$ and $\boldsymbol{\phi}_n(\mathbf{r},\omega)$ are the eigenvalues and eigenfunctions, respectively, of the Fredholm integral equation

$$\int_D \boldsymbol{\phi}_n(\mathbf{r}_1,\omega)\cdot\mathbf{W}(\mathbf{r}_1,\mathbf{r}_2,\omega)\,\mathrm{d}^3 r_1 = \lambda_n(\omega)\boldsymbol{\phi}_n(\mathbf{r}_2,\omega). \qquad (28)$$

The eigenvalues are real and nonnegative. Since the eigenfunctions are subject to a vectorial orthonormality condition in D, they represent orthogonal polarization states within the decomposition domain.

I emphasize that the coherent-mode decomposition in Eq. (27) gives the cross-spectral density matrix $\mathbf{W}(\mathbf{r}_1,\mathbf{r}_2,\omega)$ as an incoherent sum of elementary matrices that all factor in the two spatial variables. Hence, these vector-field modes are completely coherent [15] in the space-frequency domain, in terms of the degree of electromagnetic correlations given by Eq. (25).

As in scalar theory, many useful results follow from Eq. (27). For one thing, one may construct an ensemble of monochromatic vectors and show that the cross-spectral density $\mathbf{W}(\mathbf{r}_1,\mathbf{r}_2,\omega)$ can rigorously be expressed as a correlation matrix averaged over this ensemble [15] in analogy to Eq. (7). Straightforward calculations based on Eq. (27) lead to an effective degree of spectral correlations [15,33]

$$\frac{\iint_{DD} S(\mathbf{r}_1, \omega) S(\mathbf{r}_2, \omega) \zeta^2(\mathbf{r}_1, \mathbf{r}_2, \omega)\, \mathrm{d}^3 r_1\, \mathrm{d}^3 r_2}{\iint_{DD} S(\mathbf{r}_1, \omega) S(\mathbf{r}_2, \omega)\, \mathrm{d}^3 r_1\, \mathrm{d}^3 r_2} = \frac{\sum_n \lambda_n^2(\omega)}{\left[\sum_n \lambda_n(\omega)\right]^2} \geq \frac{1}{N}, \qquad (29)$$

where N is the number of modes. The average electromagnetic correlations can be zero in D only if N approaches infinity. On the other hand, the expression in Eq. (29) is equal to unity only if $N = 1$. The relation between the eigenvalues and the degree of correlations remains unchanged if scalar fields are considered. Thus, the spectral theory of electromagnetic coherence and the quantity $\zeta(\mathbf{r}_1, \mathbf{r}_2, \omega)$ in Eq. (25) are fully consistent with their classical scalar counterparts.

4.7 Conclusions

It is hoped that this overview manages to convey some impressions of the rapid pace at which the electromagnetic coherence theory and its applications are currently progressing, despite long and well-established traditions. The theory of optical coherence with vector fields is, perhaps surprisingly, far from being finalized, as is evidenced by the ongoing endeavors to address partial polarization, electromagnetic coherence, and also entropy (which was not considered here) in near-field optics and nanophotonics.

The overriding conclusions to draw are that while electromagnetic coherence theory in many ways is similar to the conventional scalar theory of coherence, at the same time it is fundamentally different from scalar coherence in many aspects. The similarities are in the overall mathematical structure. The differences primarily arise from the fact that, unlike scalar waves, electromagnetic fields contain several components and their correlations must be included in a comprehensive treatment of polarization and coherence. Recent definitions of the degree of polarization and the degree of correlations (coherence) for arbitrary random 3D electromagnetic fields were assessed, but many questions still remain unanswered.

One thing is clear, though. The preceding sections and citations demonstrate that Wolf's work has laid the foundations and he continues to influence much of today's research in the field of electromagnetic theory of optical coherence. It has been a privilege to work with him all these years.

Acknowledgments

I thank Tero Setälä and Matti Kaivola (both at Helsinki University of Technology, Finland) and Jani Tervo (University of Joensuu, Finland) for extensive collaboration and useful discussions. Financial support from the Swedish Research Council is acknowledged.

References

1. E. Wolf, "A macroscopic theory of interference and diffraction of light from finite sources. II. Fields with a spectral of arbitrary width," *Proc. Royal Soc. (London) A* **230**, 246–265 (1955).
2. E. Wolf, "Coherence properties of partially polarized electromagnetic beams," *Nuovo Cimento* **13**, 1165–1181 (1959).
3. P. Roman and E. Wolf, "Correlation theory of stationary electromagnetic fields. Part I—The basic field equations," *Nuovo Cimento* **17**, 462–476 (1960).
4. P. Roman and E. Wolf, "Correlation theory of stationary electromagnetic fields. Part II—Conservation laws," *Nuovo Cimento* **17**, 477–490 (1960).
5. R.J. Glauber, "The quantum theory of optical coherence," *Phys. Rev.* **130**, 2529–2539 (1963).
6. C.L. Mehta and E. Wolf, "Coherence properties of blackbody radiation. III. Cross-spectral tensors," *Phys. Rev.* **161**, 1328–1334 (1967).
7. L. Mandel and E. Wolf, "Spectral coherence and the concept of cross-spectral purity," *J. Opt. Soc. Am.* **66**, 529–535 (1976).
8. E. Wolf, "New theory of radiative energy transfer in free electromagnetic fields," *Phys. Rev. D* **13**, 869–886 (1976).
9. E. Wolf, "Coherence and radiometry" (1977 Ives Medal Address), *J. Opt. Soc. Am.* **68**, 7–17 (1978).
10. L. Mandel and E. Wolf, *Optical Coherence and Quantum Optics*, Cambridge University Press, Cambridge, U.K. (1995).
11. E. Wolf and D.F.V. James, "Correlation-induced spectral changes," *Rep. Prog. Phys.* **59**, 771–818 (1996).
12. E. Wolf, "New spectral representation of random sources and the partially coherent fields that they generate," *Opt. Commun.* **38**, 3–6 (1981).
13. A.T. Friberg and E. Wolf, "Relationships between the complex degrees of coherence in the space-time and in the space-frequency domains," *Opt. Lett.* **20**, 623–625 (1995).
14. F. Gori, M. Santarsiero, S. Vicalvi, R. Borghi, and G. Guattari, "Beam coherence-polarization matrix," *J. Eur. Opt. Soc. A: Pure Appl. Opt.* **7**, 941–951 (1998).
15. J. Tervo, T. Setälä, and A.T. Friberg, "Theory of partially coherent electromagnetic fields in the space-frequency domain," *J. Opt. Soc. Am. A* (in press).
16. C. Henkel, K. Joulain, R. Carminati, and J.J. Greffet, "Spatial coherence of thermal near fields," *Opt. Commun.* **186**, 57–67 (2000).
17. A.V. Shchegrov, K. Joulain, R. Carminati, and J.J. Greffet, "Near-field spectral effects due to electromagnetic surface excitations," *Phys. Rev. Lett.* **85**, 1548–1551 (2000).

18. T. Setälä, A. Shevchenko, M. Kaivola, and A.T. Friberg, "Degree of polarization for optical near fields," *Phys. Rev. E* **66**, 016615 (2002).

19. T. Setälä, M. Kaivola, and A.T. Friberg, "Degree of polarization in near fields of thermal sources: Effects of surface waves," *Phys. Rev. Lett.* **88**, 123902 (2002).

20. T. Setälä, *Spatial Correlations and Partial Polarization in Electromagnetic Optical Fields: Effects of Evanescent Waves*, Ph.D. dissertation, Helsinki University of Technology, Finland (2003).

21. T. Setälä, K. Blomstedt, M. Kaivola, and A.T. Friberg, "Universality of electromagnetic-field correlations within homogeneous and isotropic sources," *Phys. Rev. E* **67**, 026613 (2003).

22. R. Carminati and J.J. Greffet, "Near-field effects in spatial coherence of thermal sources," *Phys. Rev. Lett.* **82**, 1660–1663 (1999).

23. J.J. Greffet, R. Carminati, K. Joulain, J.P. Mulet, S. Mainguy, and Y. Chen, "Coherent emission of light by thermal sources," *Nature* **416**, 61–64 (2002).

24. B. Karczewski, "Coherence theory of the electromagnetic field," *Nuovo Cimento* **30**, 906–915 (1963).

25. E. Wolf, "Unified theory of coherence and polarization of random electromagnetic fields," *Phys. Lett. A* **312**, 263–267 (2003).

26. F. Zernike, "Diffraction and optical image formation," *Proc. Phys. Soc. (London)* **61**, 156–164 (1948).

27. W.H. Carter and E. Wolf, "Far-zone behavior of electromagnetic fields generated by fluctuating current distributions," *Phys. Rev. A* **36**, 1258–1269 (1987).

28. S. Ponomarenko and E. Wolf, "The spectral degree of coherence of fully spatially coherent electromagnetic beams," *Opt. Commun.* **227**, 73–74 (2003).

29. J. Tervo, T. Setälä, and A.T. Friberg, "Degree of coherence for electromagnetic fields," *Opt. Express* **11**, 1137–1143 (2003).

30. T. Setälä, J. Tervo, and A.T. Friberg, "Complete electromagnetic coherence in the space-frequency domain," *Opt. Lett.* **29**, 328–330 (2004).

31. T. Setälä, J. Tervo, and A.T. Friberg, "Theorems on complete electromagnetic coherence in the space-time domain," *Opt. Commun.* (in press).

32. F. Gori, M. Santarsiero, R. Simon, G. Piguero, R. Borghi, and G. Guattari, "Coherent-mode decomposition of partially polarized, partially coherent sources," *J. Opt. Soc. Am. A* **20**, 78–84 (2003).

33. P. Vahimaa and J. Tervo, "Unified measures for optical fields: degree of polarization and effective degree of coherence," *J. Opt. A: Pure Appl. Opt.* **6**, S41–S44 (2004).

Ari Friberg (left), Prof. Wolf, and Kurt Oughstun at SPIE Conference AM100: Tribute
to Emil Wolf: Engineering Legacy of Physical Optics.

Ari T. Friberg obtained his Ph.D. on the subject of radiometry and partial coherence from the Institute of Optics in Rochester, NY. He also holds a D.Sc.
(Tech.) degree in engineering physics from Helsinki University of Technology,
where, from 1983 to 1996, he worked in various positions. Since 1990 he has been
associated with the Academy of Finland. In 1987–1988 he was a Royal Society
guest research fellow at Imperial College of Science and Technology in London,
and in 1996, he served as a visiting scientist at the Optisches Institut of Tech-
nische Universität Berlin. After a brief period as professor of physics in Joensuu,
Finland, in 1997 he was appointed professor of optics at the Royal Institute of
Technology (KTH) in Stockholm, Sweden. Dr. Friberg has published more than
140 peer-reviewed research papers, presented roughly a similar number of conference reports, edited a book, and co-edited two special issues of Pure and Applied
Optics and the Proceedings of ICO Topical Meeting "Image Science" (Helsinki,
1985). He is a fellow of OSA and a senior member of IEEE. He belongs to the
editorial boards of Progress in Optics, Optics Communications, OME Information (China) and previously Pure and Applied Optics and Optical Revue (Japan).
From 1998–2003 he was a topical editor of JOSA A and served on OSA's David
Richardson Medal committee (2002–2003). Dr. Friberg was a founding member
and the first President of the Finnish Optical Society, and he is a board member
of the Swedish and European Optical Societies. Since 1996 he has served as associate secretary of the International Commission for Optics (ICO). He is chair
of the ICO Finnish territorial committee, and chair of the ETOP (Education and
Training in Optics and Photonics) long-range planning committee.

⅋CHAPTER 5⅋

PHYSICAL OPTICS AT PHYSICAL OPTICS CORPORATION

Tomasz P. Jannson

5.1 Introduction

My venture with physical optics started during my graduate studies at Warsaw Tech in Poland when my supervisor, the late Prof. Bohdan Karczewski, proposed to me as a subject of my M.S. thesis, "electromagnetic analysis of polarization states of waves diffracted on a perfectly conducting half-plane" [1], based on Emil Wolf's coherency matrix formalism [2,3]. (This so-called Sommerfeld problem [4], as well as the coherency matrix formulation of polarization states, are discussed elsewhere in the present work [5,6].) From his Rochester discussion with Emil Wolf, Prof. Karczewski also suggested to me as a subject of my Ph.D. dissertation "inverse diffraction coherence theory" [7], a subject closely connected with inverse properties and the information content of evanescent waves [8], later seen as one of the earlier attempts at nano-optics, also discussed in this book [9]. At that time in Poland considerable study was stimulated by Prof. Rubinowicz and his school into diffraction of electromagnetic and acoustic waves, including the equivalence problem of integrating the Young and Huygens approaches, first solved for spherical incidence wave by Rubinowicz [10,11], and then generalized by Miyamoto and Wolf [12].

Working as an adjunct professor at Warsaw Tech, I had directed my interest to the engineering aspects of physical optics, mostly in holography, holographic interferometry, and Fourier optics. Here again, critical to my studies were Emil Wolf's inverse diffraction problem developments [13] based on the first Born approximation [14], which was instrumental in volume (Bragg) holography and diffraction tomography, the latter developed by A.J. Devaney, also described in this book [15]. My further efforts in Poland concentrated on structural information in volume holography [16,17], planar holograms [18], and integrated optics [19].

Those efforts materialized later in the U.S. in such applications as chip-to-chip waveguide interconnects [20,21].

My real venture with physical optics, however, started when I met Joanna. We soon married, and have worked together ever since. It was a fantastic cooperation from the beginning, and I would wish for anybody to work in such a "coherent" way, when *"1 + 1 > 2."* While working together in Poland at Warsaw Tech, we had developed a new approach to Fourier optics [22,23], based on the temporal Fourier transform [22,23]. She then started her Ph.D. dissertation under Prof. Jan Petykiewicz, in which she developed the basic theoretical framework for prism coupling into an anisotropic waveguide [24], based on complex-variable Rieman spaces, as an extension of those in the Sommerfeld problem [25]. She completed her Ph.D. a year later at the University of New Mexico, in Albuquerque.

Just before leaving Poland, I spent three months in Olomunc, Czech Republic (then Czechoslovakia), where Prof. Jan Peřina [26] advised me to study Carter's and Wolf's paper on physical radiometry [27]. This paper turned out to be critical for engineering applications of physical optics, because, by introducing the concept of quasi-homogeneous sources, it opens up the theoretical framework of physical optics into a broad range of partially coherent light sources such as thermal, fluorescent, LEDs, and LDs. This paper stimulated my own studies in this new area of physical optics [28,29], later critical in our development of diffuser products at Physical Optics Corporation (POC). Today at POC and the company's subsidiaries [30] only about half of our efforts are directly related to physical optics. Other areas include electronic imaging [31] and soft computing [32,33]; small RF communication platforms [34,35] such as unmanned ground vehicles (UGVs) and unmanned aerial vehicles (UAVs); fiber sensors [30], fiber communication [30], microwave phased array antennas based on physical optics [30], remote lighting, and others [36,37]. Nonetheless, physical optics still plays a crucial role in our product efforts, and these physical optics-based commercial efforts at POC will be the main subject of this paper.

5.2 Non-Lambertian Diffusers Theory

While many semitransparent scattering media, either natural or artificial (such as milky glass), can be considered diffusers, most of them are Lambertian scatterers. The main issue then is how to control and/or modify their angular spectrum in a useful way for specific practical lighting applications such as backlighting, cellular phones displays, diffuser screens for rear-projection and front-projection TV; optical sensors and illuminators producing wide-angle uniform white light; flat-panel displays; and other lighting systems. For projection screens in particular, it would be useful to develop diffusers with broader horizontal angular divergence usually characterized by a half-width and half-maximum angle (HWHM) and a narrow vertical divergence. It is not easy to produce such non-Lambertian "elliptical" diffusers by any conventional means, such as mechanically or lithographically, since scattering microscopic centers should be random/aperiodic (to avoid rainbow/grating effects on white light) and have an anisotropic profile. Speckle optics gives us a general tool for this purpose: producing a non-Lambertian diffuser by recording anisotropic laser speckle random pattern is one solution to this problem; reducing this solution to practice is itself a problem.

In our early studies of speckle optics diffusers, we used volume holographic materials such as dichromate gelatin (DCG) as a recording medium [38], building on our volume holography expertise [39,40]. Later we began using photoresist, a surface-relief material, as a recording medium [41] in which high-relief aspect ratios (ARs) (see Fig. 1) and volume (Bragg) effects can still be observed in at least two ways: the zero-order beam can be reduced to a minimum, and diffraction efficiencies of master copies can be *higher* than those of the master itself, a strange manifestation of Kogelnik's volume holographic resonance effect [42].

A theoretical explanation of "holographic" diffusers (or rather "speckle" diffusers) should include the fact that controlled-shape speckles are recorded with laser light and then reconstructed by an incident beam of either white or laser light. White light is important for applications such as rear-projection TV and laptop/wireless handset display backlighting, yet white light optics has been rather unpopular in the scientific world since the invention of the laser.

To record a holographic diffuser, a coherent laser beam is incident on a plane original diffuser mask, as in Fig. 2. In the reconstruction process, in the linear approximation, the amplitude transmittance of the diffuser mask is

$$t = BI, \tag{1}$$

where B is a proportionality constant and I is the intensity modulation. According to the spatial representation of the Wiener-Khintchine theorem [3], the angular spectrum intensity of diffused (scattered) quasi-homogeneous light is proportional

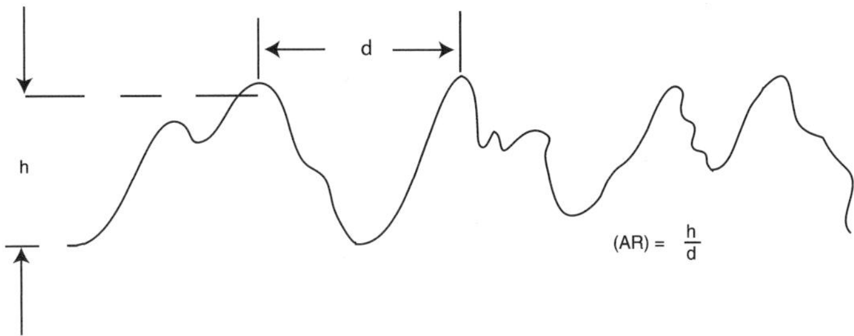

Figure 1 Illustration of diffuser relief aspect ratio (AR).

to the 2D Fourier transform of the spatial coherence autocorrelation density function of amplitude transmittance (assuming a monochromatic collimated normal incident beam). Therefore, according to Eq. (1), we need to find the autocorrelation intensity function in the form

$$W(\mathbf{r}) = C\langle I(\mathbf{r}' + \mathbf{r})I(\mathbf{r}')\rangle, \qquad (2)$$

where $\langle \cdots \rangle$ is the ensemble average, and $\mathbf{r} = (x,y)$.

This problem was solved by Goldfischer [43], whose final formula [38] is

$$\langle I(\mathbf{r} + \mathbf{r}')I(\mathbf{r}')\rangle = D|F(\mathbf{r})|^2, \qquad (3)$$

$$F(\mathbf{r}) = F(x,y) = \iint du\, dv P(u,v)\exp[-2\pi i(xu + yv)]/\lambda h, \qquad (4)$$

where D is a constant, and $P(u,v)$ is the aperture function of the original diffuser mask. (Other setups for diffusers are described in Ref. [41].)

Unfortunately, Goldfischer's derivation of Eq. (3) is rather complicated, since in his derivation in Ref. [43] he applied discrete cosine and sine functions rather than complex-variable continuous functions based on the concept of the analytic signal [3]. Therefore, we repeat this important derivation in the complete complex-variable analytic signal notation, with some variables defined in Fig. 2.

Consider a coherent collimated incident laser beam. The complex amplitude of the diffracted light on the (u,v) screen has the form (within the accuracy of the

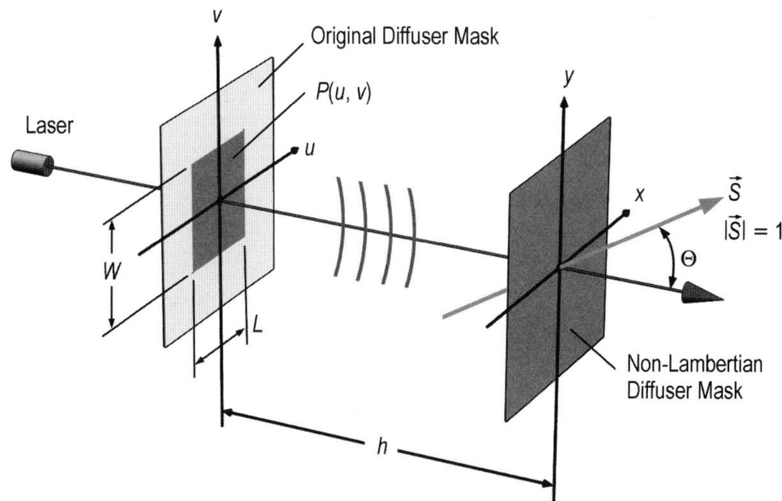

Figure 2 Configuration for recording non-Lambertian diffuser mask (screen).

x, y-dependent quadratic term)

$$A(x,y) = \iint [P(u,v)]^{1/2} \exp i\varphi \exp\left[\frac{ik(u^2 + v^2)}{2h} \exp\frac{-ik(xu + yv)}{h}\right] du\, dv, \quad (5)$$

where $k = 2\pi/\lambda$, h is the distance between the (u,v) and (x,y) planes, φ is the original diffuser's random phase, and $P(u,v)$ is the intensity aperture function, equal to 1 inside the diffuser aperture, and 0 otherwise. The intensity function is

$$I(x,y) = \iiiint [P(u,v)]^{1/2} [P(u',v')]^{1/2} \exp[i\varphi(u,v)] \exp[-i\varphi'(u',v')]$$

$$\times \exp\left[\frac{-ik}{h}(xu + yv)\right] \exp\left[\frac{ik}{h}(xu' + yv')\right] du\, dv\, du'\, dv', \quad (6)$$

where quadratic factors have been included in the random phases φ and φ'. Now we apply simplifying relations; this is equivalent to a general statement that Fresnel and Fraunhofer diffraction regions are equivalent in the context of speckle statistics; also, we have

$$\exp\left[\frac{-i2\pi x}{\lambda h}(u - u')\right] = \exp[-i2\pi xp], \quad (7a)$$

$$u - u' = p\lambda h, \text{ or } u' = u - p\lambda h \quad (7b)$$

to obtain (for x-dependent terms)

$$\exp\left[\frac{-ik}{h}(xu) + \frac{ik}{h}(xu')\right] = \exp\left[\frac{-ikx}{h}(u - u')\right] = \exp\left[\frac{-i2\pi x}{\lambda h}(u - u')\right]$$

$$= \exp(-i2\pi px), \quad (8)$$

and the intensity function formula becomes

$$I(x,y) = \iiiint [P(u,v)]^{1/2} [P(u - p\lambda h, v - q\lambda h)]^{1/2} \exp[i\Psi(u,v;p,q)]$$

$$\times \exp[i2\pi(px + qy)]\, du\, dv\, dp\, dq, \quad (9)$$

where $\Psi = \varphi - \varphi'$. Therefore, the intensity autocorrelation function is

$$\langle I(x,y)I(x + \Delta x, y + \Delta y)\rangle$$

$$= \iint \iint \iint [P(u,v)]^{1/2}[P(u',v')]^{1/2}[P(u-p\lambda h, v-q\lambda h)]^{1/2}$$

$$\times [P(u'-p'\lambda h, v'-q'\lambda h)]^{1/2}\exp[i\Psi(u,v;p,q)]\exp[-i\Psi'(u',v';p',q')]$$

$$\times \exp[-i2\pi x(p-p')]\exp[-i2\pi y(q-q')]\exp[i2\pi p'\Delta x]\exp[i2\pi q'\Delta y]$$

$$\times du\,dv\,dp\,dq\,du'\,dv'\,dp'\,dq'. \tag{10}$$

Since function $(\Psi - \Psi')$ is random, the nonzero contribution is only from those terms that include nonrandom items; i.e.,

$$\boxed{\Psi - \Psi' = 0}, \tag{11}$$

which is equivalent to $p = p', q = q', u = u'$, and $v = v'$. Thus, Eq. (10) reduces to the following relation:

$$\langle I(x,y)I(x + \Delta x, y + \Delta y)\rangle = \iint \iint P(u,v)P(u-p\lambda h, v-q\lambda h)$$

$$\times \exp(i2\pi p\Delta x)\exp(i2\pi q\Delta y)\,du\,dv\,dp\,dq, \tag{12}$$

which is equivalent to Eq. (17) in Ref. [43]. However, we have

$$\iint P^{*}(p\lambda h - u, q\lambda h - v)\exp(i2\pi p\Delta x)\exp(i2\pi q\Delta y)dpdq$$

$$= \hat{P}^{*}\left(\frac{\Delta x}{\lambda h}, \frac{\Delta y}{\lambda h}\right)\exp\left(-i2\pi\Delta x\frac{u}{\lambda h}\right)\exp\left(-i2\pi\Delta y\frac{v}{\lambda h}\right), \tag{13}$$

where $\hat{P}$ is the 2D Fourier transform of the P-function; also,

$$\iint P(u,v)\exp\left(-i2\pi\Delta x\frac{u}{\lambda h}\right)\exp\left(-i2\pi\Delta y\frac{v}{\lambda h}\right)du\,dv = \hat{P}\left(\frac{\Delta x}{\lambda h}, \frac{\Delta y}{\lambda h}\right). \tag{14}$$

Substituting Eqs. (13) and (14) into Eq. (12), we obtain

$$\langle I(x,y)I(x + \Delta x, y + \Delta y)\rangle = \left|\hat{P}\left(\frac{\Delta x}{\lambda h}, \frac{\Delta y}{\lambda h}\right)\right|^{2}, \tag{15}$$

which, with accuracy to proportionality constant, is identical with Ref. [43], Eq. (18), and our Eq. (3). In order to obtain Eq. (3), however, we need to change notation:

$$(x,y) \rightarrow (x',y') \rightarrow (\mathbf{r}'), \tag{16a}$$

$$(\Delta x, \Delta y) \rightarrow (x,y) \rightarrow (\mathbf{r}), \tag{16b}$$

$$\hat{P} \rightarrow \Gamma, \tag{16c}$$

then Eq. (15) becomes

$$\langle I(\mathbf{r} + \mathbf{r}')I(\mathbf{r}')\rangle = D|F(\mathbf{r})|^2, \tag{17}$$

where

$$F(\mathbf{r}) = F(x,y) = \iint du\,dv\,P(u,v)\exp\left[-i2\pi(xu + yv)/\lambda h\right], \tag{18}$$

which is identical to Eq. (4), confirming Goldfischer's derivation [43].

If the incident source is spherical instead of collimated, this modifies the factor

$$\exp\left[-i2\pi(xp + yq)\right] \text{ into } \exp\left[-i2\pi(p'u_{\mathrm{o}} + px)\right], \tag{19}$$

where

$$px + p'u_{\mathrm{o}} = p\left(x + u_{\mathrm{o}}\frac{h}{h'}\right).$$

Figure 3 explains the existence of this new factor, since

$$\frac{u'}{h} = \frac{u_{\mathrm{o}}}{h'} \Rightarrow u' = u_{\mathrm{o}}\frac{h}{h'}, \tag{20}$$

which coincides with Eq. (19).

Therefore, modifying the incident beam from collimated to spherical is equivalent to the lateral translation of a diffraction pattern. In addition, if the incident beam is Gaussian, we have an additional attenuation factor that is equivalent to the

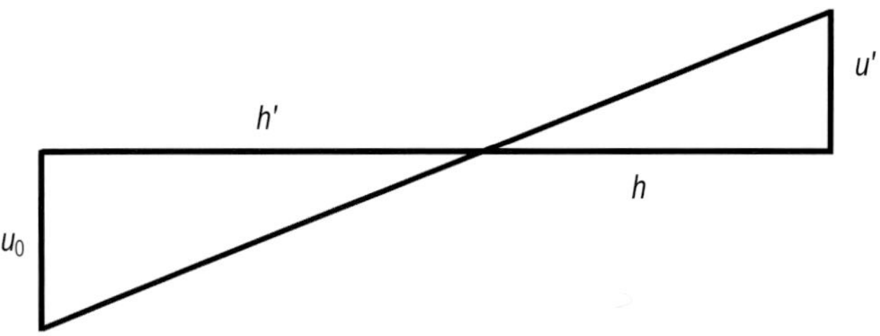

Figure 3 Illustration of Eq. (20).

following pupil function:

$$P(u,v) = \exp\left[\frac{-(u^2 + v^2)}{w^2}\right].$$ (21)

5.2.1 Angular spectrum of light scattered by non-Lambertian diffuser

In the reconstruction process, we illuminate the diffuser mask (x,y) with a collimated white-light beam; thus, for a monochromatic component of the light at a reconstruction wavelength that typically is different from the recording wavelength $\lambda_o \neq \lambda$, the radiant intensity [27] is

$$J(s) = s_z A \iint dx\, dy \exp\left[-ik_o \mathbf{r} \cdot \mathbf{s}\right] W(\mathbf{r}),$$ (22)

where $k_o = 2\pi/\lambda_o$, $W(\mathbf{r}) = \langle t(\mathbf{r}' + \mathbf{r})t^*(\mathbf{r}')\rangle$, and, from Eqs. (1) and (2), we have

$$W(\mathbf{r}) = G\langle I(\mathbf{r}' + \mathbf{r})I(\mathbf{r}')\rangle,$$ (23)

where $\mathbf{r} = (x,y)$, and G is a constant; $\mathbf{s} = (s_x, s_y)$ is the (x,y) projection of directional unit vector:

$$s_x^2 + s_y^2 + s_z^2 = 1.$$ (24)

By substituting Eq. (17) into Eq. (23) and Eq. (23) into Eq. (22), we obtain

$$J(s_x, s_y) = s_z \cdot D \iint du\, dv P(u,v) P\left(u + \frac{\lambda}{\lambda_o}hs_x, v + \frac{\lambda}{\lambda_o}hs_y\right),$$ (25)

i.e., the autocorrelation of the aperture function P, which is the original diffuser pupil function; h is the distance between the original diffuser (used for *recording* the non-Lambertian diffuser), and the *recorded* diffuser. Thus, Eq. (25) contains information about both the recording and reconstruction processes, while "speckle statistics" is defined by Eq. (4), which is the Fourier transform of the pupil function P in the following form:

$$F(f_x, f_y) = \iint du\, dv P(u,v) \exp\left[-i2\pi(f_x \cdot u + f_y \cdot v)\right],$$ (26)

$$f_x = \frac{x}{\lambda h}, \quad f_y = \frac{y}{\lambda h}.$$ (27)

In order to obtain statistically averaged speckle sizes, consider a typical case of a rectangular aperture, as illustrated in Fig. 2. Then, $P(u,v)$ is the rectus function

$$P(u,v) = \text{rect}\left(\frac{u}{L}\right) \text{rect}\left(\frac{v}{W}\right),$$

where

$$\text{rect}(x) = \begin{cases} 1 & \text{for } |x| \leq 1/2 \\ 0 & \text{for } |x| > 1/2 \end{cases}. \tag{28}$$

Then, according to Eq. (26), the F-function is

$$F(x,y) = \text{sinc}(f_x \cdot L)\,\text{sinc}(f_y \cdot W),$$

where

$$\text{sinc}(\xi) = \frac{\sin\pi\xi}{\pi\xi}. \tag{29}$$

The first zeros of this function are defined by the relations

$$f_{x_0} \cdot L = 1, \quad f_{y_0} \cdot W = 1. \tag{30}$$

Substituting Eq. (27) into Eq. (30), we obtain

$$x = x_0 = \frac{\lambda h}{L}, \quad y_0 = \frac{\lambda h}{W}. \tag{31}$$

Defining the average speckle sizes as $(2x_0, 2y_0)$, we obtain

$$\delta_x = 2x_0 = \frac{2\lambda h}{L}, \quad \delta_y = 2y_0 = \frac{2\lambda h}{W}. \tag{32}$$

This relation is illustrated in Fig. 4.

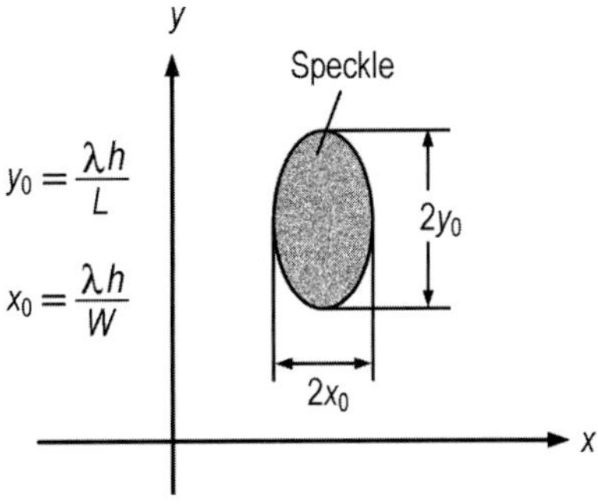

Figure 4 Illustration of Eq. (32) describing statistically averaged speckle sizes.

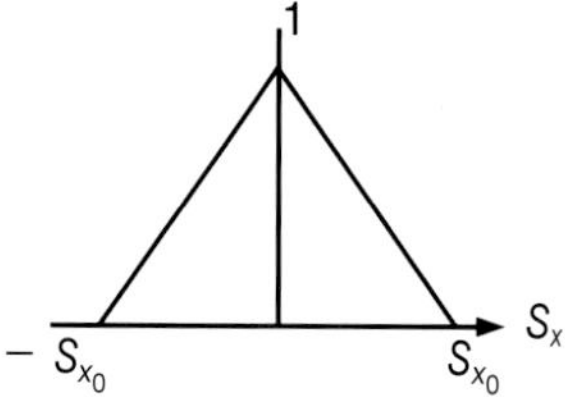

Figure 5 Illustration of 1D triangular function [Eq. (33)].

For example, for $\lambda = \lambda_o = 0.5$ μm, $h = 10$ cm, and $L = 1$ cm we obtain $\delta_x = 10$ μm.

Substituting Eq. (28) into Eq. (25), we obtain

$$J(s_x, s_y) = s_z D \Lambda\left(\frac{s_x}{s_{x0}}\right) \Lambda\left(\frac{s_y}{s_{y0}}\right), \tag{33}$$

where Λ is the triangular function shown in Fig. 5 with an amplitude of 1 and zero-intersections at points $-s_{x0}$ and s_{x0}, or $-s_{y0}$ and s_{y0}, where

$$s_{x0} = \frac{L}{h}\frac{\lambda_o}{\lambda}, \quad s_{y0} = \frac{W}{h}\frac{\lambda_o}{\lambda}. \tag{34}$$

We see that when the reconstruction wavelength λ_o, is longer than the recording wavelength λ, the angular sizes of the beam scattered by a non-Lambertian diffuser are larger by a factor of λ_o/λ. For $\lambda_o = \lambda$, we have

$$s_{x0} = \frac{L}{h}, \quad s_{y0} = \frac{W}{h}. \tag{35}$$

On the other hand, the angular sizes of the pupil (aperture) "seen" from the non-Lambertian diffuser center, according to Fig. 2, are

$$s_{x_0'} = \frac{L}{2h}, \quad s_{y_0'} = \frac{W}{h}, \tag{36}$$

i.e., twice-smaller than beam angular sizes described by Eq. (35). This angular relation is shown in Fig. 6.

Substituting Eq. (36) into Eq. (32), we obtain

$$s_{x_0}'\delta_x = \lambda, \quad s_{y_0}'\delta_y = \lambda, \tag{37}$$

or using Eq. (35) we have

$$s_{x_0}\delta_x = 2\lambda, \quad s_{y_0}\delta_y = 2\lambda. \tag{38}$$

Those "uncertainty relations" combine angular sizes with speckle sizes. They are illustrated in Table 1.

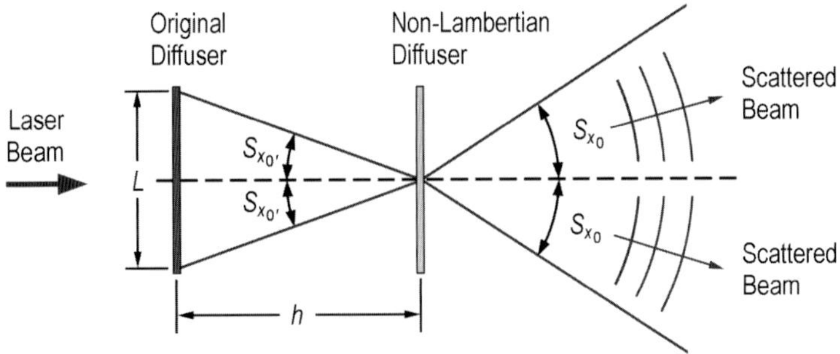

Figure 6 Illustration of the fact that the angular sizes (s_{x0}, s_{y0}) of scattered beam, are twice as large as the angular sizes of the original diffuser (s'_{x_0}, s'_{y_0}) "seen" from the non-Lambertian diffuser center.

Table 1 Speckle sizes (δ_x) versus angular size of the pupil (s'_{x_0}) and scattered beam (s_{x0}), for $\lambda = \lambda_o = 0.5$ μm.

s'_{x_0}	0.05	0.1	0.15	0.2	0.25
s_{x0}	0.1[*]	0.2	0.3	0.4	0.5
δ_x[**]	10 μm	5 μm	3.3 μm	2.5 μm	2 μm

[*]Equivalent to $\pm 5°$-angular divergence.
[**]Statistically averaged size.

The formula Eq. (22) can be generalized for slanted angles of incidence; then, instead of Eq. (22), we have (assuming a slowly varying s_z-parameter)

$$J(\mathbf{s}, \mathbf{s}_1) = s_z \cdot A I_o(\mathbf{s}_1) \iint dx\, dy\, e^{-ik_o(\mathbf{s}-\mathbf{s}_1)r} \cdot W(\mathbf{r}), \qquad (39)$$

where $\mathbf{s}_1$ is the directional vector of the incident beam illuminating the non-Lambertian diffuser. For an angular spectrum of incident beams we have

$$J(\mathbf{s}) = \iint h(\mathbf{s}-\mathbf{s}_1)I_o(\mathbf{s}_1)\,ds_{x1}\,ds_{y1}, \qquad (40)$$

where

$$h(\mathbf{s}-\mathbf{s}_1) = s_z \iint dx\, dy\, \exp[-ik_o(\mathbf{s}-\mathbf{s}_1) \cdot \mathbf{r}]W(\mathbf{r}) \qquad (41)$$

is the impulse response of the system. Equation (40) is a generalization of Eq. (22), and describes broadening of the angular spectrum scattered by the diffuser as a result of nonplane illumination. The angular spectrum of incident illumination is described by $I_o(\mathbf{s}_1)$.

In a similar way, we can generalize the monochromatic case to the nonmonochromatic case:

$$J(\mathbf{s}) \rightarrow J(\mathbf{s}_1, \lambda_o) \tag{42}$$

to obtain the following formula:

$$J(\mathbf{s}) = \iiint h(\mathbf{s} - \mathbf{s}_1, \lambda_o) I_o(\mathbf{s}_1, \lambda_o)\, ds_{x1}\, ds_{y1}\, d\lambda_o, \tag{43}$$

where $I_o(\mathbf{s}_1, \lambda_o)$ describes both the angular and the wavelength spectra of the incident beam. The angular broadening of the scattered beam as a result of wavelength broadening can, according to Ref. [38], be described by the following formula:

$$\frac{\Delta\lambda_o}{\lambda_o} = \frac{\Delta s_x}{s_x}, \tag{44}$$

where $s_x = \sin\alpha_x$; thus, we have

$$\frac{\Delta s_x}{s_x} = \frac{\Delta\alpha_x}{\tan\alpha_x}, \tag{45}$$

and for $\tan\alpha_x \approx 1$, we roughly have

$$\frac{\Delta\lambda_o}{\lambda_o} \cong \Delta\alpha_x. \tag{46}$$

For example, for $\lambda_o = 0.5$ μm and $\Delta\lambda = 20$ nm, $\Delta\alpha_x = 2$ deg. However, it should be emphasized that, because of the diffuser random structure, the angular broadening that is due to nonmonochromatic incident wave *does not* create a rainbow effect similar to that of grating diffraction.

It is well known that in the case of a fully coherent incident screen, when the diffuser is moving (e.g., rotating), the radius of spatial coherence is equal to the rms of the speckle (average speckle size), and since, according to Eq. (33), we can control the speckle sizes, we can also control the spatial coherence of the scattered beam [38].

5.2.2 New degree of freedom for communication as a result of spatial coherence

According to Ref. [38], we can transmit information through spatial coherence modulation only by changing pupil size over time, and thus proportionally increasing or reducing the intensity of incident light. Angular spectrum and wavelength spectrum of scattered light are unaffected, as is intensity. Therefore, a standard intensity detector, spectrometer, or some detector arrays *will not* detect any light modulation. Yet, information will be still sent by means of the spatial coherence modulation, which can be detected only by a Young interference detector. This effect can be compounded with Wolf's wavelength shift effect, also discussed in this book.

5.3 Non-Lambertian Diffusers Experiment

5.3.1 Applicability of the theoretical formulas

For practical purposes it is important to determine precisely the assumptions under which the theoretical formulas given in the previous chapter have been derived. It should be emphasized that similar formulas can be obtained by other means than those shown in Fig. 2 {see, e.g., Fig. 23 in Ref. [44]}. However, the methods proposed here are very practical ones because they do not require lenses. Also, note that the proposed non-Lambertian diffuser production process consists of two-steps: (1) *registration* of a coherent laser speckle pattern with wavelength λ, as described by the Goldfischer formula Eq. (3); and (2) *illumination* of a *recorded* non-Lambertian diffuser with a partially coherent (LED, thermal) nonmonochromatic (including white) incident light beam, with average wavelength λ_o as described by Eq. (25). Here, emphasis is on the application aspect of the introductory photometric formula Eq. (22), which can be considered a particular case of Schell's theorem {see, e.g., Eqs. (5.7–10) in Ref. [44]}. Equation (22) coincides with the Schell formula if we assume that the spatial coherence radius d_c is much larger than the autocorrelation radius d_a of the complex pupil function, identified here with the non-Lambertian diffuser transmittance function t, as in Eq. (22) {do not confuse the pupil function defined by Goodman [44] with our aperture function from Eq. (9)}; i.e., we obtain

$$d_a \ll d_c. \tag{47}$$

This is because, in the vast majority of practical cases, a given non-Lambertian diffuser is illuminated either with collimated laser light or with LED light. In both cases Eq. (47) is satisfied. In the second case, by applying the Van Cittert-Zernike theorem for LED with size $a = 50$ μm, distance $z = 2$ cm, and $\lambda_o = 0.63$ μm, the coherence radius is

$$d_c = \frac{\lambda_o \cdot z}{a} = \frac{0.63 \ \mu\text{m} \times 2 \ \text{cm}}{50 \ \mu\text{m}} = 252 \ \mu\text{m}, \tag{48}$$

while d_a does not typically exceed 10 μm. We see that Eq. (22) can be treated as an advanced version of the Brian Thompson formula for the coherent case {see Eqs. (5.7–13) in Ref. [44]} if we replace the typical open diffraction aperture with

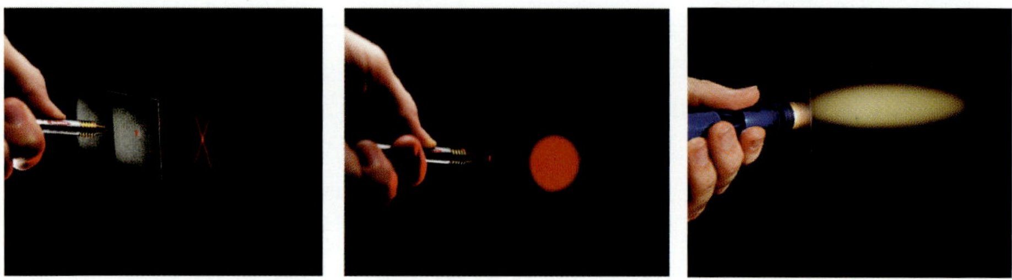

Figure 7 Typical diffuser illumination schemes described by Eq. (22) for various diffuser FWHM angular values.

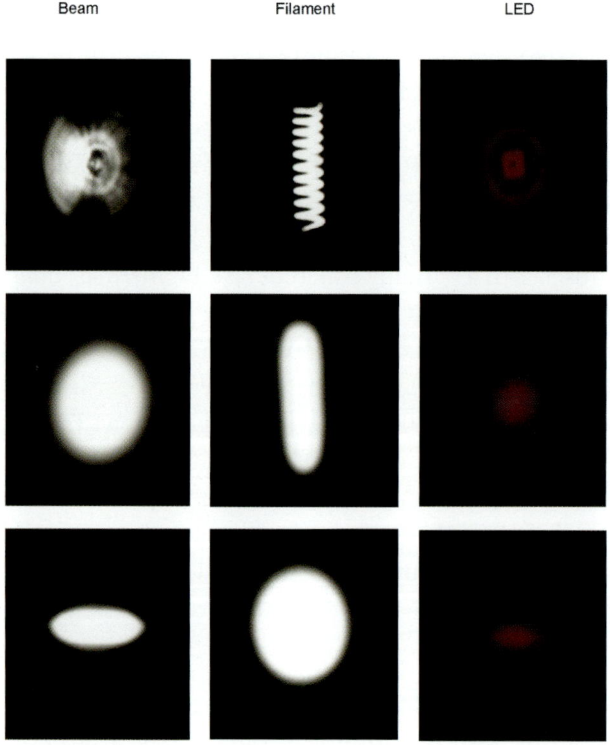

Figure 8 Homogenization by two types of non-Lambertian diffusers: circular (second row) and elliptical (third row), for various types of nonhomogeneous sources (first row).

our diffuser aperture. The typical application of Eq. (22) is shown in Fig. 7. The typical homogenization process as an effect of a non-Lambertian diffuser is shown in Fig. 8.

Figure 9 shows two scanning electron microscope (SEM) pictures of light shaping diffuser (LSD[*]) structures, for the circular and elliptical diffusers.

---

[*] Light shaping diffuser is a trademark of Physical Optics Corp.

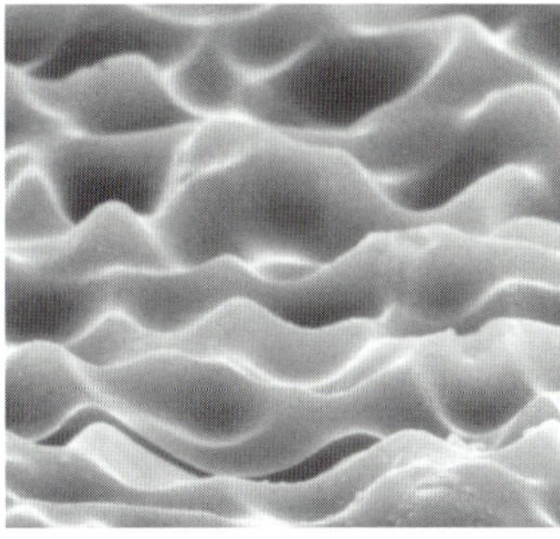

Figure 9 SEM pictures of non-Lambertian diffuser structures: circular and elliptical; 18-deg LSD and $1500\times$ magnification, as well as 0.2 deg $\times$ 11 deg LSD and $75\times$ magnification.

5.3.2 Mass production process

Non-Lambertian diffusers can be mass-produced by a replication of a single diffuser master many times. Master fabrication proceeds through: master coating, master holographic exposure, and submaster fabrication. The replication steps are: hot embossing, injection molding, and web-machine-based UV curing. The master fabrication and replication processes are shown in Fig. 10.

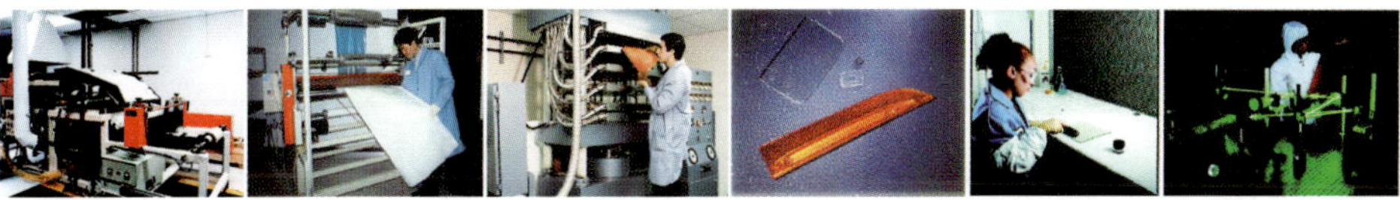

Figure 10 Mastering and replication processes at POC, including (from left to right): diffuser roll web replication; diffuser roll inspection; hot embossing; injection molding; hand replication; and diffuser master holographic recording.

5.4 Physical Radiometry at POC

5.4.1 Radiometric ray tracing (R^2T)

Developed by Wolf [27,45–47] and others [28,29,48–53], physical radiometry (photometry) extends the rules of geometric optics as weak diffraction phenomena. Such diffraction optics phenomena are within the scope of optical radiation created by quasi-homogenous sources, first defined by Carter and Wolf [27]. Based on physical photometry, spatial coherence, and Fourier optics [54], radiometric ray tracing (R^2T) [28,29,55] expands the standard ray-tracing algorithms (such as in ZEMAX) by adding information about the optical intensity and spatial coherence distribution of quasi-homogeneous sources. It should be emphasized that this R^2T formalism is *not* based on the complex wave amplitude characteristic of typical diffraction formulas (see, e.g., Refs. [3] and [54]), but rather on standard photometric quantities such as: radiance, or brightness (luminance); radiant intensity; emissivity; and optical flux; therefore, it should not be confused with diffraction optics algorithms, also included in recent versions of ZEMAX and other ray-tracing algorithms. Its ultimate goal would be quantitative ray-tracing system design based on optical properties of all system components one by one without the need for an overall system physical experiment. *In other words, by applying R^2T, we could predict weak diffraction optical system performance without making an actual optical experiment.* The optical components to be analyzed within R^2T could include all lenses, mirrors, and other imaging elements as well as any scatterers and diffusers, assuming that we know their optical characteristics. The R^2T does not, however, cover diffractive optical elements, gratings, multilayers, and other components that produce strong diffraction phenomena manifested by interference/rainbow effects.

The practical importance of the R^2T is significant: by finding an optical measure of all system components, treated separately, and by extension of the system R^2T based on the known component optical parameters, we obtain all overall radiometric (photometry) parameters of the system, *without* its actual measure. That saves significant optical design time for many optical systems of interest, especially nonimaging optical (NIO) systems [56], including ones that have scattering elements such as diffusers. Therefore, R^2T is especially useful for designing illumination and display systems based on the non-Lambertian diffusers developed at POC.

The fundamental connection between photometric quantities and 4D phase space is based on the fact that radiance (brightness, luminance) is a 4D phase-space density [55]. Thus, an arbitrary bundle of rays passing through a given plane (x,y) can be presented as a multiplicity of points in phase space (i.e., a single ray is represented homomorphically by a single point in phase space). In particular, a Lambertian source radiation is represented by a *uniform* 4D (radiance) distribution in phase space [55].

The fundamental relation between radiance and phase-space density, or 4D ray density, can be applied to evaluate photometric quantities by numerically calculating the number of rays passing through selected phase-space domains. For example, radiance can be calculated as the number of rays located in the phase-space elementary cell defined by Heisenberg's uncertainty relation and adapted to optics. Emissivity, on the other hand, can be calculated by integrating elementary cells through directional vector space (s_x, s_y). As a result, the formalism presented enables us to connect standard ray-tracing methods such as ZEMAX, for a multiplicity of rays, with basic radiometric quantities and relevant spatial coherence.

5.4.2 Physical radiometry, R^2T, and geometric optics

It was shown by Wolf and others [28,29,50,57,58] that in the short-wavelength approximation the physical radiometry preparation rules collapse to geometric-optics ray-tracing rules. In this sense R^2T is automatically applicable to related geometric optics problems.

There is yet another area of physical optics that can benefit from R^2T, namely, the area in which phenomena related to physical optics are manifested in a soft way (see Sect. 5.4.1). In these cases, the susceptibility to optical interference (and diffraction) is weak; in other words, the spatial coherence of an optical source is low and optical intensity spatial distribution varies slowly. Such weak-spatial-coherence sources are said to be quasi-homogeneous [27]. They are very common in nature and in industry, and include all thermal sources, fluorescent sources, LEDs, the vast majority of semiconductor lasers, plasma sources, and others.

The criterion for defining a source as quasi-homogeneous is as follows. Consider coherence radius, d_c, defined as the radius of the area as in Sect. 5.3.1, where the spatial coherence degree μ of the source is significantly larger than zero $(0 \leq \mu| \leq 1)$ [27]. The source is called quasi-homogeneous when: (1) its spatial coherence properties are homogeneous; and (2) within its coherence radius, the optical intensity is almost constant. It could be observed that these properties are easily satisfied for all the types of sources mentioned above. It should also be observed that the minimum value of coherence radius is not zero but a wavelength; in other words, there is no completely incoherent source assuming that inhomogeneous waves are ignored (the exception is the nano-optics domain, when the inhomogeneous (evanescent) waves cannot be ignored [9]).

Physical radiometry also exhibits a connection between Lambert's law and spatial coherence, as shown in Table 2 [55]. In particular, we see that non-Lambertian sources have *narrower* angular distribution than Lambertian ones, and consequently, their coherence radius is larger than wavelength [28]. On the other hand, no angular distribution broader than a Lambertian one can exist (see Table 2), assuming that inhomogeneous (evanescent) waves are ignored. These conclusions

Table 2 Relation between angular characteristics and spatial coherence.

Type of source	B*	J†	$\tilde{\mu}$‡
Lambertian	1§	$\cos\theta$**	$\cos^{-1}\theta$
General $\cos^n\theta$	$\cos^n\theta$♯	$\cos^{(n+1)}\theta$	$\cos^{(n-1)}\theta$

* Radiance; † Radiant intensity; ‡ Fourier transform of complex degree of spatial coherence; § for Lambertian source, the radiance angular distribution is constant; then, accordingly to the physical optics, the FWHM-angle is equal to 120 deg; ** Lambert law; ♯ general $\cos^n\theta$ angular distribution; for $n = 0$, the source is Lambertian; in general, $n \geq 0$.

have important meaning for the R^2T since phase-space $(x, y; s_x, s_y)$ density of rays is identical to radiance 4D distribution.

5.4.3 Radiometric anomalies

Physical radiometry [27–29,45–53] integrates classical radiometry with physical optics in general and with spatial coherence in particular as in Ref. [27], where radiance is shown to be proportional to the Fourier transform of the complex degree of (spatial) coherence. This is demonstrated by applying to physical radiometry three physical optics features: (1) wave-complex-amplitude definitions of radiance [48,49]; (2) the physical optics definition of quasi-homogeneous sources; and (3) the physical optics formula for generalized radiance free-space propagation. Unfortunately, there is unavoidable arbitrariness in choosing these features; therefore, only an experiment could decide which is correct. One of the necessary tests for the validity of this formalism is to prove that, in the "short-wavelength" approximation, the radiance propagation formula becomes the classical radiometric formula. This was shown in two ways in [28,29,57–59], based on: (1) Walther's first or second definition of radiance [49]; (2) the definition of quasi-homogeneous sources; and (3) the Rayleigh-Sommerfeld diffraction formula of the first kind [60]. In Refs. [28,29], however, it was shown that the "short-wavelength" approximation (called the "first-order" approximation in Ref. [29]) only holds for short z-distances, thus; the second-order approximation is needed in order for the results to cover the full range of interest. The second-order approximation leads, however, to strange behavior of radiance during propagation, referred to here as radiometric anomaly.

Specifically, the basic formula of Ref. [29] departs significantly from the conventional radiometry propagation law, which states that projecting radiance in a straight line through free space preserves its constant values. In contrast, according to Ref. [28], Eq. (42), the values of radiance projected along a straight line are *not* constant. Instead, they depend on the specific spatial distribution of radiance in the source plane. In particular, if the spatial distribution is periodic, and we decompose that distribution onto 2D Fourier series characterized by specific spatial frequency vector $\mathbf{f}$, each radiance Fourier component will be projected with a distinct modulation factor M {see Eq. (45) in Ref. [28]}. It is shown in Refs. [28] and [29] that sometimes this M factor can even be zero:

$$M = 0. \tag{49}$$

In this case, there are directions in which no modulated radiation is projected, even for nonzero modulated radiance distribution at the source plane. This new hypothetical phenomenon is called a radiometric anomaly.

In other words, for a sinusoidal distribution of radiance with a specific spatial frequency vector $\mathbf{f}$, and for a specific direction of observation defined by angle θ, the sinusoidal modulation is completely erased! Such a radiometric anomaly, still to be verified experimentally, occurs for a specific distance z, wavelength λ, spatial frequency f, observation angle θ, and specific angle α, between spatial frequency vector $\mathbf{f}$ and projection vector $\mathbf{p}$, the latter the (x, y) projection of the observation unit vector $\mathbf{s}$ (see Fig. 2 in Ref. [28]). For example, for $\lambda = 1$ μm, $f = 10$ l/mm, $\theta = 30$ deg, and $\alpha = 60$ deg, such an anomaly occurs for $z = 4$ mm, while for $f = 5$ λ/mm and $f = 2$ λ/mm, with the other parameters remaining the same, the corresponding distances are 16 mm and 10 cm, respectively. Also, for $\lambda = 1$ μm, $\theta = 15$ deg, $\alpha = 0$, and $f = 5$ λ/mm, we obtain $z = 18$ mm. All three geometric parameters values are quite common, so the radiometric anomaly should occur often, assuming the second Walther definition of radiance [49] is correct.

5.5 Optics and Imaging at POC

In addition to statistical (diffuser) optics and physical radiometry, other areas of physical optics include: holographic 3D autostereoscopic display [61]; Littrow-grating-based coarse (nondense) wavelength division multiplexing (WDM) [62]; planar optics, including planar optical interconnects [21]; and chip-to-chip channel waveguide interconnects [21]. Among the many types of optics and imaging at POC are: single-mode GaAs channel waveguide pigtailing [21]; WKB analytic modeling of nonuniform holograms [36]; stratified volume diffraction elements [36]; optical computer motherboards [36]; millimeter wave antennas based on physical optics principles [63]; spatial coherence filtering [64]; and nonlinear grating filters for laser eye-protection [65,66]. Recently, electronic imaging technologies have been in development at POC [67–72].

The most mature of POC's technologies is the non-Lambertian diffuser protected by over 35 U.S. patents and its production line is ISO-9001:2000-certified. Also very mature at POC are electronic imaging technologies for robotics, unmanned vehicles, cellular communication, and many defense and homeland security areas. POC's 3D autostereoscopic (without glasses) imaging [61] is an expansion of Dennis Gabor's 3D cinema concept [73]. The integrated optics fabrication methods are derived from analogous non-Lambertian diffuser fabrication at POC, including hot embossing. Finally, POC's capabilities in the areas of small system integration (SSI) and the theory of catastrophes [74,75] have been applied to navigation without the Global Positioning System (GPS).

5.6 Conclusions

This chapter has presented a capsule of Physical Optics Corporation, including ISO:9001-2000-certified diffuser manufacturing process and small system integration capabilities. The technical review started with spatial coherence [64] and physical optics, represented by non-Lambertian diffusers, radiometric ray tracing (R^2T), 3D displays, and others, and ended with a brief description of some exotic subjects as radiometric anomalies and optical applications of the mathematical theory of catastrophes.

References

1. T. Jannson, "Analysis of the polarization state of a wave diffracted by an ideally conducting half-plane," *Acta Physica Polonica* **36**, 803 (1969).
2. E. Wolf, *Nuovo Cimento* **13**, 1165 (1959).
3. M. Born and E. Wolf, *Principles of Optics*, 7th ed., Secs. 10.9.1 and 10.9.2, Cambridge Univ. Press, Cambridge (1999).
4. See Ref. [3], Sect. 11.
5. J.T. Foley, R.R. Letfullin, and H.F. Arnoldus, "The diffractive multifocal focusing effect," *Tribute to Emil Wolf: Science and Engineering Legacy of Physical Optics*, 287-315, SPIE Press, Bellingham, WA (2004).
6. C. Brosseau, "What polarization of light is: the contribution of Emil Wolf," *Tribute to Emil Wolf: Science and Engineering Legacy of Physical Optics*, 51–93, SPIE Press, Bellingham, WA (2004).
7. T. Jannson and B. Karczewski, "Inverse diffraction in the coherence theory," *Opt. Commun.* **2**, 5 (Oct. 1970).
8. T. Jannson and R. Janicki, "An eigenvalue formulation of inverse theory of scalar diffraction," *Optik* **56**(40), 429–441 (1980).
9. V.S. Lethokov, "Nano-optics: atoms in the near field," *Tribute to Emil Wolf: Science and Engineering Legacy of Physical Optics*, 475–487, SPIE Press, Bellingham, WA (2004).
10. A. Rubinowicz, *Ann. D. Physik* **53**(4), 257 (1917).
11. See Ref. [3], Sect. 8.9.
12. K. Miyamoto and E. Wolf, *JOSA* **52**, 615–626 (1962).
13. E. Wolf, "Three-dimensional structure determination of semi-transparent objects from holographic data," *Opt. Commun.* **1**, 153 (1969).
14. See Ref. [3], Sect. 13.
15. A.J. Devaney and P. Guo, "Digital holographic microscopy," *Tribute to Emil Wolf: Science and Engineering Legacy of Physical Optics*, 177–198, SPIE Press, Bellingham, WA (2004).
16. T. Jannson, "Structural Information in volume holography," *Opt. Applicata* **9**(3), 170 (1979).
17. T. Jannson, "Shannon number of an image and structural information capacity in volume holography," *Opt. Acta* **27**(9), 1335–1344 (1980).

18. T. Jannson, "Information capacity of Bragg holograms in planar optics," *JOSA* **71**(3), 342 (1981).

19. T. Jannson and J. Sochacki, "Primary aberrations of thin planar surface lenses," *JOSA* **70**(9), 1079 (1980).

20. T. Jannson, J. Jannson, and P. Yeung, "Holographic planar optical interconnect," U.S. Patent 4,838,620, June 13, 1989.

21. R. Shi and T. Jannson, "Integrated optical waveguide routing-micro-optics," in *Optical Interconnection, Foundation, and Applications*, C. Tocci and H.J. Caulfield, Eds., Artech House, Boston (1999).

22. T. Jannson and J. Jannson, "Temporal self-imaging effect in single-mode fibers," *JOSA* **71**, 1373–1376 (1981).

23. T. Jannson, "Real-time Fourier transformation in dispersive optical fibers," *Opt. Lett.* **8**, 237 (1983).

24. J. Jannson, "Prism coupling selectivity in anisotropic uniaxial waveguide," *Appl. Opt.* **20**, 379–380 (1981).

25. See Ref. [3], Sect. 11.5.

26. J. Perina, "Emil Wolf and optics in the Czech Republic," *Tribute to Emil Wolf: Science and Engineering Legacy of Physical Optics*, 245–263, SPIE Press, Bellingham, WA (2004).

27. W.H. Carter and E. Wolf, "Coherence and radiometry with quasi-homogeneous planar sources," *JOSA* **67**, 785–796 (1977).

28. T. Jannson, "Radiance transfer function," *JOSA* **70**, 1544–1549 (1980).

29. T. Jannson, T. Aye, I. Tengara, and D.A. Erwin, "Second-order radiometric ray tracing," *JOSA A* **13**, 1448–1455 (1996).

30. POC's major subsidiaries include Intelligent Optical Systems, Inc. (fiber optic sensors); Broadata Communications, Inc. (fiber-optic communication); and WaveBand Corporation (microwave antennas).

31. I. Ternovskiy, T. Jannson, and H.J. Caulfield, "Is catastrophe theory analysis the basis for visual perception?" in *Three-Dimensional Holographic Imaging*, C.J. Kuo and M.H. Tsai, Eds., Wiley, New York (2002).

32. T. Jannson, D.H. Kim, A.A. Kostrzewski, and I.V. Ternovskiy, "Soft computing and soft communication (SC2) for synchronized data," in *SPIE Proc.* Vol. 4120, 120–133, Bellingham, WA (2000).

33. A.A. Kostrzewski, T.P. Jannson, and S. Kupiec, "Soft computing and wireless communication," in *SPIE Proc.* Vol. 4479, 70–74, Bellingham, WA (2001).

34. T. Jannson and A.A. Kostrzewski, "Small communication platforms, PCMCIA-packaging and soft computing for UGV applications," in *SPIE Proc.* Vol. 4024, 326–340, Bellingham, WA (2000).

35. T. Jannson, A.A. Kostrzewski, G. Zeltser, and T. Forrester, "Telementoring and teleparamedic communication platforms and robotic systems for battle-

field biomedical applications," in *SPIE Proc.* Vol. 4047, Bellingham, WA (2000).

36. T. Jannson and R. Lieberman, Eds., *POC Proceedings 1997*, SPIE reprint of 39 POC papers, all published in1997 *SPIE Proc.* (1997).

37. T. Jannson, Ed., *POC Proceedings 2000*, SPIE reprint of 52 POC papers, all published in 2000 *SPIE Proc.* (2000).

38. T. Jannson, D. Pelka, and T.M. Aye, "GRIN type diffuser based on volume holographic material," U.S. Patent 5,365,354, issued Nov. 15, 1994.

39. T.P. Jannson, I. Tengara, Y. Qiao, and G.S. Savant, "Lippmann-Bragg broadband holographic mirrors, *JOSA A* (1991); also, in *Selected Papers on Fundamental Techniques in Holography*, H. Bjelkhegen and H.J. Caulfield, Eds., *SPIE Milestone Series* Vol. MS 171, 37 (2001).

40. T.P. Jannson and J.L. Jannson, "Bragg holograms and concentrator optics," *SPIE Milestone Series* Vol. MS 171, 393 (2001).

41. J. Peterson and J. Lerner, "Homogenizer formed using coherent light and a holographic diffuser," U.S. Patent 5,534,386, issued July 9, 1996.

42. H. Kogelnik, "Coupled wave theory for thick hologram gratings," *BSTJ* **48**, 2909 (1969).

43. L.I. Goldfischer, "Autocorrelation function and power spectral density of laser-produced speckle patterns," *JOSA* **55**(3), 247 (1965).

44. J.W. Goodman, *Statistical Optics*, Wiley, New York (2000).

45. E.W. Marchand and E. Wolf, "Radiometry with sources of any state of coherence," *JOSA* **64**, 1219–1226 (1979).

46. E.W. Marchand and E. Wolf, "Walther's definition of generalized radiance," *JOSA* **64**, 1273–1274 (1974).

47. W.H. Carter and E. Wolf, "Coherence properties of Lambertian and non-Lambertian sources," *JOSA* **65**, 1067–1071 (1975).

48. A. Walther, "Radiometry and coherence," *JOSA* **58**, 1256–1259 (1968).

49. A. Walther, "Radiometry and coherence," *JOSA* **63**, 1622–1623 (1973).

50. A. Walther, "Propagation of the generalized radiance through lenses," *JOSA* **68**, 1606–1611 (1978).

51. B. Steinle and H.P. Baltes, "Radiant intensity and spatial coherence for finite planar sources," *JOSA* **67**, 241–247 (1977).

52. H.P. Baltes, J. Geist, and A. Walther, "Radiometry and coherence," in *Topics in Current Physics*, Vol. 9, H.P. Baltes, Ed., Springer, Berlin (1978).

53. A.T. Friberg, "On the existence of a radiance function for finite planar sources of arbitrary states of coherence," *J. Opt. Soc. Am.* **69**, 192–199 (1979).

54. J.W. Goodman, *Introduction to Fourier Optics*, McGraw-Hill, New York (1979).

55. T. Jannson, "Phase-space formalism and ray-tracing modeling of photometric quantities," in *SPIE Proc.* Vol. 3140, 36, Bellingham, WA (1997).

56. T. Jannson and R. Winston, "Liouville's theorem and concentrator optics," *JOSA A* **3**, 7 (1986).

57. J.T. Foley and E. Wolf, "Radiometry as a short-wavelength limit of statistical wave theory with globally incoherent sources," *Opt. Commun.* **55**, 236–241 (1985).

58. K. Kim and E. Wolf, "Propagation law for Walther's first generalized radiance function and its short wavelength limit with quasi-homogeneous sources," *JOSA A*, 1233–1236 (1987).

59. J.T. Foley and E. Wolf, "Radiance functions of partially coherent fields," *J. of Med. Opt.* **38**, 10, 2053 (1991).

60. J.R. Shewell and E. Wolf, "Inverse diffraction and new reciprocity theorem," *JOSA* **58**, 1595 (1968).

61. T. Jannson, A.A. Kostrzewski, S.A. Kupiec, K.M. Yu, T.M. Aye, and G.D. Savant, "True 3D display and Beowulf connectivity," in *SPIE Proc.* Vol. 5080, Bellingham, WA (2003).

62. B. Moslehi and T. Jannson, "Fiber optic wavelength division multiplexing using volume holographic gratings," *Opt. Lett.* **14**, 1088 (1989).

63. V.A. Manasson and L.S. Sadovnik, "Millimeter wave MMIC hologram beam former," *Tribute to Emil Wolf: Science and Engineering Legacy of Physical Optics*, 167–176, SPIE Press, Bellingham, WA (2004).

64. J. Jannson, T. Jannson, and E. Wolf, "Spatial coherence discrimination in scattering," *Opt. Lett.* **13**, 1060 (1988).

65. D.A. Erwin and T. Jannson, "Transient mixed gratings induced by light in the third-order nonlinear Kerr media," in *Application and Theory of Periodic Structures*, T.P. Jannson, Ed., *SPIE Proc.* Vol. 2532, 463, Bellingham, WA (1995).

66. D.A. Erwin and T. Jannson, "Transient periodic structures in third-order nonlinear polymers," in *SPIE Proc.* Vol. 2532, 348, Bellingham, WA (1995).

67. T. Jannson, A. Kostrzewski, T. Forrester, S. Kupiec, and T. DeBacker, "Aerogels, SC2 compression, and Gbytes flash memories," in *SPIE Proc.* Vol. 4479, 96–104, Bellingham, WA (2001).

68. A. Kostrzewski, S. Kupiec, and T. Jannson, "Object-oriented soft computing video compression and digital processing," in *SPIE Proc.* Vol. 4787, Bellingham, WA (2002).

69. T. Jannson, A. Kostrzewski, and W. Wang, "Soft computing and minimization/optimization of video/imagery redundancy," in *SPIE Proc.* Vol. 5200, Bellingham, WA (2003).

70. T. Jannson and A. Kostrzewski, "GSM video communication," in *SPIE Proc.* 5071, Bellingham, WA (2003).

71. T. Jannson, A. Kostrzewski, S. Ro, and T. Forrester, "Soft computing techniques in network packet video," in *SPIE Proc.* Vol. 5200, Bellingham, WA (2003).

72. T. Jannson, A. Kostrzewski, B. Sun, W. Wang, K. Kolesnikov, and S. Kupiec, "Soft computing, visual communication, spatial/temporal events, and optimum bandwidth utilization," in *SPIE Proc.* Vol. 4787, Bellingham, WA (2003).

73. D. Gabor, "A new microscopic principle," *Nature*, **161**, 778 (1948).

74. R. Thom and H. Levin, *Singularities of Differentiable Mappings*, Mathematische Institute der Universitat, Bonn (1959).

75. R. Gilmore, *Catastrophe Theory for Scientists and Engineers*, Wiley, New York (1994).

Acknowledgments

I would like to thank once more Prof. Emil Wolf for his constant inspiration in physical optics, and my wife, Joanna, for a wonderful life in science, business, and otherwise.

I would also like to thank all of our scientific staff, which at present includes 20 Ph.Ds, for their constant support, especially Gajendra Savant for his contributions to diffuser manufacturing and volume holography; Andrew Kostrzewski, for his contributions to small system integration, electronic imaging, and 3D displays; Tin M. Aye, for his contributions to co-invention of non-Lambertian diffusers and holographic autostereoscopic display; Il'ya Agurok, for his contributions to the development of fisheye lens capability; and Igor Ternovskiy, for introducing the theory of catastrophes to POC.

I would also like to thank Sharon Peet for her significant help in preparation of this manuscript, and Robert Lundy for help in the editing of this manuscript. Figures 7–10 were obtained from POC's commercial brochures prepared by Rick Shie.

Emil Wolf (left) with Tomasz Jannson, at OSA's Annual Meeting, October 1997, Long Beach, California.

Tomasz Jannson, Chief Scientist and cofounder of Physical Optics Corporation (POC), has expertise in physical optics, fiber optics, integrated optics, video/imagery, soft computing, Fourier optics, 3D visualization, robotics, and small communication platforms. He is an SPIE Fellow.

CHAPTER 6

MICROSCOPIC ORIGIN OF SPATIAL COHERENCE AND WOLF SHIFTS

Girish S. Agarwal

6.1 Introduction

Emil Wolf [1–4] discovered how the spatial coherence characteristics of the source affect the spectrum of the radiation in the far zone. In particular, the spatial coherence of the source can result either in red or blue shifts in the measured spectrum. His predictions have been verified in a large number of different classes of systems. Wolf and coworkers usually assume a given form of source correlations and study its consequence. In this paper we consider the microscopic origin of spatial coherence and radiation from a system of atoms [5–8]. We discuss how the radiation is different from that produced from an independent system of atoms. We show that the process of radiation itself is responsible for the creation of spatial correlations within the source. We present different features of the spectrum and other statistical properties of the radiation, which show strong dependence on the spatial correlations. We show the existence of a new type of two-photon resonance that arises as a result of such spatial correlations. We further show how the spatial coherence of the field can be used in the context of radiation generated by nonlinear optical processes. We conclude by demonstrating the universality of Wolf shifts and its application in the context of pulse propagation in a dispersive medium.

We start by giving a summary of Wolf's main results [1,2]. Consider the radiation produced by two point sources P_1 and P_2 at the observation point P (Fig. 1). Let us consider for simplicity the case of scalar fields $U(P, \omega)$. The spectrum of the field at P is given by

$$S_U(P, \omega) = \langle U^*(P, \omega) U(P, \omega) \rangle, \tag{1}$$

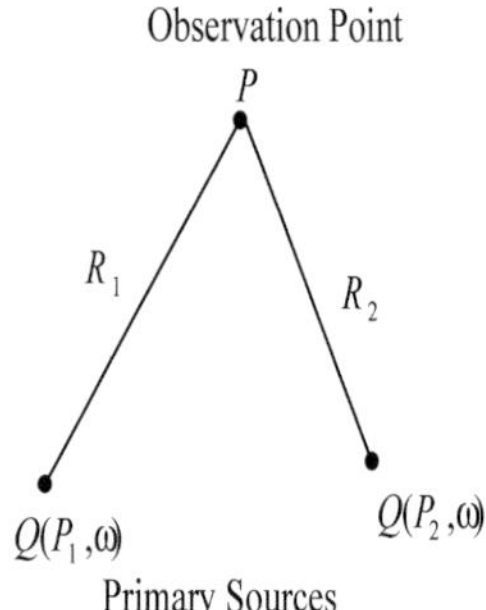

Figure 1 Radiation produced by two point sources P_1 and P_2 at the observation point P.

whereas the spectrum of the source is defined by

$$S_Q(\omega) = \langle Q^*(P_1, \omega)Q(P_1, \omega)\rangle \tag{2}$$

$$= \langle Q^*(P_2, \omega)Q(P_2, \omega)\rangle. \tag{3}$$

We assume identical spectra for the two sources. Let $\mu_Q(\omega)$ be the spectral degree of coherence between two sources

$$\mu_Q = \frac{\langle Q^*(P_1, \omega)Q(P_2, \omega)\rangle}{S_Q(\omega)}. \tag{4}$$

This is a measure of correlation between the two sources. For two coherent sources μ is 1, whereas for incoherent sources $\mu = 0$. The field U at the point P can be related to the strength of the sources via

$$U(P, \omega) = Q(P_1, \omega)\frac{e^{ikR_1}}{R_1} + Q(P_2, \omega)\frac{e^{ikR_2}}{R_2}. \tag{5}$$

Here we have ignored unnecessary numerical factors. Using Eq. (5), the spectrum of the field is related to the spectrum of the source and the degree of spatial coherence:

$$S_U(P, \omega) = S_Q(\omega)\left\{\frac{1}{R_1^2} + \frac{1}{R_2^2} + \frac{1}{R_1R_2}\left[\mu_Q(\omega)e^{ik(R_2-R_1)} + \text{c.c}\right]\right\}. \tag{6}$$

Clearly, in general, the source spectrum and the spectrum at P are not equal:

$$S_U(P, \omega) \neq S_Q(\omega). \tag{7}$$

So, the measured spectral characteristics will also be determined by μ_Q and $S_U(P, \omega)$, in general, would exhibit correlation-induced spectral shifts. Wolf used phenomenological models for S_Q and μ_Q to demonstrate a variety of spectral shifts and even the correlation-induced splitting of a line into several lines. Clearly, it is desirable to understand the origin of source correlations.

6.2 Microscopic Origin of Source Correlations

We thus examine the question of how the atoms radiate. Consider for example an atom in its excited state (Fig. 2). It interacts with the modes of a quantized electromagnetic field in vacuum state. The atom makes a transition to the ground state by the emission of a photon. The photon can be emitted in any mode of the field. The atom has an infinite number of available modes. It is known that the spectrum of the emitted radiation has Lorentzian spectrum

$$S_A(\omega) = \frac{\gamma/\pi}{(\omega - \omega_0)^2 + \gamma^2}, \tag{8}$$

where ω_0 is the frequency of the atomic transition and γ is half the Einstein A coefficient.

Next consider two atoms located at $\vec{r}_A$ and $\vec{r}_B$, and let each atom be initially in its excited state (Fig. 3). The question is whether the atoms radiate independently of each other, i.e., whether the spectrum of the emitted photons factorizes

$$S(\omega_1, \omega_2) = S_A(\omega_1)S_B(\omega_2) \tag{9}$$

or not. The correlations between the two atoms [6–8] would invalidate Eq. (9) and would also imply that

$$S_A(\omega_1) \neq \int S(\omega_1, \omega_2)d\omega_2, \tag{10}$$

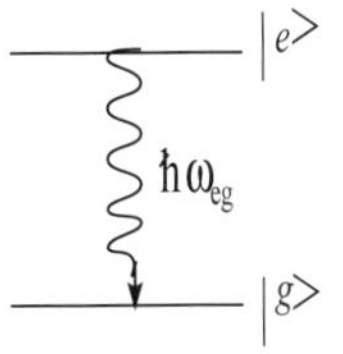

$|e\rangle$

$\hbar\omega_{eg}$

$|g\rangle$

Figure 2 Radiation from a single atom.

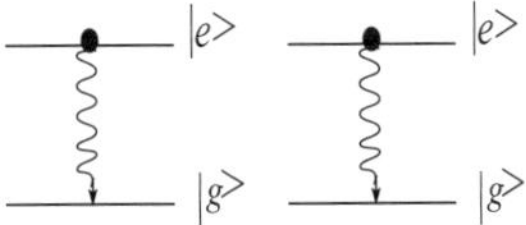

Figure 3 Two-photon emission by two atoms.

i.e., the spectrum of the emitted radiation would be different from one if the other atom was absent. Note that both atoms interact with a common quantized electromagnetic field. This interaction with a common field results in an effective interaction between two atoms even if the atoms do not interact. This can also be understood by considering, for example, the net field on the atom $B(A)$, which would consist of the vacuum field and field radiated by the atom $A(B)$. Let us denote $\chi_{ij}(\vec{r}_A, \vec{r}_B, \omega)$ as the ith component of the field at position $\vec{r}_A$ due to a unit dipole oriented in the direction j at the position $\vec{r}_B$ (Fig. 4).

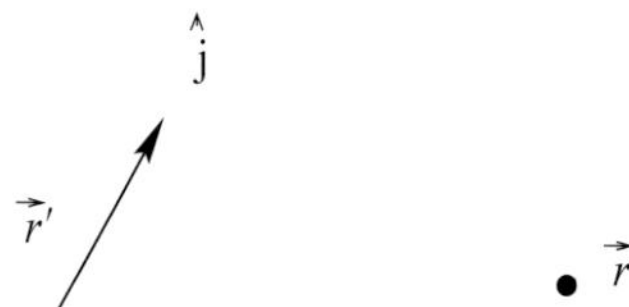

Figure 4 Meaning of the response function x.

This field [9–11] is well known from the solution of Maxwell equations

$$\chi_{ij}(\vec{r}_A, \vec{r}_B, \omega) = \left(\frac{\omega^2}{c^2}\delta_{ij} + \frac{\partial^2}{\partial r_A \partial r_B} \right) \frac{\exp(i|\vec{r}_A - \vec{r}_B|\omega/c)}{|\vec{r}_A - \vec{r}_B|}. \tag{11}$$

This function has a close connection with the spatial coherence of the vacuum of the electromagnetic field. Let us write the electric field operator in terms of its positive and negative frequency parts:

$$E = E^{(+)} + E^{(-)}. \tag{12}$$

It is well known in quantum optics that $E^{(+)}(E^{(-)})$ corresponds to the absorption (emission) of photons. Further, $E^{(+)}$ is an analytical signal. Let us consider the second-order coherence function of the electromagnetic field

$$S^A_{\alpha\beta}(\vec{r}_1, \vec{r}_2, \tau) = \left\langle E^{(+)}_\alpha(\vec{r}_1, t + \tau)E^{(-)}_\beta(\vec{r}_2) \right\rangle, \tag{13}$$

which is nonvanishing even though the field is in vacuum state. Its Fourier transform is given by [9]

$$\int d\tau e^{i\omega\tau} S^A_{\alpha\beta}(\vec{r}_1, \vec{r}_2, \tau) = 2\hbar Im \chi_{ij}(\vec{r}_1, \vec{r}_2, \omega) \quad \text{if } \omega > 0$$

$$= 0 \quad \text{if } \omega < 0. \tag{14}$$

We thus conclude that the vacuum of the electromagnetic field has spatial coherence that extends over the dimensions of wavelength. Therefore the correlation between atoms would extend over distances of the order of wavelength. Clearly in a macroscopic sample these correlations could build up over much larger distances. Explicit results for two atoms can be found in Refs. [6–8].

6.3 Source Correlation-Induced Two-Photon Resonance

We next discuss several other situations where atom-atom correlations play an important role. Consider first the case of two unidentical atoms with transition frequencies ω_A and ω_B and located within a wavelength of each other (see Fig. 5). Let both atoms start in ground state and interact with a laser field of frequency ω_l. We now study the total intensity $I(\omega_l)$ of the emitted radiation as a function of ω_l. Clearly $I(\omega_l)$ will exhibit single-photon resonance at $\omega_l = \omega_{Aeg}, \omega_{Beg}$. In principle there is also the possibility of two-photon resonance $2\omega_l = \omega_{Aeg} + \omega_{Beg}$. It turns out that in the absence of source correlations, the two-photon resonance does not occur, as the two paths

$$|g_A, g_B\rangle \to |e_A, g_B\rangle \to |e_A, e_B\rangle, \quad \text{and} \quad |g_A, g_B\rangle \to |g_A, e_B\rangle \to |e_A, e_B\rangle, \quad (15)$$

interfere destructively. Thus the source correlations are the key to the two-photon resonance. In an earlier work the effect of source correlations on such a two-photon resonance was studied in great detail [7], and recently it has been observed in experiments involving single molecules [12]. Furthermore, very recently we showed how the source correlation arises in a cavity [13].

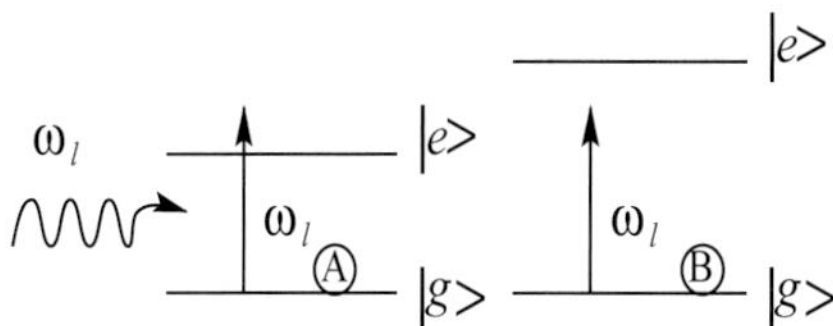

Figure 5 Schematic illustration of two-photon absorption by two atoms.

6.4 Spatial Coherence and Emission in Presence of a Mirror

Another class of systems where spatial coherence plays an important role is, for example, the emission of radiation in front of a metallic mirror [10] or in a cavity formed by metallic or dielectric mirrors (see Fig. 6). The spectrum of the emitted

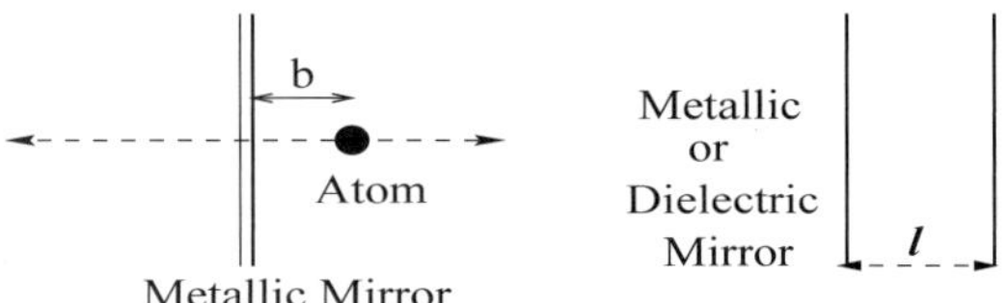

Figure 6 Emission in the presence of boundaries.

radiation depends on the distance of the atom from mirror. As a matter of fact both line width and line shift become b-dependent. If the metallic mirror is treated as a perfect conductor, then the calculations show that the line shifts, for example, are determined by the spatial coherence of the field at the location of the atom and its image. Thus the correlation of the vacuum $\langle \vec{E}^{(+)}(\vec{b},t)\vec{E}^{(-)}(-\vec{b},t')\rangle$, which is related to $\chi(\vec{b},-\vec{b},\omega)$, determines the line shifts and line widths. Explicit results for the b-dependence of shifts and widths can be found in Refs. [10,14].

6.5 Spatial Coherence-Induced Control of Nonlinear Generation

We next discuss the effects of spatial coherence in the context of nonlinear optics. We would show that the generation of radiation using nonlinear processes can be controlled by source correlations. Consider, for example, the process of second harmonic generation (SHG) with $P = \chi^{(2)}E^2$, $E \sim e^{i\vec{k}\cdot\vec{r}}$ (Fig. 7). The efficiency of the SHG depends on the phase-matching integral

$$f = \frac{1}{V}\int e^{-i\vec{q}\cdot\vec{r}}e^{2i\vec{k}\cdot\vec{r}}d^3r, \tag{16}$$

which goes to unity if $\vec{q} = 2\vec{k}$. The function f determines the direction in which second harmonic generation is dominant.

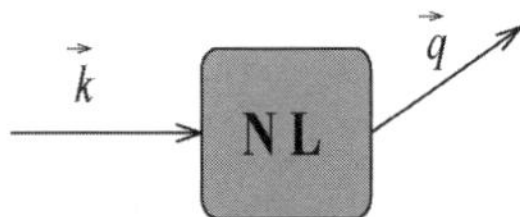

Figure 7 Generation of coherent radiation in the direction $\vec{q}$.

If, however, the field E is partially coherent, then in place of Eq. (16) we need to consider

$$f = \int d^3r' d^3r'' e^{-i\vec{q}\cdot\vec{r}'} e^{2i\vec{k}\cdot\vec{r}''} \langle P(\vec{r}')P^*(\vec{r}'')\rangle. \tag{17}$$

Note that for SHG with coherent radiation

$$\langle P(\vec{r}')P^*(\vec{r}'')\rangle \equiv \langle P(\vec{r}')\rangle\langle P^*(\vec{r}'')\rangle \equiv e^{2i\vec{k}\cdot(\vec{r}'-\vec{r}'')}, \tag{18}$$

and then

$$I \propto (vol)^2. \tag{19}$$

On the other hand, for the case of incoherent radiation

$$\langle P(\vec{r}')P^*(\vec{r}'')\rangle \equiv |\wp|^2\delta(\vec{r}'-\vec{r}''), \tag{20}$$

$$I \rightarrow |\wp|^2(vol). \tag{21}$$

For the partially coherent radiation

$$\langle P(\vec{r}')P^*(\vec{r}'')\rangle = |\chi^{(2)}|^2\langle E^2(\vec{r}')E^{*2}(\vec{r}'')\rangle, \tag{22}$$

which under the assumption of a Gaussian field will become

$$\langle P(\vec{r}')P(\vec{r}'')\rangle = 2I^2\left|\mu(\vec{r}'-\vec{r}'')\right|^2, \tag{23}$$

where $\mu(\vec{r}'-\vec{r}'')$ denotes the degree of spatial coherence of the incident field. Thus, SHG would now be determined by the integral

$$\left|f(\vec{Q})\right|^2 = \iint d^3r' d^3r'' \left|\mu(\vec{r}'-\vec{r}'')\right|^2 e^{\vec{Q}\cdot(\vec{r}'-\vec{r}'')}, \tag{24}$$

$$\vec{Q} = -\vec{q} + 2\vec{k}. \tag{25}$$

Clearly, now the direction of SHG would be determined by the spatial coherence of the field (Fig. 8). Thus spatial coherence can serve as a control parameter for the nonlinear generation. The above ideas should also find interesting applications in other areas of nonlinear optics as well.

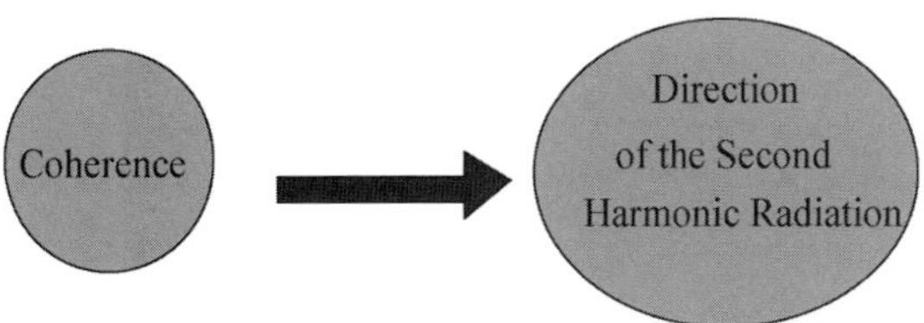

Figure 8 Role of coherence in nonlinear generation.

6.6 Universality of Wolf Shift

Before concluding the paper, we would also like to make some general remarks on the universality and applicability of Wolf shifts in the context of other systems. We know, for instance, that other standard equations of physics (such as those describing vibrations of string and heat transport) admit the following relation between the effect Φ of the source P at the observation point

$$\Phi(\vec{r}) = \int G(\vec{r},\vec{r}')P(\vec{r}')d^3r', \tag{26}$$

where G is Green's function for the underlying equation. The observed quantities are usually quadratic in Φ. Thus, observation at the point $\vec{r}$ would depend on the correlations of the source at two points. This is due to the nonlocal nature of the solution [Eq. (26)].

6.7 Fluctuating Pulses in a Dispersive Medium

As another example of this universality, we can consider the propagation of pulses in a dispersive medium that is described by the equation

$$i\frac{\partial E}{\partial z} = \frac{\tilde{k}}{2}\frac{\partial^2 E}{\partial t^2}. \tag{27}$$

The solution of this equation can be given in terms of Green's function

$$E(z,t) = \int G(z,t;0,t')E(0,t')dt'; \tag{28}$$

$$G = \frac{i}{2\pi z\tilde{k}}\exp\left[-\frac{i}{2z\tilde{k}}(t^2 - 2tt' + t'^2)\right]. \tag{29}$$

If the input pulse has fluctuations, then the intensity of the output pulse would be determined by the correlation in pulses on input plane

$$I(L,t) = \iint dt'dt''G^*(L,t;0,t')G(L,t;0,t'')\langle E(t')E^*(t'')\rangle. \tag{30}$$

Clearly, the intensity of the pulse at the output plane is not completely determined by the intensity of the pulse at the input plane.

6.8 Conclusions

In conclusion we have shown that the vacuum of an electromagnetic field has intrinsic partial spatial coherence in a frequency domain that effectively extends over regions of the order of wavelength λ. This spatial coherence leads to a dynamical coupling between atoms and is the cause of source correlations. We showed how such correlations can lead to a new type of two-photon resonance and how these are relevant for near-field optics. We further showed how the source spatial correlations can lead to new phase-matching conditions for nonlinear optical effects, leading to the possibility of using spatial coherence to produce tailor-made emissions. We also discussed the universality of source correlation effects and, as a specific example, we treated the case of the propagation of fluctuating pulses in a dispersive medium.

The author thanks Emil Wolf for many discussions on the subject of correlation-induced shifts.

References

1. E. Wolf, "Invariance of the spectrum of light on propagation," *Phys. Rev. Lett.* **56**, 1370–1372 (1986); E. Wolf, "Red shifts and blue shifts of spectral lines emitted by two correlated sources," *Phys. Rev. Lett.* **58**, 2646–2648 (1987); E. Wolf, "Correlation-induced Doppler-type frequency shifts of spectral lines," *Phys. Rev. Lett.* **63**, 2220–2223 (1989).

2. E. Wolf, "Non-cosmological redshifts of spectral lines," *Nature (London)* **326**, 363–366 (1987).

3. E. Wolf and D.F.V. James, "Correlation-induced spectral changes," *Rep. Prog. Phys.* **59**, 771–818 (1996).

4. L. Mandel and E. Wolf, *Optical Coherence and Quantum Optics*, Cambridge University Press, Cambridge, 1995.

5. G.S. Agarwal, *Quantum Optics*, Springer Tracts in Modern Physics, Vol. 70 (1974).

6. G. Varada and G.S. Agarwal, "Microscopic approach to correlation-induced frequency shifts," *Phys. Rev. A* **44**, 7626–7634 (1991).

7. G. Varada and G.S. Agarwal, "Two-photon resonance induced by the dipole-dipole interaction," *Phys. Rev. A* **45**, 6721–6729 (1992).

8. D.F.V. James, "Frequency shifts in spontaneous emission from two interacting emission," *Phys. Rev. A* **47**, 1336–1346 (1993).

9. G.S. Agarwal, "Quantum electrodynamics in the presence of dielectrics and conductors: I. Electromagnetic-field response functions and black-body fluctuations in finite geometries," *Phys. Rev. A* **11**, 230–242 (1975).

10. G.S. Agarwal, "Quantum electrodynamics in the presence of dielectrics and conductors: IV. General Theory of spontaneous emission in finite geometries," *Phys. Rev. A* **12**, 1475–1497 (1975).

11. G.S. Agarwal, "Finite boundary effects in quantum electrodynamics," in *Quantum Electrodynamics and Quantum Optics*, A. Barut, Ed., Plenum (1983).

12. C. Hettich, C. Schmitt, J. Zitzmann, S. Kuhn, I. Gerhardt, and V. Sandoghdar, "Nanometer resolution and coherent optical dipole coupling of two individual molecules," *Science* **298**, 385–389 (2002).

13. P.K. Pathak and G.S. Agarwal, "Giant two-atom two-photon vacuum Rabi oscillations in a high quality cavity," *Phys. Rev. A* (2004) (in press).

14. G.S. Agarwal and H.D. Vollmer, "Surface polariton effects in spontaneous emission," *Physica Status Solidi B* **79**, 249 (1977); G.S. Agarwal and H.D. Vollmer, "Surface polariton effects in spontaneous emission. II. Effects of spatial dispersion," *Physica Status Solidi B* **85**, 301 (1978).

Girish Agarwal received his M.Sc. in 1966 from Banaras Hindu University, he then joined the University of Rochester as a graduate student and obtained his Ph.D. in 1969. He is a leading theoretician who has contributed extensively to many subdisciplines of optics. Currently he is the Director of the Physical Research Laboratory (PRL), Ahmedabad, and, in addition, holds the Indian National Science Academy's Albert Einstein Research Professorship. Prof. Agarwal's work has been recognized by a very large number of national and international awards, including the Max Born Prize from the Optical Society of America in 1988, the

Physics Prize of the Third World Academy of Sciences in 1994. Prof. Agarwal is a Fellow of the American Physical Society, the Optical Society of America, and is a recipient of the prestigious Humboldt Research Award (1997) of Germany. During his career, Prof. Agarwal has published more than 500 papers in top international journals, including extensive review articles and a research monograph. Prof. Agarwal serves on the editorial boards of many international journals.

Roland Winston, Robert G. Littlejohn, Yupin Sun, and K. A. Snail

❧CHAPTER 7☙

PARADIGM FOR A WAVE DESCRIPTION OF OPTICAL MEASUREMENTS

7.1 Introduction

Radiance, which is the density of radiative power in phase space, has been the subject of a rich literature over the past 40 and more years [1]. Emil Wolf, whom we, along with his many students and colleagues, honor herein, has been a principal contributor to the development of this subject. It is a tribute to the work of Emil Wolf and his school that the development of a wave theory of radiance, known as "generalized radiance," continues today as exemplified by Ref. [2]. Our own work has been complementary to this line of development in that we have attempted to bridge the gap between theory and practical radiometry. Radiometric measurements are important in many branches of science and technology. For example, in illumination engineering, the visibility of displays is quantified by radiometers. In astrophysics, radiometry in the far infrared has played a critical role in understanding the large space-time structure of the universe [3]. Of course, in the short wavelength limit where diffraction effects can be neglected, geometrical optics suffices and one can dispense with the technical difficulties that the wave property of light introduces. But it is precisely in the regime where diffraction effects cannot be neglected that the properties of the measuring instrument have to be taken into account. Moreover, in the absence of a consistent formalism that does take the diffraction property of the instrument into account, it may be difficult to assess the significance of such effects. "Back-of-the-envelope" estimates of diffraction effects may not be reliable and the practical scientist carrying out radiometeric measurements is left with little guidance as to the magnitude of such effects. For example, an excellent text on radiometry famously states that diffraction effects are "beyond the scope" of the book. In recent papers [4] we showed how the measurement of radiance can be understood in terms of the statistical properties of the electromagnetic field and the properties of the instrument. However, the utility of this approach was limited by the availability of accessible instrument functions that represent the measuring apparatus. In the process, we exhibited a remarkable analogy between the result of measuring radiance and the van Cittert-Zernike theorem. In this paper we first give an overview of a wave description of the measurement of radiance, referring details to previous publications. Then we compare the theory to experiments we performed with highly sophisticated radiometers, finding excellent agreement. The excellent agreement with the analytical model suggests that while our demonstration was confined to the measurement of radiance, it is likely that similar considerations apply to a wide class of optical measurements, where diffraction effects are significant.

7.2 The van Cittert-Zernike Theorem

The well-known van Cittert-Zernike (VCZ) theorem states that for an incoherent, quasi-monochromatic source of radiation, the equal-time degree of coherence (two-point correlation function) $\Gamma(\mathbf{r}, \mathbf{r}')$ is proportional to the complex amplitude in a certain diffraction pattern: the amplitude at $\mathbf{r}$ formed by a spherical wave converging to $\mathbf{r}'$ and diffracted by an aperture the same size, shape, and location as the source [5]. The source could, for example, be a thermal blackbody followed by a filter that selects a small wavelength range. A familiar geometry is a circular source. Then, apart from a normalizing factor, $\Gamma(\mathbf{r}, \mathbf{r}')$ in a transverse plane becomes the well-known Airy diffraction amplitude:

$$\Gamma(\mathbf{r}, \mathbf{r}') = (\text{const.})F(ks\theta_s), \tag{1}$$

where $F(x) = 2J_1(x)/x$, $k = 2\pi/\lambda$, θ_s is the angle subtended by the source at $\mathbf{r}$ or $\mathbf{r}'$, and where $s = |\mathbf{r} - \mathbf{r}'|$. Recall that $\Gamma(\mathbf{r}, \mathbf{r}')$ has its first zero at $s_1 = 0.61\lambda/\theta_s$. For a numerical example, we consider terrestrial sunlight. Then θ_s is 4.7 mrad, so that for $\lambda = 0.5$ μm, s_1 is approximately 65 μm. This is the scale of the transverse correlation of sunlight.

7.3 Measuring Radiance

In a previous paper [6], we examined the relationship between the generalized radiance and the measuring process. We showed how this process can be quantified by introducing the *instrument function*, which is a property of the measuring apparatus [4]. We showed that the result of the measurement is represented by the quantity

$$Q = \mathrm{Tr}(\hat{M}\hat{\Gamma}), \qquad (2)$$

where $\hat{M}$ is a nonnegative-definite Hermitian operator that characterizes the measuring apparatus, and $\hat{\Gamma}$ is the two-point correlation function of the incident light, viewed as an operator. The instrument function itself is a coordinate representation of the measurement operator $\hat{M}$, for example, its matrix element or its Weyl transform. The Weyl transform maps an operator to a Wigner function (for a discussion of the Wigner-Weyl formalism in optics see Ref. [6]). It is appropriate to associate Q with the signal. We then derived an analytical form for the instrument function for a simple radiometer in one space dimension.

One difficulty in using Eq. (2) is that it may not be easy to compute the instrument function. Although the one-dimensional calculation in Ref. [6] was not too hard, we do not expect it to be easy to compute the instrument function for many realistic radiometers, which are two-dimensional in cross section and which may have complicated geometry. Therefore we have considered other means for determining the instrument function. In a previous publication [7] we considered the possibility that the instrument function could be measured. In this section we present an alternative approach. That is, we point out a physical interpretation of the instrument function that is similar to the VCZ theorem. We do this initially by working through the example of a simple "pinhole" radiometer, and then we comment about generalizations.

Radiance is the power per unit volume in phase space. Therefore an instrument for measuring radiance (called a radiometer) has to select a window function in phase space. For measurements close to the diffraction limit, the exact shape of the window function is not critical. For this reason we examine a simple radiometer, illustrated in Fig. 1. The dotted line in the figure is the axis of the radiometer. Light enters from the left and passes through the circular pinhole of radius a. It then passes through a drift space of length L, before passing through another circular aperture of radius b. We assume $L \gg a, b$, so the rays are paraxial. The detector is assumed to measure the total power passing through the aperture b and can be thought of as composed of tiny, densely packed, independent absorbing particles (which is a fairly good approximation to what commonly used detectors like photon detectors, thermal detectors, or photographic film do).

As explained in Ref. [6], the effect of the radiometer on the radiation field is described by the operator

$$\hat{P} = \hat{A}(b)\hat{D}(L)\hat{A}(a), \qquad (3)$$

which maps the wavefield at the entrance aperture a into the wavefield at the exit aperture b. Here $\hat{A}(a)$ is the aperture or "cookie cutter" operator representing the

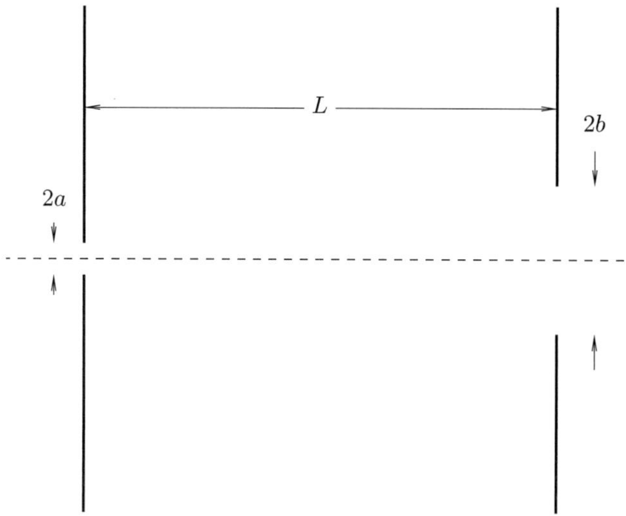

Figure 1 A pinhole radiometer. The dotted line is the axis. Light enters from the left, passing through circular pinhole a, drift space of length L, and finally circular aperture b.

pinhole, $\hat{D}(L)$ is the Huygens-Fresnel operator representing the drift space, and $\hat{A}(b)$ is the aperture operator for the aperture b. In the approximation $L \gg \lambda$ the drift operator has the kernel (or matrix element)

$$\left\langle \mathbf{r}_\perp | \hat{D}(L) | \mathbf{r}'_\perp \right\rangle = -\frac{ikL}{2\pi} \frac{e^{ikR}}{R^2},\tag{4}$$

where $\mathbf{r}_\perp = (x,y)$, $\mathbf{r}'_\perp = (x',y')$, $\mathbf{r} = (x,y,z)$, $\mathbf{r}' = (x',y',z')$, $L = z - z'$ and $R = |\mathbf{r} - \mathbf{r}'|$. Here z is the coordinate along the optical axis and it is assumed that $z > z'$. If in addition we assume rays are paraxial ($r_\perp, r'_\perp \ll L$), then the kernel can be written

$$\left\langle \mathbf{r}_\perp | \hat{D}(L) | \mathbf{r}'_\perp \right\rangle = -\frac{ik}{2\pi L} e^{ikL} \exp\left(\frac{ik}{2L} |\mathbf{r}_\perp - \mathbf{r}'_\perp|^2\right).\tag{5}$$

Following the methods of Ref. [6], the instrument is represented by the operator $\hat{M} = \hat{P}^\dagger \hat{P}$ whose matrix elements are

$$\left\langle \mathbf{r}_\perp | \hat{M} | \mathbf{r}'_\perp \right\rangle = \left(\frac{kL}{2\pi}\right)^2 \int_{|\mathbf{r}''_\perp| \leq b} d^2 \mathbf{r}''_\perp \frac{\exp\left[ik(R_1 - R_2)\right]}{R_1^2 R_2^2},\tag{6}$$

where $R_1 = |\mathbf{r}' - \mathbf{r}''|$, $R_2 = |\mathbf{r} - \mathbf{r}''|$, and the integration is over the detector area. In Eq. (6), the transverse variables $\mathbf{r}_\perp$ and $\mathbf{r}'_\perp$ are understood to lie in the entrance plane (the pinhole), so that $r_\perp, r'_\perp \leq a$. If this condition is not met, the matrix element is understood to be zero.

The expression (6) is identical (up to constants) to the mutual intensity evaluated at aperture a of a uniform, delta-correlated source at aperture b (the location of the detector). Thus, to form the VCZ interpretation of the instrument function, we replace the detector (at aperture b, in this example) by a delta-correlated source, and measure the radiation field emanating from the entrance aperture of the instrument (the pinhole in this example). The instrument function (at a given plane) is then proportional to the amplitude at $\mathbf{r}'$ formed by a spherical wave converging to $\mathbf{r}$ and diffracted by an aperture the same size, shape, and location as the detector. The detector emulates a delta-correlated source. In a sense, this model involves running the radiometer backwards (exchanging the detector for a source).

This interpretation applies also to other radiometers, for example, those with lenses, as in Fig. 2. The essential property is that the operator $\hat{P}^{\dagger}$ should serve as a propagator for wavefields traveling to the left (in the negative z direction), just as $\hat{P}$ serves as a propagator for waves traveling to the right. The situation is rather like time reversal in quantum mechanics. Not all time evolutions in quantum mechanics are time reversal invariant (only those for which the Hamiltonian commutes with time reversal). In the case of optical fields, it is a kind of "z-reversal" that we need. Lenses, drift spaces, and apertures are "z-reversal invariant," as long as evanescent waves can be ignored. We remark that the same conditions apply to the usual VCZ theorem.

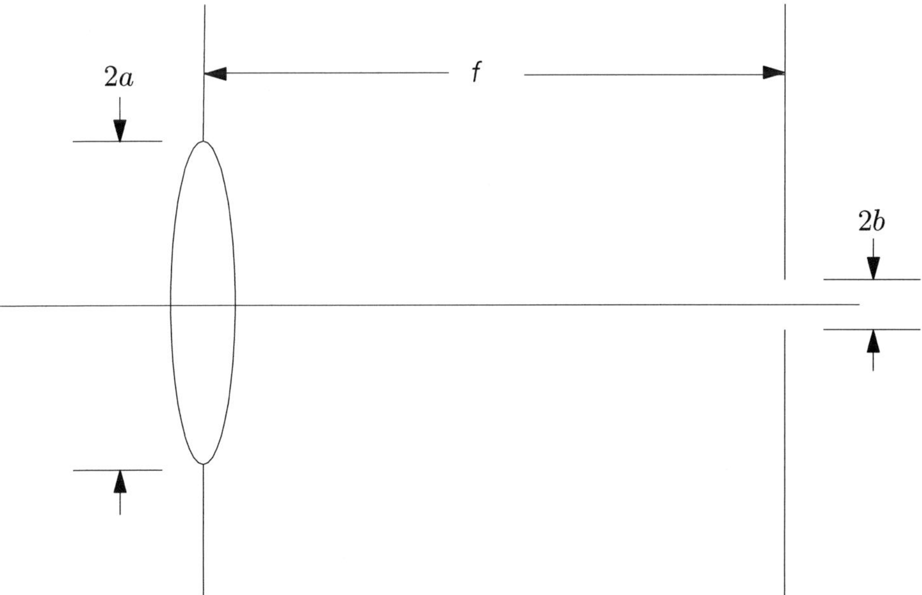

Figure 2 A practical radiometer. The pinhole is replaced by a lens. The drift space length is essentially the focal length for distant objects $L \approx f$.

7.4 Near-Field and Far-Field Limits

It is useful to examine the pinhole radiometer and the formalism we have presented in two limiting cases. First, however, we explain some notation regarding the correlation operator and function (see Ref. [6] for more details). The mutual coherence is defined by $\Gamma(\mathbf{r},\mathbf{r}') = \overline{\psi(\mathbf{r})\psi^*(\mathbf{r}')}$, where the overbar means a statistical or ensemble average, and where we use a scalar model for the wavefield ψ, which in electromagnetic applications can be loosely identified with one of the components of the electric field. When $z = z'$, we can associate the mutual coherence with an operator $\hat{\Gamma}(z)$ by $\Gamma(\mathbf{r},\mathbf{r}') = \Gamma(\mathbf{r}_\perp,z;\mathbf{r}'_\perp,z) = \langle \mathbf{r}_\perp|\hat{\Gamma}(z)|\mathbf{r}'_\perp\rangle$, so that $\hat{\Gamma}(z) = \overline{|\psi(z)\rangle\langle\psi(z)|}$. Thus,

$$\mathrm{Tr}\,\hat{\Gamma}(z) = \int d^2\mathbf{r}_\perp \overline{|\psi(\mathbf{r}_\perp,z)|^2}. \tag{7}$$

If we identify $|\psi|^2$ with 4π times the energy density, then for paraxial rays $(c/4\pi)\,\mathrm{Tr}\,\hat{\Gamma}(z)$ is the power crossing the plane z.

Now let us consider the case that a uniform, thermal source is very close to the radiometer, which is useful for normalizing the signal. Then at the entrance aperture the mutual intensity is proportional to a delta function, $\Gamma(\mathbf{r},\mathbf{r}') = I_0\lambda^2\delta(\mathbf{r}_\perp - \mathbf{r}'_\perp)$, where I_0 is a constant with dimensions of energy per unit volume. A thermal source is not really delta correlated, of course; the spatial correlation is really a sinc function with a width of the order of a wavelength. It is for this reason that we insert the factor of λ^2 into the formula for the mutual coherence, so that if we need to set $\mathbf{r}_\perp = \mathbf{r}'_\perp$ for the purposes of taking the trace, we can interpret $\lambda^2\delta(0)$ as being of order unity. In any case, when we compute the signal according to Eq. (2), we obtain

$$\mathrm{Tr}\big(\hat{M}\hat{\Gamma}\big) = I_0\lambda^2\,\mathrm{Tr}\,\hat{M} = I_0\lambda^2 N_i, \tag{8}$$

where we set

$$\mathrm{Tr}\,\hat{M} = N_i \tag{9}$$

for the number of phase space cells in the acceptance region of the instrument (this point is discussed more fully in Ref. [6]). The trace of $\hat{M}$ is easy to compute. We first make the paraxial approximation in Eq. (6), which gives

$$\left\langle \mathbf{r}_\perp|\hat{M}|\mathbf{r}'_\perp\right\rangle = \left(\frac{k}{2\pi L}\right)^2 \int d^2\mathbf{r}''_\perp \exp\left[\frac{ik}{L}\mathbf{r}''_\perp \cdot (\mathbf{r}_\perp - \mathbf{r}'_\perp)\right], \tag{10}$$

where the $\mathbf{r}''_\perp$ integration is taken over a circle of radius b. Now setting $\mathbf{r}_\perp = \mathbf{r}'_\perp$ and integrating $\mathbf{r}_\perp$ over a circle of radius a, we obtain

$$N_i = \mathrm{Tr}\,\hat{M} = \left(\frac{kab}{2L}\right)^2 = \left(\frac{\pi a \theta_0}{\lambda}\right)^2, \tag{11}$$

where $\theta_0 = b/L$.

The number N_i has a simple interpretation. A phase space cell in the four-dimensional $\mathbf{k}_\perp$-$\mathbf{r}_\perp$ phase space has volume $(2\pi)^2$. As viewed from the standpoint of the exit aperture b of the radiometer, the rays passing through each point of the aperture b occupy a solid angle of $\pi(a/L)^2$, or a region of $\mathbf{k}_\perp$-space of area $\pi(ka/L)^2$. The region of $\mathbf{r}_\perp$-space is just the exit aperture, of area πb^2. Multiplying these areas and dividing by $(2\pi)^2$ gives precisely N_i.

Next we consider the case of a very distant thermal source, which effectively produces a coherent plane wave at the entrance aperture, for example, $\sqrt{I_0}\sqrt{\pi\theta_s^2}e^{ikz}$, so that $\langle\mathbf{r}_\perp|\hat{\Gamma}|\mathbf{r}'_\perp\rangle = I_0(\pi\theta_s^2)$. The dimensionless factor $\pi\theta_s^2$ will be explained below. Using this and Eq. (10), we obtain

$$Q = \mathrm{Tr}(\hat{M}\hat{\Gamma}) = I_0\pi\theta_s^2\left(\frac{k}{2\pi L}\right)^2\int d^2\mathbf{r}_\perp\, d^2\mathbf{r}'_\perp\, d^2\mathbf{r}''_\perp \exp\left(\frac{ik}{L}\mathbf{r}_\perp\cdot\mathbf{r}''_\perp\right)$$

$$\times \exp\left(-\frac{ik}{L}\mathbf{r}'_\perp\cdot\mathbf{r}''_\perp\right). \tag{12}$$

It is easiest to do the $\mathbf{r}_\perp$ and $\mathbf{r}'_\perp$ integrals first (both of which go out to radius a). These are identical, and are given by

$$\int d^2s\exp\left(\pm\frac{ik}{L}\mathbf{s}\cdot\mathbf{r}''_\perp\right) = \pi a^2 F(kar''_\perp/L), \tag{13}$$

where $\mathbf{s} = \mathbf{r}_\perp$ or $\mathbf{r}'_\perp$. This leaves only the $\mathbf{r}''_\perp$ integration (taken out to radius b),

$$Q = \mathrm{Tr}(\hat{M}\hat{\Gamma}) = I_0\pi\theta_s^2\left(\frac{\pi a^2}{\lambda L}\right)^2\int d^2\mathbf{r}''_\perp\left[\frac{2J_1(kar''_\perp/L)}{(kar''_\perp/L)}\right]^2, \tag{14}$$

which agrees with the expected fraction of the Airy diffraction pattern contained by the detector of radius b.

7.5 A Wave Description of Measurement

We begin with the statistical properties of the incident wavefield. The two-point correlation function at the source plane is

$$\langle\mathbf{r}_\perp|\hat{\Gamma}(0)|\mathbf{r}'_\perp\rangle = I_0\delta(\mathbf{r}_\perp - \mathbf{r}'_\perp), \tag{15}$$

with $\mathbf{r}_\perp$ and $\mathbf{r}'_\perp$ inside the source, σ, and equal to 0 otherwise. The matrix elements of the Fresnel free space propagator of a distance L are given by

$$\left\langle \mathbf{r}_\perp | \hat{D}(L) | \mathbf{r}'_\perp \right\rangle = \frac{-ik}{2\pi L} e^{ikL} \exp\left(\frac{ik}{2L} |\mathbf{r}_\perp - \mathbf{r}'_\perp|^2 \right), \tag{16}$$

where $k = 2\pi/\lambda$. The two-point correlation at a plane a distance L from the source in the Fresnel diffraction regime is

$$\left\langle \mathbf{r}_\perp | \hat{\Gamma}(L) | \mathbf{r}'_\perp \right\rangle = I_0 \int d^2\mathbf{s}_\perp d^2\mathbf{s}'_\perp \left\langle \mathbf{r}_\perp | \hat{D}(L) | \mathbf{s}_\perp \right\rangle \left\langle \mathbf{s}_\perp | \hat{\Gamma}(0) | \mathbf{s}'_\perp \right\rangle \left\langle \mathbf{s}'_\perp | \hat{D}^\dagger(L) | \mathbf{r}'_\perp \right\rangle$$

$$= I_0 \int_\sigma d^2\mathbf{s}_\perp \left\langle \mathbf{r}_\perp | \hat{D}(L) | \mathbf{s}_\perp \right\rangle \left\langle \mathbf{s}_\perp | \hat{D}^\dagger(L) | \mathbf{r}'_\perp \right\rangle$$

$$= I_0 \exp\left[\frac{ik}{2L} (|\mathbf{r}_\perp|^2 - |\mathbf{r}'_\perp|^2) \right] \left(\frac{k}{2\pi L} \right)^2$$

$$\times \int_\sigma d^2\mathbf{s}_\perp \exp\left[\frac{ik}{L} \mathbf{s}_\perp \cdot (\mathbf{r}_\perp - \mathbf{r}'_\perp) \right]$$

$$= I_0 \exp\left[\frac{ik}{2L} (|\mathbf{r}_\perp|^2 - |\mathbf{r}'_\perp|^2) \right] F_r(\mathbf{r}_\perp - \mathbf{r}'_\perp). \tag{17}$$

In Eq. (17), $F_r(\mathbf{r}_\perp - \mathbf{r}'_\perp)$ is the Fourier transform of the source area; it is the VCZ result for the far-field two-point correlation function.

We will be testing two source geometries, circular and square. For the square geometry, F_r is

$$F_i(\mathbf{r}_\perp - \mathbf{r}'_\perp) = \left(\frac{1}{\pi} \right)^2 \frac{1}{(x - x')(y - y')} \sin\left[\frac{kd}{2L}(x - x') \right] \sin\left[\frac{kd}{2L}(y - y') \right], \tag{18}$$

with d as the linear dimension of the source. For the circular geometry, F_r is

$$F_i(\mathbf{r}_\perp - \mathbf{r}'_\perp) = \left(\frac{kd}{2L} \right) \frac{J_1\left(\frac{kd}{2L} |\mathbf{r}_\perp - \mathbf{r}'_\perp| \right)}{2\pi |\mathbf{r}_\perp - \mathbf{r}'_\perp|}, \tag{19}$$

with d as the diameter of the source.

7.6 Focusing and the Instrument Operator

The matrix elements for the lens operator are given by

$$\left\langle \mathbf{r}_\perp | \hat{L}(f) | \mathbf{r}'_\perp \right\rangle = \exp\left(\frac{-ik}{2f} |\mathbf{r}_\perp|^2 \right) \delta(\mathbf{r}_\perp - \mathbf{r}'_\perp). \tag{20}$$

The instrument operator is

$$\hat{M} = \hat{P}^{\dagger}\hat{P},\tag{21}$$

where $\hat{P}$ is the propagator from the aperture, $\hat{A}(a)$, to the detector $\hat{A}(b)$.

$$\hat{P} = \hat{A}(b)\hat{D}(l)\hat{L}(f)\hat{A}(a).\tag{22}$$

The matrix elements for $\hat{A}(a)$ are given by

$$\left\langle \mathbf{r}_{\perp}|\hat{A}(a)|\mathbf{r}'_{\perp}\right\rangle = \delta(\mathbf{r}_{\perp} - \mathbf{r}'_{\perp}),\tag{23}$$

with $\mathbf{r}_{\perp}$ and $\mathbf{r}'_{\perp}$ inside the aperture a, and equal to 0 otherwise. The matrix elements for $\hat{A}(b)$ are given by

$$\left\langle \mathbf{r}_{\perp}|\hat{A}(b)|\mathbf{r}'_{\perp}\right\rangle = \delta(\mathbf{r}_{\perp} - \mathbf{r}'_{\perp}),\tag{24}$$

with $\mathbf{r}_{\perp}$ and $\mathbf{r}'_{\perp}$ inside the detector b, and equal to 0 otherwise.

From Eqs. (16) and (20) we have

$$\left\langle \mathbf{r}_{\perp}|\hat{D}(l)\hat{L}(f)|\mathbf{r}'_{\perp}\right\rangle = \frac{-ik}{2\pi l}e^{ikl}\exp\left(\frac{ik}{2l}|\mathbf{r}_{\perp} - \mathbf{r}'_{\perp}|^2\right)\exp\left(\frac{-ik}{2f}|\mathbf{r}'_{\perp}|^2\right).\tag{25}$$

This gives

$$\left\langle \mathbf{r}_{\perp}|\hat{P}|\mathbf{r}'_{\perp}\right\rangle = \frac{-ik}{2\pi l}e^{ikl}\exp\left(\frac{ik}{2l}|\mathbf{r}_{\perp} - \mathbf{r}'_{\perp}|^2\right)\exp\left(\frac{-ik}{2f}|\mathbf{r}'_{\perp}|^2\right),\tag{26}$$

with $\mathbf{r}_{\perp}$ inside the detector, $\mathbf{r}'_{\perp}$ inside the aperture, and equal to 0 otherwise.

The matrix elements of $\hat{M}$ are given by

$$\left\langle \mathbf{r}_{\perp}|\hat{M}|\mathbf{r}'_{\perp}\right\rangle = \int_b d^2\mathbf{s}_{\perp}\left\langle \mathbf{s}_{\perp}|\hat{P}|\mathbf{r}_{\perp}\right\rangle^*\left\langle \mathbf{s}_{\perp}|\hat{P}|\mathbf{r}'_{\perp}\right\rangle.\tag{27}$$

The integration is over the detector area:

$$\begin{aligned}
\left\langle \mathbf{r}_{\perp}|\hat{M}|\mathbf{r}'_{\perp}\right\rangle &= \exp\left[\frac{ik}{2f}(|\mathbf{r}_{\perp}|^2 - |\mathbf{r}'_{\perp}|^2)\right]\exp\left[\frac{-ik}{2l}(|\mathbf{r}_{\perp}|^2 - |\mathbf{r}'_{\perp}|^2)\right]\\
&\quad \times \left(\frac{k}{2\pi l}\right)^2\int_b d^2\,\mathbf{s}_{\perp}\exp\left[\frac{ik}{l}\mathbf{s}_{\perp}\cdot(\mathbf{r}_{\perp} - \mathbf{r}'_{\perp})\right]\tag{28}\\
&= \exp\left[\frac{ik}{2f}(|\mathbf{r}_{\perp}|^2 - |\mathbf{r}'_{\perp}|^2)\right]\exp\left[\frac{-ik}{2l}(|\mathbf{r}_{\perp}|^2 - |\mathbf{r}'_{\perp}|^2)\right]F_i(\mathbf{r}_{\perp} - \mathbf{r}'_{\perp}),
\end{aligned}$$

$$\tag{29}$$

with $\mathbf{r}_\perp$ and $\mathbf{r}'_\perp$ inside the aperture a, and equal to 0 otherwise. Here F_i is

$$F_i(\mathbf{r}_\perp - \mathbf{r}'_\perp) = \left(\frac{k}{2\pi l}\right)^2 \int_b d^2\mathbf{s}_\perp \exp\left[\frac{ik}{l}\mathbf{s}_\perp \cdot (\mathbf{r}_\perp - \mathbf{r}'_\perp)\right]. \tag{30}$$

If the camera is focused on the source,

$$\frac{1}{l} + \frac{1}{L} = \frac{1}{f}, \tag{31}$$

where L is the distance between the camera entrance aperture and the source. In this case the nonzero matrix elements are given by

$$\left\langle \mathbf{r}_\perp | \hat{M} | \mathbf{r}'_\perp \right\rangle = \exp\left[\frac{ik}{2L}\left(|\mathbf{r}_\perp|^2 - |\mathbf{r}'_\perp|^2\right)\right] F_i(\mathbf{r}_\perp - \mathbf{r}'_\perp). \tag{32}$$

The VCZ result for the nonzero matrix elements of $\hat{M}$ are given by Eq. (30) with the camera focused at infinity, $l = f$. This results in a slight difference between the VCZ result and the general result in the phase space volume of $\hat{M}$ defined by

$$N_i = \mathrm{Tr}(\hat{M}), \tag{33}$$

with N_i being the number of phase space cells. The infrared camera we used in the experiment has a circular entrance aperature and a square detector. Thus, we have

$$F_i(\mathbf{r}_\perp - \mathbf{r}'_\perp) = \left(\frac{1}{\pi}\right)^2 \frac{1}{(x-x')(y-y')} \sin\left[\frac{kb}{l}(x-x')\right]\sin\left[\frac{kb}{l}(y-y')\right], \tag{34}$$

with $\mathbf{r}_\perp$ and $\mathbf{r}'_\perp$ inside the circular entrance aperture of radius a and equal to 0 otherwise. The linear dimension of the square detector is $2b$. From Eq. (32),

$$N_i = \pi a^2 \left(\frac{2\theta_i}{\lambda}\right)^2, \tag{35}$$

where $2\theta_i = 2b/l$ is the full acceptance angle. For all practical purposes, the difference in the values of N_i for the focused case and with $l = f$ is negligible. Henceforth, we will mean $l = f$ when refering to N_i.

7.7 Measurement by Focusing the Camera on the Source

From Eqs. (17) and (32), the detected signal, Q, is

$$Q = \mathrm{Tr}\left[\hat{M}\hat{\Gamma}(L)\right] = I_0 \int d^2\mathbf{r}_\perp d^2\mathbf{r}'_\perp\, F_i(\mathbf{r}'_\perp - \mathbf{r}_\perp) F_r(\mathbf{r}_\perp - \mathbf{r}'_\perp), \tag{36}$$

The normalized signal, Q_n, is defined as

$$Q_n = \frac{Q}{I_0 N_i}. \tag{37}$$

7.8 Experimental Test of Focusing

The matrix elements of the instrument operator modeling the infrared camera used in the experiment are given by Eq. (32). The experiment was conducted by focusing the camera on the source. The results of the experiments are compared with Eqs. (36) and (37).

The following is the protocol for processing the data. The value of the signal when the detector of the camera is flood illuminated by the blackbody radiation subtracted by the value of the signal when the detector is flood illuminated by the background is used as the normalization. The measured normalized signal is obtained by first subtracting the detected signal by the background and then divided by the normalization. The normalized signal is compared with theory.

The camera has an interference filter and a HgCdTd detector giving a wavelength window with a peak at $\lambda = 8.8$ μm and $\Delta\lambda \simeq \pm 0.75$ μm. The square HgCdTd detector is of dimensions 75 μm × 75 μm. The focal length of the camera is $f = 18.99$ mm. In Eq. (32) F_i is then given by Eq. (30), with $b = (75/2)$ μm.

Two circular aperture plates of radii $a = 0.136$ in. and $a = 0.272$ in. were placed in front of the camera aperture. We have $N_i = 1.888$ for the $a = 0.136$ in. aperture and $N_i = 7.551$ for the $a = 0.272$ in. aperture. The face of each plate facing the lens was painted with high emissive paint to provide the background. The source's size and shape were controlled by placing aluminum masks with either square or circular apertures over a blackbody source set at $T \sim 500°$C.

Define N_r as

$$N_r = \int_a d^2\mathbf{r}_\perp \langle \mathbf{r}_\perp | \hat{\Gamma}(L) | \mathbf{r}_\perp \rangle, \tag{38}$$

where the integration is over the camera aperture. N_r can be interpreted as the number of phase space cells of the radiation field intercepted by the camera aperture. Therefore, N_r can be much less than 1 without violating the uncertainty principle (but when it is, the light is coherent at the detector). For a square source

$$N_r = \pi \left(\frac{2a\theta_s}{\lambda} \right)^2, \tag{39}$$

and for a circular source

$$N_r = \left(\frac{\pi a \theta_s}{\lambda} \right)^2. \tag{40}$$

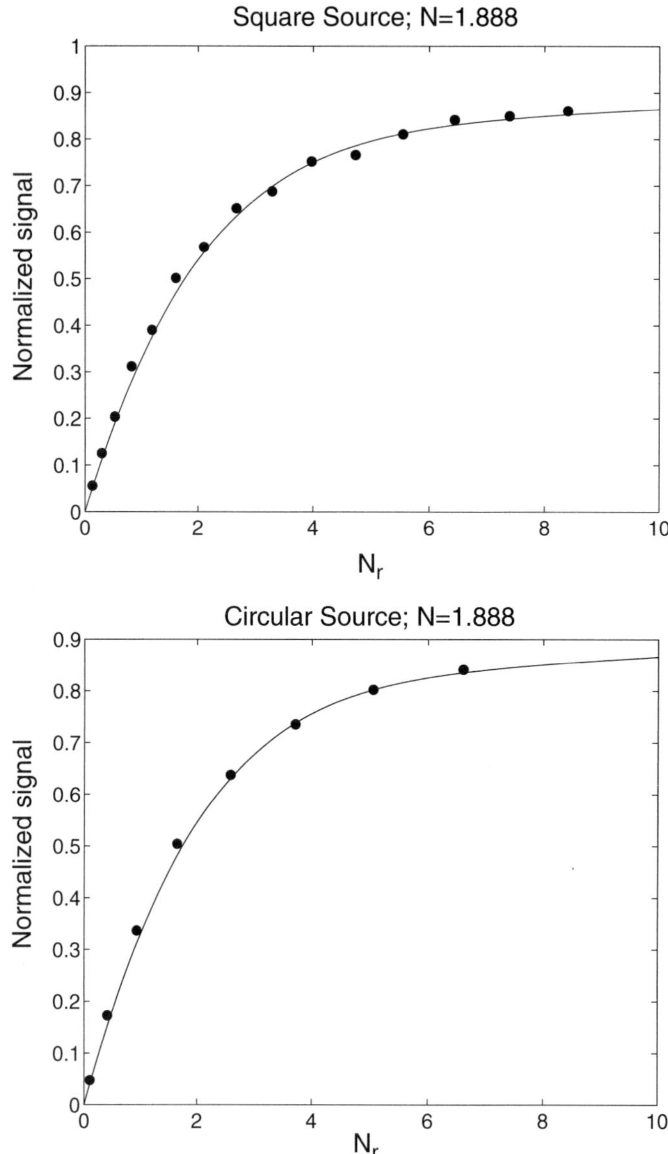

Figure 3 Comparison of experiment with theory for the $N = 1.888$ peak measurements. Filled circles are data points.

Here θ_s is the half-angle subtended by the source at the camera's entrance aperture. For the square source $\theta_s = d/2L$, with d being the linear dimension for the square source. For the circular source $\theta_s = d/2L$, with d being the diameter of the circular source.

The results of the measured normalized signal are plotted with theory versus N_r in Figs. 3 and 4. The agreement with experiment is highly satisfactory.

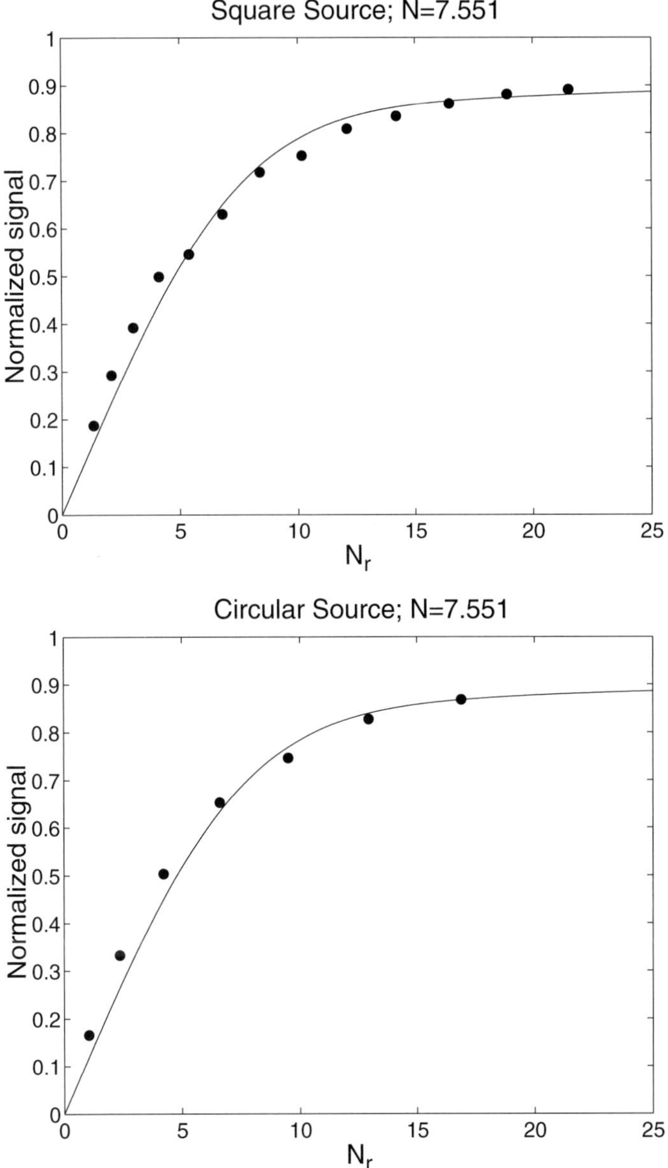

Figure 4 Comparison of experiment with theory for the $N = 7.551$ peak measurements. Filled circles are data points.

7.9 Conclusions

We have demonstrated a consistent approach to incorporating the diffraction properties of the instrument in optical measurements. We used a remarkable analogy between the result of measuring radiance and the van Cittert-Zernike theorem that exploits the symmetry between an incoherent source whose radiance is being measured, and the detector whose signal represents the measurement. It turns out that the measured radiance is represented (up to an overall constant) by the double integral over the instrument aperture of the mutual intensity of the field and the mutual intensity of a delta-correlated source of the same size, shape, and location as the detector. One may therefore form a perspective of the subject. "Generalized radiance" is a useful calculational tool that comes in a variety of forms ranging from the classical Wigner distribution [1] to the "Alonso function" [2]. "Measured radiance" is the value obtained by a particular instrument that can be calculated by the techniques described in this paper. While we have expressed our results in the context of radiometry, one would go through a similar analysis in analyzing the detection of any partially coherent wave. The signal is represented by the double integral of two mutual coherence functions. One of these is for the incident wave, the other arising from the detector considered as a source. It is likely that entirely similar considerations may apply to other signal detection processes where diffraction effects are important.

References

1. E. Wolf, "Coherence and radiometry," *J. Opt. Soc. Am.* **68**, 6–17 (1978); A. Walther, "Radiometry and coherence," *J. Opt. Soc. Am.* **58**, 1256–1259 (1968) and *J. Opt. Soc. Am.* **63**, 1622–1623 (1973). For a recent treatment, see also R.G. Littlejohn and R. Winston, "Corrections to classical radiometry," *J. Opt. Soc. Am. A* **10**, 2024–2037 (1993).

2. L.E. Vicent and M.A. Alonso, "Generalized radiometry as a tool for the propagation of partially coherent fields," *Optics Communications* **207**, 101–112 (2002).

3. J.C. Mather, M. Toral, and H. Memmati, "Heat trap with flare as multimode antenna," *Applied Optica* **25**, 2826–2830 (1986).

4. R. Winston, Y. Sun, and R.G. Littlejohn, "Measuring radiance and the van Cittert-Zernike theorem," *Optics Communications* **207**, 41–48 (2002); Y. Sun, R. Winston, J.J. O'Gallagher, and K.A. Snail, "Statistical optics and radiance measurement in the diffraction limit," *Optics Communications* **206**, 243–251 (2002).

5. For an excellent, accessible discussion of this theorem, see L. Mandel and E. Wolf, *Optical Coherence and Quantum Optics*, Cambridge University Press, Cambridge (1995); also M. Born and E. Wolf, *Principles of Optics*, 7th (Expanded edition), Cambridge University Press, Cambridge (1999).

6. R.G. Littlejohn and R. Winston, "Generalized radiance and measurement," *J. Opt. Soc. Am. A* **12**, 2736–2743 (1995).

7. R. Winston and R.G. Littlejohn, "Measuring the instrument function of radiometers," *J. Opt. Soc. Am. A* **14**, 3099–3101 (1997).

Roland Winston, Ph.D., Prof. of Physics (Chairman), University of California at Merced and former Professor (Chairman) of the Department of Physics, University of Chicago, is known as the "father of Non-Imaging Optics," with over 25 patents in this field, including the invention of the compound parabolic concentrator (CPC). His contributions to this field are extensive, including such highly praised works on the subject as his seminal book on the subject, *High-Collection Non-Imaging Optics*, which he co-authored with Dr. W.T. Welford

Robert G. Littlejohn received his B.A. in 1975 and his Ph.D. in 1980, both from the University of California at Berkeley. After postdoctoral positions at the La Jolla Institute and the University of California, Los Angeles, he joined the Berkeley faculty in 1983. He has been a Presidential Young Investigator and a Miller Professor, and he is a fellow of the American Physical Society.

Yupin Sun graduated from the University of California, Irvine, in 1995 with B.S. degrees in physics and biology. He subsequently did graduate work in physics and received his Ph.D. in physics in 2001 from the University of Chicago, where his thesis advisor was Prof. Roland Winston. Dr. Sun's Ph.D. thesis was on measuring radiance in the diffraction limit. He is now an optical scientist at LPI, LLC.

Keith A. Snail is a Section Head in the Optical Sciences Division at the Naval Research Laboratory in Washington, D.C. Dr. Snail received his Ph.D. from the University of Chicago in 1983; his Ph.D. research examined a novel nonimaging solar thermal collector with Prof. Roland Winston. While at NRL, Dr. Snail has published more than 30 refereed journal articles, 10 U.S. patents and several book chapters. Dr. Snail's research has focused on the application of nonimaging concentrators to the instrumentation used to measure directional hemispherical reflectance.

CHAPTER 8

MILLIMETER WAVE MMIC HOLOGRAM BEAM FORMER

Vladimir A. Manasson and Lev S. Sadovnik

8.1 Introduction

A physical optics approach has been found to be very fruitful in designing antennas that make use of the diffraction phenomenon. Flexibility in forming a desired wavefront is achieved by creating a complex diffraction grating, also known as holographic beam forming. Based on this notion, a new type of electronically controlled beam former has been demonstrated. The device operation is based on a real-time reconfigurable hologram formed by electron-hole plasma injected into a planar semiconductor waveguide. The device operates at millimeter-wave frequencies and is capable of forming a wavefront with an arbitrary profile. Digital control is achieved through the use of a millimeter wave/microwave integrated circuit (MMIC) design that holds the promise of unparalleled cost effectiveness. Potential applications include smart antennas, imaging radar, and communications.

8.2 Principles of Operation

The presence of free carriers in a solid body strongly affects its optical characteristics. For metals, this phenomenon has been known for over a century (Drude, 1900). However, for a given metal electron density it is a rather fixed value, and it depends only very slightly on external conditions (such as temperature, pressure, etc.). In contrast, semiconductors represent media into which electrons and holes can be easily injected. Injection of only one type of carrier (electrons or holes) usually leads to the formation of charge carrier clouds, which limits the injection level. Bipolar injection (injection of both electrons and holes) produces a quasi-neutral formation called electron-hole plasma. The achievable plasma density is sufficient to cause local changes in the real and imaginary parts of the dielectric constant of the material, generating changes in reflection, absorption, refraction, and phase velocity, and causes coherent scattering. Plasma-induced changes alter the propagation of electromagnetic waves passing through the medium. In other words, a semiconductor medium with a nonuniform distribution of electron-hole plasma acts as a hologram, with spatially and temporarily variable refractive index. As a result, the wavefront changes and its shape depends on the plasma pattern. By varying the plasma pattern, one can control the wavefront and thus shape the resulting beam(s) and aim it (them) in the desired direction(s). The injected plasma is a nonequilibrium formation and is subject to fast recombination. Typically, the plasma lifetime varies from 10^{-10} sec to 10^{-5} sec. This means that the hologram pattern can practically be updated (rewritten) in real time. Plasma can be excited with the use of various techniques such as photoconductivity [1–17] and current injection that was used to excite plasma patterns in optically controlled beam formers. Another way to create plasma patterns is to use PN-junctions, PIN structures, heterostructures, or other carrier-injecting electrodes. In this paper we present the results of a study of a plasma hologram beam former, where plasma patterns are created by carrier injection via heavily doped P- and N-electrodes.

The key element of a beam former is a semiconductor chip holding a linear array of individually controlled P- and N-electrodes that constitute an MMIC aperture. Biased electrodes inject carriers into their vicinity, while unbiased ones do not. As in the case of photoinjection, the resulting plasma pattern acts as a hologram that controls the wavefront. To change the latter one needs to alter the plasma pattern, i.e., change the biasing of the electrodes.

We have simulated beam forming by an MMIC aperture comprising 220 electrode pairs. Our simulations have demonstrated high flexibility in controlling the beam profile. Some examples are shown in Figs. 1–4. In particular, Fig. 1 shows

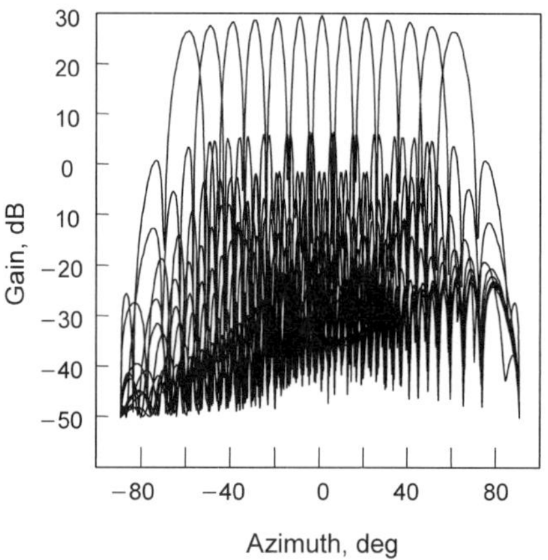

Figure 1 Simulation of continuous beam steering in azimuth plane. Frequency: 76 GHz; aperture size: 75 mm.

how an MMIC digital hologram regenerates a diffraction beam in the arbitrary direction, providing continuous beam steering. Figure 3 shows generation of a complex wave front consisting of eight main lobes. Figure 4 demonstrates two-lobe variable shape patterns, and Fig. 5 shows the ability of forming a deep null at an arbitrary position. As seen from the simulation, the device is capable of forming a single pencil beam in an arbitrary direction, i.e., it can perform a scan. It can also create several beams simultaneously, the direction and shape of each being independently controlled. It can also form wide beams with deep "nulls" blocking the reception (or transmission) from (to) undesired direction(s). All those features are very useful for "smart" antenna applications.

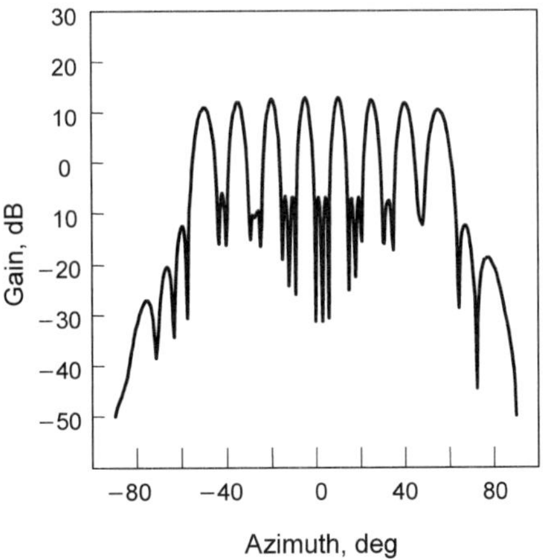

Figure 2 Beam pattern for a hologram that forms eight simultaneous pencil beams. Frequency: 76 GHz; aperture size: 75 mm.

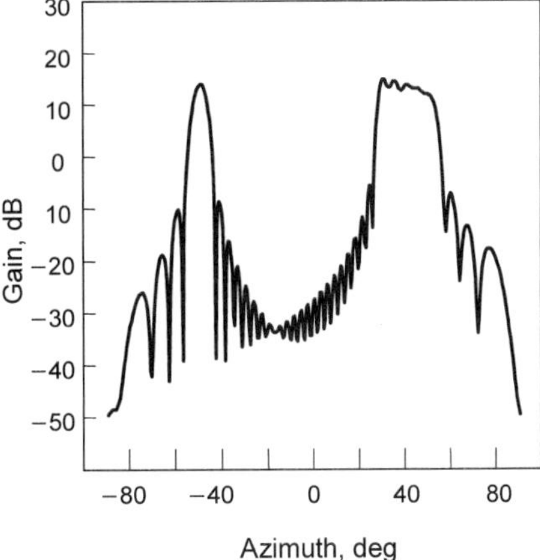

Figure 3 Beam pattern for a hologram that forms two simultaneous beams of different shapes.

The MMIC digital hologram beam former has been implemented using silicon chips as shown in Fig. 5. A photo of the beam former is shown in Fig. 6. A cylindrical lens covers the MMIC aperture and is used to form the wavefront in the elevation plane, while the hologram forms the beam shape in the azimuth plane.

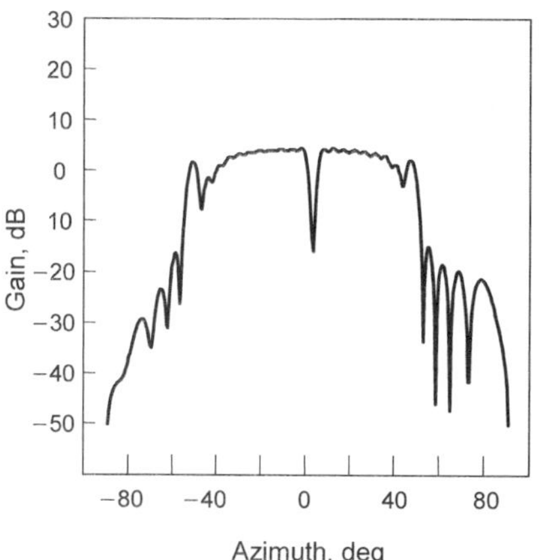

Figure 4 Beam pattern for a hologram that forms a wide beam with a deep "null" at the center.

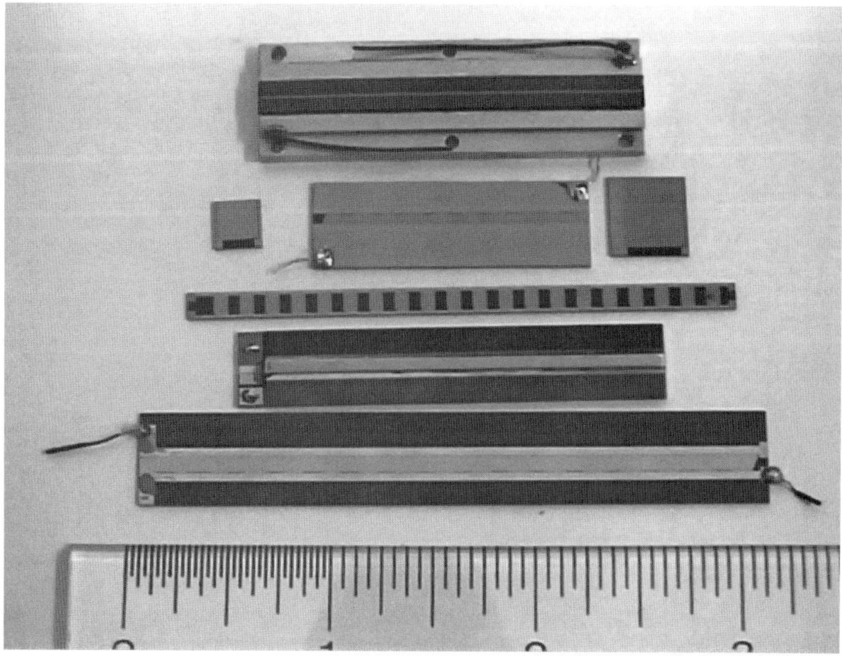

Figure 5 Examples of semiconductor chips for plasma hologram beam formers—the single chip concept.

As predicted by preliminary simulations, the antenna experimentally demonstrated continuous beam steering over a wide azimuth angle. Some of the beam patterns are shown in Fig. 7. They were measured at the frequency of 76 GHz in the transmitting mode using a near-field measurement technique. We also tested the beam former in the receiving mode. The device is capable of forming multiple simultaneous beams. Some of the measured beam patterns are shown in Fig. 8. The beam patterns comprise two beams, each controlled independently.

Figure 6 Beam former assembly in a transmitting mode. Beam former includes MMIC aperture, dielectric rod feeder, cylindrical lens, driving electronics, cables connecting MMIC aperture and driving electronics.

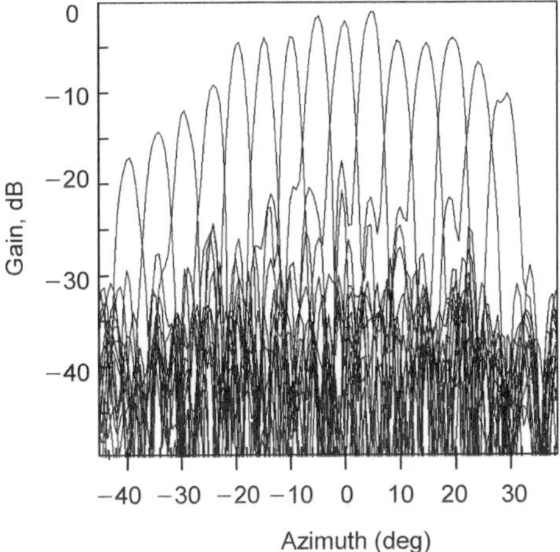

Figure 7 Fifteen overlapping beam patterns that demonstrate beam-former scanning capabilities. Operation frequency: 76 GHz. Data was obtained using a near-field antenna measurement system.

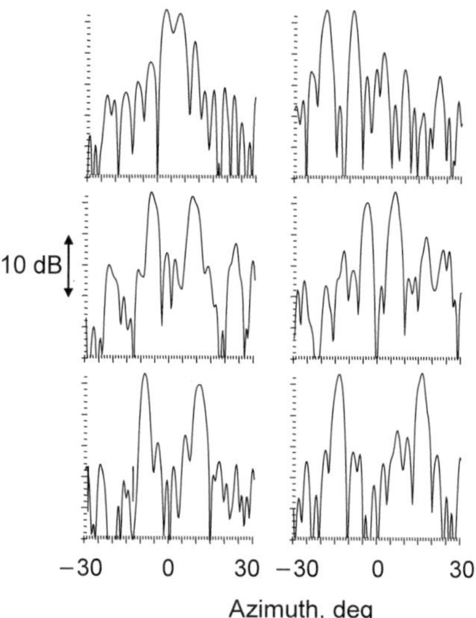

Figure 8 Examples of beam patterns for MMW holograms that generate two simultaneous beams. Each beam can be directed independently from the other. Data was obtained with near-field antenna measurement system at a frequency of 76 GHz.

Summary

We have simulated and successfully demonstrated the operation of a *single-chip* MMIC plasma hologram beam former. The device provides excellent beam control over a steering range of 70 degrees. The switching time between two consequent hologram patterns was shorter than 10 μsec. Potential applications for the new beam-forming technology include smart antennas for imaging and surveillance radars, as well as satellite, mobile, and point-to-point communications. The MMIC embodiment of the hologram aperture projects a dramatic cost reduction compared to competing technologies such as phased-array antennas.

References

1. C.H. Lee, P.S. Mark, and A.P. DeFonzo, "Optical control of millimeter-wave propagation in dielectric waveguides," *IEEE J. Quantum Electron.* **QE-16**, 277–288 (1980).
2. M. Matsumoto, M. Tsutsumi, and N. Kumagai, "Radiation of millimeter waves from a leaky dielectric waveguide with a light-induced grating layer," *IEEE Trans. Microwave Theory Tech.* **MTT-35**, 1033–1042 (1987).
3. A. Rosen, P.J. Stabile, P. Herczfeld, A. Daryoush, and J.K. Butler, "Optically controlled IMPATT diodes and subsystems," in *Proc. 1989, SBMO Int. Microwave Symp.*, Vol. 2, 589–594 (1989).

4. V.A. Manasson, L.S. Sadovnik, P.I. Shnitser, and V. Litvinov, "Guided wave antenna induced by light," *Antenna Application Symposium*, Allerton Park, Monticello, Illinois, 1–14 (1994).

5. A. Alphones and M. Tsutsumi, "Leaky wave radiation from a periodically photoexcited semiconductor slab waveguide," *IEEE Trans. Microwave Theory Tech.* **43**, 2435–2441 (1995).

6. V.A. Manasson, L.S. Sadovnik, A. Moussessian, and D.B. Rutledge, "Millimeter-wave diffraction by a photo-induced plasma grating," *IEEE Trans. Microwave Theory Tech.* **43**, 2288–2290 (Sept. 1995).

7. V.A. Manasson, L.S. Sadovnik, P.I. Shnitser, R.M. Mino, and J.S. Kruger, "MMW optically scanning antenna based on plasma-induced grating," in *Application and Theory of Periodic Structures, SPIE Proc.* Vol. 2532, 290–299, Bellingham, WA (1995).

8. K. Nishimura and M. Tsutsumi, "Scattering of millimeter waves by metallic strip gratings on an optically plasma-induced semiconductor slab," *IEEE Trans. Microwave Theory Tech.* **44**, 2231–2237 (Dec. 1996).

9. V.A. Manasson, L.S. Sadovnik, and K. Spariosu, "Compact version of an optically scanning MMW antenna," in *Twentieth Antenna Application Symposium*, Allerton Park, Monticello, IL, 1–11 (1996).

10. V.A. Manasson and L.S. Sadovnik, "Optical approaches to MMW scanning antenna," in *PSAA-6 Proc. of sixth Annual ARPA Symposium on Photonic Systems for Antenna Applications*, March 1996, Monterey, CA (1996).

11. V.A. Manasson, L.S. Sadovnik, P.I. Shnitser, R.M. Mino, and J.S. Kruger, "Millimeter-wave optically scanning antenna based on photo-induced plasma grating," *Optical Engineering* **35**(2), 357–361 (1996).

12. V.A. Manasson, L.S. Sadovnik, and V.A. Yepishin, "Optically controlled scanning antennas comprising a plasma grating," in *IEEE Antennas and Propagation Society International Symposium, 1997*, Vol. 2, 1228–1231 (1997).

13. V.A. Manasson, L.S. Sadovnik, and V.A. Yepishin, "New architectures of light-controlled MMW steering antennas," in *Optical Technology for Microwave Applications VIII, SPIE Proc.* Vol. 3160, 80–88, Bellingham, WA (1997).

14. V.A. Manasson, L.S. Sadovnik, and V.A. Yepishin, "Use of light for direct control of MMW propagation," presented at the Seventh Annual DARPA Symposium on Photonic Systems for Antenna Applications, Monterey, CA, January 14–16, 68–72 (1997).

15. V.A. Manasson, L.S. Sadovnik, V.A. Yepishin, and D. Marker, "An optically controlled MMW beam-steering antenna based on a novel architecture," *IEEE Transactions on Microwave Theory and Techniques* **45**(8), Part 2, 1497–1500 (1997).

16. V.A. Manasson, V.A. Yepishin, D. Eliyahu, L.S. Sadovnik, H. Buss, and V. Rubtsov, "Application of photo-induced plasma-grating (PIPG) technology for 2-D tracking and 1-D beam steering in x-band," *Eighth Annual DARPA Symposium on Photonic Systems for Antenna Applications*, Monterey, CA (1998).
17. V.A. Manasson, V.A. Yepishin, D. Eliyahu, L.S. Sadovnik, and V. Rubtsov, "Photo-induced plasma-grating technology for x- and w-band beam-steering antennas," in *Enabling Photonic Technologies for Aerospace Applications II, SPIE Proc.* Vol. 4042, 156–159, Bellingham, WA (2000).

A Personal Perspective

As many of us have, I, Lev Sadovnik, learned about Prof. Emil Wolf from his and Max Born's classic text *Principles of Optics*. In addition to its educational value, the book was a source of a fateful coincidence in my life.

Going back to those days in 1978, I find myself a senior at Chernovtsi State University (now Ukraine), studying optics. Professor Polyanski the head of Optics Department brought a copy of the just-translated Born and Wolf book (Fig. 9) to the classroom, as unusual as it was under the Soviet era educational system, and announced that the Theoretical Optics course would include studying some of the

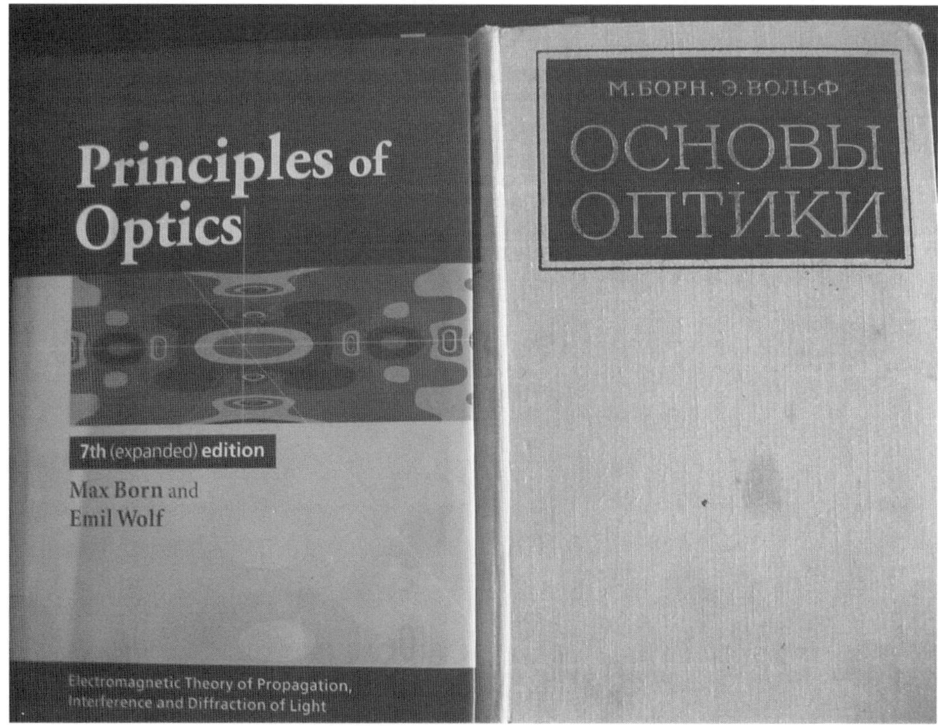

Figure 9 The book and its translation.

book's chapters. He further suggested that students in that small class would be assigned a chapter each to actually give a lecture on the subject. Yours truly was given the lengthy and tedious van Cittert-Zernike Theorem. After sweating over it for quite a while, I delivered a lecture and eventually graduated with a diploma that read "Physical Optics."

Eleven years later, I came to the U.S., and looking for my first job was knocking on the door of a company called ... Physical Optics Corporation. My job interview was with the company's Chief Technical Officer, Dr. Tomasz Jannson. Lamenting about the paucity of serious optical background among this generation's scientists, Dr. Jannson asked if the author heard of *Principles of Optics*. Hearing a positive response, Dr. Jannson inquired if I had actually read it. Feeling that my future job might be hanging on the answer to this question, I produced a positive reply. In turn, probably noting some hesitation in that reply, Dr. Jannson went on to inquire how much of the book's content the author remembered. Paralyzed by the question, what I could immediately recall was, for the most part, the book's cover, so I mumbled something about coherence, diffraction, and lens design. Dr. Jannson didn't buy that and, in an attempt to clear it once for all, asked pointedly about ... van Cittert-Zernike theorem. Needless to say, Dr. Jannson was quite impressed when I produced not only the theorem's formulation, but even the outline of its proof. As a result, I was offered the job. Some years later I was fortunate to meet Prof. Wolf himself, and was privileged to personally discuss the subject of diffraction and coherence with him.

The research described above started when the two authors, VM and LS, discussed how to apply a rewritable hologram concept to the millimeter wave, so that all the advantages of achieving the desired field distribution via the diffraction phenomena could be utilized. While all the hardware components of MMW optics are quite similar to those of conventional optics, the missing part has been the material to record a dynamic diffraction pattern. Such a material, the semiconductor has been proposed by VM and the antenna developed based on this approach is discussed below.

Vladimir Manasson received his Ph.D. degree in Physics from Chernovtsy State University and M.S. degree in Semiconductor and Electro-Vacuum Engineering from Moscow Institute for Electronic Manufacturing (MIEM). His achievements include invention and development of hologram-based millimeter-wave beam formers, microwave diffraction gratings based on solid-state plasma effects, polarimetry instruments, new mechanism of photocurrent amplification in semiconductor heterostructures, various microwave and semiconductor devices, sensors, and solar cells. He worked at the Institute for Problems in Material Science, Chernovtsy, Ukraine; Physical Optics Corporation, Torrance, California, and is currently a VP and CTO at WaveBand Corporation, Irvine, California. He is the author and coauthor of more than 100 publications in scientific journals and conference proceeding and holds 16 patents.

Lev Sadovnik received his M.S. degree in Physical Optics from the University of Chernovtsy (Ukraine) and his Ph.D. from the University of Southern California Signal and Image Processing Institute. His expertise is in diffraction optics, holography and optical image processing. Among his professional achievements, Dr. Sadovnik is the co-inventor of the Micro-Patterning System for microlithography. While at Physical Optics Corporation, Dr. Sadovnik completed a theoretical study of high-frequency anti-reflection gratings, aberration-corrected HOEs, and highly efficient concave gratings. He proposed and experimentally verified a new principle of phase coding for an all-optical processor. He co-authored a novel concept for millimeter wave beam steering and beam forming. Among his other inventions is an innovative algorithm and optical design for a distortion-invariant processor. His work experience has involved the management of system projects from conceptual design through full-scale development. Currently, Dr. Sadovnik is President of WaveBand Corporation. He is the author or co-author of 60 publications and eleven patents.

DIGITAL HOLOGRAPHIC MICROSCOPY

Anthony J. Devaney and Pengyi Guo

9.1 Introduction

More than a half century ago Dennis Gabor proposed that conventional lens-based imaging systems, such as the optical microscope, could be replaced by a new type of imaging system that is theoretically free of aberrations and can achieve numerical apertures arbitrarily close to unity [1,2]. This new idea was, of course, *holography*, which Gabor referred to as "microscopy by wavefront reconstruction." In this paper a modern version of Gabor's original scheme for lensless microscopy is presented that employs a fully coherent (laser) source and no imaging lenses, and generates images on a digital computer using algorithms that mimic (and, actually improve upon) the imaging operation of a diffraction limited lens system. This class of imaging systems, which we will refer to as "digital holographic microscopes" (DHM), generates complex-valued images of amplitude, phase, and even three-dimensional objects by employing digital holography in combination with state-of-the-art computer algorithms to implement the imaging process.

The key experimental ingredient of DHM is phase-shifting holography (PSH) [3–5], an idea originally conceived by Gabor in the 1950s and later published in 1966 [6]. PSH supplies the means of determining the complex amplitude of a coherent wavefield diffracted by an object by digitally recording multiple Gabor (in-line) holograms over an aperture that forms the entrance pupil of the microscope using, for example, a CCD array. Although other schemes exist for performing this task without the need of holography (such as the use of phase retrieval algorithms [7,8]), PSH is an easily implemented and accurate method that is ideally suited to DHM and is exclusively employed in the microscopes discussed in this paper.

The key computational ingredient of the DHM is the process of field back propagation. This mathematical operation was developed and refined in the 1960s largely through the work of Wolf and coworkers [9–11], who employed plane wave expansions [12,13] (also called angular spectrum expansions) to implement the back propagation operation. Field back propagation is used in DHM to form the coherent images of thin objects from the object's diffracted wavefields as determined by PSH and is also a key ingredient of the algorithms of diffraction tomography (see discussion below) that are employed to image 3D objects.

One of the primary advantages of DHM over conventional microscopy is that it can yield images of phase objects without the necessity of staining or of employing ad-hoc phase contrast imaging schemes that generate images not directly or quantitatively related to the object's physical properties. An even greater advantage is that a DHM can be configured to generate quantitative, high-quality images of 3D semitransparent objects. This is accomplished by configuring the microscope in such a way that multiple measurements of the wavefield diffracted by a 3D object for different orientations of the object relative to the optical axis of the microscope can be performed. The suite of diffracted field data so acquired can then be processed using the algorithms of diffraction tomography[*] [14,15] (DT) to generate a quantitative three-dimensional image (reconstruction) of the three-dimensional complex index of refraction distribution of the object. The merging of DHM with DT results in a "lensless fully digital microscope" that has the imaging capability of a modern confocal scanning microscope [17], but at a fraction of the cost or complexity.

Due to length considerations, the current paper covers only a broad review of DHM and, in particular, does not delve into the technical or engineering details of such systems. Also, only a few examples are presented and the reader is directed to the open literature cited in the paper for more examples and further details. Finally, we would like to note that this paper is especially relevant to be included in a collection of papers honoring the work of Emil Wolf since, as is evident from the list of citations at the end of the paper, he played a fundamental role in the development of the back propagation and DT algorithms on which the operation of this class of microscopes depends.

9.2 Conventional Optical Microscopy

A conventional optical microscope can be represented using the generalized model of an imaging system illustrated in Fig. 1 [12,13]. In this figure, a monochromatic

---

[*] The origins of diffraction tomography go back to a pioneering paper by Wolf [16], who showed that multiple holograms of a weakly scattering object illuminated with plane waves with varying directions of propagation can, in principle, be used to generate a 3D reconstruction of the object.

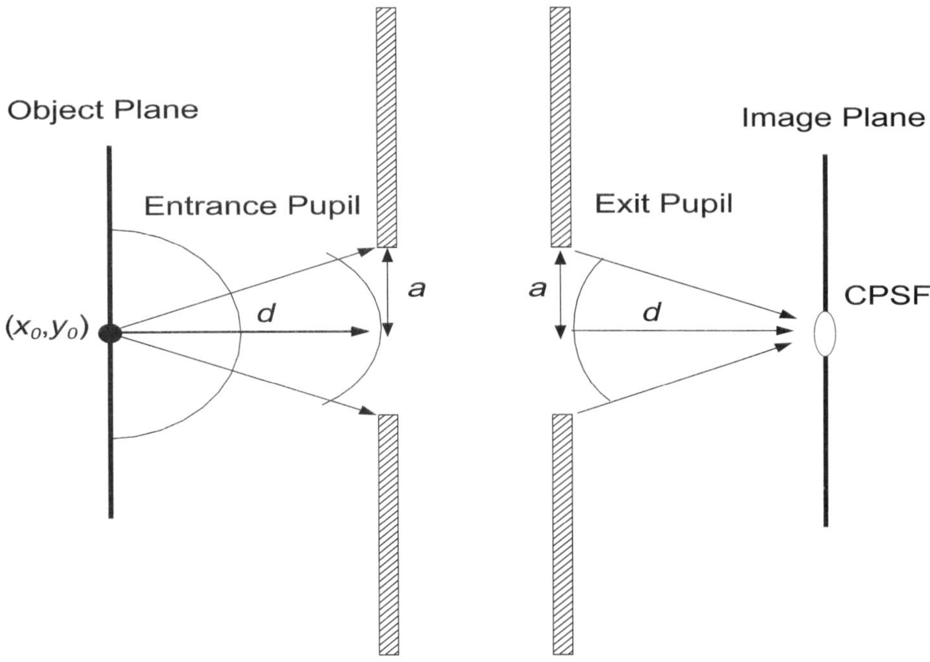

Figure 1 Generalized model of imaging system. A point source of light centered on the object plane at coordinates (x_0, y_0) generates a three-dimensional image field whose complex amplitude distribution over the image plane coherent point spread function is centered at image plane coordinates (x', y'). For a unit magnification system illustrated in the figure the distances from object plane to entrance pupil and image plane to exit pupil are equal and $x' = x_0, y' = y_0$.

point source of light located at the coordinates (x_0, y_0) on the object plane P_0 generates a diverging spherical wave that is converted by the system (microscope) to a converging spherical wave centered on the image plane at the image plane coordinates (x', y'). Since the microscope's magnification M plays no essential role in the imaging performance of the microscope[†] we will assume that $M = 1$ to simplify the following discussion. In this case the object and image distances are both equal $d_o = d_i = d$, and the converging spherical wave focuses at the point $x' = x_0, y' = y_0$ on the image plane.

The complex amplitude distribution $h(x, x'; y, y')$ of the image field as a function of image plane coordinates (x, y) is called the coherent point spread function (CPSF) of the system and can be expressed in terms of the systems pupil function

---

[†] The magnification M is only a means to scale the image and plays no role in the basic imaging performance, which is governed by the resolution of the microscope.

$\mathcal{P}$ via the equation [12,13]

$$h(x,x';y,y') = C \int d\xi d\eta \, \mathcal{P}(\xi,\eta) e^{-i\frac{k}{d}[\xi(x-x')+\eta(y-y')]}, \qquad (1a)$$

where C is an unessential constant, $k = 2\pi/\lambda$ is the free space wavenumber, and d the distance of the object and image planes from the system entrance and exit pupils, respectively (see Fig. 1). The pupil function $\mathcal{P}$ depends on the amplitude and phase of the converging spherical wave over the exit pupil as a function of the exit pupil coordinates ξ and η, and, in particular, vanishes identically when $\xi^2 + \eta^2 > a^2$, where a is the radius of the entrance and exit pupils. The phase of the pupil function models defects in the phase of the focused wave from being perfectly spherical and can be expressed in terms of the so-called *aberration function* of the system as discussed in Ref. [12].

In general the pupil function $\mathcal{P}$ will depend on the location of the object point $(x' = x_0, y' = y_0)$, but for small objects contained within a single isoplanatic patch, as can be expected in most applications of optical microscopy, this dependence is weak and can be neglected so that h is a function only of the difference coordinates $(x - x', y - y')$. Under this assumption the complex image field generated by a complex distribution of light over the object plane will then give rise to the image field

$$\psi_i(x,y) = \int dx' dy' h(x-x',y-y')\psi_0(x',y') = h \otimes \psi_0(x,y), \qquad (1b)$$

where $\psi_0(x',y')$ is the object field, i.e., the complex field distribution existing over the object plane and $\otimes$ denotes two-dimensional convolution. For optically thin objects illuminated by a monochromatic plane wave the object field can be approximated by the product of a transmittance function $T(x',y')$ with the complex amplitude of the incident plane wave so that Eq. (1b) reduces to

$$\psi_i(x,y) = \hat{T}(x,y) = \int dx' dy' h(x-x',y-y')T(x',y') = h \otimes T(x,y). \qquad (1c)$$

In this equation $\hat{T}$ is the *coherent image* of the object (transmittance function) generated by the microscope and we have assumed, for the sake of simplicity, that the illuminating plane wave has unit intensity and zero phase at the sample location.

In conventional optical microscopy, great pains are taken to make the illuminating light source as spatially incoherent as possible. In the ideal case, where the illuminating plane wave is spatially incoherent and the object being imaged is optically thin and sufficiently small so it is contained within a single isoplanatic patch, the image intensity resulting from an object having a transmittance function $T(x,y)$ is found from Eq. (1c) to be given by [12,13]

$$I_i = |\psi_i|^2 = |\hat{T}(x,y)|^2 = H \otimes |T(x,y)|^2, \tag{2}$$

where $H(x-x',y-y') = |h(x-x',y-y')|^2$ is the incoherent point spread function (PSF) of the microscope.

Even in its idealized form summarized by the imaging equation Eq. (2) the conventional microscope is limited in its performance by the quality of its optics and inherent diffraction effects that are present even if all optical components are free of aberrations and are "diffraction limited." In this latter case the ultimate resolution attainable is determined by the *numerical aperture* (NA) of the system, which is equal to the sine of the half angle θ_0 subtended by the entrance pupil, as measured from the center of the object plane, and is approximately equal to the ratio of the radius of the entrance pupil a to the object distance d. In terms of the numerical aperture $\sin\theta_0 \approx a/d$, the effective resolution of the microscope is on the order of $\lambda/(a/d)$, where λ is the wavelength of the illuminating plane wave. Although it is possible to moderately improve this diffraction-limited resolution limit via the use of oil immersion techniques, the microscope's numerical aperture sets an effective upper bound on the spatial resolution that cannot be improved upon.

The above discussion refers to an "ideal" optical microscope. A real microscope suffers from a number of drawbacks that go well beyond the limits outlined above. To begin with, in real applications the illuminating plane wave is not perfectly spatially incoherent, with the result that the simple image formation equation Eq. (2) is replaced by a complicated, nonlinear transformation between the transmittance function $T(x,y)$ and the image intensity I_i. The net result of this breakdown of spatial incoherence is to reduce the image quality of the microscope in a complicated way that is not correctable via improvements in lens quality. Perhaps the greatest limitations of conventional optical microscopy are its inability to image phase only and optically thick samples. In particular, in the case of phase-only objects for which the object intensity $I_0(x,y) = |T(x,y)|^2$ is (effectively) constant, the image I_i carries little or no structural information about the object. In order to overcome this limitation it is necessary to either *stain* the sample so as to convert the phase structure to an amplitude structure that can then be imaged or to employ phase contrast imaging techniques that have their own set of limitations. In the case of 3D objects (optically thick samples) the situation is even worse. In this case the object cannot be described via a 2D transmittance function $T(x,y)$, but rather is described via its 3D complex index of refraction distribution $\hat{n}(x,y,z)$. The coherent imaging Eq. (1b) still applies, but the object field ψ_0 is now related to the 3D index of refraction n via a three-dimensional mapping similar to a tomographic projection of the 3D object onto the 2D object plane [18]. The net result is that 3D samples cannot be successfully imaged by a conventional optical microscope and either the 3D sample has to be sectioned (thinly sliced), or a costly confocal scanning microscope must be employed.

9.3 Holographic Microscopy

A modern version of a Gabor holographic system is illustrated in Fig. 2.[‡] In this figure an incident coherent plane wave is split evenly into the upper and lower legs of a Mach–Zehnder interferometer, and the interference pattern formed between the wave diffracted by a semitransparent sample placed in the lower leg of the interferometer and the plane wave propagating in the upper leg of the interferometer is recorded by the CCD array of a high-resolution, monochrome digital camera.[§] The aperture defined by the CCD array of the monochrome camera

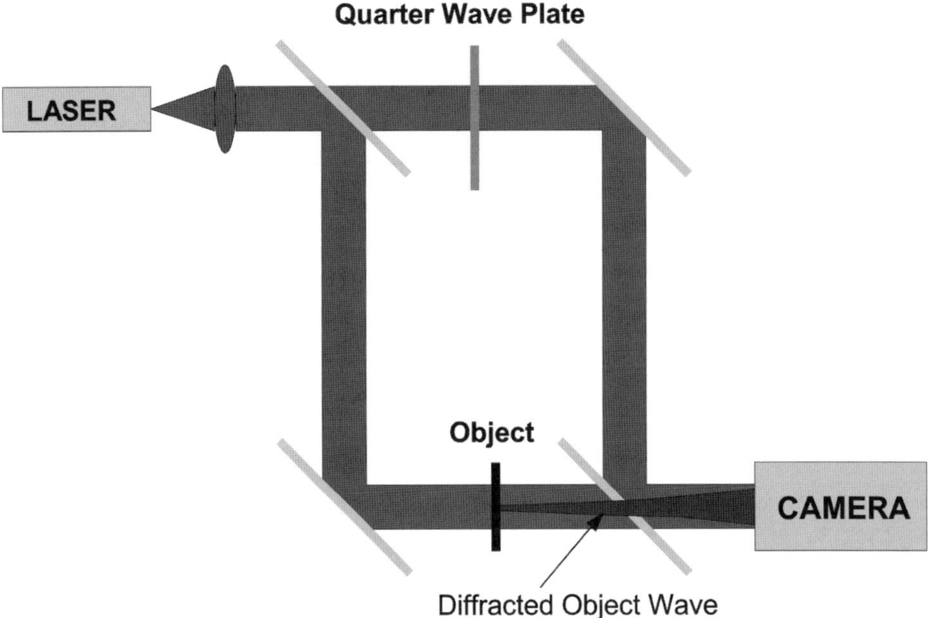

Figure 2 Mach–Zehnder interferometer used to acquire digital holograms. A quarter wave plate is included in the upper leg of the interferometer for use in phase-shifting holography (PSH).

[‡] In fact it can, in principle, achieve numerical apertures larger than unity corresponding to the measurement of *evanescent plane wave components* generated by the sample. In this paper we will not be concerned with the possibility of such *super-resolution*, although it has been achieved by a number of workers in ultrasound imaging [19].

[§] This camera is assumed to contain no imaging lenses. The camera thus records the intensity of the field across its CCD array and not a focused version of this field.

forms the hologram recording plane and the entrance pupil P_e of the holographic imaging system. The wavefield diffracted by the object will be a function of both the transverse coordinate vector $\boldsymbol{\rho} = (x,y)$ and the distance z along the optical axis as measured from the object plane P_0 so that the hologram records the intensity distribution

$$I_e(\boldsymbol{\rho}) = 1 + I_0(\boldsymbol{\rho}) + e^{-ikd}\psi(\boldsymbol{\rho},d) + e^{ikd}\psi^*(\boldsymbol{\rho},d) \tag{3}$$

over the region $|\boldsymbol{\rho}| \leq a$, where a is the radius of the entrance pupil P_e, and d the distance from object plane to hologram plane P_e. In this equation ψ is the field diffracted by the object and $I_0(\boldsymbol{\rho}) = |\psi(\boldsymbol{\rho},d)|^2$ is the intensity of the diffracted object field over the hologram plane P_e and it is assumed that the plane reference wave has unit amplitude and zero phase over the object plane P_0.

In the second step of Gabor's wavefront imaging scheme, the recorded hologram defined by the intensity distribution I_e of Eq. (3) is used to reconstruct the scattered object wave ψ. In Gabor's original work, the hologram was recorded on a high-resolution photographic plate and the reconstruction was performed experimentally by illuminating the developed photographic plate (hologram) with the same plane wave used in the first part of the procedure. In its modern form, the reconstruction is performed on a digital computer by *back propagating* the measured intensity distribution I_e (see discussion in Sect. 9.3.2 below). In either case, the resulting reconstruction has three components corresponding to the three terms comprising the hologram intensity distribution in Eq. (3). The sum of the intensities of the plane wave and scattered object wave $1 + I_0$ generate a direct (dc) beam that has uniform phase over the hologram plane and is thus an intensity-modulated plane wave that contains no object information. The other two terms generate virtual and real images of the object (so-called twin images). Of course all three image fields are superposed and in Gabor's original scheme it was not possible to untangle these terms, so the quality of the reconstructed holographic image suffered.

9.3.1 Phase-shifting holography

In the modern version of the holographic system illustrated in Fig. 2, the hologram is recorded digitally and the reconstruction is performed computationally on a digital computer. The use of a digital camera to record the hologram not only allows the image fields to be generated digitally but also introduces the possibility of removing the interfering dc field and one of the twin image fields discussed above. This has been accomplished [20] using *phase-retrieval algorithms* [7,8] and, more recently, using *phase-shifting holography* [3–5], an idea that goes back to earlier work by Gabor himself [6]. In this paper we will restrict our attention to PSH since, at the present time, it appears to offer the greatest promise for digital microscopy.

PSH is performed using the modern Gabor holographic setup illustrated in Fig. 2, where the hologram is recorded using the CCD array of a monochrome digital camera placed over the hologram plane (entrance pupil) P_e. Unlike conventional Gabor holography, a total of four separate intensity measurements are performed in PSH corresponding to (a) the intensity of the plane reference wave itself (taken to be unity here for convenience), (b) the intensity of the object scattered field alone I_0, (c) a conventional Gabor hologram formed from the interference of the object scattered wave with a plane reference wave, and (d) a conventional Gabor hologram formed from the interference of the object scattered wave with a quarter wave shifted plane reference wave. The two conventional Gabor holograms record the two interference patterns:

$$I_e^{(1)}(\boldsymbol{\rho}) = 1 + I_0(\boldsymbol{\rho}) + e^{-ikd}\psi(\boldsymbol{\rho},d) + e^{ikd}\psi^*(\boldsymbol{\rho},d) \qquad (4a)$$

$$I_e^{(2)}(\boldsymbol{\rho}) = 1 + I_0(\boldsymbol{\rho}) - ie^{-ikd}\psi(\boldsymbol{\rho},d) + ie^{ikd}\psi^*(\boldsymbol{\rho},d), \qquad (4b)$$

where the second hologram employs a quarter wave plate in the upper leg of the Mach–Zehnder interferometer to generate the $\pi/2$ phase shifts in the interference terms. Because the reference plane-wave intensity and object scattered-field intensities are known (measured), the above two equations can be solved simultaneously for the object field $\psi(\boldsymbol{\rho},d)$ boundary value over the hologram plane P_e. This quantity is then used to computationally generate the image of the object as discussed below.

9.3.2 Digital reconstruction

The interaction of the illuminating plane wave with the sample located in the lower-leg of the Mach–Zehnder interferometer generates the boundary value of the object field $\psi_0(\boldsymbol{\rho}')$ over the plane P_0 situated immediately to the right of the sample. This boundary value field then generates the diffracted object field $\psi(\boldsymbol{\rho},z)$ that propagates away from P_0 to the entrance pupil plane P_e, where its boundary value over this plane $\psi(\boldsymbol{\rho},d)$ is determined using PSH as outlined above. The goal of microscopy is to use this measured boundary value field to generate an image of the object field $\psi_0(\boldsymbol{\rho}')$, which, for an optically thin sample, is directly related to the material properties of the sample via its transmittance function $T(\boldsymbol{\rho}')$, as discussed in Sect. 9.2.[¶] For a conventional microscope employing lenses, the imaging step is performed automatically by the lenses and other optical components of the microscope, while for a fully digital microscope, which is of interest here, it is necessary to employ a reconstruction algorithm to digitally perform the imaging. This

---

[¶] For thick samples, such images are not directly useful and the methods of diffraction tomography [14,15] have to be employed, as outlined in Sect. 9.4.

is done by inverting (or back propagating) [9,10] the boundary value field $\psi(\boldsymbol{\rho},d)$ determined by PSH so as to recover the object field ψ_0 over the object plane P_0.

The back propagation of the object wave from its boundary value $\psi(\boldsymbol{\rho},d)$ on the plane P_e to the object plane P_0 is best performed using a plane-wave expansion [12,13]. The plane wave representation of the diffracted object wavefield leads to a theory of imaging for the digital microscope that is closely connected with Abbe's theory of image formation in the conventional optical microscope [12,13]. In particular, in Abbe's theory of image formation the field diffracted from the object is decomposed into a superposition of plane waves that propagate away from the sample into the entrance pupil of the microscope. The job of the imaging optics of the microscope is to then combine these various plane waves into an image of the object field ψ_0.

The plane wave expansion of the diffracted object field as it propagates away from the object toward the entrance pupil P_e can be expressed in the form [12,13]

$$\psi(\boldsymbol{\rho},z) = \frac{1}{(2\pi)^2} \int_{-\infty}^{\infty} d^2\kappa, \tilde{\psi}_0(\boldsymbol{\kappa})e^{i(\boldsymbol{\kappa}\cdot\boldsymbol{\rho}+\gamma z)}, \tag{5a}$$

where

$$\tilde{\psi}_0(\boldsymbol{\kappa}) = \int d^2\rho\psi_0(\boldsymbol{\rho})e^{-i\boldsymbol{\kappa}\cdot\boldsymbol{\rho}} \tag{5b}$$

is the spatial Fourier transform of the object field

$$\psi_0(\boldsymbol{\rho}) = \psi(\boldsymbol{\rho},z)|_{z=0},$$

with $\boldsymbol{\kappa} = (\kappa_x, \kappa_y)$ being the two-dimensional spatial transform vector conjugate to the space vector $\boldsymbol{\rho}$. The quantity γ in the plane-wave expansion Eq. (5b) is given by

$$\gamma = \begin{cases} \sqrt{k^2 - \kappa^2} & \text{if } \kappa \leq k \\ i\sqrt{\kappa^2 - k^2} & \text{if } \kappa > k \end{cases}, \tag{5c}$$

where $k = 2\pi/\lambda$ is the field wavenumber with λ being its wavelength.

A key ingredient of Abbe's treatment of the microscope is the observation that the finite size of the entrance pupil of the microscope limits the range of plane waves that enter and are processed by the microscope. In particular, only those plane waves whose transverse wavenumbers κ_x, κ_y obey the inequality

$$\kappa = \sqrt{\kappa_x^2 + \kappa_y^2} \leq k\sin\theta_0 \approx k\frac{a}{d} \tag{6}$$

will enter the entrance pupil where θ_0 is the half-angle subtended by the entrance pupil as measured from the center of the object plane P_0 and a/d the numerical aperture of the entrance pupil.[||] On making use of this result we conclude that the object field that is actually processed in PSH is not given by the full plane-wave expansion Eq. (5a) but, rather, this plane-wave expansion truncated to plane waves whose transverse wavenumbers satisfy the inequality Eq. (6). Thus, in particular, PSH will yield the boundary value

$$\hat{\psi}(\boldsymbol{\rho},d) = \frac{1}{(2\pi)^2} \int_{\kappa \leq k\frac{a}{d}} d^2K\,\tilde{\psi}_0(\boldsymbol{\kappa})e^{i(\boldsymbol{\kappa}\cdot\boldsymbol{\rho}+\gamma d)}, \tag{7a}$$

where the hat is used to denote the field boundary value determined using PSH with a finite entrance pupil and a numerical aperture a/d.

The boundary value Eq. (7a) can be Fourier inverted to yield $\tilde{\psi}_0(\boldsymbol{\kappa})$ over the spatial frequency band $|\boldsymbol{\kappa}| = \kappa \leq ka/d$; that is to say

$$\tilde{\psi}_0(\boldsymbol{\kappa}) = e^{-i\gamma d} \int d^2\rho\,\hat{\psi}(\boldsymbol{\rho},d)e^{-i\boldsymbol{\kappa}\cdot\boldsymbol{\rho}}$$

$$= e^{-i\gamma d}\tilde{\hat{\psi}}(\boldsymbol{\kappa},d), \quad \kappa \leq k\frac{a}{d}, \tag{7b}$$

where $\tilde{\hat{\psi}}(\boldsymbol{\kappa},d)$ is the Fourier transform of the boundary value field over the hologram plane P_e. If Eq. (7b) is now substituted back into the plane-wave expansion [Eq. (5a)], a band-limited approximation of the object-diffracted wave $\hat{\psi}(\boldsymbol{\rho},z)$ is obtained that is valid throughout the region $0 \leq z \leq d$ lying between, and including, the object and hologram planes P_0 and P_e:

$$\hat{\psi}(\boldsymbol{\rho},z) = \frac{1}{(2\pi)^2} \int_{\kappa \leq k\frac{a}{d}} d^2K\,\tilde{\hat{\psi}}(\boldsymbol{\kappa},d)e^{i[\boldsymbol{\kappa}\cdot\boldsymbol{\rho}+\gamma(z-d)]}. \tag{7c}$$

The "image" of the object field $\hat{\psi}_0(\boldsymbol{\rho})$ is finally obtained by evaluating the back-propagated field [Eq. (7c)] over the object plane P_0 (at $z = 0$),[**] i.e.,

$$\hat{\psi}_0(\boldsymbol{\rho}) = \frac{1}{(2\pi)^2} \int_{\kappa \leq k\frac{a}{d}} d^2K\,\tilde{\hat{\psi}}(\boldsymbol{\kappa},d)e^{i(\boldsymbol{\kappa}\cdot\boldsymbol{\rho}-\gamma d)}. \tag{7d}$$

[||] This, in turn, leads directly to the resolution limit quoted earlier of $\lambda/(a/d)$ where a/d is the numerical aperture of the entrance pupil.

[**] For an optically thin object characterized by a transmittance function $T(\boldsymbol{\rho})$ the image field is, in fact, a coherent image of the transmittance function.

9.3.2.1 Resolution and equivalent CPSF

The digital holographic microscope described in the preceding section can be characterized in terms of a coherent point spread function exactly of the form Eq. (1a), describing the imaging performance of a conventional microscope. In particular, on substituting the definition of $\tilde{\hat{\psi}}(\boldsymbol{\kappa},d)$ in terms of $\tilde{\psi}_0(\boldsymbol{\kappa})$ from Eq. (7b) into Eq. (7d), we obtain the result

$$\hat{\psi}_0(\boldsymbol{\rho}) = \frac{1}{(2\pi)^2} \int_{\kappa \leq k\frac{a}{d}} d^2K \left\{ \int d^2\rho' \psi_0(\boldsymbol{\rho}')e^{-i\boldsymbol{\kappa}\cdot\boldsymbol{\rho}'} \right\} e^{i\boldsymbol{\kappa}\cdot\boldsymbol{\rho}}$$

$$= \int d^2\rho' \psi_0(\boldsymbol{\rho}')h(\boldsymbol{\rho} - \boldsymbol{\rho}'), \tag{8a}$$

where

$$h(\boldsymbol{\rho} - \boldsymbol{\rho}') = \frac{1}{(2\pi)^2} \int_{\kappa \leq k\frac{a}{d}} d^2K \, e^{i\boldsymbol{\kappa}\cdot(\boldsymbol{\rho}-\boldsymbol{\rho}')}. \tag{8b}$$

By making the change of integration variables $\boldsymbol{\kappa} = (k/d)(\xi,\eta)$, Eq. (8b) can be expressed in the identical form as Eq. (1a), with the constant $C = k^2/(2\pi d)^2$, and where the pupil function $\mathcal{P}(\xi,\eta)$ is unity if $\xi^2 + \eta^2 \leq a^2$ and zero otherwise. Thus, in terms of its coherent imaging performance the digital holographic microscope is equivalent to an isoplanatic, aberration-free, diffraction-limited microscope of an equal numerical aperture. However, unlike a conventional microscope that operates as an incoherent imaging system, the holographic microscope is a coherent imaging system and, thus, has the capability of imaging phase objects.

9.3.3 Examples

In this section we present two examples comparing the imaging performance of DHM with digital Gabor holography and with conventional white light imaging. The experimental setup for DHM is shown in Fig. 2. A polarized He-Ne laser with a wavelength of 633 nm and power of 5 mW was used as the light source. The phase shift of the reference wave in PSH was obtained by placing a quarter wave plate in the upper arm of the Mach–Zehnder interferometer illustrated in the figure. We used a lensless 10-bit-per-pixel monochromatic CCD camera having a maximum resolution of 1024×768 pixels, with each pixel being 6.7 μm square. The camera's CCD sensor array was positioned at about 44 mm from the object. Four intensity measurements consisting of (1) the intensity of the reference wave alone, (2) the intensity of the object diffracted wave alone, (3) a Gabor hologram, and (4) a second Gabor hologram using a quarter wave shifted-reference wave were performed.

The first example used an object consisting of a 50-μm width slit and compared the performance of conventional digital Gabor holography with DHM as described above. The results are shown in Fig. 3. The conventional holographic image shown in the top of this figure was obtained by back propagating the intensity distribution corresponding to the first of the two Gabor holograms acquired in the experiment. The back propagation was performed using Eq. (7d) with $\tilde{\hat{\psi}}(\boldsymbol{\kappa},d)$ replaced by

$$\tilde{I}_e^{(1)}(\boldsymbol{\kappa}) = \int d^2\boldsymbol{\rho}\, I_e^{(1)}(\boldsymbol{\rho})e^{-i\boldsymbol{\kappa}\cdot\boldsymbol{\rho}},$$

where $I_e^{(1)}$ is defined in Eq. (4a). Because the back propagation operation is linear, a total of three component images were generated [see discussion under Eq. (3)] corresponding to a dc image, a real image, and a virtual image. The dc and virtual image components are clearly displayed in the figure and represent background noise that obscures and degrades the sought-after real image of the slit. The DHM image of the slit is shown in the bottom of the figure. The degrading dc and virtual image fields are clearly gone.

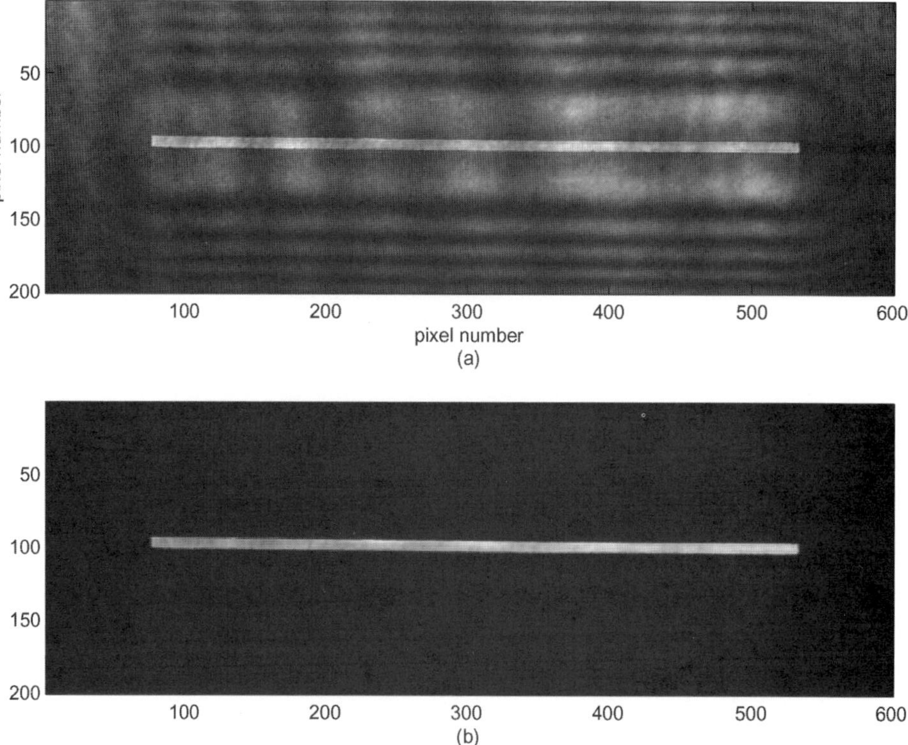

Figure 3 Reconstructed intensity images of a slit (a) using conventional Gabor holography; (b) using HDM.

The second example used an onion cell as the object and illustrates the ability of DHM to generate images of both the amplitude (intensity) and the phase of 2D objects. The DHM images were obtained using the same procedure outlined above and employed in generating the bottom image of Fig. 3. Figure 4(a) is the intensity image of the onion cell obtained using DHM and (b) is the phase image. In Fig. 4(c) we show an intensity image obtained from a conventional white light microscope. The reconstructed intensity image obtained using DHM is seen to agree closely with the image acquired by the light microscope. However, the phase image provides additional information about the onion cell's internal structure that is not available in the conventional white light image.

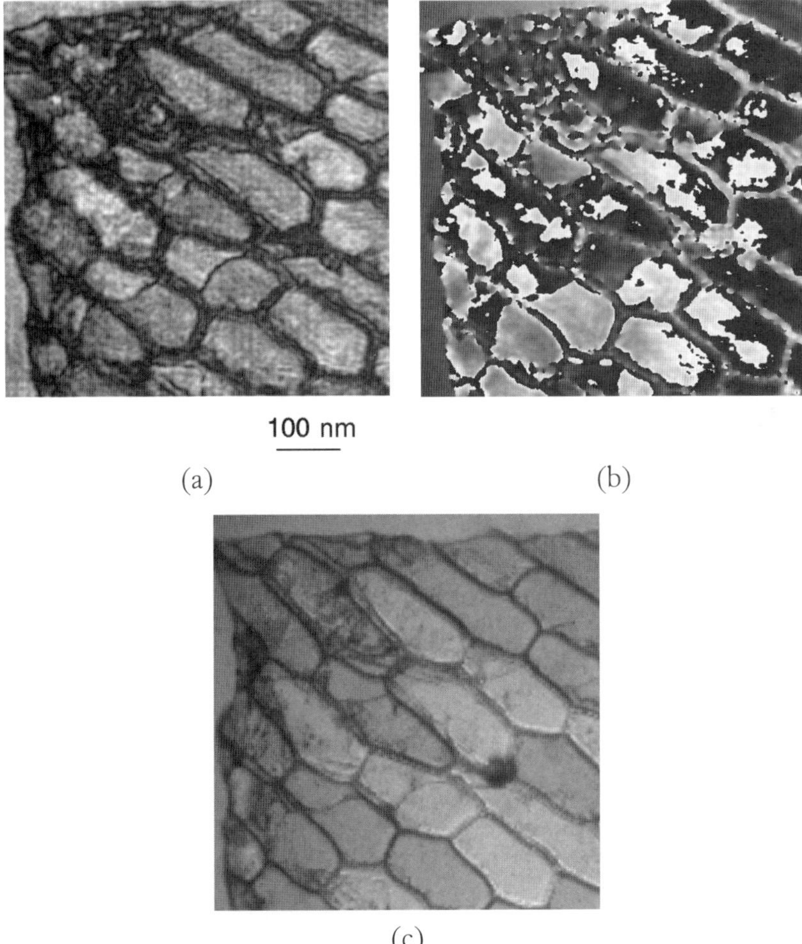

Figure 4 (a) Reconstructed intensity image of onion cells by DHM; (b) reconstructed phase image of onion cells by DHM; (c) image of onion cells viewed under a conventional white light microscope.

9.4 3D Microscopy

The problem posed by 3D (optically thick) objects is that a single coherent image generated by a digital microscope (or any other microscope for that matter) is neither simply nor unambiguously related to the internal 3D structure of the object as defined, for example, by its complex index of refraction distribution[††] $n(\mathbf{r})$. In particular, for a weak scatterer assumed to be centered on the optical axis behind the object plane P_0, the object field and 3D object are related by an integral equation of the general form [14]

$$\psi_0(\boldsymbol{\rho}) = \frac{k^2}{2\pi} \int d^3r'\, \delta n(\mathbf{r}')e^{ikz'} G_0(\boldsymbol{\rho} - \mathbf{r}'), \qquad (9)$$

where $\delta n(\mathbf{r}') = [n(\mathbf{r}') - 1]$ is the deviation of the 3D index of refraction distribution of the object from free space and G_0 is the free space Green function

$$G_0(\mathbf{R}) = \frac{e^{ikR}}{R},$$

with $R = |\mathbf{R}|$. Equation (9) can be viewed as a transformation from the 3D complex-valued function $\delta n(\mathbf{r})$ to the 2D object field $\psi_0(\boldsymbol{\rho})$. It is apparent from simply counting degrees of freedom that a coherent image of the object field $\hat{\psi}_0$ (which has two degrees of freedom) will not suffice to uniquely determine $\delta n(\mathbf{r})$ which, of course, has three degrees of freedom.

The situation described above is completely analogous to what occurs in computed tomography (CT) [18], where a single tomographic measurement of a 3D object is related to the object via a projection, which, like Eq. (9), is an integral transform of a function that describes the 3D internal structure of the object. Indeed, in the limit where the wavelength λ of the incident field used in the digital microscope is much smaller than the scale at which the internal structure of the object varies [Eq. (9)] and reduces to [21]

$$\psi_0(\boldsymbol{\rho}) \approx C \int dz'\, \delta n(\boldsymbol{\rho}, z'),$$

where C is an unessential constant. The above equation defines a tomographic projection of δn onto the object plane P_0 and, as in CT, a complete determination of the index of refraction distribution in this short wavelength limit then requires that

---

[††] We remark that although a holographic image is three-dimensional, it has, in fact, only two degrees of freedom corresponding to the two degrees of freedom of the hologram recording plane. Thus, a coherent image of a 3D semitransparent object contains, in fact, very little information about the internal structure of the object (see Sect. 9.4.1).

a number[‡‡] of such projections be available, with each projection corresponding to a different orientation of the object relative to the optical axis of the microscope. This can be accomplished by placing the 3D object on a gimbaled mount that can be rotated about both the x and y axes of the microscope, and then producing a set of coherent digital images of the object placed in varying orientations relative to the microscope optical axis. The actual determination of the index distribution $n(\mathbf{r})$ from this set of images (projections) can then be accomplished via a number of CT reconstruction algorithms, the most common being the filtered back projection algorithm [18].

In the more general case, where the wavelength cannot be assumed to be much smaller than the scale at which the object varies, it is still necessary to perform a number of experiments using, for example, the same experimental procedure outlined above. The only difference between this more general case and the short wavelength case is that the CT reconstruction algorithms can no longer be employed and the algorithms of diffraction tomography (DT) are required to process the complex image fields acquired in the sequence of digital images. Due to space limitations, we will not discuss in any detail the theory of DT here but refer the reader to the rather extensive literature on this subject (see, for example, Ref. [15] for a review of the theory of the subject). Also, we will not review early attempts at optical 3D coherent imaging that employed conventional CT algorithms or ad-hoc inversion methods that preceded DT. A rather extensive listing of this literature can be found in Ref. [23]. It is worthwhile, however, to briefly outline the steps required to generate a reconstruction of a 3D object from a set of coherent images using the so-called filtered back propagation algorithm [14,22] which is the DT generalization of the filtered back projection algorithm of CT [18].

9.4.1 Filtered back propagation algorithm

Let us assume that a 3D semitransparent object is mounted in such a way that it can achieve varying orientations relative to the optical axis of a digital holographic microscope such as illustrated in Fig. 2, and a sequence of entrance pupil boundary value fields are determined using, for example, PSH as discussed in Sect. 9.3.1. We can regard the object as being fixed (nonrotating) in space and the microscope as rotating around the object, with its optical axis (and incident plane wave direction) defined by the unit vector $\mathbf{s}_0$, which, in the ideal situation, can completely cover the unit sphere. For any given relative orientation between the object and microscope we can form a partial image of the 3D object by simply back propagating the entrance pupil boundary value field into the space region occupied by

---

[‡‡] Theoretically, an infinite number of projections are required for complete determination. Of course, in practice only a limited number are employed, which then yield an approximation to the $n(\mathbf{r})$.

the object. This back-propagated field corresponds, in fact, to the virtual image field that is generated in a holographic reconstruction procedure as discussed in Sect. 9.3.1 and is computed from the entrance pupil boundary value field using Eq. (7c). By then superimposing a set of these 3D holographic images constructed from different object/microscope orientations (as defined by the unit vector $\mathbf{s}_0$), a composite image is generated that encompasses many "views" of the object in a manner reminiscent of computed tomography [18].

The imaging procedure described above fails to account for the fact that the "source" of the diffracted object field is not the object's index of refraction deviation $\delta n(\mathbf{r})$ alone, rather it is the product of this function with the amplitude of the incident plane wave[*] $\exp(ik\mathbf{s}_0 \cdot \mathbf{r})$. Thus, any partial image formed by back propagating the diffracted object field as measured over the microscope entrance pupil will generate an image of the quantity

$$F(\mathbf{r}) = \delta n(\mathbf{r})e^{ik\mathbf{s}_0 \cdot \mathbf{r}}$$

and, hence, must be multiplied by $\exp(-ik\mathbf{s}_0 \cdot \mathbf{r})$ to isolate out the complex index of refraction deviation $\delta n(\mathbf{r})$.

A second problem with the above-described imaging scheme is that it fails to account for overlap of information from different "views" of the object. In order to remove this redundancy, it is necessary to first convolutionally filter the measured pupil boundary value fields using a universal filter that is independent of the object prior to the back-propagation step. The filtering operation corresponds to a simple multiplication in spatial frequency space so that each partial image is given by an expression of the form

$$\hat{\delta n}(\mathbf{r}; \mathbf{s}_0) = \frac{e^{-ik\mathbf{s}_0 \cdot \mathbf{r}}}{(2\pi)^2} \int_{K \leq k\frac{a}{d}} d^2K\, \tilde{H}(\boldsymbol{\kappa})\tilde{\hat{\psi}}(\boldsymbol{\kappa}, d; \mathbf{s}_0)e^{i[\boldsymbol{\kappa}\cdot\boldsymbol{\rho}+\gamma(z-d)]}, \qquad (10)$$

where $\tilde{H}(\boldsymbol{\kappa})$ is a universal filter, and where the Cartesian coordinates $(\boldsymbol{\rho}, z)$ of the vector position $\mathbf{r}$ are relative to the (rotating) coordinate system of the microscope. Finally, by linearly superimposing a full set of such partial images one can generate the composite image

$$\hat{\delta n}(\mathbf{r}) = \int_{4\pi} d\Omega_{s_0} \hat{\delta n}(\mathbf{r}; \mathbf{s}_0)$$

$$= \frac{1}{(2\pi)^2} \int_{4\pi} d\Omega_{s_0} e^{-ik\mathbf{s}_0 \cdot \mathbf{r}} \int_{K \leq k\frac{a}{d}} d^2K \tilde{H}(\boldsymbol{\kappa})\tilde{\hat{\psi}}(\boldsymbol{\kappa}, d; \mathbf{s}_0)e^{i[\boldsymbol{\kappa}\cdot\boldsymbol{\rho}+\gamma(z-d)]}. \qquad (11)$$

---

[*] This is clear from Eq. (9) and is in complete agreement with the case of a thin object where the object field ψ_0 is the product of the transmittance function with the amplitude of the illuminating plane wave on the object plane (cf., discussion in Sect. 9.2).

The algorithm Eq. (11) is the filtered back-propagation (FBP) algorithm of diffraction tomography [14,22]. Its name derives from the sequence of steps required to generate the 3D image (reconstruction): (1) convolutional filtering of the data, (2) back propagation of the filtered data, and (3) summation over "views," where each "view" of the object corresponds to a different plane wave direction $\mathbf{s}_0$. It is important to note that the back-propagation step (2) is performed relative to the rotating microscope Cartesian-coordinate system. It is necessary to interpolate the resulting partial image onto the fixed Cartesian-coordinate system of the object in order to construct the composite image Eq. (11), i.e., prior to summing over views. A detailed derivation of the FBP algorithm is presented in Ref. [14] and its computer implementation is presented in Ref. [22].

Coherent imaging of 3D weakly scattering objects using DT has been performed by a number of workers. The first work was performed using phase retrieval techniques [7,8] applied to intensity scans obtained using a scanning photodetector [23,24]. Later work employed a monochrome digital camera, but again used phase retrieval rather than PSH to deduce the phase of the optical field from its measured intensity distribution [25]. An excellent example of the use of DT in this class of applications is given in Ref. [26] and is reproduced in the latest edition of Born and Wolf (Ref. [12], p. 716).

9.5 Concluding Remarks

We have presented a general description of digital holographic microscopy but we have not delved into the practical and engineering details of DHM systems nor explored their potential use in industrial, medical, and biological applications. Although DHM has a number of attractive attributes such as simplicity and low cost, it also suffers from some practical deficits such as pixel depth (in bits per pixel) and size, and the (limited) array size of current state-of-the-art CCD arrays, all of which translate into degraded performance of the DHM. Also, in order to make such systems commercially viable the entire process of acquiring the multiple holograms required by phase-shifting holography (PSH) has to be efficiently automated so that the microscope can generate images in real or, at least, near real time. Although this is probably easily accomplished for 2D samples, it presents a much more formidable barrier for 3D objects where multiple "views" of the object are required, corresponding to multiple orientations of the object relative to the optical axis of the microscope.

In addition to the practical considerations outlined above, we have not discussed in any depth the details of the diffraction tomographic (DT) reconstruction algorithms without which 3D DHMs cannot function. In order to produce high-quality and quantitative reconstructions of 3D objects, these algorithms must account for diffraction within the object and also must account for multiple scattering between the object and the physical cell in which they are placed. Typically, this cell is a test tube filled with an index-matching fluid [20,23], and the multiple scattering between the test tube and sample has not, as yet, been adequately taken into account in the DT reconstruction algorithms.

Finally, we mention that other alternatives exist for implementing a DHM. First, phase-retrieval algorithms [7,8] can be used to deduce the complex amplitude of a diffracted object field from multiple intensity measurements without the use of PSH [20,23,25]. An alternative approach was suggested by Wolf [27] that employs a single off-axis reference beam hologram, rather than a Gabor hologram, to deduce low spatial frequency components of the diffracted object field. In the case of 3D objects it is also possible to use single, Gabor (digital) holograms [25] since the degrading effects of the zero-order and real image fields are reduced significantly in the 3D reconstruction process [28]. Finally, it is possible to use PSH in conjunction with a lens-based optical imaging system to generate high-quality coherent images of 2D amplitude, phase objects, or 3D highly reflecting objects [4]. The advantage of this latter procedure is that it can yield high numerical apertures for limited-sized CCD arrays but has a number of other disadvantages such as cost of the optics, increased speckle noise, and limited application to 3D semitransparent objects.

References

1. D. Gabor, "Microscopy by reconstructed wavefronts," *Proc. R. Soc. London Ser. A* **197**, 454 (1949).
2. D. Gabor, "Microscopy by reconstructed wavefronts: II," *Proc. Phys. Soc. London Ser. B* **64**, 449 (1951).
3. S. Lai, B. King, and M.A. Neifeld, "Wave front reconstruction by means of phase-shifting digital in-line holography," *Opt. Commun.* **173**, 155 (2000).
4. T. Zhang and I. Yamaguchi, "Three-dimensional microscopy with phase-shifting digital holography," *Opt. Lett.* **23**, 1221 (1998).
5. I. Yamaguchi, J.-i. Kato, S. Ohta, and J. Mizuno, "Image formation in phase-shifting digital holography and applications to microscopy," *Appl. Opt.* **40**, 6177 (2001).
6. D. Gabor and W.P. Goss, "Interference microscope with total wavefront reconstruction," *J. Opt. Soc. Am.* **56**, 849 (1966).
7. R.W. Gerchberg and W.O. Saxton, "A practical algorithm for the determination of phase from image and diffraction plane pictures," *Optik* **35**, 237 (1972).
8. R.A. Gonsalves, "Phase retrieval from modulus data," *J. Opt. Soc. Amer.* **66**, 961 (1976).

9. J.R. Shewell and E. Wolf, "Inverse diffraction and a new reciprocity theorem," *J. Opt. Soc. Amer.* **58**, 1596 (1968).

10. G. Sherman, "Integral transform formulation of diffraction theory," *J. Opt. Soc. Amer.* **57**, 1490 (1967).

11. E. Lalor, "Inverse wave propagator," *J. Mathematical Phys.* **9**, 2001 (1968).

12. M. Born and E. Wolf, *Principles of Optics*, Cambridge University Press, Cambridge, U.K. (1999).

13. J.W. Goodman, *Introduction to Fourier Optics*, McGraw-Hill, New York (1968).

14. A.J. Devaney, "A filtered backpropagation algorithm for diffraction tomography," *Ultrasonic Imaging* **4**, 336 (1982).

15. E. Wolf, "Principles and development of diffraction tomography," in *Trends in Optics*, A. Consortini, Ed., Academic Press, New York (1996).

16. E. Wolf, "Three-dimensional structure determination of semi-transparent objects from holographic data," *Opt. Commun.* **1**, 153 (1969).

17. B. Masters, "Three-dimensional confocal microscopy of the human optic nerve in vivo," *Optics Express* **3**, 356 (1998).

18. G.T. Herman, *Image Reconstruction from Projections: The fundamentals of Computerized Tomography*, New York, Adademic Press (1980).

19. E.G. Williams and J.D. Maynard, "Holographic imaging without the wavelength resolution limit," *Phys. Rev. Lett.* **45**, 554 (1980).

20. M. Maleki and A.J. Devaney, "Non-iterative reconstruction of complex valued objects from two intensity measurements," *Opt. Eng.* **33**, (1994).

21. A.J. Devaney, "Inverse scattering theory within the Rytov approximation," *Optics Letts.* **6**, 374 (1981).

22. A.J. Devaney, "A computer simulation study of diffraction tomography," *IEEE Trans. Biomed. Eng.* **BME-30**, 377 (1982).

23. A.J. Devaney, M. Maleki, and A. Schatzberg, "Tomographic reconstruction from optical scattered intensities," *J. Opt. Soc. Am. A* **9**, 1356 (1992).

24. T.C. Wedberg and J.J. Stamnes, *J. Opt. Soc. Amer. A* **12**, 493 (1995).

25. M. Maleki and A.J. Devaney, "Phase retrieval and intensity-only reconstruction algorithms for optical diffraction tomography," *J. Opt. Soc. Amer.* **10**, 1086 (1993).

26. T.C. Wedberg and W.C. Wedberg, *J. Microsc.* **11**, 53 (1995).

27. E. Wolf, "Determination of the amplitude and the phase of scattered fields by holography," *J. Opt. Soc. Amer.* **60**, 345 (1970).

28. A.J. Devaney, "Structure determination from intensity measurements in scattering experiments," *Phys. Rev. Letts.* **62**, 2385 (1989).

Wolf Anecdotes

When I was a graduate student with Emil we had a number of "excited technical conversations" that some would describe as "fights." On one of these occasions Joe Eberly who, at that time, had an adjoining office to Emil, came running into Emil's office thinking that we had physically attacked each other. Many years later Emil had a post doc, Yagin Lee, from China who was not accustomed to arguing with his professors or interacting with them in an "active" manner as is common between Emil and most of his students. During one of their early technical discussions Emil became somewhat "excited," whereupon Yagin became very upset, thinking that he had somehow offended Emil and would be sent back to China. Emil, seeing this response from Yagin assured him that everything was okay and that in America, and especially within his research group, this type of interaction was normal and not to be taken personally. He went on to mention the terrible sounding fights that he and I had while I was a graduate student, but that we were best of friends and see each other at least once every year. A few months later Emil mentioned to Yagin that I was going to be visiting him in a few weeks time. Upon learning this, Yagin blurted out: "Professor Wolf, can I watch you and Dr. Devaney have a fight?"

Upon his first visit to Schlumberger Doll Research Labs, where Professor Wolf was hired as a consultant, he was invited to lunch by the Director of the Laboratory and various Department heads. Unfortunately, I did not attend this luncheon due to a previous commitment and only have the following story secondhand from Emil himself. The luncheon was held at the *Elms Inn* in Ridgefield, Connecticut, which is an historic and well-known Inn and European restaurant. That particular day they were serving Wiener Schnitzel as a special entree. Now, anyone who knows Emil knows that this is absolutely his favorite dish, and that he will search for this item on the menu of every restaurant he visits. Unfortunately (for Emil), he, as the guest of honor at this luncheon, was given the privilege of ordering first. The problem was that the Wiener Schnitzel dish was very expensive and in order not to appear greedy Emil ordered some other less expensive item (probably a hamburger). To Emil's great dismay everyone else at the table then proceeded to order the Wiener Schnitzel. The Director of the lab even commented on how excellent the dish was and that Emil certainly should have gotten it rather than his hamburger. Emil made me take him to the Elms at least once on each of his many visits to Schlumberger after that day but, alas, Wiener Schnitzel never appeared again on the menu.

On one of his subsequent visits to Schlumberger, Emil and his wife Marlies, my wife Amy, and I had dinner at *Touchstones*, which is a bar/restaurant in Ridgefield that has decorated their dining room with wall-to-wall old books. Emil commented on how lousy these books were. He had heard that they bought them for about a penny a pound since no one wants them. A few days later Emil, some of our colleagues from Schlumberger, and I had lunch at this restaurant. Prior to our arrival I hid a copy of Born and Wolf among the "crappy" books lining the walls of the restaurant. During lunch I asked Emil to recite this story of his about the nature and value of the books on the wall. In Emil's typical fashion he gave a lively and spirited discourse on these valueless books, whereupon I reached up and pulled down Born and Wolf from the shelf immediately beside our table. You can guess the rest.

Emil and Marlies Wolf with Tony Devaney at the AFOSR annual meeting in San Antonio, Texas, 2003.

Anthony Devaney is a Professor of Electrical and Computer Engineering at Northeastern University in Boston, MA and a "Distinguished Professor" of the College of Engineering at Northeastern. Prof. Devaney received his Ph.D. from Institute of Optics at the University of Rochester in 1971, an M.S. in Engineering and Applied Science from Yale University in 1966, and a B.S. in Electrical Engineering from Northeastern University in 1964. Before joining Northeastern University in 1987, Prof. Devaney was with the Schlumberger Doll Research Center in Ridgefield, CT. Prof. Devaney is a member of Acoustical Society of America

Anthony J. Devaney (left) and Pengyi Guo.

and the Institute of Electronics Engineers, and is a Fellow of the Optical Society of America. He has been a chairman or cochairman of a large number of society meetings that include the SPIE as well as the Optical and Acoustical Societies of America. He has also organized a number of national and international symposiums. Prof. Devaney has been a member of the Editorial Board of *Ultrasonic Imaging* since 1984 and was Topical Editor of the *Journal of the Optical Society of America* from 1984–1986. He has been a feature editor for a special issue on *Inverse Problems in Propagation and Scattering, Journal of the Optical Society of America*, and a member of the board of editors for the journals *Wave Motion* (1988–1992), *Inverse Problems*, 1988–1993, and *Electronic Imaging*, 1988–1993.

Pengyi Guo received his B.S. in electronics and information systems in 1997 and his M.S. in Optical Engineering in 2000 from Nankai University, China. He then earned his M.S. and Ph.D. in electrical engineering in 2002 and 2004, respectively, from Northeastern University. He is a member of the Eta Kappa Nu honor society. His research interests include tomography, scattering, inverse problems, medical imaging, dosimetry, image reconstruction, holography, and optics.

CHAPTER 10

COLORED SHADOWS: DIFFRACTIVE-OPTICAL CROSS-CORRELATIONS IN THE HUMAN EYE: THE MISSING LINK BETWEEN PHYSICS AND PSYCHOLOGY, NEWTON AND GOETHE

Norbert Lauinger

The human eye is a powerful illuminant-adaptive trichromatic optical sensor. The phenomenon of colored shadows in twilight, described by Goethe and leading to a controversy among Newton and the community of physicists, shows the full circle of opponent and complementary colors adding to white. The colored shadows can be explained by von Laue interferences in the visible spectrum and diffractive-optical cross-correlations between global and local information in the diffractive-optical hardware of the human eye. They illustrate the most elementary—the spectral—transformations from physics into psychology in human vision. Scattering of global information in aperture space and diffraction of local information in image space of optical imaging systems leads to spectral/4D-spatio-temporal optical transformations into reciprocal grating space (Fourier/photoreceptor space) in the near field behind the retina. Three-dimensional (3D) grating-optical chromatic resonance—following von Laue's equation—is governing adaptation to varying illuminants. The rebalancing of RGB diffraction orders toward a new white norm by shifting the chromatic resonance guarantees color constancy in human vision. At physically unbalanced stages in twilight the human eye "does not see what physically is real at a shadow area, but what the eye optically has calculated": the hues of the colored shadows. The spectral transformations from physics into psychology, from the objective into the subjective world in vision, are based on wave/interference-optical transformations linking complementary and reciprocally interrelated worlds. Emil Wolf's prominent statements about the central role of von Laue's equation in optics have largely encouraged research on these modern aspects of physiological diffractive-optical correlators.

10.1 Introduction

Emil Wolf said in his introductory remarks to a Workshop on Physical Optics and Human Vision at the University of Rochester (June 21–22, 1993): "Now I know nothing about physiological optics, but I became interested in the possibility that some of the physical optics phenomena that we have been studying in Rochester—coherence effects and diffraction on three-dimensional gratings—might perhaps be relevant to problems of human vision." And "I will show that the von Laue's equations are coded into the field in all planes at any distance from the diffracting medium. This result follows from the basic equations of diffraction tomography ... . Diffraction tomography, holography, and the von Laue equations have a good deal in common."

The human eye traditionally is considered as a camera, imaging the visible world onto a flat 2D array of photoreceptors: the cones in daylight with photochemical RGB spectral brightness sensitivities (Fig. 1), and the rods at dim light

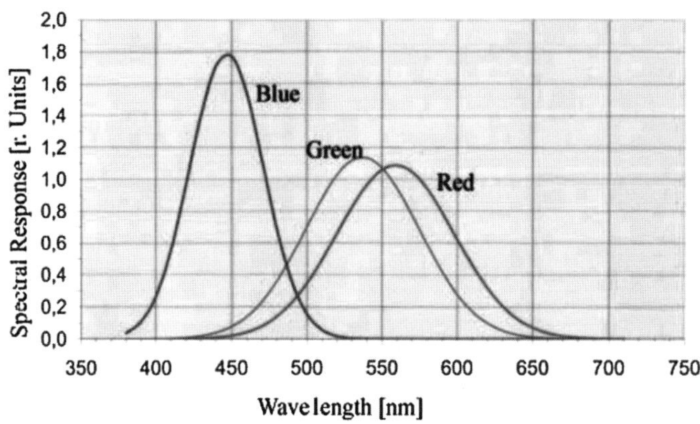

Figure 1 Color in human vision is based on physics, spectral intensity distributions in only one octave of the visible spectrum (380–760 nm). Sunlight registered under the atmosphere statistically varies around an equi-energy white spectrum, a "balanced state" in physics. Color in the human retina at photopic vision is processed on a trichromatic photochemical basis in RGB space. The RGB Gaussian spectral brightness sensitivity curves with relatively large half-widths are peaked at approx. λmax = 560 nmR, 540 nmG, 450 nmB (measurements of Wald/Rushton). At a white sensation, an RGB "balanced state" in psychology, the areae covered by the three Gaussians are identical.

with a single spectral brightness sensitivity curve. Each pixel in this model absorbs "local" information and each photoreceptor in a photochemical cascade transforms optical into electrical signals. The integration of local RGB triplets in color vision into coherent groups of data and the reduction of data into receptive fields (from 110 Mio rods and 6.5 Mio cones down to 1.1 million cables in the optic nerve) seem to be accomplished by neuronal nets in the associative layers of the retina of the human eye and at later cortical stages. Psychological phenomena in this model only start in the neuronal software programs of the retina and the cortex, based on electrical and chemical data processing. This interpretation of the human eye—and cortex—essentially is due to the histological heritage of Ramon y Cajal with Golgi staining of cortical tissue, making axons and dendrites visible and leaving nuclear (multi)layers and cellular phase gratings out of view.

10.2 Diffractive-Optical Hardware of the Human Eye: The Basis for Spectral Transformations, Cross-Correlations, and Adaptations in Color Vision

A closer histological study of the prenatal development of the optical hardware of the human eye (Fig. 2) [1,2] has shown that aperiodic diffractive-optical

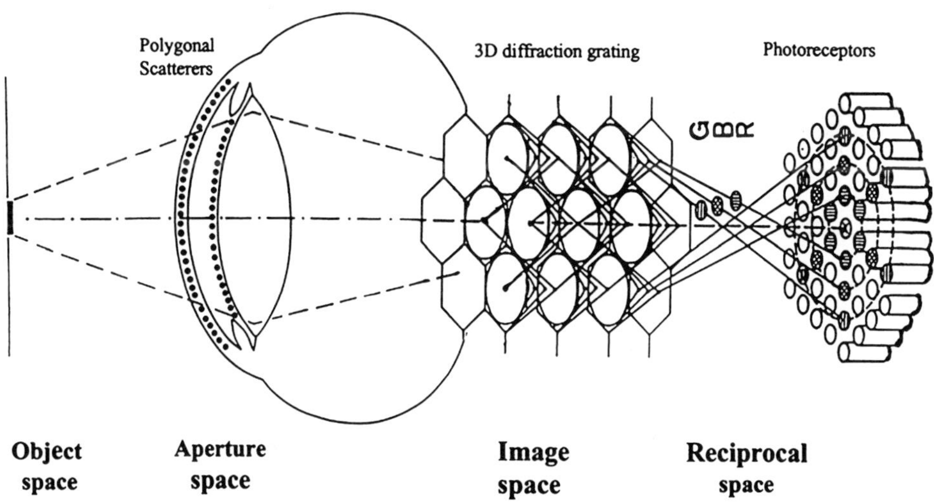

Figure 2 Diffractive-optical hardware of the human eye with diffusely scattering polygonal gratings in aperture space (where global information about objects and illuminants is available), and with a retinal hexagonal 3D diffraction-grating in image space (with local information about objects), where light is considered to become diffracted into three—RGB—diffraction orders (DOs) in reciprocal space. In the near field behind the retina, the trichromatic RGB signals (von Laue interference maxima) are located in concentric zones and absorbed by photoreceptors. (Schematic drawing not to scale.)

cellular multilayers—together with a subwavelength periodic 3D grating in the cornea [3]—are inserted into aperture space and a periodic diffractive-optical cellular 3D grating into image space (Figs. 3–4).

Fully transparent for incident light, they all represent optical phase gratings. The cellular multilayers located in image space are positioned lightward before the photoreceptors. The photoreceptors themselves are positioned in the near field behind the retinal 3D grating, in Fourier/Fresnel, or "reciprocal grating" space. Together they form the so-called inverted retina design of the human eye. Optical grouping of the data can lead to the reduction of physical data in the single octave of visible light into concentric RGB triplets in reciprocal space, and to orientational tuning of the information available to pixels (Fig. 5).

The spectral transformations at this diffractive-optical hardware in color vision (from intensity and wavelength in physics into brightness, hue, and saturation of colors in psychology) can be described by three main operations, together representing the missing link between the outer and inner visible world, physics and

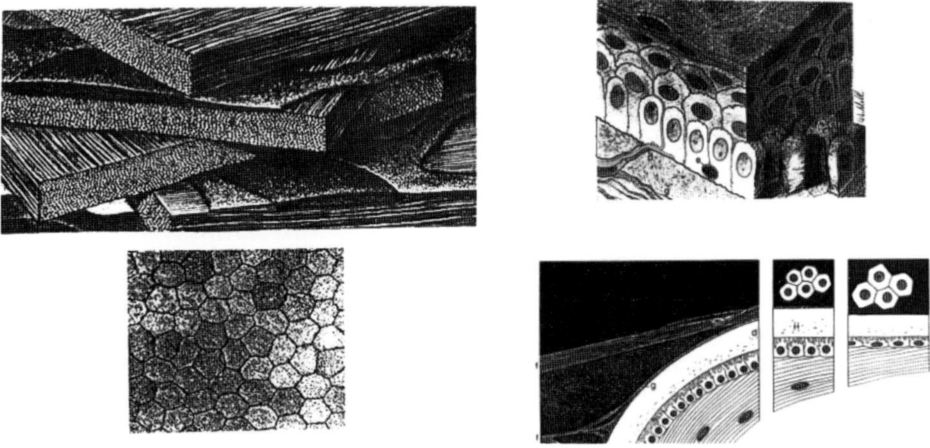

Figure 3 Polygonal diffractive-optical cellular layers in aperture space of the human eye: the fibrillar 3D cornea grating (above left) contains about 200–250 lamellae that are superimposed on each other, each having a thickness of 2.0 μm; the collagen fibrils (with diameters of 320–360 Å) of the stromal lamellae form a three-dimensional array of diffraction gratings. [Reprinted from D.M. Maurice, "The structure and transparency of the cornea," *J. Physiol.* **136**, 262–286 (1957). Copyright (1957) by Blackwell Publishing.] The multilayer polygonal cornea epithelium (above right) has a three-dimensional cellular structure with 5–6 overlapping layers, with cell-to-cell spacing of 10–30 μm. The cornea endothelium (below left) has a single layer of polygonal cells with a spacing of 18–20 μm. The lens epithelium (below right) has a single layer of polygonal cells with a spacing of 5–17 μm. [Reprinted from M. Carbon, N. Lauinger, and J. Schwab, "Self-imaging of three-dimensional phase gratings," *Pure Appl. Opt.* **7**, 1103–1120 (1998). Copyright (1998) by Institute of Physics Publishing.]

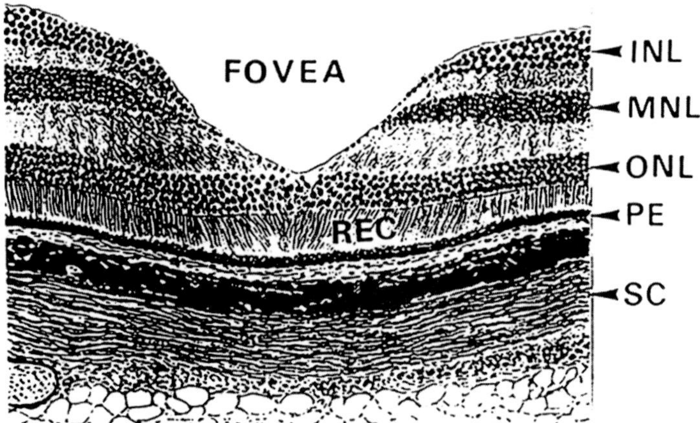

Figure 4 Hexagonally densest packed 3D diffraction grating in the "inverted" retina of the human eye (fovea region). Histological retina section from a rhesus monkey, prepared by hematoxylin-eosin coloration [5]. REC = photoreceptors; INL = inner; MNL = middle; ONL = outer nuclear layer; PE = pigment epithelium; SC = sclera. Light incidence from above. Cell nuclei and cell bodies are shifted sideways in the INL and MNL only in the center of the fovea. Thus, in the so-called foveola only 5–6 layers of cell nuclei in the ONL are in front of the cones, which are oriented toward the incident light.

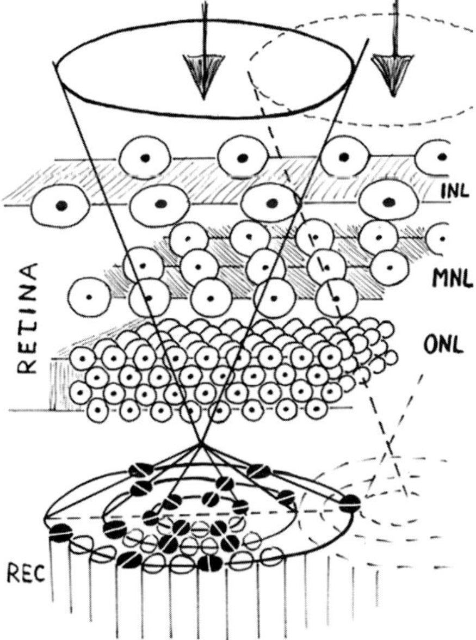

Figure 5 Optical grouping of local data by hierarchically packed hexagonal nuclear layers with decreasing grating constants: INL (greater) and MNL (smaller grating constants) and ONL (smallest grating constants) together realize the optical grouping of three RGB-DOs with orientational tuning of pixel information. Optical grouping can be considered as the basis of receptive-field formation. REC = photoreceptors located in reciprocal grating space.

psychology, Newton and Goethe. All three operations are needed to fully explain the phenomenon of colored shadows.

10.2.1 Diffractive-optical transformation of data in image space into local triplets of RGB diffraction orders in reciprocal space

This transformation can be realized by a small number of layers of hexagonally densest packed cellular elements in the retinal 3D grating and is described by von Laue's equation in crystal optics (Table 1). The three Gaussian transmission curves of the diffraction orders (DOs) for perpendicular incident white light are centered to the RGB $-\lambda$max triplet with $h_1 h_2 h_3 = 111 = 559$ nmR, $123 = 537$ nmG, $122 = 447$ nmB (25:24:20). The von Laue interference maxima in reciprocal

Table 1 Von Laue's equation describing the spectral transformations of visible light cones incident on a 3D diffraction grating into three RGB diffraction orders (reduction of spectral data in the visible range into RGB space).

---

1. Direction cosine of von Laue's equation ($\alpha\beta\gamma$ = aperture of diffraction orders; $\alpha°\beta°\gamma°$ = aperture of incident light cones):

$$(\cos\alpha - \cos\alpha°)^2 + (\cos\beta - \cos\beta°)^2 + (\cos\gamma - \cos\gamma°)^2 = 1.$$

2. Direction cosine for perpendicular light incidence: ($\alpha° = \beta° = 90°; \gamma° = 0°$):

$$(\cos^2\alpha + \cos^2\beta + \cos^2\gamma) - 2\cos\gamma\cos\gamma° = 0.$$

3. Spectral transformation at perpendicular light incidence: $\lambda = \lambda$max of a $h_1 h_2 h_3$ diffraction order.

$$\lambda = \frac{2\cos\gamma\cos\gamma°}{\cos^2\alpha + \cos^2\beta + \cos^2\gamma}.$$

4. Resonant von Laue equation with optical resonance factor $V_p = n\nu\lambda$ = phase velocity [11] (n = refractive index), with $\lambda_{h1h2h3} = \lambda_{111}$

$$\lambda_{111} = \frac{2\cos\gamma\cos\gamma°}{\cos^2\alpha + \cos^2\beta + \cos^2\gamma} x\lambda_{111}; \quad V_p = \nu_{111}\lambda_{111} = c = 1.$$

5. Triplet of RGB diffraction orders ($h_1 h_2 h_3 = 111(R); 123(G); 122(B)$ with g_x, g_y, g_z = grating constants x, y, z-axes: $g_x = 2\lambda_{111}; g_y = 4\lambda_{111}/\sqrt{3}; g_z = 4\lambda_{111}; \lambda_{111} = \sqrt{5}/4 = 559$ nm

$$\cos\alpha = \frac{h_1\lambda_{h1h2h3}}{g_x}; \quad \cos\beta = \frac{h_2\lambda_{h1h2h3}}{g_y}; \quad \cos\gamma = \frac{h_3\lambda_{h1h2h3}}{g_z};$$

$$\lambda_{111} = 0.25\sqrt{5} = 559 \text{ nm } (R)$$

$$\lambda_{123} = 0.24\sqrt{5} = 537 \text{ nm } (G)$$

$$\lambda_{122} = 0.20\sqrt{5} = 447 \text{ nm } (B)$$

$$\lambda_{111}:\lambda_{123}:\lambda_{122} = 25:24:20 = \lambda_R:\lambda_G:\lambda_B.$$

---

space are positioned—following the laws of hyperbolic geometry—on intersections between circles, ellipses, and hyperbolas, representing for concentric light cones the loci of Fourier terms of equal weights.

10.2.2 Diffractive-optical cross-correlation between RGB data of local and global information in reciprocal space

Local information is available in image space, global illumination in aperture space. Emil Wolf at his traditional workshop on modern coherence theory always quotes Zernike: "The image that will be formed in a photographic camera—i.e., the distribution of intensity on the sensitive layer—is present in an invisible, mysterious way in the aperture of the lens, where the intensity is equal at all points [6]." Global information is scattered by the aperiodic cellular layers in aperture space onto the retinal 3D grating in image space. It clearly has been documented that "the human eye exhibits considerable scatter [7]." Global information optically is superposed to local information in image space. (In addition, the inner human eye is a rotational ellipsoid [4], therefore a good optical resonator for global information). Convolution in real space leads into a multiplication in reciprocal space [8] and to the optically calculated psychological facts in the spectral domain. "Whatever exists as reality for psychology is a product of inductive inference .... These inferred realities of psychology are relations. A fact is a relation, and the simple basic fact in psychology is a correlation of a dependent variable upon an independent one [9]." The diffractive-optical serial product calculations in vector matrices predict the hues of colored shadows and explain why—at physically unbalanced spectra of illuminants (lights unlike mean sunlight)—"we cannot see, what physically is real at a shadow area, but only what our eyes optically have calculated." They also explain in a specific way why "Human color vision is a spatial calculation involving the whole image [10]."

10.2.3 Chromatic resonance between the retinal 3D grating and global information

Chromatic resonance between 3D grating constants and variable spectral compositions of the overall illumination adaptively recenter color space to a new RGB white norm. The trichromatic RGB triplet shifts in a direction, where a psychological rebalancing can compensate an "out-of-balance" state in physics. With this mechanism our eyes reach good performances of color constancy. Phase velocity in crystal optics is the resonance factor (Table 2) in x-rays as well as in the visible [11]. Chromatic resonance tuning of the retinal 3D grating constants is inherent to von Laue's equation (Table 1). Only after birth is the optical tuning of the RGB photoreceptors accomplished by means of sunlight and the training of chromatic adaptation begins.

Table 2 Ewald's article on phase velocity as the resonance factor in von Laue's Equation. [Copyright (1965) by the American Physical Society.]

REVIEWS OF MODERN PHYSICS   VOLUME 37, NUMBER 1   JANUARY 1965

Crystal Optics for Visible Light and X Rays*

P. P. EWALD

The Polytechnic Institute of Brooklyn, Brooklyn, New York

This summary of much of the author's work stresses the unity of approach to the understanding of the optical properties of a crystalline medium for visible light as well as for x rays. The scattering elements of the crystal are supposed to act as dipoles, and for the interior of the medium the problem is that of finding a balanced, or self-consistent state of optical field and dipole vibration. This is an eigenvalue problem, with the phase velocity as the eigenvalue.

What we are looking for is, in fact, a self-consistent system of resonator vibrations and emissions, so that one maintains the other. The field provides the coupling between the resonator oscillations, and the resonators the coupling between the spherical wavelets. The two kinds of coupling must harmonize—this is the condition which determines the refractive index n. To express this in an equivalent way: given the frequency ν, the self-consistency determines the wavelength λ, or the velocity q of the optical wave in the body $(n = c/q = \lambda_0/\lambda; \nu\lambda = q)$.

5. THE OPTICAL PHASE VELOCITY AS A REGULATING DEVICE

In the optical case again the phase velocity q is the regulating device between the wavelets emitted by each resonator and the strength of the field surrounding the resonator which prompts it to emit its wavelet.

From here one can easily proceed to the geometrical theory of diffraction by a three-dimensional lattice. This is built up on three translations a_1, a_2, a_3 which form the edges of the parallelopipedon of repeat, the "cell." If we make sure that the three atoms at the ends of the translations emit into the direction of observation with the same phase as the corner atom, then *all* atoms of the lattice will contribute wavelets of equal phases to a plane wave of wave vector K. We have thus to apply the condition of reinforcement three times and obtain the famous equations of Laue for the determination of the vector K:

$$(K - K_0) \cdot a_1 = h_1; \quad (K - K_0) \cdot a_2 = h_2;$$

$$(K - K_0) \cdot a_3 = h_3. \quad (2)$$

Here, h_1, h_2, h_3 are three integers which characterize the "order" of the spectrum or interference ray, and the wave vector taken from these equations will therefore be called $K_{h_1 h_2 h_3}$, or in short K_h.

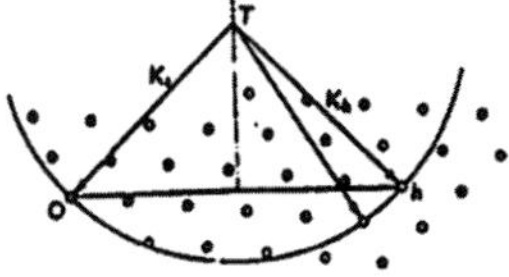

FIG. 5. Geometrical theory of space lattice diffraction in Fourier space.

10.3 The Missing Links between Physics and Psychology in the Spectral Domain

Colored shadows are observed when two lights of different color are combined in a twilight. Goethe in Ref. [12] describes the interplay of the full moon and a candle, which produce colored shadows: "One holds a board against the light of the full moon, with the candlelight a little to one side; and at a relevant distance one holds an opaque object in front of the board, creating a double shadow. The shadow from the moon (that) shines on the candlelight becomes a powerful reddish-yellow; and, in reverse, the shadow created by the candle and illuminated by moonlight will be seen as almost beautiful blue. Where both shadows meet, and unite as one, the shadow is black."

Three categories of experiments with colored shadows, limited only for reasons of brevity to the blue-yellow opponent color axis, but valuable in the full circle of opponent colors adding to white, will be differentiated to illustrate the reciprocal interdependence of physics and psychology in human color vision:

(1) Two colored lights add to an equi-energy white, physically balanced light: physics and psychology are reciprocally in balance.
(2) A colored and a white light (or two colored lights) add to a physically unbalanced light: physics and psychology are out of balance. The human eye does not see what physically is real at a shadow area, but what it has optically calculated.
(3) A colored and a white light (or two colored lights) add to a physically unbalanced light, but physics and psychology become rebalanced by optical adaptation, i.e., by 3D grating-optical chromatic resonance with the changed illumination.

10.3.1 Two colored lights add to an equi-energy white, physically balanced light: physics and psychology are reciprocally in balance

A light with an equi-energy white "balanced" spectrum (B1 + B2) can, in multiple ways, be divided into two white or two colored lights B1 and B2, forming a twilight (Fig. 6). One extreme partitioning is the horizontal division of the equi-energy spectrum into two white lights, where the two shadows become gray. At the other extreme, the vertical division of the equi-energy spectrum into two colored lights, the two shadows reach the fully saturated hues of a pair of opponent colors adding to white (blue + yellow, etc.). What holds on a blue-yellow axis, also holds for all other axes of the full circle of opponent colors centered to white. With both ends of the spectrum brought together in a purple light and the middle part of the spectrum concentrated on a green light, shadow hues not present in the physical

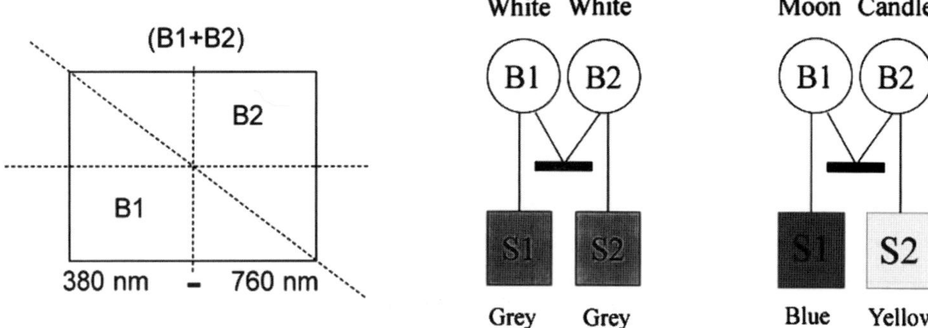

Figure 6 Two lights (B1, B2) in a twilight adding to an equi-energy white, physically balanced light (B1 + B2). A white light can be divided into two lights in many ways. When horizontally divided into two white lights, the two shadows (S1, S2) will be gray. When divided into two colored lights or vertically divided, the shadows reciprocally show opponent colors. In Goethe's experiment, the more bluish moonlight (B1) and the more yellowish candlelight (B2) show shadows with opponent colors reciprocally.

spectrum are seen in opponent color pairs (purple + green). Goethe seems to have been right when he stated that the linear spectrum in physics in human color vision is transformed into a color circle of opponent colors adding to white. The same rule holds if, instead of two lights, three colored lights adding to a white, or two pairs of lights with opponent colors, are chosen to form a white illumination. They all produce shadows of opponent colors.

Two intimately linked mathematical operations describe the transformations of the data into the observable results. The calculation is shown in Table 3 for data similar to Goethe's experiment with blue-yellow shadows, but is valuable for all other twilights adding to an equi-energy spectrum, as shown with more details in Ref. [13]. The operations are:

(1) Transform all spectral intensity data of the lights B1, B2, and (B1 + B2) into RGB space by multiplying the spectral intensities of the lights with the three Gaussian curves of spectral brightness sensitivities (SBV) of human color vision, as in Fig. 1. As a result the areas illuminated by (B1 + B2), the global information, in RGB space leads to a white (33% for R, G, and B), B1 to a local blue (61% B, 26% G, 13% R), B2 to a local yellow (0% B, 43% G, 57% R).

(2) Relate local information B1 and B2 onto global information (B1 + B2) in RGB space. The results of this optical cross-correlation determine the hues of the colored shadows. Shadow S1 is a blue (+27% B, –7% G, –20% R), S2 a yellow (–34% B, +10% G, +24% R).

Table 3 Calculation of the blue (S1) and the yellow (S2) shadow for a blue (B1) and a yellow light (B2) adding to an equi-energy white light (B1 + B2). Example with vertical division of B1 and B2 at 520 nm (arbitrary intensity units). Multiplication of B1, B2, and (B1 + B2) with the Gaussians from Fig. 1 centered to 559 nmR, 537 nmG and 447 nmB (columns 5–7) leads to the transformations into RGB space (columns 9–11): B1 + B2 = white = 33%-RGB balance; B1 = blue (61% B, 26% G, 13% R), B2 = yellow-red (0% B, 43% G, 57% R). Cross-correlation of local data to global data in RGB space leads to a blue shadow S1 (+27% B, –7% G, –20% R) and to a yellow-red shadow S2 (–34%, +10% G, +24% R). The same holds, with less saturation of the opponent colors of the shadows, for all other possible divisions of the equi-energy white light into a blue and a yellow light.

1	2	3	4	5	6	7	8	9	10	11
Wave-length (nm)	B1	B2	B1 + B2	B	G	R		B	G	R
380	450		450	37	–	–	B1 + B2	34%	33%	33%
400	450		450	264	1	–			White	
420	450		450	950	9	2				
440	450		450	1712	40	10				
460	450		450	1543	138	43	B1	61%	26%	13%
480	450		450	696	357	138	- - - - -	- - - - -	- - - - -	- - - - -
500	450		450	157	698	344	S1	+27%	–7%	–20%
520	450		450	18	1025	657			Blue	
540		450	450	1	1133	965				
560		450	450		941	1087	B2	0%	43%	57%
580		450	450		588	940	- - - - -	- - - - -	- - - - -	- - - - -
600		450	450		277	624	S2	–34%	+10%	+24%
620		450	450		98	318			Yellow	
640		450	450		26	124				
660		450	450		5	37				
680		450	450		1	9				

10.3.2 Physically unbalanced light: optically calculated vs. physically real

When a blue (B1) and a white light (B2) are combined in a twilight (Table 4), transformation of (B1 + B2) into RGB space leads to a bluish-white (49% B, 28% G, 23% R) for the global information, transformation of B1 to a local blue (60% B, 24% G, 16% R) and of B2 to a local white (33% for B, G, and R). With a white light B2, the only light illuminating the shadow area S2, this shadow "logically" should be white, but it is not. Transformation into RGB space alone is not the only logic in the transformation rules. When we now apply the second rule, the cross-correlation of local onto global information, shadow S1 becomes blue

Table 4 Calculation of the colored shadows S1, S2 for a blue (B1) and a white light (B2) adding to a physically unbalanced bluish light (B1 + B2). Multiplication of B1, B2, and (B1 + B2) with the Gaussians from Fig. 1 centered to 559 nmR, 537 nmG and 447 nmB (columns 5–7) leads to the transformations into RGB space (columns 9–11): B1 + B2 = blue (49% B, 28% G, 23% R); B1 = blue-violet (60% B, 24% G, 16% R), B2 = white (33%-RGB balance). Cross-correlation of local data to global data in RGB space leads to a blue shadow S1 (+11% B, –4% G, –7% R) and to a yellow-red shadow S2 (–15% B, +5% G, +10% R), where, without the cross-correlating, a white S2 should be seen. At S2 the human eye sees what it has optically calculated.

1	2	3	4	5	6	7	8	9	10	11
Wave length (nm) B1	Blue light B1	White light B2	B1 + B2	B	G	R		B	G	R
380	630	240	870	37	–	–	B1 + B2	49%	28%	23%
400	630	240	870	264	1	–			Blue	
420	670	240	910	950	9	2				
440	570	240	810	1712	40	10	B1	60%	24%	16%
460	640	240	880	1543	138	43	-----	-----	-----	-----
480	510	240	750	696	357	138	S1	+11%	–4%	–7%
500	380	240	620	157	698	344			Blue	
520	260	240	500	18	1025	657				
540	220	240	460	1	1133	965	B2	34%	33%	33%
560	140	240	380		941	1087	-----	-----	-----	-----
580	80	240	320		588	940	S2	–15%	+5%	+10%
600	60	240	300		277	624			Yellow	
620	40	240	280		98	318				
640		240	240		26	124				
660		240	240		5	37				
680		240	240		1	9				
700		240	240			2				

(+11% B, –4% G, –7% R) and shadow S2 becomes yellow (–15% B, +5% G, +10% R), the opponent color of blue. The same rules hold for all other combinations of colored lights with a white light. The more intense the colored light is, the more intense and saturated the opponent color is seen at shadow S2. It is now that our eyes "do not see what is physically real at the shadow S2 area, but what they have optically calculated." This psychological fact is not an optical illusion, but the result of an optical calculation. Could it be the result of a photochemical instead of an optical calculation? It is optical, because absorption of local information in an image plane alone, as stated by the camera model of the eye, cannot explain the psychological facts, which per se are relations [9]. The Gaussian curves of spectral brightness sensitivities in Table 1 therefore inevitably represent diffractive optical RGB transmission curves of the retinal 3D grating [14], which at mean sunlight are identical with the absorption characteristics of the photopigments in color vision.

10.3.3 Experimental setup to demonstrate ordinary and paradoxically colored shadows

The main aspects of the phenomenon of colored shadows analyzed in the text have been shown with a simple demonstrator on the blue-yellow and purple-green opponent color axes. What holds for the blue-yellow and purple-green axes also holds for all other opponent color pairs in the color circle.

A single white light source in a box through two optical fibers divides the intensities into two equal parts [Fig. 7(a)], representing twilight with two white light spots B1 and B2 on the whitish screen. The two white lights B1 and B2 by superposition of their spots add to (B1 + B2) white [Fig. 7(b)]. The introduction of a shadow casting stick shows two equally gray shadows S1 and S2 [Fig. 7(c)]. In Figs. 7(d) and 7(e), both optical fibers are equipped with opponent color filters adding to white: in Fig. 7(d) a violet-blue light (B1) and a yellow (B2) light in the twilight show a violet-blue (S2) and a yellow ordinary colored shadow (S1). In Fig. 7(e), a green (B1) and a purple (B2) light in the twilight (B2 adds the violet and the red parts of the white spectrum) show a green (S2) and a purple (S1) ordinary colored shadow.

The paradoxically colored shadows S2 are shown in Figs. 7(f)–(i). The inserted graphics illustrate the opponents of the colored shadows S1 and S2. Only one optical fiber (B2—shown on the left) has a color filter, the other fiber (B1—on the right) always projects white light. Shadow S1 regularly shows the hue of the colored light (B2), shadow S2, instead of showing a white or a gray as could be expected from B1, always shows the opponent hue of the color at B2, manifesting the paradoxical colored shadows.

The colors of the shadows S2 can only be photographed when a white balance system is operating in the camera. A spectral photometer will always register a white spectrum at the S2 shadow area.

10.3.4 Unbalanced light rebalanced by adaptation to the colored illumination

Adaptive chromatic grating-optical resonances with "balanced" or "unbalanced" spectral intensity distributions are described by the *resonant von Laue equation* (Table 1), where phase velocity $V_p = n\nu\lambda$, as the crystal optical resonance factor (Table 2 and Ref. [11]) becomes relevant (n = refractive index). The grating constants of the retinal 3D grating become tuned or "transformed" to a specific

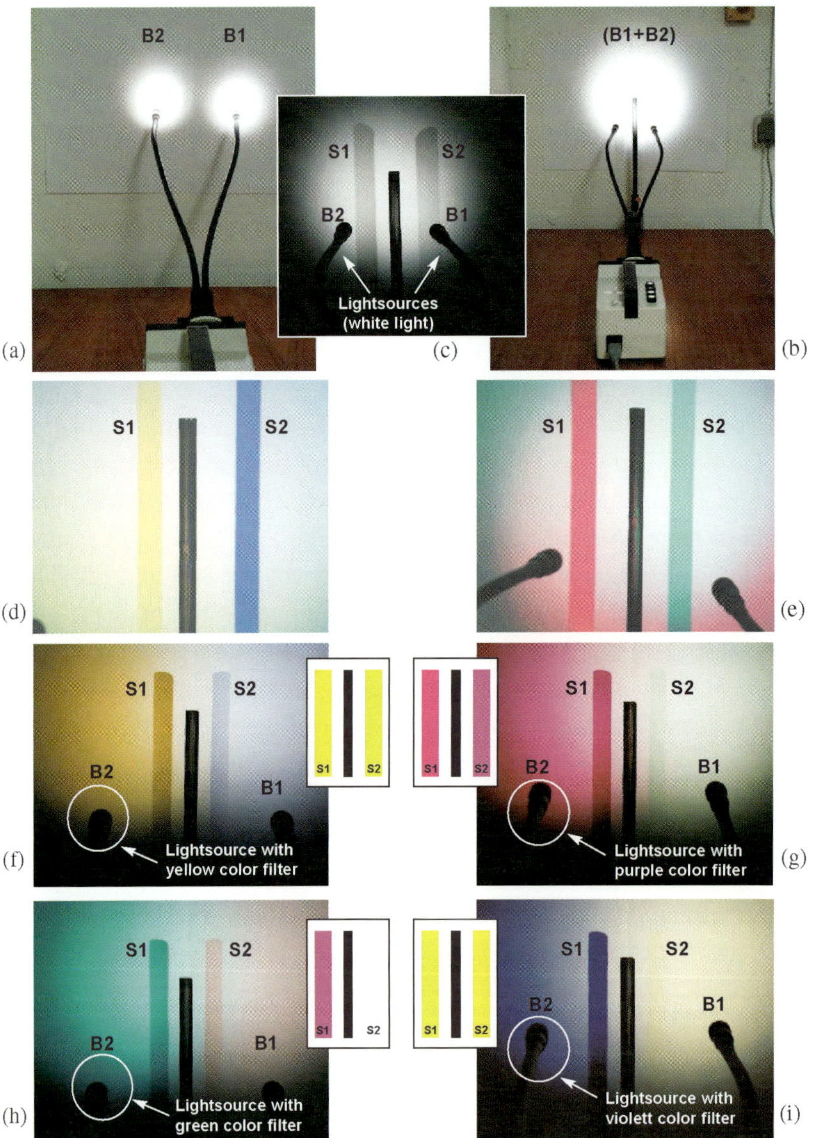

Figure 7 Experimental setup demonstrating ordinary and paradoxically colored shadows. (a) Twilight represented by two white-light spots (B1 and B2) on a white screen. (b) Superposition of white spots add to a white (B1 + B2). (c) Shadow casting stick shows two equally gray shadows (S1 and S2). (d) A violet-blue light (B1) and a yellow light (B2) show ordinary violet-blue and yellow shadows (S1 and S2). (e) Green (B1) and purple (B2) light show ordinary green (S2) and purple (S1) shadow. (f)–(i) Paradoxically colored shadows are represented by S2. One optical fiber projects color while the other projects white light. S1 shows the line of the colored light, S2 shows the opponent here of the color at B1, manifesting paradoxically colored shadows. The inserted graphics illustrate the opponents of S1 and S2.

λmax of the 111R-DO in the visible octave of the spectrum, the fundamental resonant wavelength in the RGB-triplet. Soon after birth, but only when adaptation to variable light intensities by the appropriate training of pupil reflexes has at first been learned, chromatic adaptation of the eye is trained and the DO-triplet—like a comb-filter—learns to shift along the unbalanced global spectrum (B1 + B2) up to a position where a good new psychological RGB balance—a white sensation— is reached. In an adaptive instrument of sight this optical (tri)chromatic resonance mechanism helps to stabilize psychological data in color vision. The rule for chromatic resonance adaptation is: at diffuse scatter of an unbalanced spectrum (B1 + B2)—the global information—in aperture space, the 111R-DO in the retinal 3D grating shifts the RGB-triplet toward a new white norm in RGB space to recalibrate and renormalize color space (Table 5).

Table 5 Calculation of the colored shadow S2 for a blue (B1), and a white light (B2), adding to a physically unbalanced blue light (B1 + B2) as in Table 4 (column 4), but with progressive adaptation to the dominant blue component B1 in (B1 + B2). With the Gaussians from Fig. 1 at first centered to 559 nmR, (B1 + B2) = blue (49% B, 28% G, 23% R) in RGB space, B2 = white (33%-RGB balance) and shadow S2 = yellow (–15% B, +5% G, +10% R) as before (Table 4, columns 9–11). With the Gaussians progressively shifting toward 499 nmR, (B1 + B2) in RGB space more and more approaches a new white norm (reaching an optimum at about 499 nm = 33% B, 36% G, 31% R) and B2 drifts toward a yellowish white (finally at 499 nm = 28% B, 36% G, 36% R = yellow), S2 progressively loses its yellow color and finally becomes a highly unsaturated whitish yellow (at 499 nm = –5% B, +0% G, +5% R). This bleaching out of the yellow hue of S2 toward a highly unsaturated whitish yellow experimentally is observed, when adaptation to the isolated B1 is reached and the white light B2 suddenly becomes reintroduced to form the twilight.

Adaptation to		B	G	R	
559 nmR	B1 + B2	49%	28%	23%	Blue
	B2	34%	33%	33%	White
	S2	–15%	+5%	+10%	Yellow
539 nmR	B1 + B2	46%	30%	25%	Blue
	B2	33%	33%	33%	White
	S2	–13%	+3%	+8%	Yellow
519 nmR	B1 + B2	36%	34%	29%	Bluish-White
	B2	32%	34%	34%	White
	S2	–4%	+0%	+5%	Whitish-Yellow
499 nmR	B1 + B2	33%	36%	31%	White
	B2	28%	36%	36%	Yellowish-White
	S2	–5%	+0%	+5%	Whitish-Yellow

When the sun at early morning hours radiates more in the blue part of the spectrum, and in the evening more in the red parts, adaptation simply follows the respective direction. When (B1 + B2) suddenly changes from an equi-energy white into a colored (blue + white) light (column 4 in Table 4), the three Gaussians centered at first to 559 nmR, 537 nmG, 447 nmB will progressively shift to 499 nmR, 477 nmG, 387 nmB and reach a new white norm in RGB space. When (B1 + B2) then splits into a twilight with B1 (blue) largely dominating over B2 (white) in the global light, so that adaptation can be maintained at the 499 nmR-resonance state, the critical shadow S2 will not show a saturated yellow hue, but a highly unsaturated whitish yellow. The same is true for all other opponent colors. At a perfect resonant adaptation the system is driven back to the situation, where a white light can, by a horizontal line, be divided into two "white" lights and where both shadows in the twilight become gray again (Fig. 6). At perfect adaptation, color space is recentered to a new white norm and the physically unbalanced state is compensated by a psychological rebalancing, guaranteeing invariants in color vision under variable illuminants and thereby, as shown in Ref. [15], allowing relatively good color constancy. Again it becomes evident that the three Gaussian curves in Table 1 not only reflect photochemical characteristics adapted to perpendicularly incident equi-energy white light, but also spectral transmission curves of a small number of diffractive layers in the retinal multilayer 3D grating [14]. Only a postnatal training of color vision under unbalanced spectra differentiates the resonant diffractive-optical adaptation mechanism. The initial trichromatic spectral tuning of the photo-pigments in Fig. 1 seems to be programmed by optical diffraction of perpendicularly incident mean sunlight as the ordinary global illumination early after birth. But it is the training of adaptation that differentiates the grating-optical resonance mechanism.

10.4 Conclusions

The analysis of the phenomenon of colored shadows, typically illustrated on the most critical shadow S2 at the blue-yellow axis, reveals the diffractive-optical spectral transformations and cross-correlations in human color vision, representing the most elementary psychophysical links between variable objective and adaptable subjective reciprocal worlds. The colored shadow S2 reaches fully saturated chroma when the colored lights in a twilight add to a balanced white; the hue of S2 desaturates when the lights add to an unbalanced illumination; S2 finally adopts a more and more unsaturated whitish tint when chromatic adaptation to the colored illuminant is reached. Color can be considered as an interference-optical construction, an optical calculation by the resonant diffractive-optical hardware of the human eye, a psychological fact reciprocally related to physical parameters in the outside visible world.

With the diffractive-optical hardware of the human eye, what holds for spectral transformations and brings the Goethe-Newton controversy [16] to an end by introducing modern optics also holds for the full range of 4D spatio-temporal transformations in image preprocessing. Only the final products of the diffractive-optical correlator calculations are available to photoreceptors for absorption. This is the price won by putting back absorption of photons behind diffractive/wave-interference-optical image preprocessing in optics. It makes the diffractive-optical illuminant-adaptive hardware of the human eye more intelligent, opening the door to invariants in an ever-changing world through optically resonant adaptive psychological rebalancing of unbalanced states in physics. It also makes human vision more risky by opening the door to constructive interpretation of reality, to illusions and hallucinations, pure psychological facts.

Emil Wolf summarized the main results of "Von Laue's equations and scattering from finite distances" at the Rochester Workshop in 1993: "I showed that much information about diffracting and scattering objects can be determined from measurements at finite distances from the object, without the use of any optical system. This follows from the main result that underlies the theory of diffraction tomography; namely, that the 3D Fourier components of a scattering potential are mapped, in a rather remarkable and relatively simple manner, onto the two-dimensional Fourier components of the scattered field in arbitrary cross sections of the field. I also showed that if the object is periodic, the von Laue equations clearly manifest themselves in the two-dimensional mapping of the object. All this suggests that diffraction tomography [17], together with holography, might offer a new technique for the study of the human visual system."

This work has been supported by the German Government Research Program [18].

References

1. N. Lauinger, "Learning from biotechnology; milestones in the prenatal engineering of an intelligent optical sensor: the human eye," *SPIE Proc.* Vol. 4572, 231–240 (2001).
2. N. Lauinger, "Prenatal development of the human eye: a starting point for innovation in optics?" *OPTO 2002 Proceedings*, 7–17 (2002).
3. D.M. Maurice, "The structure and transparency of the cornea," *J. Physiol.* **136**, 262–286 (1957).
4. M.J. Hogan, J.A. Alvarado, and J.E. Weddell, *Histology of the Human Eye*, Saunders, Philadelphia-London-Toronto (1971).
5. W. Bargmann, *Histologie und mikroskopische Anatomie des Menschen*, Georg Thieme Verlag, Stuttgart (1967).
6. F. Zernike, *Proc. Phy. Soc. London* **61**, 158 (1948).

7. W.A. Stiehl, J.J. McCaim, and R.L. Savoy, "Influence of intraocular scattered light on lightness-scaling experiments," *JOSA* **73**(9), 1143–1148 (1983).

8. E.G. Stewart, *Fourier Optics. An Introduction*, Ellis Horwood Series in Physics (1989).

9. E.G. Boring, *The Physical Dimensions of Consciousness*, Dover Publications, Inc., New York (1963).

10. J.J. McCann, "Color imaging systems and color theory: past, present and future," *SPIE Proc.* Vol. 3299, 38–46 (1998).

11. P.P. Ewald, "Crystal optics for visible light and x rays," *Reviews of Modern Physics* **37**(1), 46–56 (1965).

12. W.v. Goethe, *Zur Farbenlehre*, Goethes Werke Bd. 12, 314–536, Christian Wegner Verlag, Hamburg (1958).

13. N. Lauinger, "Illuminant-adaptive diffractive-optical RGB-color sensor: 3D grating-optical cross-correlator calculating colored shadows in human vision," *SPIE Proc.* Vol. 5267 (2003).

14. M. Carbon, N. Lauinger, and J. Schwab, "Self-imaging of three-dimensional phase gratings," *Pure Appl. Opt.* **7**, 1103–1120 (1998).

15. N. Lauinger and B. Badenhop, "Diffractive 3D grating-optical sensor with trichromatic color constancy at adaptation to variable illuminants," *SPIE Proc.* Vol. 4197, 352–357 (2000).

16. *Goethe's Way of Science. A Phenomenology of Nature*, D. Seamon and A. Zajonc, Eds., State University of New York Press (1998).

17. E. Wolf, "Determination of the amplitude and the phase of scattered fields by holography," *JOSA* **60**(1), 18–20 (1970).

18. BMBF-Project 03C0300A on "Maßgeschneiderte Polymerlatices und ihre Selbstorganisation zu Partikelarrays für Anwendungen in der optischen Informationsverarbeitung und Sensorik" (1999–2003).

My Encounters with Emil and Marlies Wolf

I first met Emil and Marlies Wolf at the OSA meeting in Albuquerque (September 1992), where I was an assistant to Emil's traditional workshop on "Modern Coherence Theory." After the course, together we made a private excursion to a nearby mountain via a cable railway. We spent a few hours passionately discussing historical discoveries and highlights in x-ray optics, tomography and holography. This was an immediate start into the heart of our main topic for the years to come: the potential role played by the von Laue Equations in human vision.

Emil and Marlies Wolf and Norbert Lauinger in Albuquerque, NM, 1992. (Courtesy of N. Lauinger, copyright 2004.)

In 1993, Prof. Wolf invited me, and two of my collaborators, Drs. M. Carbon and Jochen Schwab, to a "Workshop on Physical Optics and Human Vision" at the University of Rochester (June 21 – June 22, 1993). At this event we were not only fascinated by the quality of the other speakers, but also by the charming ambiance of Emil's and Marlies's home and garden, and the very friendly atmosphere between the Wolf's and their students and disciples.

Later we met at many congresses (ICO in Budapest, 1993; OSA Conference in Toronto, 1993; OSA at Rochester, 1996; OSA at Long Beach, 1997; EOS in Germany, 2003; SPIE at San Diego in 2003), and, in short intervals, I had the great pleasure to meet Emil and Marlies at their home in Rochester in 1994, 1995, and 2002. In 1995, I welcomed Emil and Marlies to the newly founded company Corrsys and at my home in Wetzlar, Germany, which has a rich history of pioneers in optics (Berek, Barnak, Bergmann, and others).

During all these encounters the topic of the von Laue's Equations was the unbroken thread encouraging our minds, and Prof. Wolf always excelled in precious knowledge of books and publications hidden in pre-Internet times and sold only in antique bookshops.

I have never met a man like Emil Wolf, who continually encouraged other scientists and I have always forgotten to express my thanks to him. This time, I would like to thank him for all he has done for my group of colleagues, and for me.

A garden party at Rochester with Emil and Marlies Wolf, Margarita Carbon, and Norbert Lauinger. (Courtesy of N. Lauinger, copyright 2004.)

Emil Wolf and Norbert Lauinger at Corrsys, Wetzlar, Germany. (Courtesy of N. Lauinger, copyright 2004.)

Norbert Lauinger began his professional career with 12 years of general management and research and development in optics, microscopy, and industrial metrology at Ernst Leitz, Paris, France, from 1963 to 1975. Then followed a period of 18 years as freelance consultant at a Management Institute in Cologne, Germany (industrial psychology), and research and development on modern aspects of diffractive optics and histology in pre- and postnatal human vision. In 2003, with the foundation of Corrsys Datron Sensorsysteme GmbH in Wetzlar, Germany, and together with a small team of optical and electrical engineers he started the technological development and industrial production of innovative optical correlators and sensor systems. He was general manager from the company's beginning through today with 60 employees. He has not only promoted this company, but has also chaired many SPIE conferences at Photonics East and published many articles about intelligent optical correlators at SPIE conferences during the last several years. He is a member of the board of Optence e.V., part of the German network for the promotion of photonics and optical technologies, supported by the German government.

Chapter 11

The Wolf Effect in Rough Surface Scattering

Zu-Han Gu, Tamara A. Leskova, Alexei A. Maradudin, and Mikael Ciftan

11.1 Introduction

The literature dealing with theoretical optics contains so many effects associated with the name Wolf that the selection of one or another of them to address in a paper written in honor of Prof. Emil Wolf is difficult. We have finally chosen to study changes in the spectrum of light due to its scattering from a randomly rough surface, an investigation that has its origins in work done by Prof. Wolf more than 15 years ago.

In the work that prompted our study, Prof. Wolf considered radiation from a three-dimensional quasi-homogeneous source, and showed that if the degree of spectral coherence of the source is appropriately chosen, the spectrum of the emitted radiation can be redshifted or blueshifted with respect to that of the source, even when the source is at rest with respect to the observer, and the radiation propagates in free space [1–4]. Unlike Doppler shifts, these correlation-induced shifts are frequency independent, and are restricted in their magnitudes to values that are smaller than the widths of the spectral lines.

In the scattering of polychromatic light from a static random medium, the different frequency components of the incident light, which are scattered in any particular direction, will be scattered with different strengths. Consequently, the spectrum of the scattered light will differ from that of the incident light, even though the different frequency components are uncorrelated. The possibility of generating a spectral redistribution by scattering is analogous to the possibility of generating a spectral redistribution in light emitted by a source caused by correlations in the fluctuations of the source. The only difference is that in scattering one is dealing with secondary sources, namely with the polarization induced in the scattering medium by the incident field. The induced polarization, in general, is correlated over finite distances of the scattering medium, and thus imitates correlations in primary sources. This analogy between scattering and radiation prompted several theoretical investigations of spectral redistributions of light scattered by volume random media [5–7]. These investigations were based on the first Born approximation, a single-scattering approximation, and were therefore applicable to media with a very low density of scatterers. The first calculations of spectral changes of light scattered by volume random media in which multiple-scattering effects are important, were carried out by Lagendijk [8,9]. He pointed out that the enhanced backscattering of light from a strongly scattering disordered medium, which is due to the coherent interference between multiply-scattered optical paths and their reciprocal partners, can be regarded as due to the reemission of light from an extended source in the random medium that possesses just the type of source correlation needed to produce a Wolf redshift. The magnitudes of the redshifts calculated by Lagendijk for scattering angles in the vicinity of the backscattering direction, however, were very small. Lagendijk's work was later followed by other studies of spectral changes in the scattering of light from volume-disordered systems that took multiple scattering into account [10,11].

Our own interest in the scattering of light from randomly rough surfaces led to several theoretical [12,13] and experimental [14,15] studies of the change in the spectrum of polychromatic light scattered from such surfaces. This change can be regarded as due to the emission of light by a secondary source, namely the random surface, whose profile function possesses the type of source correlation that leads to the spectrum of the scattered light in the far zone being different from that of the light at the secondary source.

In this chapter we theoretically and experimentally address the spectral changes occuring in the scattering of light from a system that consists of a dielectric film deposited on the planar surface of a metallic substrate, when the illuminated surface of the film is randomly rough. It has been known for some time that coherent light scattered from a slightly rough two-dimensional random surface of a dielectric film deposited on the planar surface of a reflecting substrate consists of speckle spots that arrange themselves into concentric interference rings that are centered at the normal to the mean surface. The angular positions of these rings (intensity maxima) are independent of the angle of incidence, but depend strongly on the wavelength of the incident light and the mean thickness of the film. These rings have been named Selényi rings [16]. If the roughness of the dielectric surface is strong, the angular positions of these rings depend on the angle of incidence and on the wavelength of the incident light and the mean thickness of the film. The ring corresponding to the zero order of interference passes through both the specular and retroreflection directions. These rings have been named Quételet rings [17,18]. The dielectric films in the scattering systems that will be studied experimentally in this paper are sufficiently rough and sufficiently thick to give rise to Quételet rings. Those that will be studied theoretically are either weakly rough and give rise to Selényi rings, or are strongly rough and give rise to Quételet rings. Our interest in such systems derives from the fact that in order to obtain changes in the spectrum of light scattered from a randomly rough surface that are large enough to be observed experimentally, the spectrum of the scattered light should be measured at a scattering angle that is in the near vicinity of some feature in the scattering pattern whose angular position depends strongly on the frequency of the incident light [12,13]. Selényi and Quételet rings are just such features, and it is expected that large changes in the spectrum of the light scattered at the angle at which one of these rings occurs will be observed. This expectation is borne out by our theoretical and experimental results.

The outline of this paper is as follows. In Sect. 11.2 we theoretically examine the changes in the spectrum of light scattered from a dielectric film deposited on the planar surface of a metallic substrate, in the case where the illuminated surface of the film is a one-dimensional randomly rough surface, and the scattering angle is close to the angular position of one of the Selényi or Quételet fringes produced by the scattering system, depending on the roughness of the film. In Sect. 11.3 we present experimental results for the changes in the spectrum of the scattered light in the case of the in-plane copolarized scattering of s-polarized light from a dielectric film deposited on the planar surface of a metal substrate, when the illuminated surface of the film is a two-dimensional strongly rough random surface, and the scattering angle is close to the angular position of one of the Quételet rings produced by the scattering system. A brief discussion of the results obtained, in Sect. 11.4, concludes this paper.

11.2 Theoretical Study of Changes in the Spectrum of Light Scattered from a Rough Dielectric Film on a Metallic Substrate

In this section we theoretically study the scattering of s-polarized light from a dielectric film deposited on the planar surface of a metallic substrate. The illuminated surface of the film is a one-dimensional randomly rough surface, and the plane of incidence is perpendicular to the generators of the surface. The analogs of the Selényi and Quételet rings in this case are called Selényi and Quételet fringes [19]. What will be calculated is the spectrum of the scattered light at scattering angles at which these fringes occur. The results of this study should be qualitatively similar to those obtained in the experiments on the in-plane copolarized scattering of s-polarized light from a two-dimensional randomly rough surface described in the following section.

Thus, the system we study in this section consists of vacuum in the region $x_3 > \zeta(x_1)$, a dielectric medium characterized by a dielectric constant ϵ in the region $-D < x_3 < \zeta(x_1)$, and a metal characterized by an isotropic, complex, frequency-dependent dielectric function $\epsilon(\omega)$ in the region $x_3 < -D$. Since the sample in the experimental study described in Sect. 11.3 is a smooth aluminum plate that was coated with a dielectric film whose upper surface was weakly rough, and whose thickness was slowly varying around the mean thickness, we assume that the surface profile function $\zeta(x_1)$ is a sum of two profiles

$$\zeta(x_1) = s_1(x_1) + s_2(x_1), \tag{1}$$

where both $s_1(x_1)$ and $s_2(x_1)$ are single-valued functions of x_1 that are differentiable as many times as is necessary, and constitute zero-mean, stationary, uncorrelated, Gaussian random processes defined by

$$\langle s_i(x_1)s_j(x_1') \rangle = \delta_{ij}\delta_i^2 W_i(|x_1 - x_1'|), \quad i,j = 1,2, \tag{2}$$

where δ_{ij} is the Kronecker symbol. The angle brackets in Eq. (2) denote an average over the ensemble of realizations of the surface profile function, and $\delta_i = \langle s_i^2(x_1)\rangle^{1/2}$ is the rms height of the surface. The surface height autocorrelation functions $W_i(|x_1|)$ are assumed to have the Gaussian form

$$W_i(x_1) = \exp(-x_1^2/a_i^2), \tag{3}$$

where the characteristic lengths a_i are the transverse correlation lengths of the surface roughness. The power spectrum of the surface roughness associated with each component of the surface profile function is defined by ($i = 1,2$)

$$g_i(|Q|) = \int_{-\infty}^{\infty} dx_1\, W_i(|x_1|)\exp(-iQx_1) \tag{4a}$$

$$= \sqrt{\pi}a_i\exp(-a_i^2 Q^2/4). \tag{4b}$$

This system is illuminated from the vacuum side by an s-polarized electromagnetic field whose plane of incidence is the x_1x_3 plane. The single nonzero component of the electric vector of the incident field is written as a superposition of incident monochromatic components weighted by a random function of frequency $F(\omega)$,

$$E_2(x_1,x_3;t)_{inc} = \int_{-\infty}^{\infty} \frac{d\omega}{2\pi}F(\omega)\exp\left[ikx_1 - i\alpha_0(k,\omega)x_3 - i\omega t\right], \tag{5}$$

where $\alpha_0(k,\omega) = [(\omega/c)^2 - k^2]^{1/2}$, with $\mathrm{Re}\,\alpha_0(k,\omega) > 0$, $\mathrm{Im}\,\alpha_0(k,\omega) > 0$. The wavenumber k is related to the angle of incidence θ_0, measured counterclockwise from the positive x_3 axis, by $k = (\omega/c)\sin\theta_0$. The random function $F(\omega)$ possesses the properties

$$\langle F(\omega)F^*(\omega')\rangle_F = 2\pi\delta(\omega - \omega')S_0(\omega), \tag{6a}$$

$$\langle F(\omega)F(\omega')\rangle_F = 0, \tag{6b}$$

where the angle brackets $\langle\cdots\rangle_F$ denote an average over the ensemble of realizations of the field. The spectral density of the incident light $S_0(\omega)$ is normalized to unity,

$$\int_{-\infty}^{\infty} d\omega S_0(\omega) = 1, \tag{7}$$

and will be described by a particular form characteristic to the source used in our experiments.

The single nonzero component of the electric vector of the scattered field in the region $x_3 > \zeta(x_1)_{\max}$ is

$$E_2(x_1, x_3; t) = \int_{-\infty}^{\infty} \frac{d\omega}{2\pi} F(\omega) \int_{-\infty}^{\infty} \frac{dq}{2\pi} R_\omega(q|k) \exp[iqx + i\alpha_0(q, \omega)x_3 - i\omega t], \tag{8}$$

where $R_\omega(q|k)$ is the amplitude for the scattering of an incident plane wave of frequency ω, $\exp[ikx_1 - \alpha_0(k, \omega)x_3 - i\omega t]$, whose wave vector has a projection k on the x_1 axis, into a scattered plane wave of frequency ω, whose wave vector has a projection q on the x_1 axis.

The fraction of the total time-averaged incident flux that is scattered into the angular interval $d\theta_s$ about the scattering angle θ_s, measured clockwise from the x_3 axis, and into the frequency interval $(\omega, \omega + d\omega)$, is given by $\langle\langle \partial^2 P/\partial\omega\partial\theta_s\rangle\rangle d\theta_s d\omega$, where

$$\left\langle\!\!\left\langle \frac{\partial^2 P}{\partial\omega\partial\theta_s} \right\rangle\!\!\right\rangle = S_0(\omega)\left\langle \frac{\partial P_\omega}{\partial\theta_s} \right\rangle, \tag{9}$$

and the double brackets $\langle\langle\cdots\rangle\rangle$ denote an average over both the ensemble of realizations of the surface profile and the ensemble of realizations of the incident field. The function $\langle\partial P_\omega/\partial\theta_s\rangle$, given by

$$\left\langle \frac{\partial P_\omega}{\partial\theta_s} \right\rangle = \frac{1}{L}\frac{\omega}{2\pi c}\frac{\cos^2\theta_s}{\cos\theta_0}\langle|R_\omega(q|k)|^2\rangle, \tag{10}$$

is the mean differential reflection coefficient. In Eq. (10), L is the length of the x_1 axis covered by the random surface, $k = (\omega/c)\sin\theta_0$, and $q = (\omega/c)\sin\theta_s$. As our interest is in the spectral distribution of light scattered at an angle other than that of the specular direction, we base our calculations of the spectral distribution on the contribution to the mean differential reflection from the light that has been scattered diffusely, namely

$$\left(\frac{\partial P_\omega}{\partial \theta_s}\right)_{\text{diff}} = \frac{1}{L}\frac{\omega}{2\pi c}\frac{\cos^2\theta_s}{\cos\theta_0}\left[\langle|R_\omega(q|k)|^2\rangle - |\langle R_\omega(q|k)\rangle|^2\right]. \tag{11}$$

Thus, Eq. (9) is replaced by

$$\left\langle\!\!\left\langle\frac{\partial^2 P}{\partial\omega\partial\theta_s}\right\rangle\!\!\right\rangle_{\text{diff}} = S_0(\omega)\left\langle\frac{\partial P_\omega}{\partial\theta_s}\right\rangle_{\text{diff}}. \tag{12}$$

Equations (11) and (12) define the function that used as the spectral density of light scattered in the direction defined by the scattering angle θ_s. The presence of the second factor on the right-hand side of Eq. (12) [and of Eq. (9)] shows that the spectrum of the scattered light differs from that of the incident light. We now turn to the determination of the mean differential reflection coefficient $\langle\partial P_\omega/\partial\theta_s\rangle_{\text{diff}}$.

For the system considered in this paper, the scattering amplitude $R_\omega(q|k)$ satisfies a reduced Rayleigh equation, which has the form

$$\int_{-\infty}^{\infty}\frac{dp}{2\pi}M_\omega(q|p)R_\omega(p|k) = -N_\omega(q|k), \tag{13}$$

where

$$M_\omega(q|p) = \frac{I[\alpha_1(q) - \alpha_0(p)|q - p]}{\alpha_1(q) - \alpha_0(p)} - r_1(q)\frac{I\{-[\alpha_1(q) + \alpha_0(p)]|q - p\}}{\alpha_1(q) + \alpha_0(p)}, \tag{14a}$$

$$N_\omega(q|k) = \frac{I[\alpha_1(q) + \alpha_0(k)|q - k]}{\alpha_1(q) + \alpha_0(k)} - r_1(q)\frac{I\{-[\alpha_1(q) - \alpha_0(k)]|q - k\}}{\alpha_1(q) - \alpha_0(k)}, \tag{14b}$$

with

$$\alpha_1(q) = \left(\epsilon\frac{\omega^2}{c^2} - q^2\right)^{1/2}, \qquad \text{Re}\,\alpha_1(q) > 0, \quad \text{Im}\,\alpha_1(q) > 0, \tag{15a}$$

$$\alpha(q) = \left[\epsilon(\omega)\frac{\omega^2}{c^2} - q^2\right]^{1/2}, \qquad \text{Re}\,\alpha(q) > 0, \quad \text{Im}\,\alpha(q) > 0, \tag{15b}$$

while

$$r_1(q) = \frac{\alpha_1(q) - \alpha(q)}{\alpha_1(q) + \alpha(q)}e^{2i\alpha_1(q)D} \tag{16}$$

is the Fresnel reflection coefficient at the planar dielectric film-metal interface, and

$$I(\gamma|Q) = \int_{-\infty}^{\infty} dx_1 e^{-iQx_1} e^{-i\gamma\zeta(x_1)}. \tag{17}$$

We now introduce the Green's function $G_\omega(q|k)$ for the surface and guided electromagnetic waves supported by the system with the rough vacuum-dielectric film interface as the solution of the Dyson equation

$$G_\omega(q|k) = 2\pi\delta(q-k)G_0(k,\omega) + \int_{-\infty}^{\infty} \frac{dr}{2\pi} G_\omega(q|r)V_\omega(r|k)G_0(k,\omega), \tag{18}$$

where $G_0(k,\omega)$ is a Green's function for the surface and guided waves supported by the system with planar surfaces

$$G_0(k,\omega) = i\frac{1 + r_1(k)}{[\alpha_1(k) + \alpha_0(k)] - [\alpha_1(k) - \alpha_0(k)]r_1(k)}, \tag{19}$$

and $V_\omega(q|k)$ is the scattering potential due to the surface roughness.

We assume a solution of Eq. (13) of the form

$$R_\omega(q|k) = -2\pi\delta(q-k) - 2iG_\omega(q|k)\alpha_0(k). \tag{20}$$

This form for $R_\omega(q|k)$ is suggested by the fact that when $G_\omega(q|k)$ is replaced by $2\pi\delta(q-k)G_0(q|k)$, i.e., in the case where the system has planar surfaces, Eq. (20) yields the Fresnel reflection coefficient for the latter system.

By substituting Eq. (20) into Eq. (13), and then using Eq. (18) to eliminate $G_\omega(q|k)$ from the resulting equation, we obtain the following integral equation for the scattering potential $V_\omega(q|k)$ in terms of the Green's function $G_0(q,\omega)$:

$$\int_{-\infty}^{\infty} \frac{dq}{2\pi} [N_\omega(p|q) - M_\omega(p|q)]\frac{V(q|k)}{2i\alpha_0(q)}$$
$$= \frac{N_\omega(p|k) - M_\omega(p|k)[1 + 2i\alpha_0(k)G_0(k,\omega)]}{2i\alpha_0(k)G_0(k,\omega)}. \tag{21}$$

At this point we replace the function $I(\gamma|Q)$ defined by Eq. (17) by

$$I(\gamma|Q) = 2\pi\delta(Q) + J(\gamma|Q), \tag{22}$$

where

$$J(\gamma|Q) = \int_{-\infty}^{\infty} dx_1 \exp(-iQx_1)\{\exp[-i\gamma\zeta(x_1)] - 1\}. \tag{23}$$

Then with the use of Eq. (19) we can rewrite Eq. (21) for the scattering potential in the form

$$V_\omega(p|k) = V_\omega^{(0)}(p|k) + \int_{-\infty}^{\infty} \frac{dq}{2\pi} \mathcal{K}_\omega(p|q) V_\omega(q|k), \tag{24}$$

where

$$
\begin{aligned}
V_\omega^{(0)}(q|k) &= \frac{i(\epsilon-1)(\omega^2/c^2)}{1+r_1(q)} \\
&\times \left\{ \left[\frac{J[\alpha_1(q)+\alpha_0(k)|q-k]}{\alpha_1(q)+\alpha_0(k)} - r_1(q)\frac{J\{-[\alpha_1(q)-\alpha_0(k)]|q-k\}}{\alpha_1(q)-\alpha_0(k)} \right] \right. \\
&\quad \times \{[\alpha_1(k)+\alpha_0(k)] - [\alpha_1(k)-\alpha_0(k)]r_1(k)\} \\
&\quad - \left[\frac{J[\alpha_1(q)-\alpha_0(k)|q-k]}{\alpha_1(q)-\alpha_0(k)} - r_1(q)\frac{J\{-[\alpha_1(q)+\alpha_0(k)]|q-k\}}{\alpha_1(q)+\alpha_0(k)} \right] \\
&\quad \left. \times \{[\alpha_1(k)-\alpha_0(k)] - [\alpha_1(k)+\alpha_0(k)]r_1(k)\} \right\} \frac{1}{1+r_1(k)}\frac{1}{2\alpha_0(k)},
\end{aligned}
\tag{25}
$$

and

$$
\begin{aligned}
\mathcal{K}_\omega(q|p) &= \frac{(\epsilon-1)(\omega^2/c^2)}{1+r_1(q)} \left\{ \left[\frac{J[\alpha_1(q)+\alpha_0(p)|q-p]}{\alpha_1(q)+\alpha_0(p)} - \frac{J[\alpha_1(q)-\alpha_0(p)|q-p]}{\alpha_1(q)-\alpha_0(p)} \right] \right. \\
&\quad - r_1(q) \left[\frac{J\{-[\alpha_1(q)-\alpha_0(p)]|q-p\}}{\alpha_1(q)-\alpha_0(p)} \right. \\
&\quad \left. \left. - \frac{J\{-[\alpha_1(q)+\alpha_0(p)]|q-p\}}{\alpha_1(q)+\alpha_0(p)} \right] \right\} \frac{1}{2\alpha_0(p)}.
\end{aligned}
\tag{26}
$$

Since we assumed that the surface is weakly rough in the sense that either the rms height is small compared to the wavelength, or the rms slope is very small, we solve Eq. (24) for the scattering potential iteratively. In order to have a clear physical picture of the different contributions to the scattering potential we rearrange the solution so that we can rewrite the expression for the scattering potential in the following form, which is exact through terms of $O(\zeta^2)$:

$$
\begin{aligned}
V_\omega(q|k) &= \frac{i(\epsilon-1)(\omega^2/c^2)}{1+r_1(q)} \left\{ \frac{J[\alpha_1(q)+\alpha_1(k)|q-k]}{\alpha_1(q)+\alpha_1(k)} \right. \\
&\quad - r_1(q)\frac{J\{-[\alpha_1(q)+\alpha_1(k)]|q-k\}}{\alpha_1(q)+\alpha_1(k)} r_1(k)
\end{aligned}
$$

$$+ \frac{J\alpha_1(q) - \alpha_1(k)|q-k]}{\alpha_1(q) - \alpha_1(k)} r_1(k)$$

$$-r_1(q) \frac{J[-(\alpha_1(q) - \alpha_1(k))]\,|q-k)}{\alpha_1(q) - \alpha_1(k)} \Bigg\} \frac{1}{1 + r_1(k)}. \tag{27}$$

The Green's function for the rough film on the metal surface is then obtained iteratively from Eq. (18) to second order in the scattering potential $V_\omega(q|k)$

$$G_\omega(q|k) = 2\pi\delta(q-k)G_0(q,\omega) + G_0(q,\omega)V_\omega(q|k)G_0(k,\omega) + G_0(q,\omega)$$

$$\times \int_{-\infty}^{\infty} \frac{dp}{2\pi} V_\omega(q|p)G_0(p)V_\omega(p|k)G_0(k,\omega). \tag{28}$$

The scattering amplitude $R_\omega(q|k)$ can be then determined from Eq. (20), and the mean differential reflection coefficient $\langle \partial P_\omega/\partial\theta_s\rangle_{\text{diff}}$, given by Eq. (11), can be calculated as a function of the frequency ω of the incident light and the angle of scattering θ_s:

$$\left\langle \frac{\partial P_\omega}{\partial\theta_s} \right\rangle_{\text{diff}} = \frac{1}{2\pi L} \frac{\omega^3}{c^3} \cos\theta_s \cos\theta_0 |G_0(q,\omega)|^2$$

$$\times \Bigg[\langle V_\omega(q|k)V_\omega^*(q|k)\rangle - \langle V_\omega(q|k)\rangle\langle V_\omega^*(q|k)\rangle$$

$$+ \left\langle \int_{-\infty}^{\infty} \frac{dp}{2\pi} \int_{-\infty}^{\infty} \frac{dp'}{2\pi} V_\omega(q|p)G_0(p,\omega)V_\omega(p|k) \right.$$

$$\times V_\omega^*(q|p')G_0^*(p',\omega)V_\omega^*(p'|k) \Big\rangle$$

$$- \left\langle \int_{-\infty}^{\infty} \frac{dp}{2\pi} V_\omega(q|p)G_0(p,\omega)V_\omega(p|k) \right\rangle$$

$$\times \left\langle \int_{-\infty}^{\infty} \frac{dp'}{2\pi} V_\omega^*(q|p')G_0^*(p',\omega)V_\omega^*(p'|k) \right\rangle$$

$$+ 2\mathrm{Re}\left\langle V_\omega^*(q|k) \int_{-\infty}^{\infty} \frac{dp}{2\pi} V_\omega(q|p)G_0(p,\omega)V_\omega(p|k) \right\rangle$$

$$- 2\mathrm{Re}\langle V_\omega^*(q|k)\rangle \left\langle \int_{-\infty}^{\infty} \frac{dp}{2\pi} V_\omega(q|p)G_0(p,\omega)V_\omega(p|k) \right\rangle \Bigg]$$

$$\times |G_0(k,\omega)|^2. \tag{29}$$

When calculating the averages appearing in Eq. (29), it is convenient to express them through cumulant averages [25]. Then Eq. (29) will contain the cumulant averages of products of two, three, and four scattering potentials. However, the cumulant averages of the products of more than two potentials can be neglected [26] in view of the effective weakness of both surface profile functions: $s_1(x_1)$ due to its small rms height, and $s_2(x_1)$ due to its long transverse correlation length. Thus, Eq. (29) takes the form

$$\left\langle \frac{\partial P_\omega}{\partial \theta_s} \right\rangle_{\text{diff}} = \frac{1}{2\pi L} \frac{\omega^3}{c^3} \cos\theta_s \cos\theta_0 |G_0(q,\omega)|^2 \left[\langle V_\omega(q|k) V_\omega^*(q|k) \rangle_c \right.$$

$$+ \int_{-\infty}^{\infty} \frac{dp}{2\pi} \int_{-\infty}^{\infty} \frac{dp'}{2\pi} \langle V_\omega(q|p) G_0(p,\omega) V_\omega^*(q|p') \rangle_c$$

$$\times \langle V_\omega(p|k) G_0^*(p',\omega) V_\omega^*(p'|k) \rangle_c$$

$$+ \int_{-\infty}^{\infty} \frac{dp}{2\pi} \int_{-\infty}^{\infty} \frac{dp'}{2\pi} \langle V_\omega(q|p) G_0(p,\omega) V_\omega^*(p'|k) \rangle_c$$

$$\times \langle V_\omega(p|k) G_0^*(p',\omega) V_\omega^*(q|p') \rangle_c$$

$$+ 2\text{Re} \int_{-\infty}^{\infty} \frac{dp}{2\pi} \langle V_\omega^*(q|k) V_\omega(q|p) \rangle_c G_0(p,\omega) \langle V_\omega(p|k) \rangle_c$$

$$\left. + 2\text{Re} \int_{-\infty}^{\infty} \frac{dp}{2\pi} \langle V_\omega^*(q|k) \langle V_\omega(q|p) \rangle_c G_0(p,\omega) V_\omega(p|k) \rangle_c \right]$$

$$\times |G_0(k,\omega)|^2. \tag{30}$$

In calculating the averages $\langle V_\omega(q|k) \rangle$ and $\langle V_\omega(q|k) V_\omega^*(q'|k') \rangle_c$ we utilize the results that [26]

$$\langle J(\gamma|Q) \rangle = 2\pi\delta(Q) \left\{ \exp\left[-(\delta_1^2 + \delta_2^2) \gamma^2/2 \right] - 1 \right\}, \tag{31a}$$

$$\langle J(\gamma_1|Q_1) J(\gamma_2|Q_2) \rangle_c = 2\pi\delta(Q_1 + Q_2) \exp\left[-(\gamma_1^2 + \gamma_2^2)(\delta_1^2 + \delta_2^2)/2 \right]$$

$$\times \int_{-\infty}^{\infty} du \exp(-iQ_1 u) \left(\exp\left\{ -\gamma_1\gamma_2 \left[\delta_1^2 W_1(|u|) \right. \right. \right.$$

$$\left. \left. \left. + \delta_2^2 W_2|u| \right] \right\} - 1 \right), \tag{31b}$$

so that the cumulant average of the product $V_\omega(q|p) V_\omega^*(r|k)$, namely $\langle V_\omega(q|p) V_\omega^*(r|k) \rangle_c = \langle V_\omega(q|p) V_\omega^*(r|k) \rangle - \langle V_\omega(q|p) \rangle \langle V_\omega^*(r|k) \rangle$, is found to be

$$\langle V_\omega(q|p) V_\omega^*(r|k) \rangle_c$$

$$= 2\pi\delta(q - p - r + k) \frac{(\epsilon - 1)^2 (\omega/c)^4}{[1 + r_1(q)][1 + r_1(p)][1 + r_1^*(r)][1 + r_1^*(k)]}$$

$$\times \sum_{n=1}^{\infty} \frac{1}{n!} \int_{-\infty}^{\infty} du\, e^{-i(q-p)u} \left[\delta_1^2 W_1(|u|) + \delta_2^2 W_2(|u|)\right]^n$$

$$\times \left\{ e^{-(\delta_1^2+\delta_2^2)[\alpha_1(q)+\alpha_1(p)]^2/2} [\alpha_1(q) + \alpha_1(p)]^{n-1} [1 - (-1)^n r_1(q) r_1(p)] \right.$$

$$\left. + e^{-(\delta_1^2+\delta_2^2)[\alpha_1(q)-\alpha_1(p)]^2/2} [\alpha_1(q) - \alpha_1(p)]^{n-1} [r_1(p) - (-1)^n r_1(q)] \right\}$$

$$\times \left\{ e^{-(\delta_1^2+\delta_2^2)[\alpha_1(r)+\alpha_1(k)]^2/2} [\alpha_1(r) + \alpha_1(k)]^{n-1} [1 - (-1)^n r_1(r) r_1(k)] \right.$$

$$\left. + e^{-(\delta_1^2+\delta_2^2)[\alpha_1(r)-\alpha_1(k)]^2/2} [\alpha_1(r) - \alpha_1(k)]^{n-1} [r_1(k) - (-1)^n r_1(r)] \right\}^{*} \tag{32}$$

In the limit of a small slope roughness, when only single scattering processes survive, the differential reflection coefficient is well described by the contribution $\langle |V_\omega(q|k)|^2 \rangle_c$ in Eq. (30). However, as soon as one of the components of the surface roughness contains high slopes (small transverse correlation length), the other contributions in Eq. (30) become important. In particular, the one described by the third term contains the enhanced backscattering effect. The appearance of the interference fringes in the intensity of the light scattered diffusely is well described by the contribution to $\langle \partial P_\omega / \partial \theta_s \rangle_{\text{diff}}$ from only $\langle |V_\omega(q|k)|^2 \rangle_c$:

$$\left\langle \frac{\partial P_\omega}{\partial \theta_s} \right\rangle_{\text{diff}} = \frac{\omega^7}{c^7} \cos\theta_s \cos\theta_0 (\epsilon - 1)^2 \left| \frac{1}{\alpha_1(q) + \alpha_0(q) - [\alpha_1(q) - \alpha_0(q)] r_1(q)} \right|^2$$

$$\times \sum_{n=1}^{\infty} \frac{1}{n!} \int_{-\infty}^{\infty} du\, e^{-i(q-p)u} \left[\delta_1^2 W_1(|u|) + \delta_2^2 W_2(|u|)\right]^n$$

$$\times \left| e^{-(\delta_1^2+\delta_2^2)[\alpha_1(q)+\alpha_1(k)]^2/2} [\alpha_1(q) + \alpha_1(k)]^{n-1} [1 - (-1)^n r_1(q) r_1(k)] \right.$$

$$\left. + e^{-(\delta_1^2+\delta_2^2)[\alpha_1(q)-\alpha_1(k)]^2/2} [\alpha_1(q) - \alpha_1(k)]^{n-1} [r_1(k) - (-1)^n r_1(q)] \right|^2$$

$$\times \left| \frac{1}{\alpha_1(k) + \alpha_0(k) - [\alpha_1(k) - \alpha_0(k)] r_1(k)} \right|^2 . \tag{33}$$

In the numerical calculations based on the results presented, it was assumed that the dielectric constant of the film is $\epsilon = 3.648 + i0.0075$, and the dielectric function of the substrate is given by the Drude expression with the plasma frequency of Al, $\omega_p = 15.3$ eV, and the electron collision time $\tau = 6 \cdot 10^{-15}$ s. Color-level plots of the logarithm of the contribution to the mean differential reflection coefficient from the light scattered diffusely as a function of the angles of incidence and scattering calculated in the case where the film thickness is $D = 8.9$ μm, the wavelength of the incident light is $\lambda = 632.8$ nm are presented in Fig. 1.

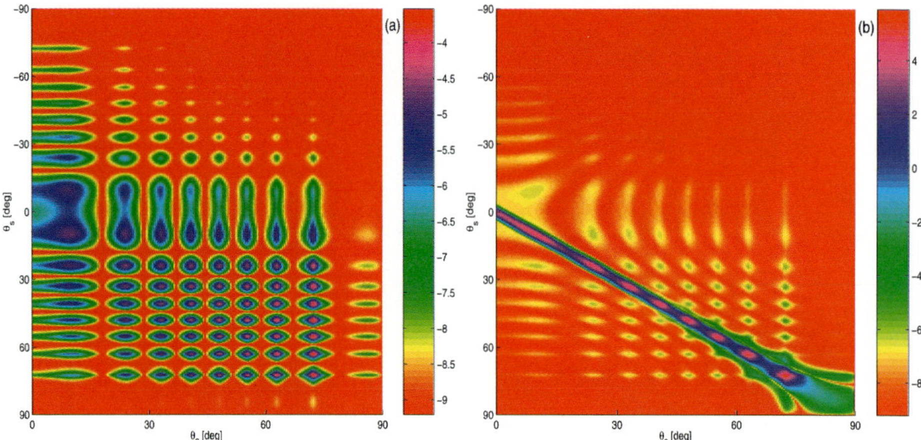

Figure 1 Color-level plots of the logarithm of the mean differential reflection coefficient of the scattered light calculated as a function of the angles of incidence and scattering. (a) $\delta_1 = 6$ nm, $a_1 = 300$ nm, $\delta_2 = 0$ (short scale roughness); (b) $\delta_1 = 6$ nm,
$a_1 = 300$ nm, $\delta_2 = 220$ nm, $a_2 = 10$ μm.

In the case illustrated in Fig. 1(a), the film surface is weakly rough, and
the dominant contribution to the mean differential reflection coefficient comes
from the first term of the sum in Eq. (33), which is governed by $|[1 + r_1(q)]$
$[1 + r_1(k)]|^2$ because the exponential factors are sensibly equal to unity. Thus,
the dependence of the mean differential reflection coefficient on the angles
of scattering and incidence factorizes. This leads to a static fringe pattern
(the Selényi fringes) that is modulated when the angle of incidence is varied.
The positions of the maxima of the intensity are determined by the maxima
of $|[1 + r_1(q)][1 + r_1(k)]|^2$. In the case of a highly reflective metallic sub-
strate the maxima occur when $\sin\alpha_1(q)D = 1 - \alpha_1^2(q)/[|\epsilon(\omega)|\omega/c]$, or when
$\sin\alpha_1(k)D = 1 - \alpha_1^2(k)/[|\epsilon(\omega)|\omega/c]$, while in the case of a perfectly conducting substrate $(r(q) = -\exp[2i\alpha_1(q)D])$ these conditions reduce to $\sin\alpha_1(q)D = 1$
and $\sin\alpha_1(k)D = 1$, respectively. The maxima therefore occur at the angles of
scattering and incidence given by

$$\theta_{s,0}^{(m)} = \sin^{-1}\left\{\epsilon - \left[\frac{\lambda}{2D}\left(m + \frac{1}{2}\right)\right]^2\right\}^{1/2}. \tag{34}$$

When the surface is moderately rough, the dominant term in the mean differential reflection coefficient is governed by the factor

$$\left|e^{-\delta_1^2+\delta_2^2[\alpha_1(q)+\alpha_1(k)]^2/2}[1 + r_1(q)r_1(k)] + e^{-(\delta_1^2+\delta_2^2)[\alpha_1(q)-\alpha_1(k)]^2/2}[r_1(k) + r_1(q)]\right|^2.$$

In this case the fringe pattern (the Quételet fringes) is produced by the second term in this sum, since $e^{-\delta^2[\alpha_1(q)-\alpha_1(k)]^2/2} \gg e^{-\delta^2[\alpha_1(q)+\alpha_1(k)]^2/2}$. Thus, the positions of the maxima of the interference fringes are determined by the maxima of the function $|r_1(k) + r_1(q)|^2$ or, if the substrate is perfectly conducting, by the condition $\cos^2[\alpha_1(q) - \alpha_1(k)]D = 1$, i.e., at the angles

$$\theta_s^{(m)} = \sin^{-1}\left[\epsilon - \left(\frac{m\lambda}{2D} + \sqrt{\epsilon - \sin^2\theta_0}\right)^2\right]^{1/2}. \tag{35}$$

In the approach taken in this paper, to model the scattering system used in the experimental work described in the following section, we have assumed that the surface roughness has two components. One of the components, $s_1(x_1)$, has a small rms height and a short correlation length, and only this component of the surface roughness was taken into account in calculations whose results are presented in Fig. 1(a). The second component of the surface roughness, $s_2(x_1)$, is characterized by a large rms height and a very large correlation length, so that the Rayleigh hypothesis, on which our work is based, remains valid for $s_1(x_1) + s_2(x_1)$. The fringe pattern in this case is shown in Fig. 1(b). The fringes move as the angle of incidence is varied, and there is always a maximum in the specular and retroreflection directions.

In Fig. 2 we present a color-level plot of the logarithm of the contribution to the mean differential reflection coefficient from the light scattered diffusely as a function of the angle of scattering and the wavelength of the incident light, calculated in the case where the film thickness is $D = 8.9$ μm, $\delta_1 = 6$ nm, $a_1 = 300$ nm,

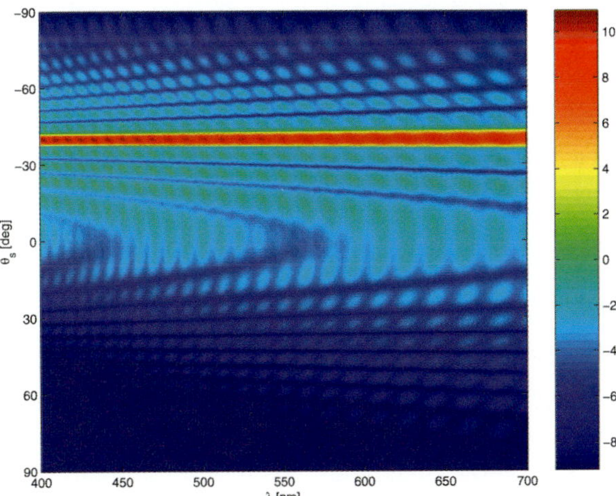

Figure 2 Color-level plot of the logarithm of the mean differential reflection coefficient of the scattered light calculated as a function of the wavelength and the angle of scattering. The angle of incidence is $\theta_0 = 40$ deg. The parameters of the scattering system are the same as in Fig. 1(b).

$\delta_2 = 220$ nm, $a_2 = 10$ µm, and the angle of incidence is $\theta_0 = 40$ deg. The Quételet fringes seen in Fig. 2 are modulated as the wavelength is varied. It is clearly seen, however, that for forward scattering, i.e., when the scattering angle is in the vicinity of the specular direction, the maxima of the intensity oscillations increase with a decrease of the wavelength, while they decrease with the decrease of the wavelength for backward scattering, i.e., when the scattering angle is in the vicinity of the retroreflection direction. The strong dependence of the mean differential reflection coefficient on the wavelength of the incident light allows us to expect a strong redistribution of the spectrum of the scattered light if the incident light is polychromatic.

To show that this expectation is confirmed, we assume that the spectral density of the incident light $S_0(\omega)$ has a Gaussian form with a central frequency $\omega_0 = 0.474 \times 10^{15}$ s^{-1} (which corresponds to the maximum at $\lambda = 632.8$ nm) and a half-width $\Delta\omega = 0.01\omega_0$,

$$S_0(\omega) = \frac{1}{\sqrt{\pi}\Delta\omega}\exp\left[-(\omega - \omega_0)^2/(\Delta\omega)^2\right]. \tag{36}$$

In this case the spectral density of the scattered light can be calculated from Eqs. (12) and (33). In Fig. 3 we present numerical results for the normalized spectral density of the scattered light (presented as a function of the wavelength rather than of the frequency) for the case where the light is incident at $\theta_0 = 40$ deg and is scattered into directions around $\theta_s = 69$ deg. To illustrate the effects of the scattering on the spectral distribution of the incident light we present in Fig. 3(a) the spectral density of the scattered light in the absence of any correlation-induced spectral changes. The spectral density of the light scattered from the surface of the dielectric film that is weakly rough and described by a single-scale roughness [the case illustrated in Fig. 1(a)] is shown in Fig. 3(b), and that of the light scattered from a surface described by a two-scale roughness [the case illustrated in Fig. 1(b)] is shown in Fig. 3(c). It should be noted that the maximum of the Selényi ring in the case illustrated in Figs. 1(a) and 3(b) at $\lambda = 632.8$ nm is at $\theta_s = 68.8$ deg, and the maximum of the Quételet ring in the case illustrated in Figs. 1(b) and 3(c) at $\lambda = 632.8$ nm is at $\theta_s = 69$ deg. The strong frequency dependence of the mean differential reflection coefficient results not only in a shift, but also in a strong redistribution of the spectral density of the scattered light, as can easily be seen in Fig. 3.

In Fig. 4 we present the analogous results for the scattering into backward directions around $\theta_s = -43$ deg. The maximum of the Selényi ring in the case illustrated in Figs. 1(a) and 4(b) at $\lambda = 632.8$ nm is at $\theta_s = -39.8$ deg, and the maximum of the Quételet ring in the case illustrated in Figs. 1(b) and 4(c) at $\lambda = 632.8$ nm is at $\theta_s = -39$ deg. The shift of the position of the maximum from the retroreflection direction is due to the fact that the Quételet rings are on the tail of the strong peak in the contribution to the mean differential reflection coefficient from the light scattered diffusely into the directions around the specular direction that arises due to the scattering by the long-scale surface roughness and mimics the power spectrum of this roughness, $g_2(|q-k|)$.

As is seen from the plots in Figs. 3 and 4, the spectral density of the scattered light changes considerably in scattering from both weakly rough and strongly rough surfaces that give rise to Selényi fringes and to Quételet fringes, respectively. The overall shifts of the maximum, however, have different signs. In the case of the Selényi rings, the scattering is stronger for wavelengths larger than the wavelength λ_0, while in the case of the Quételet rings it is stronger for wavelengths smaller than λ_0.

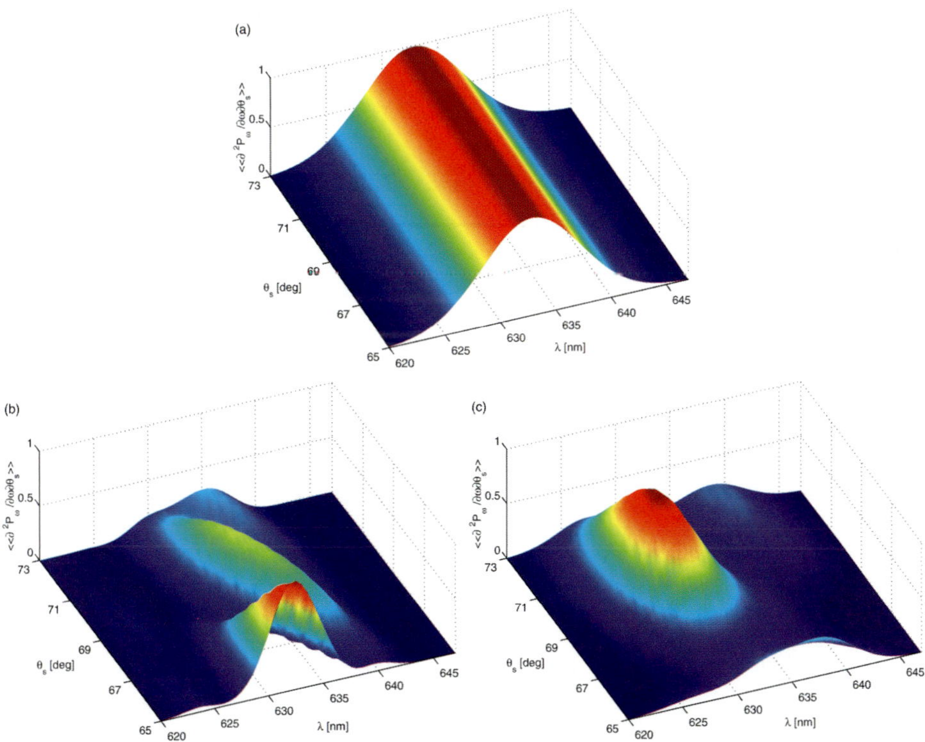

Figure 3 The spectral density of the light scattered into directions around $\theta_s = 69$ deg; (a) in the absence of the roughness-induced correlations, (b) a weakly rough surface with a single-scale roughness, and (c) a strongly rough surface with a two-scale roughness. The angle of incidence $\theta_0 = 40$ deg. The roughness parameters are the same as in Figs. 1(a) and 1(b), respectively.

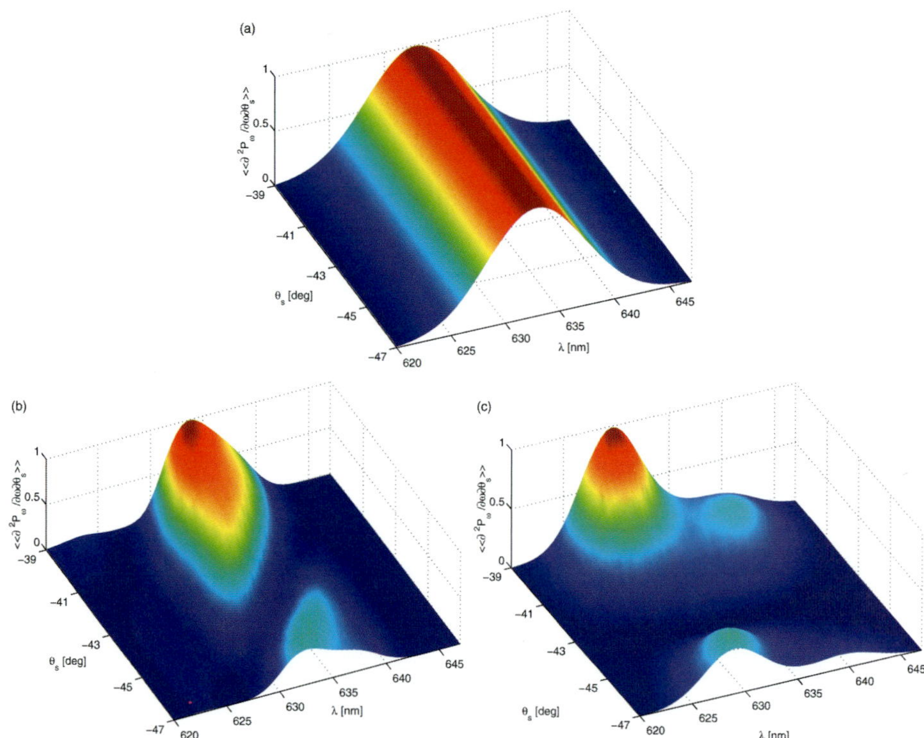

Figure 4 The spectral density of the light scattered into directions around $\theta_s = -43$ deg; (a) in the absence of the roughness-induced correlations, (b) a weakly rough surface with a single-scale roughness, and (c) a strongly rough surface with a two-scale roughness. The angle of incidence $\theta_0 = 40$ deg. The roughness parameters are the same as in Figs. 1(a) and 1(b), respectively.

11.3 Experimental Studies of Spectral Changes in the Scattering of Light from a Rough Dielectric Film on a Metallic Substrate

In this section we describe measurements of the in-plane copolarized scattering of s-polarized light from a dielectric film deposited on the planar surface of a metallic substrate, when the illuminated surface of the film is a two-dimensional randomly rough surface. The measured differential intensity of the scattered light displays a set of bright rings, whose positions depend on the angle of incidence, and there is always an interference ring present in the specular and backscattering directions. Thus, the scattering pattern displays the Quételet rings. The spectrum of the scattered light is measured at scattering angles that are very close to the angular positions of two of the Quételet rings present in the scattering pattern, and is found to differ from the spectrum of the incident light.

In our experiments, as a sample we used a smooth aluminum plate that was coated with a teflon film for high performance and protection. The upper surface of the dielectric film is weakly rough, and its surface profile constitutes a good approximation to a Gaussian random process with a Gaussian surface height autocorrelation function defined by an rms height of the surface $\delta = 6$ nm and a transverse correlation length of the surface roughness $a = 300$ nm. The thickness of the film is slowly varying around the mean thickness $D \cong 8.9$ μm, and its refractive index is almost frequency independent, $n = 1.91$ (the dielectric constant $\epsilon = 3.648$). Since the dielectric film is smooth, most of the energy goes into the specular direction; thus, a sensitive photomultiplier is used.

A fully automated bidirectional reflectometer was used to measure the fraction of incident light reflected by the sample into incremental angles over its field of view. It uses illumination by combination of a 150-W xenon lamp, a 75-W tungsten-halogen lamp, and a laser, and enables measurements to be taken for any combination of angles of incidence and scattering over the entire plane, except for a small angular range (about 0.5 deg away from the retroreflection direction) in which the source and detector mirrors interfere. A laser beam passes through a polarizer and is interrupted by a chopper and a half-wavelength plate, which rotates the polarization of the beam. Then it is directed toward the sample by a folded beam system that collimates it into a parallel beam up to 25 mm in diameter. For measurements, the beam size W is set to $W = 10$ mm. The sample is viewed by a movable off-axis paraboloid that projects the light reflected by the sample onto the detector via a polarizer and a folding mirror. Four different polarization combinations of input and receiving beams are recorded. The signal is recorded and digitized at each angular setting of interest throughout the angular range by an ITHACO lock-in amplifier, and the data are stored in the memory of a personal computer. The sample and the receiving telescope arm are separately mounted on two rotational stages run by two independent stepper motors that are controlled by a PC via a two-axis driver.

In Fig. 5 the mean differential intensity of the in-plane copolarized scattered light is plotted as a function of the angle of scattering for the case where s-polarized light of wavelength $\lambda = 632.8$ nm is incident on the film surface at two angles of incidence: (a) $\theta_0 = 0$ deg and (b) $\theta_0 = 40$ deg. The differential intensity of the scattered light displays a set of bright rings, which are positioned at the scattering angles $\theta_s = 0$ deg, $\theta_s \cong \pm 21.5$ deg, $\theta_s \cong \pm 31.1$ deg, $\theta_s \cong \pm 38.9$ deg, $\theta_s \cong \pm 46.3$ deg, $\theta_s \cong \pm 53.7$ deg, and $\theta_s \cong \pm 62$ deg in the case where $\theta_0 = 0$ deg, and at $\theta_s \cong -53$ deg, $\theta_s \cong -47$ deg, $\theta_s \cong -40$ deg, $\theta_s \cong -32$ deg, $\theta_s \cong 11$ deg, $\theta_s \cong 24$ deg, $\theta_s \cong 34$ deg, $\theta_s \cong 40$ deg, $\theta_s \cong 46$ deg, $\theta_s \cong 54$ deg, $\theta_s \cong 62$ deg, $\theta_s \cong 69$ deg, and $\theta_s \cong 79$ deg in the case where $\theta_0 = 40$ deg. A strong enhanced backscattering peak on top of the ring in the retroreflection direction is present in Fig. 5(b). The positions of the maxima in the

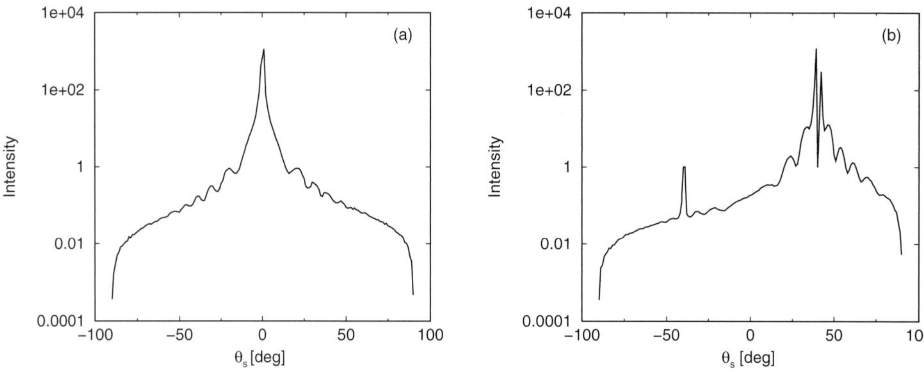

Figure 5 The experimental results for the mean differential intensity of the scattered light. (a) $\theta_0 = 0$ deg and (b) $\theta_0 = 40$ deg.

angular distribution of the intensity of the scattered light depend on the angle of incidence, and rings in the specular and backscattering directions are present; thus, the scattering patterns display Quételet rings.

To study the spectral redistribution of the scattered light, the wavelength dependences of the intensity of the light scattered into the directions of $\theta_s = 69$ deg and $\theta_s = -43$ deg have been measured. At the wavelength of the incident light $\lambda = 632.8$ nm the maximum of one of the Quételet rings is at $\theta_s = 69$ deg. At the same wavelength the angle of scattering $\theta_s = -43$ deg is almost at the minimum between two neighboring rings.

In Fig. 6 the normalized measured wavelength dependences of the intensity of the light scattered into the directions given by (a) $\theta_s = 69$ deg and (b) $\theta_s = -43$ deg are presented. In the same figures the wavelength dependences of the intensity of the incident light (red curves) and the calculated intensities of the scattered light (blue curves) are also presented. The intensities plotted in Fig. 6 are normalized by the maximum of the strongest peak in the spectrum of the incident and scattered radiation, respectively, so that the changes in the wavelength dependence can be clearly seen. The theoretical curve was obtained by multiplying the experimental data for the incident light (after converting them into a function of frequency) by numerical data for the mean differential reflection coefficient obtained from Eq. (33), and subsequently transforming the result into a function of the wavelength and normalizing it by the maximum value of the resulting distribution. As is seen from Fig. 6(a) in the case where the angle of scattering is $\theta_s = 69$ deg, the experimental results and the results of theoretical calculations are in good agreement, showing that stronger scattering occurs in the blue part of the spectrum. In the case where $\theta_s = -43$ deg the scattering is stronger in the red part of the spectrum, as was expected from the results of our numerical calculations. The agreement between the experimental and theoretical curves in the blue wing

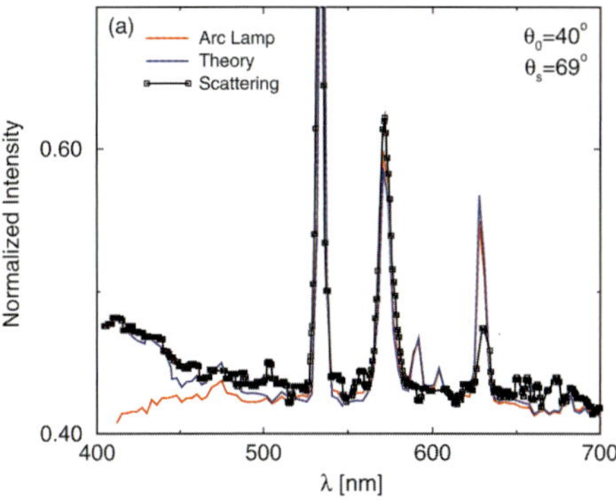
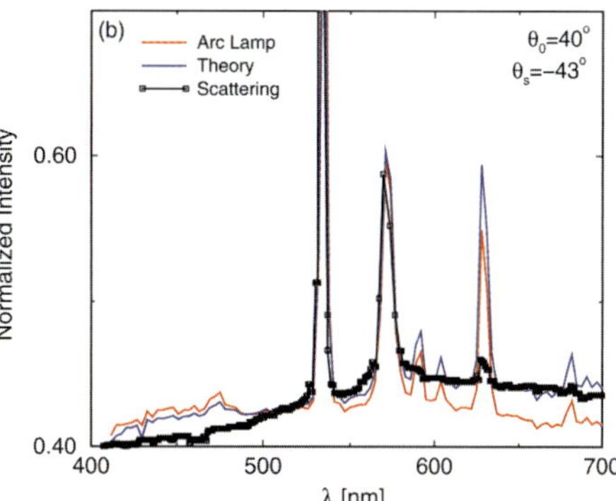

Figure 6 The wavelength dependence of the differential intensity of the light scattered into the directions (a) $\theta_s = 69$ deg and (b) $\theta_s = -43$ deg. The angle of incidence $\theta_0 = 40$ deg. The red curves give the intensity of the incident light (arc lamp), the black rectangles give the measured intensity of the scattered light, and the blue curves are the calculated wavelength dependences of the scattered light in the case where the spectral density $S_0(\omega)$ is obtained from the spectrum of the incident light. The parameters of the scattering system are the same as in Fig. 2.

of the spectrum can be improved by a slightly different choice of the value of the mean thickness of the film.

11.4 Conclusions

In this paper we have presented theoretical and experimental results for the change in the spectrum of the light scattered from a randomly rough dielectric film deposited on a metallic substrate, when the scattering angle is one of the angles at which a Selényi or a Quételet ring supported by the scattering system occurs, depending on the roughness of the film. We have shown theoretically that because the angular positions of these rings depend strongly on the wavelength of the incident light, large changes are expected in the spectrum of the light scattered at the position of one of these rings. Using a broadband source of the incident light we observed strong spectral changes in the wavelength dependence of the intensity of the light scattered at an angle at which a Quételet ring occurs. Thus, theory and experiment agree in demonstrating the occurrence of large, observable, spectral changes in the scattering of light from the randomly rough surface of a suitably chosen scattering system.

Acknowledgments

It is a pleasure to dedicate this paper to Prof. Emil Wolf, whose many pioneering contributions to theoretical optics have stimulated so much subsequent work, including ours. We wish him health, happiness, and many more years of creative energy.

This research was supported in part by Army Research Office Grant Nos. DAAD19-02-C-0056 and DAAD 19-02-1-0256.

References

1. E. Wolf, "Invariance of the spectrum of light on propagation," *Phys. Rev. Lett.* **56**, 1370–1382 (1986).
2. E. Wolf, "Non-cosmological redshifts of spectral lines," *Nature* (London) **326**, 363–365 (1987).
3. E. Wolf, "Red shifts and blue shifts of spectral lines emitted by two correlated sources," *Phys. Rev. Lett.* **58**, 2646–2648 (1987).
4. E. Wolf, "Redshifts and blueshifts of spectral lines caused by source correlations," *Opt. Commun.* **62**, 12–16 (1987).
5. E. Wolf, J.T. Foley, and F. Gori, "Frequency shifts of spectral lines produced by scattering from spatially random media," *JOSA* **6**, 1142–1149 (1989).
6. E. Wolf and J.T. Foley, "Scattering of electromagnetic fields of any state of coherence from space-time fluctuations," *Phys. Rev. A* **40**, 579–587 (1989).
7. J.T. Foley and E. Wolf, "Frequency shifts of spectral lines generated by scattering from space-time fluctuations," *Phys. Rev. A* **40**, 588–598 (1989).
8. A. Lagendijk, "Comment on 'Invariance of the spectrum of light on propagation,'" *Phys. Rev. Lett.* **65**, 2082 (1990).
9. A. Lagendijk, "Terrestrial redshifts from a diffuse light source," *Phys. Lett. A* **147**, 389–392 (1990).
10. T. Shirai and T. Asakura, "Spectral changes of light induced by scattering from spatially random media under the Rytov approximation," *JOSA A* **12**, 1354–1363 (1995).
11. T. Shirai and T. Asakura, "Multiple light scattering from spatially random media under the second-order Born approximation," *Opt. Commun.* **123**, 234–249 (1996).
12. T.A. Leskova, A.A. Maradudin, A.V. Shchegrov, and E.R. Méndez, "Spectral changes of light scattered from a bounded medium with a random surface," *Phys. Rev. Lett.* **79**, 1010–1013 (1997).
13. T.A. Leskova, A.A. Maradudin, A.V. Shchegrov, and E.R. Méndez, "Spectral changes of light scattered from a random metal surface in the Otto attenuated total reflection configuration," in *SPIE Proc.* Vol. 3426, 14–31, Bellingham, WA (1998).
14. Zong-Qi Lin and Zu-Han Gu, "Wolf effect from a real image as a secondary source," in *SPIE Proc.* Vol. 4100, 78–83, Bellingham, WA (2000).

15. Zu-Han Gu, "Angular spectrum redistribution from rough surface scattering," in *SPIE Proc.* Vol. 4447, 109–114, Bellingham, WA (2001).

16. P. Selényi, "Über lichtzerstreuung im raume wienerscher interferenzen and neue, diesen reziproke interferenzerscheinungen," *Ann. Phys. Chem.* **35**, 440–460 (1911).

17. L. Schläffi, "Über eine durch zerstreutes licht bewirkte interferenzerscheinung," *Mitt. Natureforsch. Ges* (Bern) **131–132**, 177–183 (1848).

18. C.V. Raman and G.L. Datta, "On Quételet's rings and other allied phenomena," *Philos. Mag.* **42**, 826–840 (1921).

19. Jun Q. Lu, José A. Sánchez-Gil, E.R. Mendez, Zu-Han Gu, and A.A. Maradudin, "Scattering of light from a rough dielectric film on a reflecting substrate: diffuse fringes," *JOSA A* **15**, 185–195 (1998).

20. Yu.S. Kaganovskii, V.D. Freilikher, E. Kanzieper, Y. Nafcha, and I.M. Fuks, "Light scattering from slightly rough dielectric films," *JOSA A* **16**, 331–338 (1999).

21. Sir Isaac Newton, *Optics* (originally published in London, 1704) 289, Dover, New York (1952).

22. T. Young, "The Bakerian lecture. On the theory of light and colors," *Philos. Trans. Roy. Soc. London*, Part I, 12–46 (1802).

23. Sir John Herschel, *On the Theory of Light*, Art. 676, Encyclopedia Metropolitana, London (1828).

24. A.A. Maradudin, A.R. McGurn, and V. Celli, "Localization effects in the scattering of light from a randomly rough grating," *Phys. Rev. B* **31**, 4866–4871 (1985).

25. R. Kubo, "Generalized cumulant expansion method," *Phys. Soc. Japan* **17**, 1100–1120 (1962).

26. T.A. Leskova, A.A. Maradudin, and I.V. Novikov, "Scattering of light from the random interface between two dielectric media with low contrast," *JOSA A* **17**, 1288–1300 (2000).

Professor Wolf and Zu-Han Gu at SPIE Conference AM100: Tribute to Emil Wolf:
Engineering Legacy of Physical Optics.

Zu-Han Gu received his B.S. degree from Jiaotong University, Shanghai, and his M.S. and Ph.D. from the University of California at San Diego in Electric Engineering. Since joining Surface Optics Corporation, he has been a chief scientist for more than 20 years. He has engaged in experimental and practical aspects of scattering from rough surfaces, especially enhanced backscattering, enhanced transmission in LIDAR signature, and memory effect of multiple light scattering from rough surfaces. Dr. Gu's current research interest includes the diffraction and scattering from remote targets, coherence and quantum optics, optical micro-surface chip design, fabrication and processing, and surface analysis. His interest also includes 3D holographic memory, optical pattern recognition and optical instruments. He is the author or coauthor of more than 90 technical papers, six chapters in books, and the editor for seven books. He has been the Principal Investigator for many government projects. He was an adjunct professor of the University of California at Los Angles. Dr. Gu is a Fellow of OSA and SPIE.

Tamara A. Leskova received her Ph.D. degree from the Institute of Spectroscopy of the Russian Academy of Sciences in Troitsk, Russia, in 1978, and has been a member of the research staff of that institute since then. Her research interests deal mainly with optical properties of nonideal surfaces. She is currently an assistant researcher at the University of California, Irvine.

Alexei A. Maradudin received his Ph.D. degree from the University of Bristol (U.K.) in 1957. After a faculty position at the University of Maryland and a staff position at the Westinghouse Research Laboratories, he joined the faculty of the University of California (UCI) in 1965 as Professor of Physics. Since 2002 he has been a Research Professor of Physics at UCI. His research interests include optical interactions at rough surfaces.

Chapter 12

Emil Wolf and Optics in the Czech Republic

Jan Peřina

12.1 Introduction

Emil Wolf was born in Prague, and after holding various positions he began work at the University of Rochester. He was there when Prof. B. Havelka contacted him during the ICO Congress in Paris in 1963. After that we had various levels of interaction, less fragmented before 1990 and more intense after 1990. However, from the earliest contacts, Prof. Wolf has strongly influenced modern optics in the Czech Republic (formerly Czechoslovakia), particularly its development in the branches of physical optics, optical physics, and quantum optics. Before 1990, a more systematic collaboration began after the International Optical Conference held in Jena in 1979, especially as related to contributions to *Progress in Optics*. After 1990 he visited the Czech Republic to cooperate with the Charles University, Academy of Sciences of the Czech Republic, and with the Palacký University in Olomouc, which awarded him a Gold Medal and an honorary doctorate. He also received a Gold Medal from the Czechoslovak Academy of Sciences. He became a member of Friends of the Palacký University, supporting scientific competitions of students, and an honorable member of Societas Scientiarum Bohemica. His most recent collaboration involved cooperating in the preparation of new volumes of *Progress in Optics*. This was also a period when he strongly supported us by providing copies of fundamental books. We should also mention his support in publishing the fundamental book in thin film optics by A. Vašíček [1]. In this contribution, we would like to summarize and recall some earlier results achieved under his influence. In particular we would like to mention imaging with partially coherent beams of arbitrary order, the phase problem, arbitrary ordering of field operators in multimode formulation, and the related inverse problem of reconstruction of the statistical properties of radiation from photocount measurements. The latter includes nonclassical states, application to generalized multimode superpositions of signal and quantum noise describing nonclassical behavior of systems, the quantum Zeno effect, spectral coherence, and nonlinear optical couplers. These results were further continued by systematic studies of the propagation of radiation in random and nonlinear media and of quantum theory of measurements, quantum cloning, quantum information, etc. Some of these results were reviewed earlier in several *Progress in Optics* articles.

12.2 Imaging with Partially Coherent Light of Arbitrary Order

A general formulation of imaging with partially coherent light of any statistical behavior is based on the $2n^{\text{th}}$ order correlation function $\Gamma(\mathbf{x}_1,\ldots,\mathbf{x}_{2n})$, which is considered as an object and image function and imaging is described by the complex amplitude diffraction function $K(\mathbf{x}'-\mathbf{x})$. Then the relation between the object and image functions is given by

$$\Gamma^{(i)}(\mathbf{x}'_1,\ldots,\mathbf{x}'_{2n}) = \int \ldots \int K^*(\mathbf{x}'_1 - \mathbf{x}_1)\ldots K(\mathbf{x}'_{2n} - \mathbf{x}_{2n})$$

$$\times \, \Gamma^{(o)}(\mathbf{x}_1,\ldots,\mathbf{x}_{2n})\,d\mathbf{x}_1 \ldots d\mathbf{x}_{2n}, \tag{1}$$

where the object and image correlation functions [2,3] are designated by the indices (o) and (i), respectively, and the asterisk signifies complex conjugation.

We can now distinguish linear imaging, provided that correlations of the instantaneous intensities are measured in the image, and we can distinguish nonlinear imaging, provided that the nth order intensity is measured. In the former case, we have the multidimensional spatial analysis quite analogous to the standard spatial analysis of imaging with partially coherent light. In the latter case Eq. (1) reads

$$\Gamma^{(i)}(\mathbf{x}',\ldots,\mathbf{x}') = \int \ldots \int K^*(\mathbf{x}' - \mathbf{x}_1)\ldots K(\mathbf{x}' - \mathbf{x}_{2n})$$

$$\times \, \Gamma^{(o)}(\mathbf{x}_1,\ldots,\mathbf{x}_{2n})\,d\mathbf{x}_1 \ldots d\mathbf{x}_{2n}, \tag{2}$$

where $\Gamma^{(i)}(\mathbf{x}',\ldots,\mathbf{x}')$ represents the nth order intensity $I^{(n)}(\mathbf{x}')$ at an image point $\mathbf{x}'$.

Performing the spatial Fourier analysis, we arrive at Ref. [4] (Chapter 9 and references therein)

$$\tilde{G}^{(o)}(\mu) = \int_{-\infty}^{+\infty} \cdots \int_{-\infty}^{+\infty} \mathcal{L}(\eta_1, \ldots, \eta_{2n})$$
$$\times \delta(\eta_1 + \cdots + \eta_n - \eta_{n+1} - \cdots - \eta_{2n} + \mu)$$
$$\times \tilde{u}^{(o)*}(\eta_1) \ldots \tilde{u}^{(o)}(\eta_{2n}) d\eta_1 \ldots d\eta_{2n}, \tag{3}$$

where $\tilde{u}$ is the object spatial frequency complex-field amplitude,

$$\mathcal{L}(\eta_1, \ldots, \eta_{2n}) = \int_{-\infty}^{+\infty} \cdots \int_{-\infty}^{+\infty} \tilde{g}^{(o)}(\mu_1', \ldots, \mu_{2n}')$$
$$\times \tilde{K}^{*}(\mu_1' + \eta_1) \ldots \tilde{K}(\mu_n' + \eta_{2n}) d\mu_1' \ldots d\mu_n' \tag{4}$$

is the transfer function, $\tilde{g}^{(o)}$ is the spatial Fourier transform of the degree of spatial coherence $\gamma^{(o)}(\mathbf{x}_1, \ldots, \mathbf{x}_{2n})$, and $\tilde{K}$ is the spatial Fourier transform of the diffraction function (pupil function). The well-known cases of spatial coherent or incoherent light are obtained as special cases. Nontrivial imaging is obtained only when the spatial frequencies are related as $\eta_1 + \cdots + \eta_n - \eta_{n+1} - \cdots - \eta_{2n} + \mu = 0$.

This general description of optical imaging can serve as a basis for reconstruction of the object from its image and for solving the problem of the similarity between the object and its image. We have considered two-point resolving power using a superposition of two Gaussian functions and the slit as a system. The information in the image for filtered spatial frequencies was reconstructed by means of analytic continuation. We were able to resolve two points separated 8% below the Rayleigh limit up to a 5% noise level in the image.

Considering real amplitudes $I^{(n)1/2n}(\mathbf{x}) = u(\mathbf{x})$ in imaging law [Eq. (2)], we can define the similarity between an object and its image as $u^{(o)}(\mathbf{x}) = \lambda u^{(i)}(\mathbf{x})$, which leads to the algebraic integral equation

$$u^{(o)2n}(\mathbf{x}') = \lambda^{2n} \int \cdots \int K^{*}(\mathbf{x}' - \mathbf{x}_1) \ldots K(\mathbf{x}' - \mathbf{x}_{2n})$$
$$\times \gamma^{(o)}(\mathbf{x}_1, \ldots, \mathbf{x}_{2n}) u^{(o)}(\mathbf{x}_1) \ldots u^{(o)}(\mathbf{x}_{2n}) d\mathbf{x}_1 \ldots d\mathbf{x}_{2n}. \tag{5}$$

This nonlinear equation can be solved numerically to obtain structures imaged similarly, in addition, some qualitative conclusions can be obtained, e.g., branching of a structure imaged similarly can arise for an eigenvalue λ into several structures imaged similarly with the same similarity coefficient λ.

12.3 Phase Problem

A review of the phase problem initiated in optics by Emil Wolf can be found in Ref. [4]. An interesting feature of the problem seems to be its formulation in terms of a singular integral equation following from the Hilbert transformation:

$$f(x) = -\frac{\tan\phi(x)}{\pi} P \int_{-\infty}^{+\infty} \frac{f(x')}{x'-x} dx', \tag{6}$$

where $f(x)$ is the imaginary part of the degree of coherence $\gamma(x) = |\gamma(x)| \times \exp[i\phi(x)]$, and P denotes the Cauchy principal value of the integral. This singular integral equation can be solved by means of the Sokhotski-Plemelj formulas, provided that the function $|\gamma(x)|$ tends to zero at least as $|x|^{-1}$ for large $|x|$, and that it possesses no zero in the complex plane, thus we obtain the well-known expression for the phase

$$\phi(x) = \frac{1}{\pi} P \int_{-\infty}^{+\infty} \frac{\ln|\gamma(x')|}{x'-x} dx'. \tag{7}$$

It is well known that this minimal phase may be drastically changed by the zero points of the degree of coherence lying in the complex plane, but little is known about a relation among the position of zero points and physical properties. However, if the degree of coherence has zeros in the complex plane, we can consider a small bright background $C > 0$ so that the function $\gamma'(x) = \gamma(x) + C$ has no zeros and the phase is uniquely determined [5]:

$$\tan\phi(x) = \frac{|\gamma'(x)|\sin\alpha(x)}{|\gamma'(x)|\cos\alpha(x) - C}, \tag{8}$$

where α is the phase of $|\gamma'(x)|$ obtained as the minimal phase.

12.4 Unified Correlation Tensors

As shown in Ref. [4], Chapter 7, a number of equations of motion and conservation laws (Ref. [3], Sect. 6.5), for correlation tensors $\mathcal{E}, \mathcal{H}, \mathcal{M}, \mathcal{N}$ can strongly be simplified, which is important for applications of the correlation theory in particular, if a super matrix is introduced:

$$\hat{\mathcal{K}} = \begin{pmatrix} \hat{\mathcal{E}} & \hat{\mathcal{M}} \\ \hat{\mathcal{N}} & \hat{\mathcal{H}} \end{pmatrix}, \tag{9}$$

together with the matrix

$$\hat{\sigma} = \begin{pmatrix} 0 & -1 \\ 1 & 0 \end{pmatrix}, \tag{10}$$

by which the supermatrix $\hat{\mathcal{K}}$ can successively be multiplied in various connections and its trace can be used. It holds that $\hat{\sigma}^{\dagger} = -\hat{\sigma}$. For instance, instead of eight equations of motion for single correlation tensors in vacuo, we have two equations:

$$\sum_{k,l} \epsilon_{jkl} \frac{\partial}{\partial x_{1k}} \hat{\mathcal{K}}_{lm}(x_1, x_2) - \frac{\hat{\sigma}}{c} \frac{\partial}{\partial t_1} \hat{\mathcal{K}}_{jm}(x_1, x_2) = \hat{0},$$

$$\sum_{k,l} \epsilon_{jkl} \frac{\partial}{\partial x_{2k}} \hat{\mathcal{K}}_{lm}(x_1, x_2) + \frac{1}{c} \frac{\partial}{\partial t_2} \hat{\mathcal{K}}_{jm}(x_1, x_2) \hat{\sigma} = \hat{0}, \tag{11}$$

where ϵ_{jkl} is a Levi-Civita-unit antisymmetric tensor and x denotes spatial and temporal variables. From these equations of motion we can derive wave equations in vacuo:

$$\Delta_{1,2} \hat{\mathcal{K}}_{jm} = \frac{1}{c^2} \frac{\partial^2}{\partial t_{1,2}^2} \hat{\mathcal{K}}_{jm},$$

$$\sum_{k,l,i,n} \epsilon_{jkl} \epsilon_{min} \frac{\partial}{\partial x_{1k}} \frac{\partial}{\partial x_{2i}} \hat{\mathcal{K}}_{ln} = \frac{1}{c^2} \frac{\partial^2}{\partial t_1 \partial t_2} \hat{\sigma} \hat{\mathcal{K}}_{jm} \hat{\sigma}^{-1}. \tag{12}$$

Similarly, we can write in a compact form equations for stationary fields, conservation laws and equations for spectral correlation tensors [4].

12.5 Arbitrary Ordering of Field Operators

Fundamental works in this field were published by Agarwal and Wolf [6] and Cahill and Glauber [7]. In Ref. [4] nontrivial multimode generalization leading to the generalized photodetection equation and an interesting inverse problem for determination of wave properties of radiation from photocount measurements are described (Chapter 16 and references therein).

Describing an M-mode field by the number operator $\hat{n} = \sum_{\lambda} \hat{a}_{\lambda}^{\dagger} \hat{a}_{\lambda}$ and the corresponding wave-integrated intensity $W = \sum_{\lambda} |\alpha_{\lambda}|^2$ (α_{λ} being the complex amplitude in the mode λ corresponding to the photon annihilation operator $\hat{a}_{\lambda}$), we can derive the following relation between the s-ordered and number-generating functions

$$\langle \exp(iy\hat{n}) \rangle_s = [1 - (1-s)iy/2]^{-M} \left\langle \left[\frac{1 + (1+s)iy/2}{1 - (1-s)iy/2} \right]^n \right\rangle, \tag{13}$$

where iy is a parameter of the generating function; or, we can obtain the relation between s_1- and s_2-ordered generating functions

$$\langle \exp(iy\hat{n}) \rangle_{s_2} = [1 + (s_2 - s_1)iy/2]^{-M} \left\langle \exp\left[\frac{iy\hat{n}}{1 + (s_2 - s_1)iy/2} \right] \right\rangle_{s_1}. \tag{14}$$

In particular, from Eq. (9) we can derive the generalized photodetection equation [8,9] that relates the photon number distribution $p(n)$ and quasi-distribution $P(W,s)$ of the integrated intensity related to s-ordering

$$p(n) = \frac{1}{\Gamma(n+M)} \left(\frac{2}{1+s} \right)^M \left(\frac{s-1}{s+1} \right)^n$$

$$\times \int_0^\infty P(W,s) L_n^{M-1}\left(\frac{4W}{1-s^2} \right) \exp\left(-\frac{2W}{1+s} \right) dW, \tag{15}$$

where L_n^{M-1} is the Laguerre polynomial and Γ is the Gamma function. In the limit $s \to 1$ for the normal operator ordering related to photodetection, we obtain the standard Mandel photodetection equation as the average of the Poisson kernel. With the exception of certain cases, we can state the general relations between integrated intensities distributions for two orderings and for their moments

$$P(W,s_2) = \frac{2}{s_1 - s_2} \int_0^\infty \left(\frac{W}{W'} \right)^{(M-1)/2} \exp\left[-\frac{2(W + W')}{s_1 - s_2} \right]$$

$$\times I_{M-1}\left[4\frac{(WW')^{1/2}}{s_1 - s_2} \right] P(W',s_1) dW', \quad \mathrm{Re}\, s_1 > \mathrm{Re}\, s_2, \tag{16}$$

where I_{M-1} is the modified Bessel function, and

$$\langle W^k \rangle_{s_2} = \frac{k!}{\Gamma(k+M)} \left(\frac{s_1 - s_2}{2} \right)^k \left\langle L_k^{M-1}\left(\frac{2W}{s_2 - s_1} \right) \right\rangle_{s_1}. \tag{17}$$

It is worth noting that this formulation provides a basis for solving various inverse problems, for example, to determine $P(W,s)$ from the photocount measurements giving $p(n)$. We can obtain, for example, for normal ordering

$$P_\mathcal{N}(W) = \exp[-(\zeta - 1)W] \sum_{j=0}^\infty c_j L_j^0(\zeta W), \tag{18}$$

where

$$c_j = \zeta \sum_{s=0}^{j} p(s) \frac{(-\zeta)^s}{(j-s)! s!} \tag{19}$$

or more generally for M-mode systems

$$P_{\mathcal{N}}(W) = W^{M-1} \sum_{j=0}^{\infty} c_j L_j^{M-1}(W), \tag{20}$$

where

$$c_j = \frac{j!}{\Gamma(j+M)} \sum_{s=0}^{j} \frac{(-1)^s p(s)}{(j-s)! \Gamma(s+M)}. \tag{21}$$

These reconstruction formulas have the advantage that they can be used to reconstruct approximate quasi-distributions from finite number measurements and also provide some regularized forms of quasi-distributions [when phase should also be included, a basis $\{\alpha^\lambda L_n^\lambda(|\alpha|^2)\}$ for a field complex amplitude α can be used, which is orthogonal with the weight $\exp(-|\alpha|^2)$].

The above formulas are also able to reconstruct nonclassical behavior of the quasi-distribution, e.g., its negative values, as applied, for instance, by Klyshko [10]. This can be done when defining a set of "moments" $M_n = p(n)n!$ and quadratic forms

$$q_m = \sum_{j,k=0}^{m} M_{j+k} u_j u_k \tag{22}$$

for an arbitrary nontrivial real vector $\{u_j\}$ and $m = 0, 1, \ldots$. If q_m are larger than zero for every m, then it is possible to construct the nonnegative distribution $P_{\mathcal{N}}(W)$ from a given sequence of moments M_n in a unique way. However, if for some m this quadratic form equals zero, there exists a unique quasi-distribution, the support of which is composed of a finite numbers of points equal to the minimal number of the m's for which the form equals zero. The criterion can be expressed in terms of determinants $\mathcal{D}_m = \mathrm{Det}\, M_{j+m}, j = 0, 1, \ldots, m$, because if these determinants are positive, then the forms are also positive (Ref. [11], Sect. 3.8).

12.6 Generalized Superposition of Signal and Quantum Noise

The above formalism can be applied for obtaining multimode superposition of coherent signal and noise, giving a generalization of the well-known Mandel-Rice formula and permitting a quantum generalization. Assuming a rectangular broadband spectrum of noise, the M-mode generating function is the generating function again for the Laguerre L_n^{M-1} polynomials (a polarized superposition is assumed),

$$\langle \exp(isW) \rangle_{\mathcal{N}} = \left(1 - is\frac{\langle n_{\mathrm{ch}} \rangle}{M} \right)^{-M} \exp\left[\frac{is\langle n_{\mathrm{c}} \rangle}{1 - is\dfrac{\langle n_{\mathrm{ch}} \rangle}{M}} \right], \tag{23}$$

where is is the parameter of the generating function, and $\langle n_{\mathrm{c}} \rangle$ and $\langle n_{\mathrm{ch}} \rangle$ are the mean numbers of coherent signal and chaotic noise photons (Ref. [11], Sect. 5.3 and references therein).

The corresponding distribution of the integrated intensity is expressed in terms of modified Bessel functions,

$$P_{\mathcal{N}}(W) = \frac{M}{\langle n_{\mathrm{ch}} \rangle} \left(\frac{M}{\langle n_{\mathrm{c}} \rangle} \right)^{(M-1)/2} \exp\left(-\frac{M + \langle n_{\mathrm{c}} \rangle}{\langle n_{\mathrm{ch}} \rangle} M \right)$$

$$\times I_{M-1}\left(2M\frac{[\langle n_{\mathrm{c}} \rangle W]^{1/2}}{\langle n_{\mathrm{ch}} \rangle} \right), \quad W \geq 0,$$

$$= 0, \quad W < 0. \tag{24}$$

For the photon number distribution we have

$$p(n) = \frac{1}{\Gamma(n + M)} \left(1 + \frac{\langle n_{\mathrm{ch}} \rangle}{M} \right)^{-M} \left(1 + \frac{M}{\langle n_{\mathrm{ch}} \rangle} \right)^{-n}$$

$$\times \exp\left(-\frac{\langle n_{\mathrm{c}} \rangle M}{\langle n_{\mathrm{ch}} \rangle + M} \right) L_n^{M-1}\left(-\frac{\langle n_{\mathrm{c}} \rangle M^2}{\langle n_{\mathrm{ch}} \rangle(\langle n_{\mathrm{ch}} \rangle + M)} \right). \tag{25}$$

The corresponding factorial moments are

$$\langle W^k \rangle_{\mathcal{N}} = \frac{k!}{\Gamma(k + M)} \left(\frac{\langle n_{\mathrm{ch}} \rangle}{M} \right)^k L_k^{M-1}\left(-\frac{\langle n_{\mathrm{c}} \rangle M}{\langle n_{\mathrm{ch}} \rangle} \right). \tag{26}$$

In special cases we obtain expressions for fully coherent or fully chaotic light, e.g., Poisson or Mandel-Rice distributions.

For nonlinear two-photon processes these formulas can be generalized so that they can describe nonclassical effects. The generating function "splits" as follows:

$$\langle \exp(isW) \rangle_{\mathcal{N}} = [1 - is(E-1)]^{-M/2} [1 - is(F-1)]^{-M/2}$$

$$\times \exp\left[\frac{isA_1}{1 - is(E-1)} + \frac{isA_2}{1 - is(F-1)}\right], \tag{27}$$

where $E = B - |C|$, $F = B + |C|$ represent quantum noise; $B = \langle \Delta\hat{a}^\dagger \Delta\hat{a} \rangle$ and $C = \langle (\Delta\hat{a})^2 \rangle$ are phase-independent and phase-dependent amplitude fluctuations, respectively; number 1 subtracted from E, F represents the subtraction of vacuum fluctuations in the normal operator ordering; and $A_{1,2}$ are signal components

$$A_{1,2} = \frac{1}{2}\left[\sum_{j=1}^{M} |\xi_j(t)|^2 \mp \frac{1}{2|C|}\left(C^* \sum_{j=1}^{M} \xi_j^2(t) + \text{c.c.}\right)\right], \tag{28}$$

where $\xi_j(t)$ are time-developed complex amplitudes and c.c. denotes the complex conjugate terms. For example, for the degenerate optical parametric process we obtain

$$B = \cosh^2(gt),$$

$$C = \frac{i}{2}\sinh(2gt)\exp(i\phi),$$

$$E - 1 = \frac{1}{2}[\exp(-2gt) - 1] \leq 0, \tag{29}$$

$$F - 1 = \frac{1}{2}[\exp(2gt) - 1] \geq 0,$$

$$A_{1,2} = \frac{1}{2}|\xi|^2\exp(\mp 2gt)[1 \mp \sin(2\theta - \phi)] \geq 0,$$

where g is a nonlinear coupling constant proportional to the quadratic susceptibility, ϕ is the phase of pumping, and θ is the initial phase of the signal.

We see that the quantum noise component $E - 1$ is negative for all times and we have a superposition of a signal with negative quantum noise. If the initial phases are suitably related, e.g., $2\theta - \phi = -\pi/2$, the first factor in the generating function [Eq. (27)] involving A_1 and $E - 1$ is dominating and the photon distribution reduces to sub-Poissonian distribution for short times. For later times oscillations in $p(n)$ occur. The photon number distribution and its factorial moments are expressed in terms of the Laguerre polynomials as follows:

Figure 1 The characteristic function of the generalized superposition of coherent signal and quantum noise, illustrated by Jitka Brůnová.

$$p(n) = \frac{1}{(EF)^{M/2}}\left(1 - \frac{1}{F}\right)^n \exp\left(-\frac{A_1}{E} - \frac{A_2}{F}\right)$$
$$\times \sum_{k=0}^{n} \frac{1}{\Gamma(k + M/2)\Gamma(n - k + M/2)}\left(\frac{1 - 1/E}{1 - 1/F}\right)^k$$

$$\times L_k^{M/2-1}\left(-\frac{A_1}{E(E-1)}\right)L_{n-k}^{M/2-1}\left(-\frac{A_2}{F(F-1)}\right), \tag{30}$$

$$\langle W^k\rangle = k!(F-1)^k \sum_{l=0}^{k}\frac{1}{\Gamma(l+M/2)\Gamma(k-l+M/2)}\left(\frac{E-1}{F-1}\right)^l$$

$$\times L_l^{M/2-1}\left(-\frac{A_1}{E-1}\right)L_{k-l}^{M/2-1}\left(-\frac{A_2}{F-1}\right). \tag{31}$$

Denoting $p_1(n)$ and $p_2(n)$ as partial virtual distributions related to factors containing A_1, E and A_2, F, respectively, the resulting distribution is a discrete convolution,

$$p(n) = \sum_{k=0}^{n} p_1(n-k)p_2(k). \tag{32}$$

In the above case of phase relations, the partial distribution p_1 is oscillating and takes on negative values, whereas the partial distribution p_2 is geometric and nonnegative. As a result of convolution [Eq. (32)], nonnegative resulting photon number distribution arises as an oscillating distribution.

Thus we have a generalization of the classical superposition of signal and noise, extending it to negative quantum noise components, which makes it possible to include nonclassical effects.

12.7 Quantum Zeno Effect

The quantum Zeno effect refers to the inhibition of the isolated temporal evolution of a dynamical system when the observation of such evolution is attempted (see a review in Ref. [12]). This observation is usually described by frequent measurements on the system performed in order to discover whether the initial system has changed or not. In the limit of very frequent measurements, continuous observation, or arbitrary high resolution, it may happen that the system is locked on its initial state, and the evolution, which is the aim of the observation, is in fact inhibited and does not occur. The effect was demonstrated using atomic transitions, neutron spin dynamics, etc. We have shown it using nonlinear effect of parametric down-conversion and Raman scattering [13–15].

We can consider a nonlinear crystal of length L that is pumped by a strong, classical, and coherent field to produce pairs of signal and idler photons via spontaneous parametric down-conversion. Using the interaction picture, this interaction is described by the effective Hamiltonian

$$\hat{H} = \hbar g\left(\hat{a}_s^\dagger \hat{a}_i^\dagger + \hat{a}_s\hat{a}_i\right), \tag{33}$$

Figure 2 Quantum Zeno effect, illustrated by Jitka Brůnová.

where $\hat{a}_s$ and $\hat{a}_i$ are the slowly varying annihilation operators for signal and idler beams respectively, and g is a coupling parameter depending on the pump field and the quadratic susceptibility of the medium. We have also assumed the frequency resonance condition $\omega_p = \omega_s + \omega_i$, where ω_p, ω_s, and ω_i are the frequencies of the pump, signal, and idler beams, respectively. We will denote by τ the interaction time associated with the length L of the crystal. We focus on the generation of the signal from the vacuum. The interaction Hamiltonian, together with the standard quantum lossy mechanism [11], produce, after the interaction time τ, the following general relation between the output operator $\hat{a}'_s$ for the signal field and the input signal and idler operators $\hat{a}_s$ and $\hat{a}_i$:

$$\hat{a}'_s = \mu\hat{a}_s + \nu\hat{a}_i^{\dagger} + \hat{L}_s, \tag{34}$$

where

$$\mu = \exp(-\gamma_s\tau/2)\cosh(g\tau), \quad \nu = -i\exp(-\gamma_s\tau/2)\sinh(g\tau), \tag{35}$$

$$\hat{L}_s = \sum_l w_{sl}\hat{b}_{sl} \tag{36}$$

is an operator related to the Langevin force for signal losses and it holds that

$$\sum_l |w_{sl}|^2 = 1 - \exp(-\gamma_s \tau) \tag{37}$$

here $\hat{b}_{sl}$ are initial reservoir operators, γ_s is a signal damping coefficient, and we have neglected rotating reservoir terms, which give negligible contribution in the optical region.

Now we assume that the crystal is divided into N equal parts of length $\Delta L = L/N$, with associated interaction time $\Delta \tau = \tau/N$ within each part. We can assume that the signal beams of each part are perfectly superimposed and aligned, and that reflection at each piece can be avoided or made negligible, for instance, embedding the N pieces in a linear medium with the same refractive index. On the other hand, the idler path is interrupted after each piece by means of mirrors, for instance. The output idler beams behind each piece are removed from the idler path, being replaced by new idler beams that are in vacuum. This modification makes it possible to observe the N output idler beams to detect the emission when it occurs, for instance, by means of N photodetectors. Then, the moment of emission can be inferred with accuracy $\Delta \tau$, and the relative resolution is given by the number of pieces $\Delta \tau / \tau = 1/N$. All losses related to various imperfections can be included in the lossy reservoirs.

Now we can examine the influence of this arrangement on the single-photon emission. The signal output operator after N pieces reads

$$\hat{a}'_{sN} = \mu^N \hat{a}_{s1} + \sum_{k=1}^{N} \mu^{N-k} \nu \hat{a}_{ik}^{\dagger} + \frac{1-\mu^N}{1-\mu} \hat{L}_{sN}, \tag{38}$$

where the coefficients in Eq. (35) are considered for $\Delta \tau$. Now the probability to have one signal photon is given by

$$\langle \hat{a}'^{\dagger}_{sN} \hat{a}'_{sN} \rangle_{\text{vac}} = N(g\Delta\tau)^2 + N^2 \langle n_{rsN} \rangle \gamma_s \Delta\tau$$

$$= \frac{(g\tau)^2}{N} + \langle n_{rs} \rangle \gamma_s \tau, \tag{39}$$

where we have considered $g\Delta\tau \ll 1$, $\gamma_{s,i}\Delta\tau \ll 1$ so that $\mu \simeq 1$, $\nu \simeq -ig\Delta\tau$, $\sum_l |w_{sl}|^2 \simeq \gamma_s \Delta\tau$, $\langle n_{rs} \rangle$ are mean numbers of reservoir oscillators, which are negligibly small in the optical region for room temperatures, and the signal beam is also initially in vacuum. We have two terms here: the first one arises from nonlinear dynamics, the second one from the signal lossy mechanism. The first term exhibits the quantum Zeno effect because no signal photons are radiated if the accuracy of the observation is increased by increasing N. In the limit of N tending to infinity, the probability of signal photon emission tends to zero and there is no emission at all. We can note that whether the attempted measurement on the idler modes is actually made or not appears to make no difference. It is sufficient that it could be made. In general, the losses in the signal beam degrade the quantum Zeno effect. This lossy effect is nonlocal because the measurement on the idler beam reduces the Zeno effect in the signal beam through its losses.

We see that for the unobserved system the N emitters are stimulated by the same vacuum, imparting phase correlations between them. On the observed system, the pieces are influenced by different and statistically independent vacuum fields leading to mutually incoherent emissions. This refers rather to the idler beam; however, the signal beam is also controlled through the strong quantum nonlocal correlations. Alternatively, the probability of emission on the unobserved system can be considered as the constructive interference between N possible and intrinsically indistinguishable ways for the emission to occur. When we interrupt the idler path N times, these ways become distinguishable by the possible detection of the idler photon. This possibility wipes out the interference, and the emission is modified. In the optical region, for room temperatures the obtained lossy effects in the signal beam are not critical for observation, N_{max} being of about 10^{20}.

If phase mismatch in the nonlinear process is also taken into account, the signal photon emission can be supported and the anti-Zeno effect may arise [14,16] in which signal photon emission is increased by the measurement. The above arrangement can be modified using the Kerr effect [13]. The influence of losses is similar.

12.8 Spectral Coherence

Further, we can mention the stimulating influence of the Wolf effect [17–19] on the research of Czech physicists in spectral coherence. In particular, original results were obtained for spherical symmetric fields [20–23] and partially coherent fields diffracted on gratings [24,25] and propagating in optical waveguides [26–29]. Some earlier experimental results in x-ray physics are also related [30].

12.9 Nonlinear Optical Couplers

The above results were continued by systematic studies of the propagation of radiation in random and nonlinear media [31] and of the evolution of quantum statistics in nonlinear optical processes [11,32,33]. Much effort has been devoted to the research of the quantum statistical properties of nonlinear optical couplers composed of two or three linear and nonlinear waveguides connected by evanescent

Figure 3 Evanescent waves in a nonlinear optical coupler, illustrated by Jitka Brůnová.

waves [34,35] used for generation and propagation of nonclassical light. Nonlinear waveguides operated by degenerate as well as nondegenerate optical parametric processes, by Raman or Brillouin scattering and Kerr effect, and also a bandgap coupler, were considered. Squeezing of vacuum fluctuations, sub-Poissonian photon behavior, collapses, and revivals of oscillations and properties of quantum phase were examined in single and compound modes in a fully quantum way in short-length approximations or in a parametric approximation of strong pumping. In some cases, symbolic computations to obtain higher-order fully quantum solutions were applied. Both regimes of codirectional and contradirectional propagation were considered. Substituting schemes and stability problems were also investigated.

Such composed nonlinear devices can be applied not only as sources for the generation and propagation of light exhibiting nonclassical properties and as switching devices, but also as elements for quantum measurements using a linear waveguide as a continuous probe device [36] and for the investigation of quantum coherence [37].

12.10 Conclusions

We have illustrated some results stimulated by scientific contacts with Emil Wolf. These results form a basis for recent studies in the quantum theory of measurements [38,39], in quantum propagation [40], and in quantum cloning and quantum information [41].

Two further contributions to *Progress in Optics* from the Czech Republic were in the field of scattering of light from multilayer systems with rough boundaries [42] and in ellipsometry of real thin film systems [43].

We are very happy that our life trajectory crossed the life trajectory of Emil Wolf several times.

Acknowledgment

The author acknowledges the support of the grant LN00A015 of the Czech Ministry of Education and Dr. Jitka Brůnová for interesting illustrations.

References

1. A. Vašíček, *Optics of Thin Films*, North-Holland, Amsterdam (1960).
2. E. Wolf, "Basic concepts of optical coherence theory," in *Proc. Symp. Opt. Masers*, 29–42, J. Wiley, New York (1963).
3. L. Mandel and E. Wolf, *Optical Coherence and Quantum Optics*, Cambridge Univ. Press, Cambridge (1995).
4. J. Peřina, *Coherence of Light*, 1st ed., Van Nostrand, London (1972); 2nd ed., D. Reidel, Dordrecht (1985).
5. R.E. Burge, M.A. Fiddy, A.H. Greenaway, and G. Ross, "The application of dispersion relations (Hilbert transforms) to phase retrieval," *J. Phys. D: Appl. Phys.* **7**, L65–L68 (1974).
6. G.S. Agarwal and E. Wolf, "Calculus for functions of noncommuting operators and general phase-space methods in quantum mechanics, I, II, III," *Phys. Rev. D* **2**, 2161–2186, 2187–2205 and 2206–2225 (1970).

7. K.E. Cahill and R.J. Glauber, "Ordered expansions in boson amplitude operators;" "Density operators and quasiprobability distributions," *Phys. Rev.* **177**, 1857–1881 and 1882–1902 (1969).

8. J. Peřina and J. Horák, "On arbitrary ordering of M-mode field operators in quantum optics," *Opt. Commun.* **1**, 91–94 (1969).

9. A. Zardecki, "Functional formalism for ordered electromagnetic field operators," *J. Phys. A* **7**, 2198–2210 (1974).

10. D.N. Klyshko, "Observable signs of nonclassical light," *Phys. Lett. A* **213**, 7–15 (1996).

11. J. Peřina, *Quantum Statistics of Linear and Nonlinear Optical Phenomena*, 1st ed., D. Reidel, Dordrecht (1984); 2nd ed., Kluwer, Dordrecht (1991).

12. P. Facchi and S. Pascazio, "Quantum Zeno and inverse quantum Zeno effects," in *Progress in Optics* **42**, E. Wolf, Ed., 147–217, Elsevier, Amsterdam (2001).

13. A. Luis and J. Peřina, "Zeno effect in parametric down-conversion," *Phys. Rev. Lett.* **76**, 4340–4343 (1996).

14. K. Thun and J. Peřina, "Zeno effect in optical parametric process with quantum pumping," *Phys. Lett. A* **249**, 363–368 (1998).

15. K. Thun and J. Peřina, "Quantum Zeno effect in Raman scattering," *Phys. Lett. A* **299**, 19–30 (2002).

16. A. Luis and L.L. Sánchez-Soto, "Anti-Zeno effect in parametric down-conversion," *Phys. Rev. A* **57**, 781–787 (1998).

17. E. Wolf, "Red shifts and blue shifts of spectral lines emitted by two correlated sources," *Phys. Rev. Lett.* **68**, 2646–2648 (1987).

18. E. Wolf, "Non-cosmological redshifts of spectral lines," *Nature* **326**, 363–365 (1987).

19. E. Wolf, "Influence of source correlations on spectra of radiated fields," *International Trends in Optics*, J.W. Goodman, Ed., 221–232, Academic Press, San Diego (1991).

20. M. Dušek, "Wolf effect in fields of spherical symmetry-energy conservation," *Opt. Commun.* **100**, 24–30 (1993).

21. M. Dušek, "Changes in the radiation spectrum arising on propagation in case of spherical symmetry," *Opt. Commun.* **95**, 189–191 (1993).

22. M. Dušek, "Spectral properties of stochastic electromagnetic fields with spherical symmetry," *Phys. Rev. E* **49**, 1671–1676 (1994).

23. M. Dušek, "The Wolf effect," *Fine Mech. Opt.* **41**(3), 70–76 (1996).

24. M. Dušek, "Diffraction grating illuminated by partially coherent beam," *Opt. Commun.* **111**, 203–208 (1994).

25. M. Dušek, "Diffration of partially coherent beams on three-dimensional periodic structures and the angular shifts of the diffraction maxima," *Phys. Rev. E* **52**, 6833–6840 (1995).

26. P. Hlubina, "Coherence of light at the exit face of a fibre waveguide analysed in the space-frequency domain," *J. Mod. Opt.* **42**, 1407–1426 (1995).

27. P. Hlubina, "The mutual interference of modes of a few-mode fibre waveguide analysed in the frequency domain," *J. Mod. Opt.* **42**, 2385–2399 (1995).

28. P. Hlubina, "Far zone spectral shifts in light propagation from the exit face of a few-mode fibre waveguide guiding the frequency-dependent linearly polarized modes," *J. Mod. Opt.* **43**, 1999–2008 (1996).

29. P. Hlubina, "Experimental demonstration of the spectral interference between two beams of a low-coherence source at the output of a Michelson interferometer," *J. Mod. Opt.* **44**, 1049–1059 (1997).

30. M. Čerňanský, "Broadening and shift of diffraction lines due to partial coherence of X-rays," *Phys. Stat. Sol. B* **114**, 365–372 (1982).

31. J. Peřina, "Photocount statistics of radiation propagating through random and nonlinear media," *Progress in Optics* **18**, E. Wolf, Ed., 127–203, North-Holland, Amsterdam (1980).

32. J. Peřina, Z. Hradil, and B. Jurčo, *Quantum Optics and Fundamentals of Physics*, Kluwer, Dordrecht (1994).

33. V. Peřinová, A. Lukš, and J. Peřina, *Phase in Optics*, World Scientific, Singapore (1998).

34. J. Peřina, Jr. and J. Peřina, "Quantum statistics of nonlinear optical couplers," *Progress in Optics* **41**, E. Wolf, Ed., 361–419, Elsevier, Amsterdam (2000).

35. J. Fiurášek and J. Peřina, "Quantum statistics of light propagating in nonlinear optical couplers," *Coherence and Statistics of Photons and Atoms*, J. Peřina, Ed., 65–110, J. Wiley, New York (2001).

36. J. Řeháček, J. Peřina, P. Facchi, S. Pascazio, and L. Mišta, Jr., "Quantum Zeno effect in a probed down-conversion process," *Phys. Rev. A* **62**, 013804-1–013804-6 (2000).

37. J. Řeháček, L. Mišta, Jr., J. Fiurášek, and J. Peřina, "Continuously induced coherence without induced emission," *Phys. Rev. A* **65**, 043815-1–043815-10 (2002).

38. V. Peřinová and A. Lukš, "Continuous measurements in quantum optics," *Progress in Optics* **40**, E. Wolf, Ed., 115–269, Elsevier, Amsterdam (2000).

39. A. Luis and L.L. Sánchez-Soto, "Quantum phase difference, phase measurements and Stokes operators," *Progress in Optics* **41**, E. Wolf, Ed., 421–481, Elsevier, Amsterdam (2000).

40. A. Lukš and V. Peřinová, "Canonical quantum description of light propagation in dielectric media," in *Progress in Optics* **43**, E. Wolf, Ed., 295–431, Elsevier, Amsterdam (2002).

41. S.Ya. Kilin, "Quanta and information," in *Progress in Optics* **42**, E. Wolf, Ed., 1–91, Elsevier, Amsterdam (2001).

42. I. Ohlídal, K. Navrátil, and M. Ohlídal, "Scattering of light from multilayer systems with rough boundaries," in *Progress in Optics* **34**, E. Wolf, Ed., 249–331, Elsevier, Amsterdam (1995).
43. I. Ohlídal and D. Franta, "Ellipsometry of thin film systems," in *Progress in Optics* **41**, E. Wolf, Ed., 181–282, Elsevier, Amsterdam (2000).

Emil Wolf and Jan Peřina in the front of the Rector Office at the Palacký University in Olomouc at the occasion of Dr. H.C. obtained by Emil Wolf, 1992. (Courtesy of Jan Perina, copyright 2004.)

Jan Peřina, a Fellow of OSA and Societas Scientiarum Bohemica, and a member of APS, EOS, SPIE, and Union of Czech Mathematicians and Physicists, is a professor at the Palacký University in Olomouc, Czech Republic. His main fields of research are quantum, statistical, and nonlinear optics. He is the author of about 300 papers and 6 books from these fields, and has contributed to development of theory of coherence and nonclassical light. He served as a member of international editorial boards of *Journal of Modern Optics (Optica Acta)*, *Optics Letters*, *Journal of Optics B: Quantum and Semiclassical Optics (Quantum Optics)*, *Acta Physica Polonica*, *Czechoslovak Journal of Physics*, *Acta Physica Slovaca*, *Fine Mechanics and Optics*, *Optica Applicata* and *Progress in Optics* and as a vice-president of the International Commission for Optics. He obtained various scientific awards and was honored by the City of Olomouc and by President of the Czech Republic.

CHAPTER 13

OPTICAL PATHLENGTH SPECTROSCOPY

Aristide Dogariu

13.1 Introduction

For a long time, the intensity and phase fluctuations determined by multiple light scattering were regarded as "optical" noise that degrades the radiation properties. Recently, remarkable advances in fundamental understanding and experimental methodologies proved that light propagation in random media is a source of unexplored physics with a wide range of potential applications. Among them, the medical applications occupy a special place, since it has been proven that scattering of optical radiation can be successfully used as a noninvasive investigation technique.

In many cases of practical interest, light propagating in dense scattering media can be described by the diffusion of scalar photons and therefore is fully characterized by their distribution of optical pathlengths. This is a comprehensive quantity that describes the statistics of photon random walk through many scattering events in a random medium. Experimental techniques such as diffusive wave spectroscopy [1] and coherent backscattering [2] rely on different models of the probability density of optical pathlengths in order to describe the measurements. The theoretical models typically use the time-resolved diffusion equation, which provides a satisfactory description of the photon transport phenomenon in scattering media whenever the absorption is not significant and the medium can be considered as infinite. Refined boundary conditions for the diffusion equation extend its applicability closer to the interfaces of finite-sized media. Nevertheless, there are many situations in which the scattering process cannot be treated as diffusive and, therefore, a direct way to *measure* the pathlength distribution is highly desirable when approximate theoretical values are not available anymore. For quite some time, direct time-of-flight measurements were the only experimental techniques able to provide direct information about the pathlength distribution of scattered light. However, low dynamic range and limited resolution impose severe limitations in using time-resolved measurements to characterize light propagation through highly scattering media.

Due to characteristics such as beam directionality and intensity, the use of highly coherent radiation produced by lasers has been the undisputed choice for many light scattering procedures. Recent developments in light sources and detection techniques offer new, more sophisticated experimental possibilities. By adjusting the coherence properties of light, one can use interferometric approaches to select specific orders of scattering and, therefore, directly infer the pathlengths distribution of photons scattered by a random medium. The background of this approach is introduced, different implementations are presented, and several applications are discussed in the next sections.

13.2 Multiple Light Scattering in Random Media

13.2.1 Radiative transfer

For random media with volume disorder often encountered in practice, the scattering phenomena depend essentially on the ratio between the characteristic length scales of the system and the radiation wavelength, suggesting that a statistical description in terms of such characteristic scattering lengths could be sufficient. In general, the particular location, orientation, and size of a scattering center are irrelevant, and the underlying wave character of the propagating field seems to be washed out. Because energy is transported through multiple scattering processes, what matters is only the energy balance. Of course, this approach cannot account for subtle interference and correlation effects, but refinements can be developed on the basis of a microscopic interpretation of radiative transfer [3].

A comprehensive mathematical description of the nonstationary radiative transport is given by Chandrasekhar [4]. The net effect of monochromatic radiation flow through a medium with a number density ρ of scattering centers is expressed in terms of the specific intensity $\mathcal{I}(\mathbf{r}, \mathbf{s}, t)$. This quantity is sometimes called radiance and is defined as the amount of energy that, at the position $\mathbf{r}$, flows per second and per unit area in the direction $\mathbf{s}$. When radiation propagates over the distance ds, there is a loss of specific intensity due to both scattering and absorption: $d\mathcal{I} = -\rho(\sigma_{sc} + \sigma_{abs})\mathcal{I}ds$. In the meantime, there is a gain of specific intensity due to scattering from a generic direction $\mathbf{s}'$ into the direction $\mathbf{s}$ quantified by the scattering phase function $P(\mathbf{s}', \mathbf{s})$. In general, there could be an increase, $\varepsilon(\mathbf{r}, \mathbf{s}, t)$, of specific intensity due to emission within the volume of interest and the net loss-gain balance represents the nonstationary radiative transfer equation [5]:

$$\left[\frac{1}{c}\frac{\partial}{\partial t} + \mathbf{s} \cdot \nabla + \rho(\sigma_{sc} + \sigma_{abs})\right]\mathcal{I}(\mathbf{r}, \mathbf{s}, t) = \rho\sigma_{sc}\int P(\mathbf{s}', \mathbf{s})\mathcal{I}(\mathbf{r}, \mathbf{s}, t)d\omega_{\mathbf{s}} + \varepsilon(\mathbf{r}, \mathbf{s}, t).$$

$$(1)$$

No analytical solution exists for the transfer equation and, in order to solve specific problems, one needs to assume functional forms for both the phase function and the specific intensity. Successive orders of approximation are obtained by spherical harmonic expansion of the specific intensity. One approximation, for instance, is obtained by expressing the diffuse radiance as a linear combination of an isotropic radiance and a cosine-modulated term.

13.2.2 Diffusion approximation

Perhaps one of the most widely used treatments for multiple light scattering is the diffusion approximation. When the absorption is small compared to scattering, scattering is almost isotropic, and if the radiance is not needed close to the source or boundaries, then the diffusion theory can be used as an approximation following from the general radiative transfer theory. To get insight into the physical meaning of this approximation it is convenient to define measurable quantities such as the diffuse energy density (average radiance) $U(\mathbf{r},t) = \int_{4\pi} \mathcal{I}(\mathbf{r},\mathbf{s},t)d\omega_{\mathbf{s}}$ and the diffuse flux $\mathbf{J}(\mathbf{r},t) = \int_{4\pi} \mathcal{I}(\mathbf{r},\mathbf{s},t)\mathbf{s}d\omega_{\mathbf{s}}$. In the diffusion approximation, the diffuse radiance is approximated by the first two terms of a Taylor's expansion [6]:

$$\mathcal{I}(\mathbf{r},\mathbf{s},t) \simeq U(\mathbf{r},t) + \frac{3}{4\pi}\mathbf{J}(\mathbf{r},t) \cdot \mathbf{s}, \tag{2}$$

and the following differential equation can be written for the average radiance:

$$D\nabla^2 U(\mathbf{r},t) - \mu_a U(\mathbf{r},t) - \frac{\partial U(\mathbf{r},t)}{\partial t} = S(\mathbf{r},t). \tag{3}$$

The isotropic source density is denoted by $S(\mathbf{r},t)$ and D is the diffusion coefficient, which is defined as

$$D = \frac{1}{3[\mu_a + \mu_s(1-g)]} \tag{4}$$

in terms of the absorption μ_a and scattering μ_s coefficients and the scattering asymmetry (anisotropy factor) g. Because the phase function associated with the scattering process is characterized by a single anisotropy factor, the diffusion approximation provides mathematical convenience. Through renormalization, an asymmetry-corrected scattering cross section that depends only on the average cosine of scattering angle defines the diffusion coefficient in Eq. (4) and, therefore, an essentially anisotropic propagation problem is mapped into an almost isotropic, diffusive model.

The photon migration approach based on the diffusion approximation has been very successful in describing the interaction between light and complex fluids or biological tissues [7]. It is instructive to note that three length scales characterize the light propagation in this regime: the absorption length $l_a = \mu_a^{-1}$, which is the distance traveled by a photon before it is absorbed, the scattering length $l_s = \mu_s^{-1}$, which is the average distance between successive scattering events, and the transport mean-free path $l^* = l_s/(1-g)$ that defines the distance traveled before the direction of propagation is randomized. In experiments that are interpreted in the frame of the diffusion approximation, l^* is the only observable quantity and, therefore, the spatial and temporal resolution are limited by l^* and l^*/c, respectively.

The following observation is worth making regarding the time of flight t. If one assumes a photon propagation with constant group velocity, this time can be immediately related to a characteristic length scale, the optical pathlength $s = v \cdot t$. Moreover, a pathlength probability density $P(s) = J(s) \int_0^\infty P(s)\,ds$ can be defined that represents the probability that photons traveled an equivalent optical pathlength in the interval $(s, s + ds)$. The pathlength distribution $P(s)$ fully describes various situations of strong multiple light scattering and can be considered as the solution of the diffusion equation for the specific boundary conditions and source specifics.

13.2.3 Various solutions for $P(s)$

In practice, diffusion Eq. (3) is solved subject to boundary conditions and source specifics. Once the energy density U is calculated, the current density $\mathbf{J}$ can be obtained using Fick's law $\mathbf{J}(\mathbf{r},t) = -D\nabla U(\mathbf{r},t)$ [8]. It is worth noting that in most scattering experiments, the measurable quantity is actually the diffuse flux $\mathbf{J}(\mathbf{r},t)$. For instance, the total optical power measured is proportional with the integral of $\mathbf{J}(\mathbf{r},t)$ over the area of the detector. Most appealing, however, is the fact that analytical solutions can be obtained for reflectance and transmittance calculations.

The first example is that of a scattering medium that is infinitely extended and has homogeneous scattering and absorption properties. In this case, it was shown that the diffuse energy density at a distance r away from a pointlike source is [9]

$$U(r,s) = (4\pi Ds)^{-3/2}\exp\left(-\frac{r^2}{4Ds} - \mu_a s\right). \tag{5}$$

Using this result, the corresponding pathlength distribution can be evaluated for a particular detection geometry by accounting for the acceptance angle of the detection system.

A situation often encountered in practice is that of a point source placed at the surface of a semi-infinite random medium. Setting an appropriate boundary condition such that U vanishes on the plane $z = -z_e$, the diffusion equation can be solved using Fick's law, and the pathlength resolved reflectance can be evaluated to be

$$P(r,s) = (4\pi Dc)^{-3/2} z_e s^{-5/2} \exp(-\mu_a s) \exp\left(-\frac{r^2 + z_e^2}{2Ds}\right), \tag{6}$$

where r is now the distance between the injection and collection points measured across the surface of the semi-infinite random medium. In the next section we will use this result for the particular case when the points of injection and detection coincide, i.e., $r = 0$.

Another situation of practical interest is that of a slab of finite thickness d. Using the method of images and appropriate boundary conditions, one finds that the pathlength dependence of the energy flux is given by [10]

$$P_R(r,s,d) = (4\pi Dc)^{-3/2} s^{-5/2} \exp(-\mu_a s) \exp\left(-\frac{r^2}{4Ds}\right) F_R(d,z_e), \tag{7}$$

for the reflection geometry, while a similar calculation for transmission leads to

$$P_T(r,s,d) = (4\pi Dc)^{-3/2} s^{-5/2} \exp(-\mu_a s) \exp\left(-\frac{r^2}{4Ds}\right) F_T(d,z_e), \tag{8}$$

where F_R and F_T are model-specific functions depending on the extrapolation length z_e and the slab thickness. Note that in all examples presented here the energy flux corresponding to large optical pathlengths behaves like $s^{-5/2}$, which is a general feature of the diffusive behavior and does not depend on the experimental geometry.

It is also worth mentioning that the extrapolation length z_e depends on the effective reflectivity R_{eff} at the boundaries. The extrapolation length ratio z_e/l^* is critical for describing light diffusion inside bounded media and it will be discussed further in Sect. 13.4.2.

13.3 Using Coherence to Isolate Scattering Orders

13.3.1 Low-coherence interferometry in random media

Developed initially in the field of fiber optics, low-coherence interferometry (LCI) [11] has become a widely used technique for various applications involving biomedical imaging. The use of light sources with short temporal coherence offers the depth resolution needed for optical imaging; the method is generally referred to as optical coherence tomography (OCT) [12]. So far, LCI has been used as a filter that suppresses the multiple light-scattering contribution and preserves the single scattering component characterized by well-defined scattering angles and polarization. Not long ago, LCI was also used to characterize the multiple scattering regime of wave propagation in random media [13].

In an LCI setup as depicted in Fig. 1, light from a broad-bandwidth source is first divided into probe and reference beams that are retroreflected from a targeted scattering medium and from a reference mirror, respectively, and are subsequently recombined to generate an interference signal. Assuming quasi-monochromatic optical fields ($\Delta\lambda/\lambda \ll 1$), the detected intensity I_d has the simple form of

$$I_d = I_0 + I_{\text{ref}} + 2 \cdot \sqrt{I_0} \cdot \sqrt{I_{\text{ref}}} \cdot \gamma(\Delta s)\cos(2\pi \cdot \Delta s/\bar{\lambda}), \qquad (9)$$

where I_0 and I_{ref} are the intensities scattered by the multiple scattering medium and by the reference mirror, respectively, whereas $\gamma(\Delta s)$ is the complex degree of coherence associated with the light source. The optical path difference between the scattered and reference fields is denoted by Δs and $\bar{\lambda}$ is the central wavelength. In Eq. (9), two conditions are needed in order to obtain interference maxima: (1) Δs to be a multiple of wavelength, and (2) $\Delta s < L_c$, where $L_c = \bar{\lambda}^2/\Delta\lambda$ is the coherence length of the source. Also, it is worth noting that the simple form of the interference law described by Eq. (9) assumes that the two beams returning from the target and the reference mirror maintain their full spatial coherence. If this not the case, the concept of interferometric phase must be carefully examined. It is known that monochromaticity is not a necessary condition for interference and it is rather a restriction on the spatial coherence that is required [14]. It has been pointed out that if a strict condition of spatial coherence is imposed, the statistically averaged behavior of a polychromatic field can be described by an associated

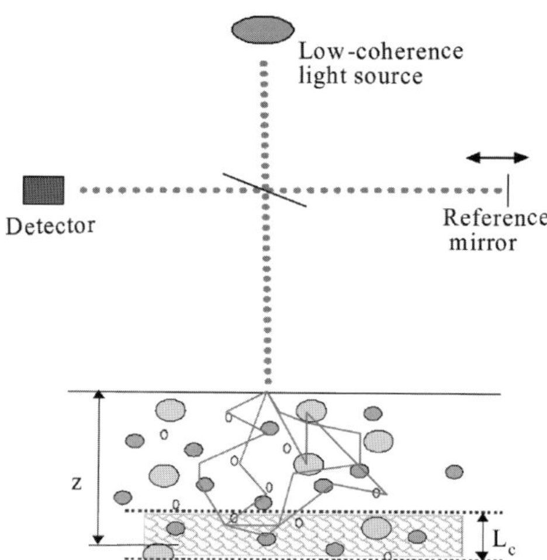

Figure 1 In a typical Michelson interferometer, the use of low-coherence radiation allows isolating scattering contributions with desired pathlengths.

monochromatic wave [15]. In a recent experiment, it has also been shown that for spatially coherent polychromatic optical fields, the measurement of the phase of the second-order correlations determines the phase of this associated, spatially coherent field [16].

In practice, the signal intensity I_d can be processed through conventional envelope detection to determine the total reflectance corresponding to the backscattered field I_{ref}. When the propagation through the random medium is determined by a multitude of optical paths with different lengths as depicted in Fig. 1, one can adjust the position of the reference mirror such that a specific pathlength s in the medium is matched (within the coherence length L_c). Accordingly, the detected signal measures only the contribution $I_{\mathrm{ref}}(s)$ of the class of waves that have traveled an optical distance s. In the optical pathlengths domain, the interferometer acts as a bandpass filter with a bandwidth given by the coherence length of the source. Accordingly, the shorter the coherence length, the narrower the optical pathlengths interval that is recorded. Now, if we let the reference mirror sweep the reference arm, waves with different optical pathlengths are detected and an optical pathlength distribution is reconstructed.

The procedure outlined here describes, of course, the scattering process at the central wavelength $\bar{\lambda}$. This assumption is often justified in practice when relatively narrow-band sources of radiation are used and dispersion effects can be neglected. In high-resolution experiments, however, the monochromatic approach fails and a full acount of the actual spectral density is required [17].

13.3.2 Optical pathlength spectroscopy

By scanning the position of the reference mirror in the standard Michelson interferometer of Fig. 1, one effectively determines the pathlength-resolved contributions $I_{\mathrm{ref}}(s)$ to the total reflection. After appropriately normalizing the detected signal, the probability density $P(s)$ of optical pathlength through the medium is obtained. The procedure is called optical pathlength spectroscopy (OPS) [18] and allows the direct measurement of the pathlength distribution $P(s)$ of backscattered waves in the particular geometry of the experiment.

In a fiber optics configuration of the interferometer, the probability density $P(s)$ is measured for the situation when the point source and the point detector coincide, as shown in Fig. 2. In this case, a typical OPS signal consists of backscattered intensity contributions corresponding to waves scattered along closed loops that have the same optical pathlengths and, in addition, have the total momentum transfer equal to $4\pi/\lambda$ (backscattering). In the following, we will restrict ourselves to this geometry, but it is straightforward to imagine other source-detector configurations in which the points of photon injection and detection do not coincide. Using fiber optics for launching and collecting the light, $P(s)$ can then be determined interferometrically following a procedure similar to the one described here.

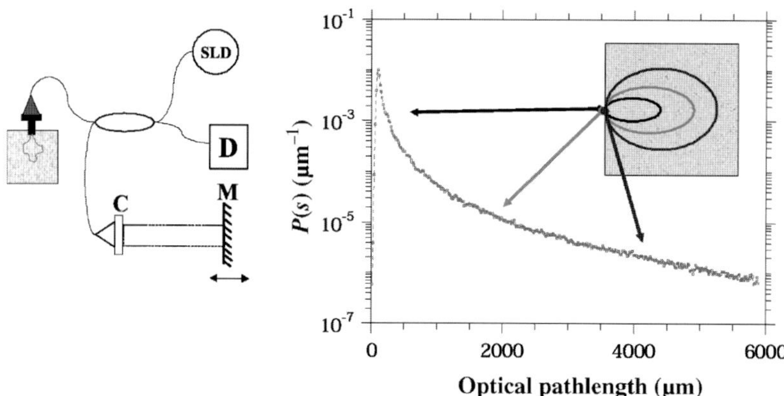

Figure 2 Fiber optic implementation of the OPS measurement setup that provides the direct measurement of the pathlength distribution.

It should be noted that somewhat similar information could be obtained by time-resolved measurements. As long as the waves propagate with a constant average velocity, the waves characterized by the same optical pathlengths can be considered as being emitted at the same moment t_0. Thus, the steady-state source could be replaced by a short pulse emitted at t_0, and the pathlength probability density could be determined using experimentally more demanding time-of-flight measurements.

In practice, the OPS approach is limited by the fact that the signal corresponding to long paths within the medium is weak and a large dynamic range is needed for accurate measurements of pathlength distribution tails. However, as opposed to dynamic techniques, the measurements can be extended over longer periods of time and there is no need for sophisticated time-of-flight configurations.

An example of an OPS signal is illustrated in Fig. 3, where the pathlength-resolved backscattered intensities corresponding to water suspensions of polystyrene microspheres with particle diameter of 0.46 μm and various volume fractions are shown. The values of normalized backscattered intensities have been compared with the corresponding solutions of the diffusion approximation given in Eq. (6). In obtaining these results, absorption effects have been neglected because the absorption lengths of the media are roughly two orders of magnitude longer than the corresponding scattering lengths. As can be seen, the diffusion theory makes a good description of the experimental data corresponding to different volume fractions, i.e., different scattering properties. Applying the Mie scattering theory, one can calculate the values of transport mean-free paths for the media examined in Fig. 3. Excellent agreement is obtained between the Mie-based estimations of l^* and the results obtained by fitting the measurement data within the diffusion approximation. For example, in the cases shown in Fig. 3, the measured l^* values of 197, 101, and 49.2 μm are to be compared with 206, 103,

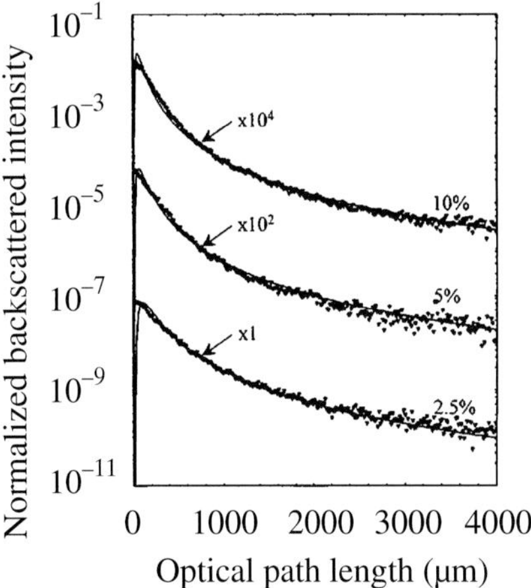

Figure 3 Probability density functions for optical pathlengths measured in water solution of polystyrene microspheres 0.46 μm in diameter and having the volume fractions of 2, 2.5, and 10%.

and 51.5 μm, respectively, obtained from Mie theory. The remarkable agreement proves that reliable measurements of photon transport mean-free path can be based on OPS.

Another example is presented in Fig. 4, where OPS is applied to characterize the structure of porous media such as thin membranes (mixtures of cellulose esters, polyvinylidene fluoride, and polycarbonate). The membranes consist of interconnected networks of pores that are strong light scatterers, with the scattering strength depending on the refractive index contrast and therefore on the membrane water content. As can be seen in Fig. 4, the presence of water diminishes the refractive index contrast, which reduces the scattering ability and leads, on average,

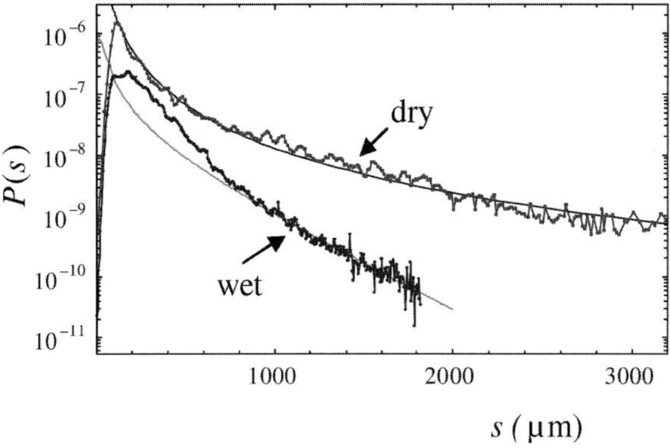

Figure 4 Pathlength distribution measured on porous membranes in dry and wet states, as indicated. The measurements are based on the experimental arrangement of Fig. 1.

to a shortened average optical pathlength in comparison with the highly scattering (dry) case [19].

The measured OPS signals are well described, at least the tails of the pathlength distributions are, by solutions of the diffusion approximation applied to slabs of finite thickness. The results of fitting the measured OPS with the prediction of Eq. (7) are also shown in Fig. 4. Once a reliable structural model is available, the measured pathlength spectra, or just moments of the distribution, can be further used to determine various characteristics related to pore sizes or overall porosity of random media [20].

13.4 Applications of OPS

13.4.1 Dynamic light scattering

Dynamic light scattering (DLS) has been established as a powerful technique for investigating dynamic processes. Measuring the temporal fluctuations of the scattered light, detailed information about the dynamics of the scattering medium can be extracted [21]. Originally, the applications were limited to weakly scattering media, where light propagation could be described by single-scattering models. An important breakthrough in the field of DLS is represented by the extension to strongly scattering media. The technique is referred to as diffusing wave spectroscopy (DWS) and has been used to study particle motion in concentrated fluids such as colloids, microemulsions, and other systems that are characterized by strong multiple scattering [22].

In a regime of complete diffusion of light, the DWS technique exploits the variations of light arising from phase and amplitude fluctuations determined by the dynamic scattering centers. To analyze the temporal autocorrelation, DWS relies on accurate knowledge of the optical pathlength distribution $P(s)$ of waves propagating through the multiple scattering medium. The pathlength distribution is usually obtained by solving a diffusionlike equation [such as Eq. (3)] with appropriate boundary conditions. As we discussed before, the approach is suitable for situations in which the characteristic scattering lengths are much smaller than the geometric dimensions of the scattering medium and the absorption length is much longer than both scattering length and sample size. Situations of practical interest are often at odds with these assumptions; subdiffusive photon regimes are encountered in systems with localized dynamics or flows, inhomogeneous absorption, etc. Accordingly, in order to accurately assess the dynamic structure, it is highly desirable to provide a simultaneous and independent measurement of photon pathlength distribution.

In treating the temporal fluctuations of multiply scattered light, one evaluates the scattered electric field as a sum over all scattered photon trajectories. The system's dynamics can be described as an incoherent summation of contribution corresponding to different pathlengths. For a specific photon path of length s, the electric field autocorrelation function is

$$G_1(\tau,s) \sim P(s)\frac{\langle E^*(t,s)E(t+\tau,s)\rangle}{\langle |E(t,s)|^2\rangle}, \tag{10}$$

where $E(t,s)$ is the electric field corresponding to photons that have traveled a distance s and $P(s)$, and the pathlength distribution is the fraction of total scattered intensity that traverses a path of length s from the point of injection to the point where it is detected. It has been shown that the autocorrelation of the electric field $g^{(1)}$ can be related to the characteristic diffusion time τ_0 of a noninteracting system of colloidal particles through [23]

$$g^{(1)}(\tau) = \int_0^\infty P(s)\exp\left(-\frac{2\tau}{\tau_0}\frac{s}{l^*}\right)ds, \tag{11}$$

where $\tau_0 = (Dk_0)^{-1}$; D is the diffusion coefficient of the scatterers in the suspending fluid and k_0 is the wave vector associated with the optical field. For Brownian particles of diameter ϕ, the diffusion coefficient relates to the temperature T and the viscosity η of the medium through the well-known Stokes-Einstein expression $D = k_B T/(3\pi\eta\phi)$, where k_B is the Boltzmann's constant. Note that there is no assumption of photon diffusion implied in the derivation of this result. It is evident from Eq. (11) that the autocorrelation function still features a negative exponential behavior, as in conventional DLS, but now it also depends on the optical pathlength s. It is also apparent that longer paths will decorrelate faster in time, while shorter paths will decorrelate more slowly. This fact can be easily explained by recognizing that the fluctuations of light undergoing a certain trajectory are due to a cumulative effect of the total number of scattering events. Finally, it is worth mentioning that the temporal autocorrelation function depends on the geometry of the experiment only through the boundary conditions as applied to obtain the probability density function $P(s)$.

In practice, the intensity correlation $g^{(2)}(\tau)$ rather than the field autocorrelation function is measured and it can be shown that the relationship between the two autocorrelation functions is

$$g^{(2)}(\tau) = 1 + 2\beta\,\mathrm{Re}\{g^{(1)}(\tau)\}, \tag{12}$$

where $\beta = \langle I_s\rangle/I_F \ll 1$, with I_s and I_F being the scattered and background components of the detected intensity.

Thus, in order to determine the parameter τ_0 describing the dynamic system under test, one has direct, experimental access to $g^{(2)}(\tau)$ and then determines

$g^{(1)}(\tau)$ via Eq. (12). If the distribution $P(s)$ is known, one can then use Eq. (11) to determine τ_0. Using OPS, the pathlength distribution can be directly measured, rather than calculated based on diffusion models, and, therefore, the dynamic properties of the dynamic system can be inferred without imposing any assumptions about the diffusive transport of light. This procedure has been implemented for different experimental configurations and some results are reviewed here.

Typical photon pathlength distributions are illustrated in Fig. 5 for the case of an aqueous suspension of spherical polystyrene particles, as indicated. As can be seen, these media belong to scattering regimes in which the photon diffusion as described by Eq. (3) is not appropriate. In subdiffusive cases like this, neither classical dynamic light scattering nor the DWS formalism in which $P(s)$ is determined by diffusion arguments are appropriate, and the direct evaluation of Eq. (11) should be used to determine τ_0. The procedure is exemplified in Fig. 6, where $P(s)$ functions obtained experimentally are used in Eq. (11) to calculate $g^{(1)}(\tau)$, which is compared with the measured values $g^{(1)}(\tau)$.

Starting from Eq. (11) and using the formalism of Laplace transform, one can evaluate values for the characteristic time τ_0. Figure 7 summarizes the comparison between these values and the ones obtained by using the solutions for $P(s)$, as provided by the diffusion approximation for the specific experimental situations. The zero-error limit—denoted by the horizontal line—is approached as the volume fraction of scattering centers increases. This is the limit of strong scattering, where the diffusion approximation is expected to provide an accurate description of the optical scattering. On the other hand, when the concentration of scatterers decreases, the procedure based on the solution of the diffusion approximation leads to increasing errors.

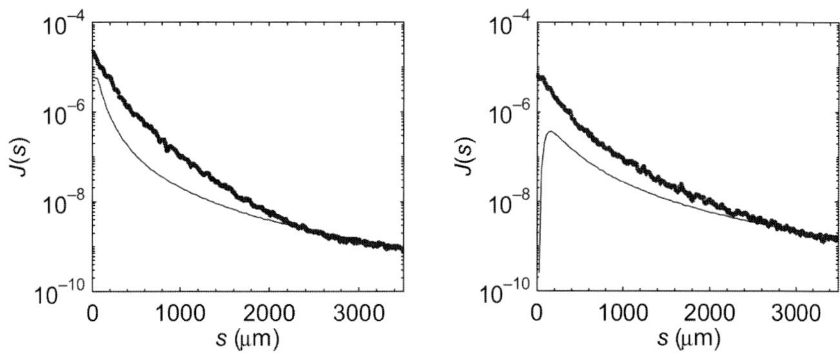

Figure 5 Measured $P(s)$ for suspensions with l^* values of 250 (left panel) and 1000 (right panel) μm, respectively. Also shown with continuous lines are the $P(s)$ solutions obtained from the diffusion approximation applied to the corresponding cases of semi-infinite media.

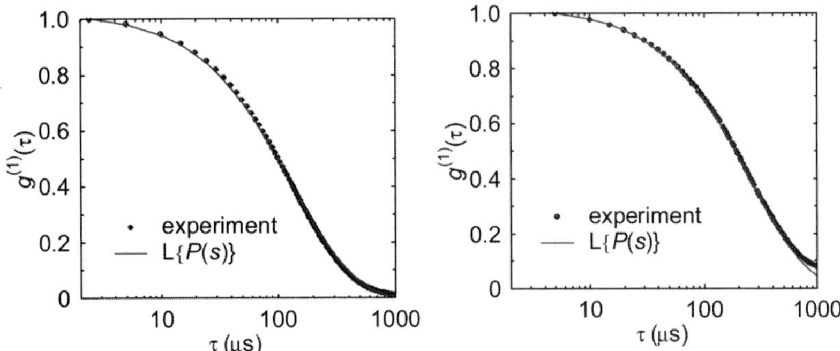

Figure 6 Temporal field autocorrelation functions for the same colloidal suspensions as in Fig. 4. The dots represent the experimental data and the continuous lines are the best fit obtained by evaluating numerically the Laplace transform of Eq. (11) with measured $P(s)$ distributions.

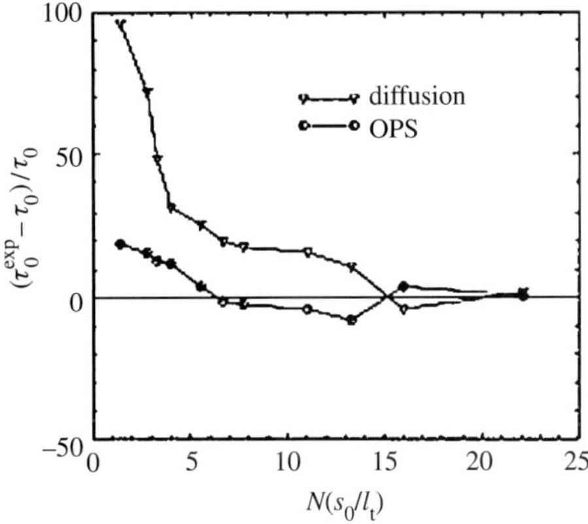

Figure 7 The relative error of the experimental value $\tau_0^{\exp}$ with respect to the theoretical one τ_0, for a series of colloidal suspensions characterized by different optical density parameters s/l^*.

As can be seen, in this limit the OPS approach brings a significant improvement over the conventional treatment based on the diffusion model. Of course, as the scattering strength decreases even more and the single scattering regime is approached, the signal level reduces and the experimental uncertainties start to become important. This can be understood by noting that Eq. (11), which is used in combination with the measured $P(s)$ to describe the data, fails to describe the single scattering limit [24].

It is interesting to remember that in OPS, the measured pathlength distribution is smeared by the coherence function of the source. This effect is, of course, more important for small values of the pathlength s. In dynamic multiple light scattering, on the other hand, small s corresponds to long correlation time [as seen in

Eq. (10)]. This observation makes the OPS-based approach especially appealing for the study of fast dynamics.

13.4.2 Diffusion at the boundaries

The ability to measure directly the photon pathlength distribution allows one to study phenomena beyond the classical diffusion approximation. For instance, $P(s)$ measured experimentally can be used to investigate subtle properties of multiple light scattering phenomena such as the process of photon diffusion close to interfaces. Closed-form solutions to the diffusion equation can be easily obtained for infinite media and different source-detector geometries. However, when the scattering medium is bounded, special conditions are needed to quantify the energy density at the interfaces. Through the phenomenon of reflection at the interface, light is reinjected into the medium and forced to travel a new diffusive path inside the scattering medium. As the reflectivity increases, this process becomes gradually important and it has to be taken into account in the overall diffusive description of light propagation in the bounded scattering medium. The diffusion process at the boundary introduces a new scale length that characterizes the specific medium interface. This is the *extrapolation length* that measures the distance outside the medium where the energy density vanishes linearly. Accurate values of this length are needed to interpret any experiment based on diffusion of light, and it is expected that the boundary inhomogeneity should directly affect the photon pathlength distribution. An appropriate boundary condition is, therefore, desirable and may extend the applicability of the diffusion model close to the interface. In general, it is well understood that the extrapolation length depends on the refractive index mismatch at the boundary but, so far, little has been done to clarify the potential influence of scattering anisotropy on boundary phenomena [25]. Our experiments have consistently shown that the extrapolation length ratio depends not only on the refractive index mismatch, but also on the anisotropy of individual scattering events.

The problem of semi-infinite random media that has been described in Sect. 2.3 becomes more complex when the reflection at the boundary is taken into account. This is not possible using the separate Dirichlet-Neuman boundary conditions because setting the energy density to zero at the interface is incorrect and does not correspond to the physical situation. When a modified Green's function was used to take into account the reflection at the boundary, it has been shown that the effect of reflection is to lower the effective diffusion coefficient of the medium [26].

The most general approach is to use a mixed boundary condition, which for a semi-infinite medium can be written as $(U - z_0 l^* \partial U/\partial z)_{z=0} = 0$, where z_0 is called the extrapolated length ratio, since $z_0 l_t$ is the distance outside the medium, where U extrapolates to zero. Using a partial current technique, it was suggested that z_0 in Eq. (6) depends only on the reflection phenomenon at the boundaries and is given by [27]

$$z_0 = \frac{2}{3}\left(\frac{1 + R_{\text{eff}}}{1 - R_{\text{eff}}}\right), \tag{13}$$

where R_{eff} is the effective reflectivity at the interface and can be easily calculated using the first two moments of the Fresnel coefficient. For a boundary without reflections, i.e., a totally absorbing interface, z_0 takes the value 2/3, which is consistent with the diffusion theory. However, as R_{eff} increases, in situations where the total internal reflection is present, the value of the extrapolation length ratio predicted by Eq. (13) can be 2 to 3 times larger. The parameter z_0 is of utmost importance since it solely defines the boundary condition for a given system; therefore, an experimental way to determine the extrapolation length for a given random medium is highly desirable.

Using OPS, this problem can be systematically investigated. It has been found that the more refined boundary conditions described in Eq. (13) are still not sufficient to describe the diffusion phenomenon close to the interface [28]. Using the modifed pathlength distribution, we demonstrated that z_0, which uniquely defines the relationship between the energy density at the boundary and its gradient normal to the surface, depends not only on the reflectivity at the boundary, but also on the anisotropy of the single-scattering process. We found that z_e, the additional length scale describing interfacial phenomena, decreases for larger anisotropy factors when light is detected in a medium of lower refractive index than that of the scattering medium.

Two types of OPS measurements were performed on water suspension of polystyrene microspheres with three different scattering anisotropies. First, the reflectance was recorded at the water/air interface of the sample. In this case, the condition for total internal reflection is present, since the backscattered light encounters a medium of lower refractive index at the boundary. Then, a layer of oil was added on top of the scattering medium in order to change the refractive index contrast at the interface. Typical experimental results are presented in Fig. 8, where it can be easily observed that the reflection at the boundary changes dramatically the optical pathlength distribution $P(s)$ of waves in the medium in a sense that the probability of long optical pathlengths is increased for a highly reflective boundary. As can be seen, the average penetration depth of light in the medium is considerably increased. The continuous curves in Fig. 8 are the result of data fitting with the normalized version of the diffusion model given by Eq. (6). Because the transport mean-free path l^* and the asymmetry parameter g are known, the only fitting parameter used is the extrapolation length ratio z_0.

The values of z_0 calculated with Eq. (13) are 0.68 for the medium/oil interface and 1.67 for the medium/air interface. Therefore, the oil interface simulates

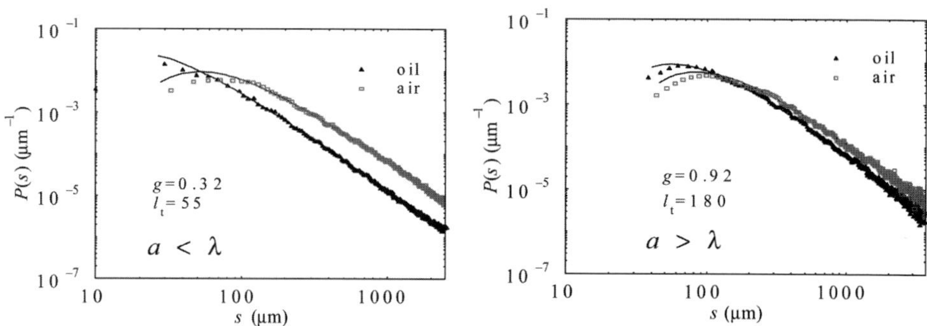

Figure 8 Measured photon pathelength distributions for media with scattering asymmetry $g = 0.32$ (a) and $g = 0.92$ (b) and for two different interfaces, as indicated. The solid curves represent the fit with the diffusion.

an almost perfect absorbing plane at the boundary of the scattering medium for which $z_0 = 0.67$. The experimental values obtained for the medium/oil boundary are 0.74, 0.73, and 0.76, corresponding to the anisotropy factors of 0.32, 0.49, and 0.92, respectively. These values are slightly different from that of 0.67 predicted by Eq. (13), and closer to the value of 0.7104 given by the Milne problem for isotropic scattering. As expected, for this index-matched interface, the anisotropy does not have a significant effect on z_0. However, for the medium/air interface, the experimental values are 1.57, 1.43, and 1.12, respectively, which are significantly different from that of 1.67 calculated with Eq. (13). Thus, the measured trend for z_0 is to decrease as g increases. We obtained experimentally a minimum value of 1.12 corresponding to $g = 0.92$. This value is considerably smaller than the calculated one and it suggests that the anisotropy dependence of the extrapolated length ratio becomes stronger at higher values of g. All the values of the extrapolation length corresponding to media with different asymmetry parameters are summarized in Fig. 9. The experimental data clearly illustrate the influence of the boundary properties on the optical pathlength distribution of the backscattered light.

Also included are the results of a model that accounts for the fact that, although the light transport is essentially diffusive in the bulk, the angular distribution of waves at the boundary is shaped by the last scattering event. As a result, extrapolation length predicted in Eq. (13) is modified to be

$$z_0 = \frac{C_1(F_g) + C_2(F_g, R)}{C_3(F_g) - C_4(F_g, R)}, \tag{14}$$

where the functions C_i depend on R, the angularly dependent Fresnel reflection coefficient at the interface, and on $F_g(\theta, \varphi)$, which represents the angular distribution of light emerging from one last scatterer that is exposed to the incident

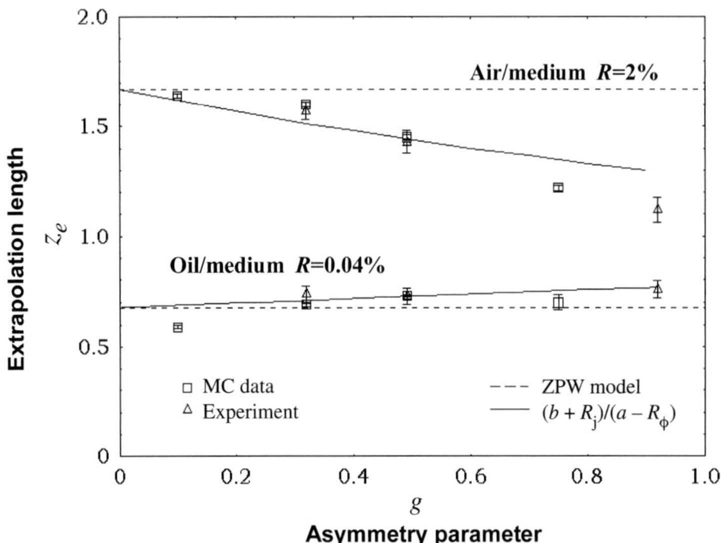

Figure 9 Values for z_0 as obtained from (1) the OPS experiment (triangles), (2) a numerical simulation (squares), and (3) from Eq. (13) for reflective boundary conditions (dotted line) and the z_0 as predicted by last scattering-dependent boundary conditions (continuous line). The data are separated in two groups corresponding to the two interfaces, as indicated.

diffusive field. The new extrapolation length ratio depends on the anisotropy factor in a manner shown in Fig. 9. In the case of isotropic scattering, $F_g(\theta, \varphi)$ is constant and the values for z_0 predicted by Eq. (13). In the situation where the reflection at the boundary is close to zero, z_0 has a slight dependence on g in such a way that increases for higher anisotropy factors. The monotonic behavior of the function $z_0(g)$ for medium/oil interface is reversed with respect to the medium/air case because now the light exits the sample into a medium with higher index of refraction. As can be seen in Fig. 9, the z_0 values predicted by this new model are fairly close to the measured and simulated ones.

Due to the high dynamic range of the fiber-optic-based OPS, a broad domain of the optical pathlengths is resolved, which allows for a thorough investigation of the diffusive backscattering light. The pathlength-resolved distribution of backscattered light, which contains the global effect of reflection at the boundary, is used to determine the value of z_0. In the overall description of the angular distribution of the backscattered waves, the contribution of the last scattering event is particularly important. For larger values of the asymmetry parameter g, the total angular flux impinging on the physical boundary is more peaked forward and undergoes less reflection than in the case of isotropic scatterers, which corresponds to a smaller value of z_0.

13.5 Conclusions

Multiple scattering of light is encountered in many cases when the dielectric properties of media are randomly varying in time or space. When the optical radiation propagates through such inhomogeneous media, the direction and the phase of the wave are randomized. A detailed description of multiple scattering phenomena accounting for the full vector characteristics of the electromagnetic waves is quite a challenging task. In many practical situations, however, the relevant properties of multiply scattered light can be obtained by simply considering a scalar field of diffusive photons that satisfy a transportlike equation. In this approach, the scalar photons travel along various random paths between the source and detector and accumulate an optical phase that is proportional with the geometrical pathlength s of each specific trajectory. All the macroscopic, measurable properties of the multiple scattering phenomenon are therefore determined by the distribution of photon pathlengths. One can use a random walk description of photon paths and, if a characteristic length scale—the transport mean-free path l^*—is introduced, multiple scattering phenomena can then be alternatively discussed in terms of scattering orders s/l^*.

The probability density $P(s)$ of photon pathlength can be obtained by solving the transport equation numerically or analytically within various approximations. When absorption losses are negligible and the average optical path is sufficiently long, it is straightforward to evaluate $P(s)$ by solving the transport equation in the diffusion approximation. This is the procedure of choice for most experimental techniques that rely on multiple light scattering to infer different properties of the random media. Of course, this approach will fail when scattering is strong, i.e., when the characteristic scattering length reduces and becomes of the order of radiation's wavelength. In this case, the interference effects between multiply scattered waves cannot be neglected anymore and the concept of optical pathlength should be carefully examined.

In a number of situations of practical interest, the scattering regime can be classified as subdiffusive. This is the class of multiple scattering phenomena in which, due to finite size, absorption, or experimental geometry, the scattering is not sufficiently strong and the isotropic diffusion behavior is not reached before the photons are detected. A typical example is any backscattering experiment in which a significant contribution originates from single or low-order scattering events. In these situations, a direct, experimental determination of $P(s)$ is desirable in order to correctly interpret the measurements. We have shown here that the optical pathlength distribution can be obtained in a variety of situations by using the coherence properties of light to select successive orders of scattering. The technique, called optical pathlength spectroscopy (OPS), enables us to obtain the probability density $P(s)$ by normalizing the path-resolved reflectivity obtained in an interferometric measurement. Subsequently, characteristics such as the average optical pathlength or higher-order moments can be inferred for a variety of random media. For example, when OPS was applied in the context of dynamic scattering, the measured $P(s)$ permitted to bridge the gap between single and diffusive light scattering and to extend the applicability of dynamic light scattering techniques to regimes relevant to many applications in biology, colloidal, and polymer sciences.

The ability to directly determine the photon pathlength distribution enables us to refine certain aspects of the photon diffusion model. For instance, a problem of practical relevance in many applications relates to the photon diffusion near the physical boundaries of the random medium. The OPS approach allowed us to elucidate the subtle effects of the scattering phase function and to establish the correct boundary conditions, which in turn permitted quantitative description of backscattering experiments. Besides giving a detailed description of different aspects related to multiple scattering phenomena, OPS could also provide a measure of the field penetration depth that is important in various light delivery applications, especially for medical techniques like photodynamic therapy.

References

1. G. Maret and E. Wolf, *Z. Phys.* **B65**, 409 (1987); D.J. Pine, D.A. Weitz, P.M. Chaikin, and E. Herbolzheimer, *Phys. Rev. Lett.* **60**, 1134 (1988).
2. M.P. Van Albada and A. Lagendijk, *Phys. Rev. Lett.* **55**, 2692 (1985); E. Wolf and G. Maret, *Phys. Rev. Lett.* **55**, 2696 (1985).
3. M.C.W. van Rossum and Th.M. Nieuwenhuizen, *Rev. Mod. Phys.* **71** (1999).
4. S. Chandrasekhar, *Radiative Transfer*, Oxford Univ. Press, Oxford (1950).
5. A. Ishimaru, *Wave Propagation and Scattering in Random Media*, Academic Press, New York (1971).
6. S. Chandrasekhar, *Radiative Transfer*, Oxford Univ. Press, Oxford (1950).
7. S.R. Arridge, *Inverse Problems* **15**, R41 (1999).
8. J.J. Duderstadt and L.J. Hamilton, *Nuclear Reactor Analysis*, John Wiley & Sons Inc., New York (1976).
9. S. Chandrasekhar, *Radiative Transfer*, Oxford Univ. Press, Oxford (1950).
10. M.S. Patterson, B. Chance, and B.C. Wilson, *Appl. Opt.* **28**, 2331 (1989).
11. R.C. Youngquist, S. Carr, and D.E. Davies, *Opt. Lett.* **12**, 158 (1987).
12. B. Masters, *Selected Papers on Optical Low-Coherence Reflectometry and Tomography*, SPIE Milestone series, Vol. MS165, SPIE, Bellingham, WA (2001).
13. G. Popescu and A. Dogariu, *Opt. Lett.* **24**, 442 (1999).
14. L. Mandel and E. Wolf, *Optical Coherence and Quantum Optics*, Cambridge University Press, Cambridge (1995).
15. E. Wolf, *Opt. Lett.* **28**, 5 (2003).
16. A. Dogariu and G. Popescu, *Phys. Rev. Lett.* **89**, 243902 (2002).

17. G. Popescu and A. Dogariu, *Appl. Opt.* **39**, 4469 (2000); G. Popescu and A. Dogariu, *Phys. Rev. Lett.* **88**, 183902 (2002).

18. G. Popescu and A. Dogariu, *Opt. Lett.* **24**, 442 (1999).

19. C. Mujat, L. Denney, and A. Dogariu, *Mat. Res. Soc. Symp. Proc.* **613**, E.6.10.1 (2000).

20. C. Mujat, M. van der Veen, J.L. Rueben, J. Ten Bosch, and A. Dogariu, *Appl. Opt.* **42**, 2979 (2003).

21. B.J. Berne and R. Pecora, *Dynamic Light Scattering with Applications to Chemistry, Biology, and Physics*, John Wiley & Sons, Inc., New York (1976).

22. D.J. Pine, D.A. Weitz, P.M. Chaikin, and E. Herbolzheimer, *Phys. Rev. Lett.* **60** (1988).

23. G. Maret and P.E. Wolf, *Z. Phys. B* **65** (1987); D.J. Pine, D.A. Weitz, P.M. Chaikin, and E. Herbolzheimer, *Phys. Rev. Lett.* **60** (1988); D.J. Pine, D.A. Weitz, J.X. Zhu, and E. Herbolzheimer, *J. Phys. France* **51** (1990).

24. G. Popescu and A. Dogariu, *Appl. Opt.* **40**, 4215 (2001).

25. R.C. Haskell, *J. Opt. Am. A* **11**(10), 2727 (1994); N.G. Chen and J. Bai, *Phys. Rev. Lett.* **80**(24), 5321 (1998).

26. A. Lagendijk, R. Vreeker, and P. De Vries, *Phys. Lett. A* **136**, 81 (1989).

27. J.X. Zhu, D.J. Pine, and D.A. Weitz, *Phys. Rev. A* **44**(6), 3948 (1991).

28. G. Popescu, C. Mujat, and A. Dogariu, *Phys. Rev. E* **61**, 4253 (2000).

Emil and Marlies Wolf, May 2004. (Courtesy Aristide Dogariu, copyright 2004.)

Aristide Dogariu's main interests lie in the general field of physical optics and, in particular, in the area of optical waves propagation and scattering. Professor Dogariu received his Ph.D. at Hokkaido University, Japan, and he is currently with the College of Optics and Photonics at the University of Central Florida. He leads the *Laboratory for Photonic Diagnostics of Random Media*, which is involved in exploring different light-scattering principles for sensing, optical metrology, diagnostics, and various applications of optical waves propagation in random media. Professor Dogariu is a fellow of OSA and the division editor of *Applied Optics—Optical Technology and Biomedical Optics*.

CHAPTER 14

THE DIFFRACTIVE MULTIFOCAL FOCUSING EFFECT

John T. Foley, Renat R. Letfullin, and Henk F. Arnoldus

14.1 Introduction

It is well known that when a monochromatic plane wave of intensity I_0 is normally incident upon a circular aperture, the intensity at on-axis observation points behind the aperture oscillates between the values $4I_0$ and zero as the distance from the aperture is increased. The reason for this is that the various Fresnel zones in the aperture contribute either constructively or destructively to the amplitude of the field at the observation point in question, causing the amplitude to oscillate between zero and twice the incident field value. For an incident wavelength λ, aperture radius a, and aperture-plane to observation-plane distance z, the number of zones that contribute is given by the Fresnel number, $N = a^2/\lambda z$. The maxima and minima occur at observation points where the Fresnel number is an odd or even integer, respectively.

What is commonly not recognized is the fact that in the region near the z-axis, as z is increased, the light is repeatedly focusing and defocusing, over and over again due to diffraction. The focal points occur at positions where the Fresnel number is an odd integer. This was pointed out in a series of papers by Lit and coworkers [1–3] and most recently by Letfullin and George [4], who referred to this phenomenon as the diffractive multifocal focusing of radiation (DMFR) effect.

Letfullin and George proposed to use a system of two circular apertures for which the on-axis intensity of an incident monochromatic plane wave would increase dramatically due to the DMFR effect. In their system the second aperture was located where the Fresnel number of the first aperture was unity. They analyzed this system theoretically, and showed that the on-axis intensity behind the second aperture oscillates between maximum values of the order of 10 times that of the incident wave and minimum values that are very small, but not zero. These predictions were verified experimentally [5,6] and extended theoretically to incident fields with a Gaussian amplitude distribution [7]. In Refs. [4–7] the phase of the field was not investigated.

In this chapter we investigate the intensity and phase of the diffracted field behind a circular aperture when a monochromatic plane wave is incident upon it, and when a monochromatic Gaussian beam is incident upon it. We also investigate the intensity and phase of the diffracted field behind a system of two circular apertures for the same two incident fields. In each case we substantiate the focusing, defocusing and refocusing interpretation mentioned above, and investigate the intensity at the focal points.

In Sect. 14.2.1 we show that when a plane wave is incident upon a single circular aperture, in the neighborhood of a focal point (where the Fresnel number is odd) the phase of the wave approaching the focal point is that of a converging wave, the phase front in the focal plane is planar, and the phase of the wave exiting the focal point is that of a diverging wave. We also show that the wave becomes more and more divergent as the distance from the focal point is increased, until a position is reached where the Fresnel number is even. At such a point the intensity of the wave is zero, and the phase of the wave is undefined, i.e., singular. We show that as the on-axis observation point moves away from the aperture and passes through a singular point, the nature of the wave in the neighborhood of the axis changes from that of a diverging wave to that of a converging wave, i.e., the wave refocuses. In Sect. 14.2.2 we show that when the incident field is a Gaussian beam, the phase behaves similarly, and the intensities at the focal points decrease as the ratio of the radius of the aperture to the spot size of the incident beam is increased.

In Sect. 14.3.1 we use the results of Sect. 14.2 to investigate the intensity and the phase of the field after the second aperture in a system of two circular apertures when a plane wave is incident. In this case the ratio of the radii of the two apertures is a key parameter, and we discuss the effect of varying this ratio. In Sect. 14.3.2, we do the same for a Gaussian beam incident upon a two aperture system.

We use scalar wave theory throughout this paper. Unlike in Refs. [4] and [7], where the monochromatic wave equation (the Helmholtz equation) was integrated numerically, we use the Fresnel approximation and the paraxial approximation in our calculations.

14.2 Fresnel Diffraction by a Circular Aperture

14.2.1 Incident plane wave

14.2.1.1 Basic equations

Consider a monochromatic plane wave of amplitude U_o and angular frequency ω, propagating in the positive z-direction, and normally incident upon an opaque screen in the plane $z = 0$, containing an aperture of radius a. The aperture is centered about the origin. Let $P' = (x', y', 0)$ be a point inside the aperture, and let $P = (x, y, z)$ be a point in an observation plane $z = $ constant > 0 (see Fig. 1). In cylindrical polar coordinates we have $P' = (\rho', \theta', 0)$ and $P = (\rho, \theta, z)$. We assume a time dependence of $\exp(-i\omega t)$ for the field. The complex amplitude, $U^{(i)}(\rho, \theta, z)$, of the incident field is given by

$$U^{(i)}(\rho, \theta, z) = U_o e^{ikz}, \tag{1}$$

where U_o is a positive constant and $k = \omega/c$ is the wave number of the light.

We make the following assumptions. First, that the wavelength λ is much smaller than the distance z from the aperture plane to the observation plane. Second, that the Fresnel number of the observation plane is small, and that the trans-

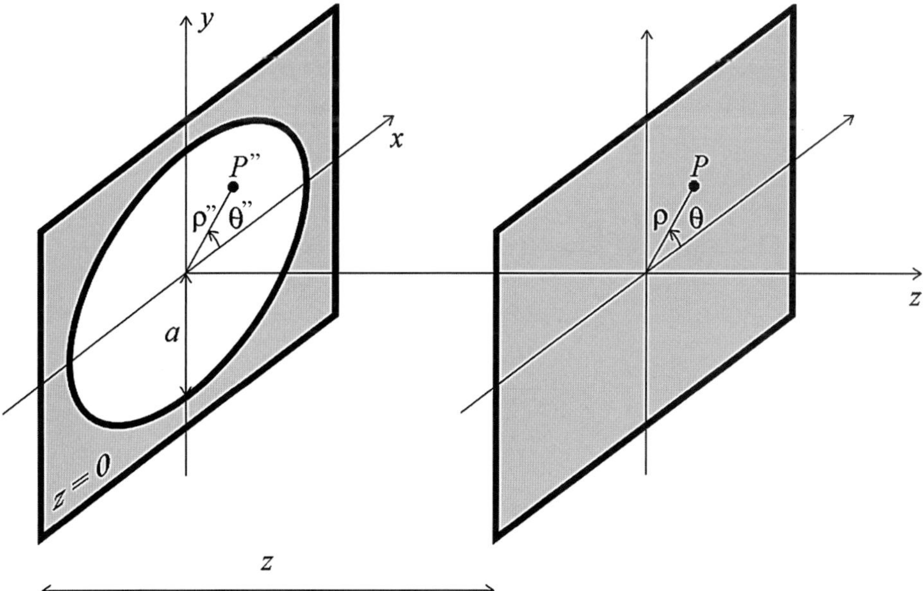

Figure 1 Geometry for the diffraction of a plane wave from a circular aperture with radius a. Point $P' = (\rho', \theta', 0)$ is a point inside the aperture and $P = (\rho, \theta, z)$ is a point in the observation plane $z = $ constant.

verse distance ρ is less than the aperture radius a and much less than the distance z. In this case the paraxial form of the Fresnel approximation to the Rayleigh-Sommerfeld diffraction formula [8] is appropriate for describing the field. The complex amplitude, $U(\rho, \theta, z)$, of the diffracted field can then be written as

$$U(\rho, \theta, z) = \frac{k}{2\pi i z} U_o e^{ikz} e^{ik\rho^2/2z}$$

$$\times \int_0^a \int_0^{2\pi} U^{(i)}(\rho', \theta', 0) e^{ik\rho'^2/2z} e^{-ik\rho\rho'\cos(\theta-\theta')/z} \rho' d\rho' d\theta'. \qquad (2)$$

Equation (2) is equivalent to the formula used by Lommel [9] for the case in which the incident field is a diverging spherical wave.

Let us now simplify this equation. Upon substituting the value of $U(\rho', \theta', 0)$ from Eq. (1) into this equation and performing the angular integration, we find that the complex amplitude of the field is independent of the angle θ and is given by the expression

$$U(\rho, z) = \frac{k}{iz} U_o e^{ikz} e^{ik\rho^2/2z} \int_0^a e^{ik\rho'^2/2z} J_0(k\rho\rho'/z) \rho' d\rho', \qquad (3)$$

where $J_0(x)$ is the zero-order Bessel function of the first kind. Let us now make the change of variables $\rho' = \xi a$. Upon making this change, Eq. (3) can be rewritten in terms of two dimensionless variables u and v as

$$U(\rho, z) = -iu U_o e^{ikz} e^{iv^2/2u} \int_0^1 e^{iu\xi^2/2} J_0(v\xi) \xi d\xi, \qquad (4)$$

where

$$u = 2\pi N, \quad v = 2\pi N\rho/a, \qquad (5)$$

and N is the Fresnel number of the aperture at the on-axis observation point,

$$N = a^2/\lambda z. \qquad (6)$$

Let us now put Eq. (4) into a form more suitable for calculations. It is shown in Sect. 14.A that in the lit region (where $\rho < a$ and hence $v < u$) the integral on the right-hand side of Eq. (4) can be expressed as

$$\int_0^1 e^{iu\xi^2/2} J_0(v\xi) \xi d\xi = \frac{i}{u} \left\{ e^{-iv^2/2u} - e^{iu/2} [V_0(u,v) - iV_1(u,v)] \right\}, \qquad (7)$$

where $V_0(u,v)$ and $V_1(u,v)$ are Lommel functions of two variables:

$$V_n(u,v) = \sum_{s=0}^{\infty} (-1)^s \left(\frac{v}{u}\right)^{2s+n} J_{2s+n}(v). \tag{8}$$

Upon substituting the right-hand side of Eq. (7) into Eq. (4), the field in the lit region attains the form

$$U(\rho,z) = U_o e^{ikz} M(\rho,z), \tag{9}$$

where

$$M(\rho,z) = 1 - e^{iu/2} e^{iv^2/2u} \left[V_0(u,v) - iV_1(u,v)\right]. \tag{10}$$

Equation (9) shows that the diffracted complex amplitude is the product of the incident field amplitude $U_o e^{ikz}$ (i.e., the total field if no aperture were present), and the function $M(\rho,z)$. We shall therefore refer to $M(\rho,z)$ as the *modifier* function, since it describes how the presence of the aperture modifies the field.

The intensity, $I(\rho,z)$, of the field in the lit region is given by

$$I(\rho,z) = U(\rho,z)^* U(\rho,z) = I_o|M(\rho,z)|^2, \tag{11}$$

where * denotes the complex conjugate and $I_o = |U_o|^2$ is the intensity of the incident field. It follows from Eq. (9) that the phase $\phi(\rho,z)$ of the field is given by

$$\phi(\rho,z) = kz + \psi(\rho,z), \tag{12}$$

where

$$\psi(\rho,z) = \arg M(\rho,z), \tag{13}$$

and arg denotes the argument of the complex-valued function $M(\rho,z)$. We shall refer to $\psi(\rho,z)$ as the reduced phase.

14.2.1.2 On-axis intensity and phase

Let us first examine the intensity and phase at on-axis observation points. At such points, $\rho = 0$ and hence $v = 0$. It follows directly from Eq. (8) that $V_0(u,0) = 1$ and $V_1(u,0) = 0$. We then find from Eq. (10) that the on-axis modifier function is

$$M(0,z) = 1 - e^{iu/2} = 1 - e^{iN\pi}, \tag{14}$$

where the z-dependence enters through the Fresnel number N. It then follows from Eq. (11) that the on-axis intensity is given by

$$I(0,z) = I_o \left| 1 - e^{iN\pi} \right|^2 = 4I_o \sin^2(N\pi/2). \tag{15}$$

This function is plotted in Fig. 2 over the interval $1 \leq N \leq 6$. When the Fresnel number is odd, the intensity is maximum with a value of $I(0,z) = 4I_o$. We will refer to the corresponding observation points as *focal points*. When the Fresnel number is even we have $I(0,z) = 0$, and we will refer to the corresponding observation points as *singular points* because the modulus of the complex amplitude $U(0,z)$ is zero at such points, and hence its phase is undefined there [10,11].

The expression for the reduced phase can be obtained from Eqs. (13) and (14), and we find that

$$\psi(0,z) = \arg(1 - e^{iN\pi}). \tag{16}$$

This function is plotted in Fig. 3 over the interval $1 \leq N \leq 6$. Note that the phase jumps by π as we pass through the singular points $N = 2, 4$ and 6.

14.2.1.3 General case

In this section, the intensity and reduced phase in observation planes at a variety of distances from the aperture will be investigated. For the sake of comparison, let us first recall the paraxial form for a diverging spherical wave. A spherical wave emanating from the origin and arriving at the position P in Fig. 1 is described by the wave function $\exp(ikr)/r$, where $r = (\rho^2 + z^2)^{1/2}$. The paraxial approximation to this function is $\exp(ikr)/r \approx \exp[ik(z + \rho^2/2z)]/z$. Upon comparing this equation to Eq. (9), we see that the paraxial approximation to the reduced phase of this

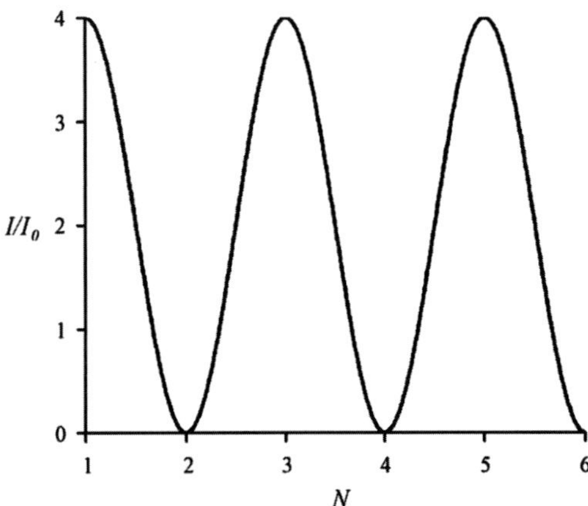

Figure 2 The on-axis intensity I, in units of I_o, as a function of the Fresnel number N.

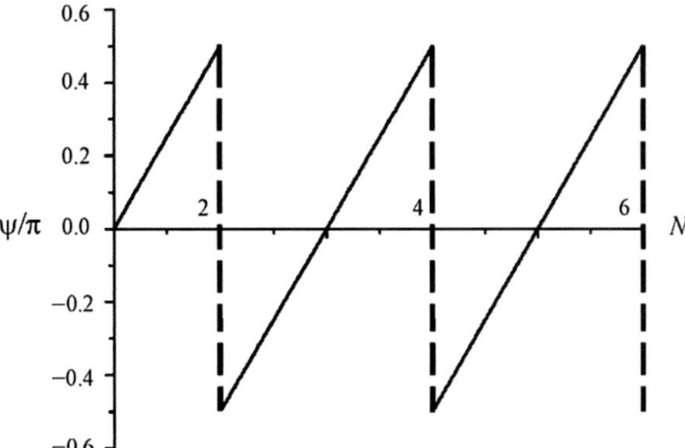

Figure 3 The on-axis reduced phase ψ, in units of π, as a function of the Fresnel number N.

wave is

$$\psi(\rho, z) \approx k\rho^2/2z. \tag{17}$$

The intensity and reduced phase of the diffracted field along the x-axis in several observation planes at different distances from the aperture plane are plotted in Figs. 4 and 5. The behavior of the field as we travel outward from the aperture plane and pass through a focal point is depicted in Figs. 4(a) through 4(c). In Fig. 4(a) the Fresnel number is 3.5, and in this plane we see that the reduced phase near the axis has a curvature with the opposite sign of that of the phase in Eq. (17). Hence the wave in this plane corresponds to a converging wave. The on-axis intensity value is approximately 2.0. In Fig. 4(b) the Fresnel number is 3. This is a focal plane, and we see that the reduced phase is constant near the axis, i.e., the wave is behaving like a plane wave in this region. The on-axis intensity value in this case is 4.0. In Fig. 4(c) the Fresnel number is 2.5, and we see that in this plane the reduced phase near the axis has a curvature with the same sign as the phase in Eq. (17). Hence the wave corresponds to a diverging wave. The on-axis intensity value is approximately 2.0.

As we move further away from the aperture plane, the wave near the axis diverges more strongly, until we reach the on-axis point where the Fresnel number is 2.0. At this point the intensity of the field is zero, and its phase is undefined. Such a point is referred to as a singular point of the field. Figure 4(d) shows the intensity and phase of the wave in the plane where the Fresnel number is 2.01, i.e., just before we reach the singular point. The phase near the axis has a steep upward curvature, corresponding to a strongly diverging wave. The value of the phase on-axis is approximately $-\pi/2$. Figure 5(a) shows the intensity and phase of the wave in the plane where the Fresnel number is 1.99, i.e., just after we have passed through the singular point. The phase near the axis has a steep downward

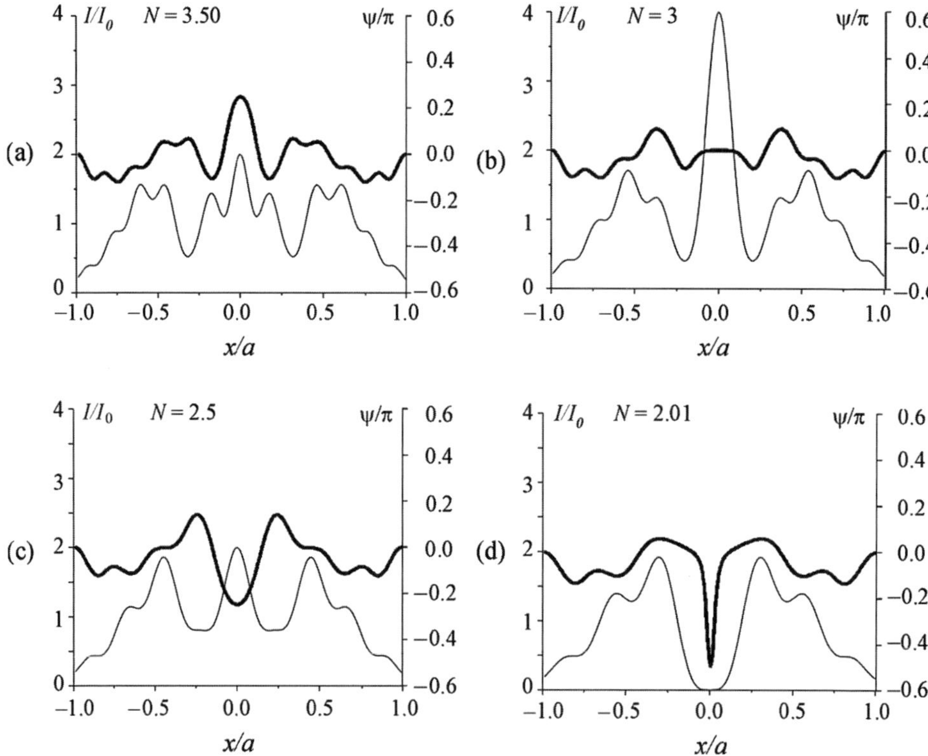

Figure 4 Plots of I/I_0 (thin line) and ψ/π (thick line) as functions of position along the x-axis for observation planes with Fresnel numbers: (a) $N = 3.5$, (b) $N = 3$, (c) $N = 2.5$, and (d) $N = 2.01$.

curvature, corresponding to a strongly converging wave. The on-axis value of the phase here is approximately $\pi/2$, so as we pass through the singular point, the phase jumps by π. Figure 5(b) shows that as we continue to move further away from the aperture plane, the field near the axis starts to converge less strongly. Finally, at the next focal point, the plane where $N = 1.0$, we see that the phase near the axis is again constant; and hence the wave is again behaving like a plane wave. This is shown in Fig. 5(c). In addition, by comparing Fig. 5(c) to Fig. 4(b), we see that for focal points further from the aperture plane, the wavefront is planar over a larger area in the observation plane.

　　　Figure 6 shows the lines of constant phase, ϕ, in the xz-plane near the $N = 2$ singular point, for the case $a/\lambda = 50$. It is evident from the figure that as the wave approaches the singular point, it is a diverging wave whose radius of curvature becomes smaller and smaller. It also follows from the figure that after the wave has passed through the singular point, it is now a *converging* wave whose radius of curvature is increasing.

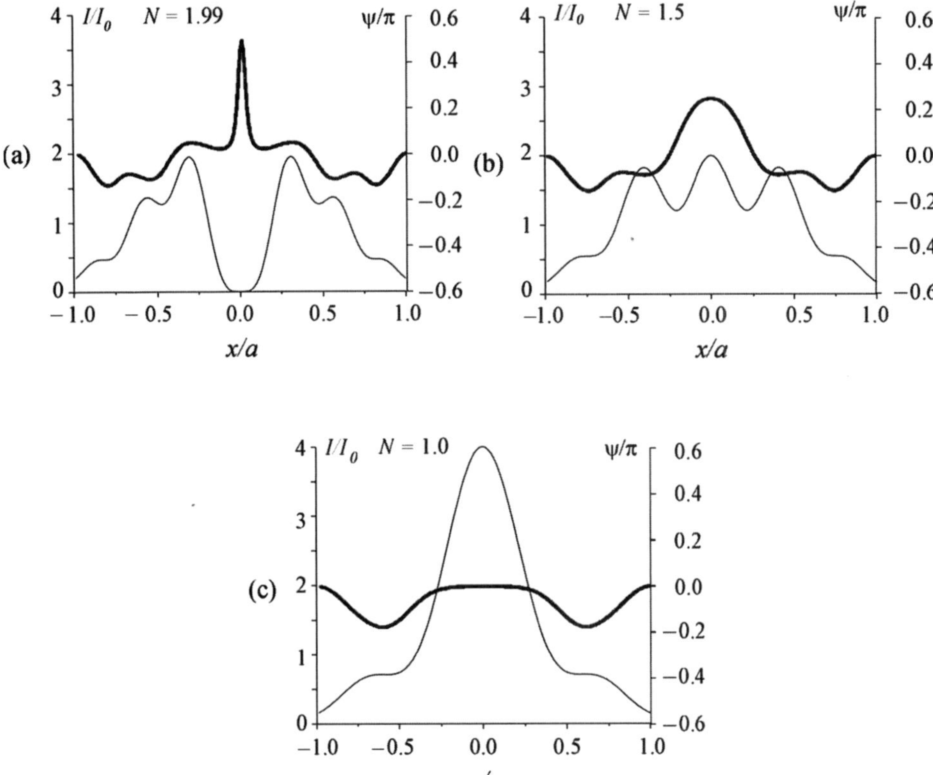

Figure 5 Plots of I/I_0 (thin line) and ψ/π (thick line) as functions of position along the x-axis for observation planes with Fresnel numbers: (a) $N = 1.99$, (b) $N - 1.5$, and (c) $N = 1.0$.

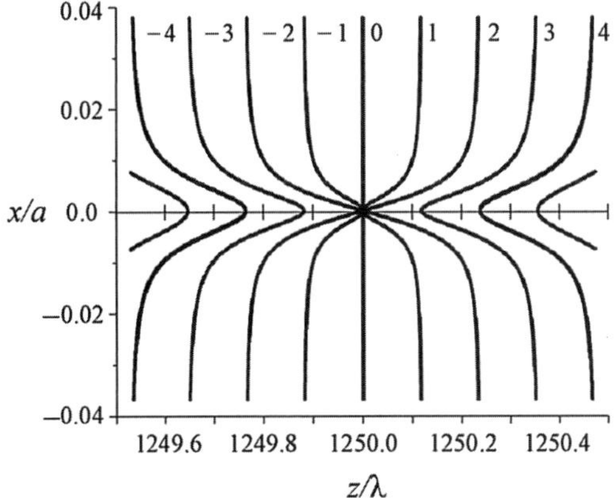

Figure 6 Lines of constant phase, ϕ, in the xz-plane near the $N = 2$ singular point when $a/\lambda = 50$. The transverse coordinate x is in units of a, and the z-coordinate is in units of the wavelength λ. The value of the phase for each line is labeled in units of $\pi/4$.

14.2.2 Incident Gaussian beam

14.2.2.1 Basic equations

Let the incident field be a Gaussian beam whose waist occurs at the aperture plane $z = 0$. The complex amplitude of the incident field is

$$U^{(i)}(\rho,\theta,z) = U_o \frac{w_o}{w} e^{i(kz-\chi)} e^{ik\rho^2/2R} e^{-\rho^2/w^2}, \tag{18}$$

where U_o is a positive constant and w_o is the spot size of the beam waist. Here, w is the spot size in the plane $z = $ constant,

$$w = w_o\sqrt{1 + (z/z_R)^2}, \tag{19}$$

R is the radius of curvature in that plane,

$$R = z + \frac{z_R^2}{z}, \tag{20}$$

χ is the Gouy phase

$$\tan\chi = \frac{z}{z_R}, \tag{21}$$

and $z_R = \pi w_o^2/\lambda$ is the Rayleigh length of the beam. It follows from Eqs. (18)–(21) that the complex amplitude in the plane $z = 0$ is

$$U^{(i)}(\rho',\theta',0) = U_o e^{-\rho'^2/w_o^2}. \tag{22}$$

In order to find the field at the observation point $P = (\rho,\theta,z)$, we substitute the incident field into Eq. (2). After performing the angular integration, we find that the complex amplitude of the field is independent of θ and is given by the expression

$$U(\rho,z) = \frac{k}{iz} U_o e^{ikz} e^{ik\rho^2/2z} \int_0^a e^{-\rho'^2/w_o^2} e^{ik\rho'^2/2z} J_0(k\rho\rho'/z)\rho'\,d\rho'. \tag{23}$$

Let us now make the change of variables $\rho' = \xi a$. Upon making this change, Eq. (23) can be rewritten in terms of two dimensionless variables u and v as

$$U(\rho,z) = -iU_o 2\pi N e^{ikz} e^{iN\pi(\rho/a)^2} \int_0^1 e^{iu\xi^2/2} J_0(v\xi)\xi\,d\xi, \tag{24}$$

where

$$u = 2\pi N + 2i\beta^2, \quad v = 2\pi N(\rho/a). \tag{25}$$

Here, N is the Fresnel number of the aperture at an on-axis observation point, given by Eq. (6), and

$$\beta = a/w_o. \tag{26}$$

We will refer to β as the aperture-spot ratio since its numerical value specifies how many beam spots would fit across the aperture.

If we now substitute the right-hand side of Eq. (7) into Eq. (24), we find that the complex amplitude is

$$U(\rho,z) = U_o e^{ikz} G(\rho,z), \tag{27}$$

where

$$G(\rho,z) = \frac{2\pi N}{u} e^{iN\pi(\rho/a)^2} e^{-iv^2/2u} M(\rho,z), \tag{28}$$

and $M(\rho,z)$ is given by Eq. (10). Note, however, that in the present case the variable u on the right-hand side of Eq. (10) is a complex value [see Eq. (25)]. The intensity of the field in the lit region is given by

$$I(\rho,z) = |U(\rho,z)|^2 = I_o|G(\rho,z)|^2, \tag{29}$$

where $I_o = |U_o|^2$ is the on-axis incident intensity in the plane $z = 0$. It follows from Eq. (28) that the phase of the field is

$$\phi(\rho,z) = kz + \psi(\rho,z), \tag{30}$$

where

$$\psi(\rho,z) = \arg G(\rho,z). \tag{31}$$

As before, we shall refer to $\psi(\rho,z)$ as the reduced phase.

Equation (27) is a useful way of writing the complex amplitude of the diffracted field, especially for calculations, but there is an alternative representation that is interesting. It is shown in Sect. 14.B that

$$\frac{2\pi N}{u} = \frac{w_o}{w} e^{-i\chi}, \tag{32}$$

and that

$$e^{iN\pi(\rho/a)^2}e^{-iv^2/2u} = e^{ik\rho^2/2R}e^{-\rho^2/w^2}. \tag{33}$$

After substituting the right-hand side of this equation into Eq. (28), and then substituting that result into Eq. (27) we obtain

$$U(\rho,z) = U_o\frac{w_o}{w}e^{i(kz-\chi)}e^{ik\rho^2/2R}e^{-\rho^2/w^2}M(\rho,z), \tag{34}$$

which is

$$U(\rho,z) = U^{(i)}(\rho,\theta,z)M(\rho,z), \tag{35}$$

with $U^{(i)}(\rho,\theta,z)$ given by Eq. (18). Equation (35) shows that the diffracted complex amplitude is the product of the incident field complex amplitude and the function $M(\rho,z)$. Hence, as in the previous section, we shall refer to $M(\rho,z)$ as the modifier function, since it describes how the presence of the aperture modifies the field.

14.2.2.2 On-axis intensity and phase

Let us now examine the intensity and phase at on-axis observation points. At such points, $\rho = 0, v = 0$, and the corresponding modifier function is given by

$$M(0,z) = 1 - e^{iu/2} = 1 - e^{iN\pi}e^{-\beta^2}. \tag{36}$$

It then follows from Eqs. (29) and (28) that

$$I(0,z) = I_o|G(0,z)|^2 = I_o\frac{4\pi^2N^2}{|u|^2}|M(0,z)|^2 = I_o\frac{4\pi^2N^2}{|u|^2}\left|1-e^{iN\pi}e^{-\beta^2}\right|^2, \tag{37}$$

where u is given by Eq. (25).

Figure 7 depicts the on-axis intensity as a function of the Fresnel number for four different values of the aperture-spot ratio. In all four cases the intensity

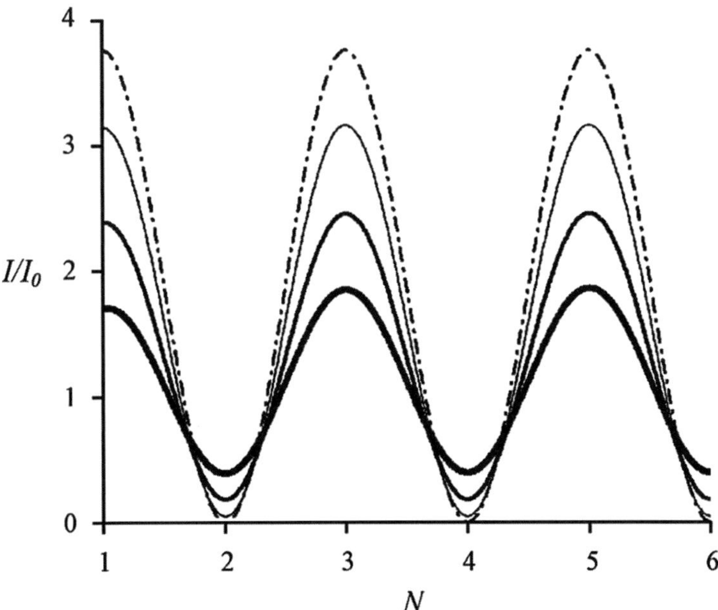

Figure 7 The on-axis intensity I, in units of I_o, for $\beta = 0.25$ (dashed line), $\beta = 0.50$ (thin solid line), $\beta = 0.75$ (medium solid line), and $\beta = 1.0$ (thick solid line).

is maximum (minimum) when the Fresnel number is approximately odd (even). These plots show that as the aperture-spot ratio increases, the focusing effect becomes weaker: the values of the maximum intensity decrease (as compared to the value of 4 for the case of an incident plane wave) and the values of the minimum intensity increase (with respect to the value of zero for the incident plane wave).

It follows from Eqs. (35) and (36) that the on-axis field is

$$U(0,z) = U_o \frac{w_o}{w} e^{i(kz-\chi)} \left(1 - e^{iN\pi} e^{-\beta^2}\right). \tag{38}$$

The on-axis reduced phase is therefore

$$\psi(0,z) = \arg\left[e^{-i\chi}\left(1 - e^{iN\pi} e^{-\beta^2}\right)\right]. \tag{39}$$

It is interesting to notice that when the Fresnel number is an integer, we have $\psi(0,z) = -\chi$. Figure 8 shows the on-axis reduced phase as a function of the Fresnel number for the same four values of the aperture-spot ratio as in Fig. 7. As the aperture-spot ratio is increased, the size of the phase change decreases.

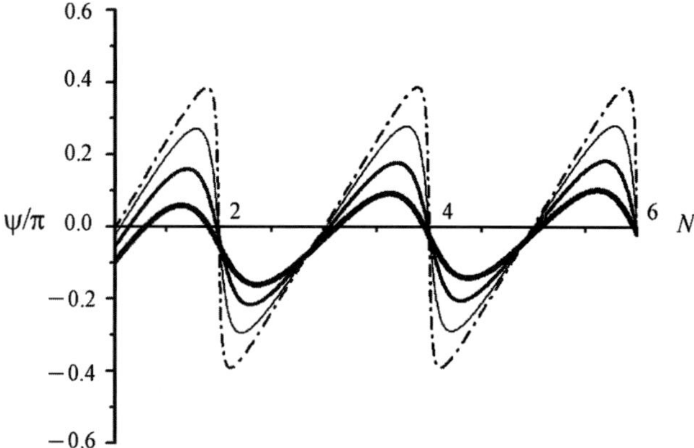

Figure 8 The on-axis reduced phase ψ, in units of π, for $\beta = 0.25$ (dashed line), $\beta = 0.50$ (thin solid line), $\beta = 0.75$ (medium solid line), and $\beta = 1.0$ (thick solid line).

14.2.2.3 General case

The intensity and reduced phase of the diffracted field along the x-axis in several observation planes are plotted in Fig. 9, for the case $\beta = 0.57$, i.e., $a = 0.57w_o$. The behavior of the field as we travel outward from the aperture plane and pass through a focal plane is depicted in Figs. 9(a) through 9(c). In Fig. 9(a) the Fresnel number is 3.5. The downward curvature of the phase near the axis shows that the wave is converging in that region. In Fig. 9(b) the Fresnel number is 3. The phase is constant near the axis, and hence the wavefront is planar in this region. In Fig. 9(c) the Fresnel number is 2.5. The upward curvature of the phase shows that the wave is diverging in this region. The qualitative behavior of the intensity and phase pictured in Figs. 9(a)–(c) is very similar to the behavior of the intensity and phase in the plane wave case [see Figs. 4(a)–(c)]. The main difference is that the peak intensity in the $N = 3$ focal plane is 2.96 in the present case, and was 4 in the plane wave case.

As we move further away from the aperture plane, the field continues to diverge, until we approach the on-axis position where $N = 2$ [Fig. 9(d)]. The intensity then goes through a minimum, and the character of the wave changes from diverging to converging as we pass through this point. The behavior is similar to that seen at the position where $N = 2$ in the plane wave case. The present behavior is different in that the phase varies continuously as we pass through this point (see Fig. 8), instead of discontinuously, as it did in the plane wave case (see Fig. 3).

As we get further away from the aperture plane the wave becomes converging again. This is illustrated by the plot of the phase in Fig. 9(e), where $N = 1.5$. The downward curvature of the phase near the axis means that the wave is converging in this region. As we approach the point where $N = 1$, the wavefront flattens out.

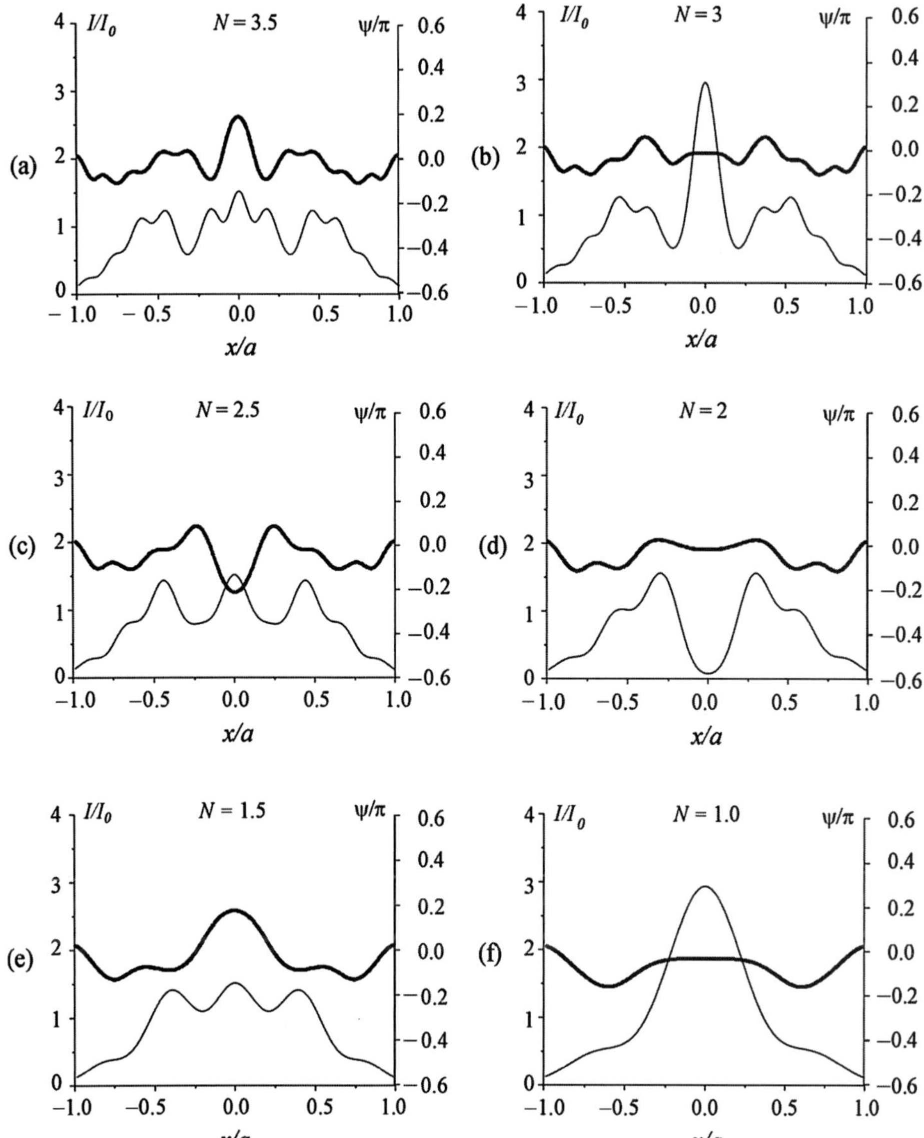

Figure 9 Plots of I/I_0 (thin line) and ψ/π (thick line) as functions of position along the x axis for observation planes with Fresnel numbers: (a) $N = 3.5$, (b) $N = 3$, (c) $N = 2.5$, (d) $N = 2$, (e) $N = 1.5$, and (f) $N = 1$. The parameter β is 0.57.

Figure 9(f) shows that the phase is constant near the axis when $N = 1$; therefore the wavefront is planar in this region. The character of the wave as we move from the plane $N = 1.5$ to the plane $N = 1$ is similar to that in the plane wave case. The major difference is that the peak intensity at the $N = 1$ plane here is 2.94, whereas for the case of a plane wave it was 4.

14.3 Fresnel Diffraction by a Bicomponent System of Apertures

14.3.1 Incident plane wave

14.3.1.1 Basic equations

Let us now consider the two-aperture system depicted in Fig. 10. The radius of the first aperture is a_1, the radius of the second aperture is a_2, and the distance between the planes containing the apertures is L. Let $P'' = (x'', y'', L)$ be a point inside the second aperture, and $P = (x, y, z)$ be an observation point in the plane $z = \text{constant} > L$. In cylindrical polar coordinates we then write $P'' = (\rho'', \theta'', L)$ and $P = (\rho, \theta, z)$. It follows from Eqs. (9) and (10) that the complex amplitude of the field incident upon the second aperture is given by

$$U(\rho'', \theta'', L) = U_o e^{ikL} \left\{ 1 - e^{iu_1/2} e^{iv_1^2/2u_1} \left[V_0(u_1, v_1) - iV_1(u_1, v_1) \right] \right\}, \qquad (40)$$

where $u_1 = 2\pi N_1, v_1 = 2\pi N_1 \rho''/a_1$ and $N_1 = a_1^2/\lambda L$.

In order to compare our results for the intensity to those of Ref. [4], we now assume that the distance L is such that N_1, the Fresnel number of the first aperture at the center of the second aperture, is equal to unity. In this case, $u_1 = 2\pi$,

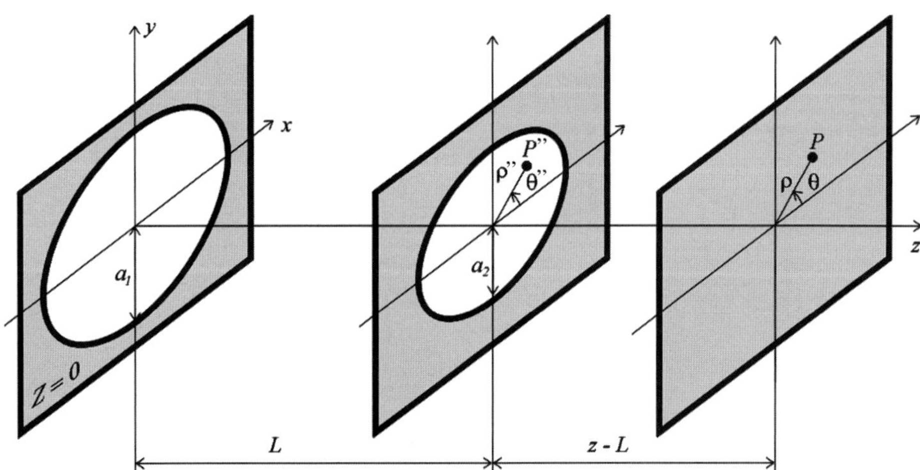

Figure 10 Geometry for the diffraction of a plane wave by a system of two circular apertures with radii a_1 and a_2, respectively. The distance between the two aperture planes is L. Point $P'' = (\rho'', \theta'', L)$ is a point inside the second aperture and $P = (\rho, \theta, z)$ is a point in the observation plane $z = \text{constant}$.

$v_1 = 2\pi\rho''/a_1$ and Eq. (40) can be written as

$$U(\rho'', \theta'', L) = U_o e^{ikL}\left[1 + D(\rho''/a_1)\right], \tag{41}$$

where

$$D(w) = e^{i\pi w^2}\left[V_0(2\pi, 2\pi w) - iV_1(2\pi, 2\pi w)\right]. \tag{42}$$

Let us now investigate the complex amplitude, $U(\rho, \theta, z)$, of the diffracted field in the region $z > L$. The paraxial form of the Fresnel approximation to the Rayleigh-Sommerfeld diffraction formula tells us that this field is given by the expression

$$U(\rho, \theta, z) = \frac{k}{2\pi i(z - L)} e^{ik(z-L)} e^{ik\rho^2/2(z-L)}$$

$$\times \int_0^{a_2}\int_0^{2\pi} U(\rho'', \theta'', L) e^{ik\rho''^2/2(z-L)} e^{-ik\rho\rho''\cos(\theta-\theta'')/(z-L)} \rho'' d\rho'' d\theta''. \tag{43}$$

Upon substituting the right-hand side of Eq. (41) into Eq. (43) and performing the angular integration, we find that the field is independent of the angle θ and given by

$$U(\rho, z) = \frac{k}{i(z - L)} U_o e^{ikz} e^{ik\rho^2/2(z-L)}$$

$$\times \int_0^{a_2}\left[1 + D(\rho''/a_1)\right] e^{ik\rho''^2/2(z-L)} J_0\left[k\rho\rho''/(z-L)\right]\rho'' d\rho''. \tag{44}$$

Let us now make the change of variable $\rho'' = \xi a_2$. After using this relation in the right-hand side of Eq. (44), we find that the field can be described in terms of the dimensionless variables u_2 and v_2 as

$$U(\rho, z) = -iu_2 U_o e^{ikz} e^{iv_2^2/2u_2}\int_0^1\left[1 + D(\alpha\xi)\right] e^{iu_2\xi^2/2} J_0(v_2\xi)\xi\, d\xi, \tag{45}$$

where

$$u_2 = 2\pi N_2, \quad v_2 = 2\pi N_2\rho''/a_2, \tag{46}$$

with N_2 the Fresnel number of the second aperture at the on-axis point in the observation plane,

$$N_2 = a_2^2/\lambda(z - L), \tag{47}$$

and α the ratio of the radii of the two apertures,

$$\alpha = a_2/a_1. \tag{48}$$

By analogy with the results of Sect. 14.2, let us write Eq. (45) as

$$U(\rho,z) = U_o e^{ikz} M(\rho,z), \tag{49}$$

where

$$M(\rho,z) = -iu_2 e^{iv_2^2/2u_2} \int_0^1 [1 + D(\alpha\xi)] e^{iu_2\xi^2/2} J_0(v_2\xi)\xi\,d\xi. \tag{50}$$

It follows from Eq. (49) that the intensity and phase of the field in the lit region are given, respectively, by Eqs. (11) and (12), with $M(\rho,z)$ given by Eq. (50) and the reduced phase defined as in Eq. (13). The function $M(\rho,z)$ can be evaluated by numerical integration.

14.3.1.2 On-axis intensity and phase

The on-axis intensity and reduced phase of the field after the second aperture were calculated by the method described above, and are plotted as a function of the Fresnel number N_2 for $\alpha = 0.1$ in Fig. 11, and for $\alpha = 0.5$ in Fig. 12. One general comment is in order before discussing the results. There are no true singular points behind the second aperture, because even at points where the intensity is minimum, its value is not exactly zero.

Figures 11 and 12 show the results for $\alpha = 0.1$ and 0.5, respectively. In both cases the values of the intensity and the reduced phase oscillate as functions of N_2. Upon comparing the two sets of curves, we see that three changes occur when α is increased from 0.1 to 0.5. First, the maximum value of the on-axis intensity decreases (from approximately $15I_o$ to $8I_o$), and the minimum value increases (from approximately zero to I_o). Second, the amplitude of the oscillation of the reduced phase decreases and does not occur so suddenly. Finally, at the smaller value of α the maxima (minima) of the intensity occur when the Fresnel number is odd (even), but at the larger value they are shifted to slightly higher values of N_2; likewise the reduced phase curve is also shifted toward higher values of N_2.

The explanation for this behavior is as follows. When $\alpha = 0.1$, the radius of the second aperture is 10 times smaller than that of the first. In this case the results of Sect. 14.2 show that the phase of the field incident upon the second aperture is constant across it [see Fig. 5(c)], and that the value of the intensity incident upon it varies by only 10% across it. Therefore the field incident upon the second aperture is very similar to the field incident upon the first aperture (a constant

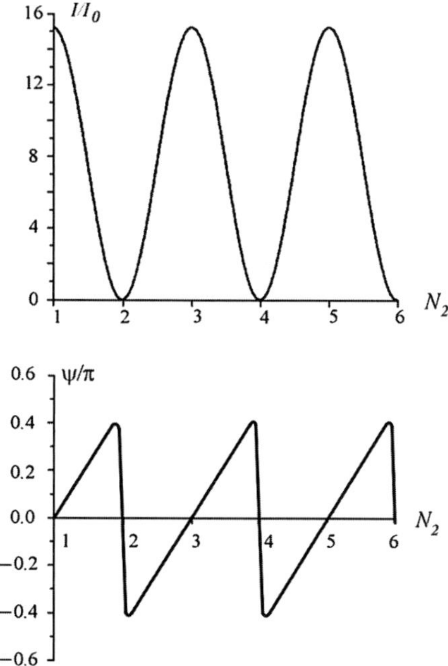

Figure 11 Plots of I/I_o and ψ/π for on-axis observation points as functions of the Fresnel number N_2 for $\alpha = 0.1$.

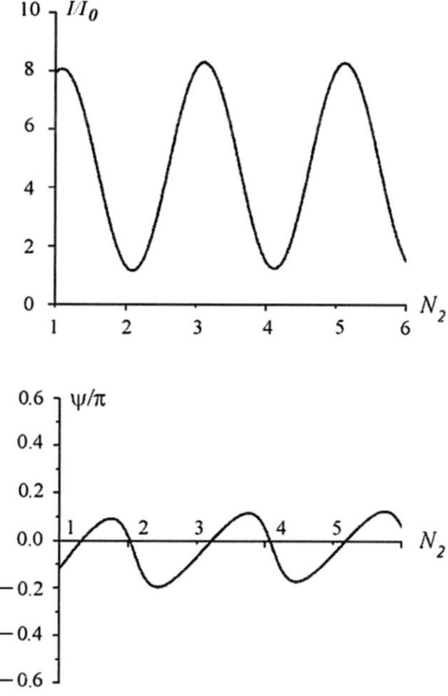

Figure 12 Plots of I/I_o and ψ/π for on-axis observation points as functions of the Fresnel number N_2 for $\alpha = 0.5$.

amplitude, normally incident plane wave), and this explains why the second aperture increases the on-axis intensity at the focal points by a factor of close to four, why the intensities at the minima are approximately zero, and why the phase jumps are approximately equal to π. As α increases from the value 0.1, the phase of the field incident upon the second aperture remains approximately constant across it, but the intensity begins to vary considerably. As a result, the effect of the second aperture becomes less ideal. When $\alpha = 0.5$, as in Fig. 12, both the phase and the intensity of the field incident upon the second aperture vary significantly across it [see Fig. 5(c)], and the effect of the second aperture is correspondingly less ideal.

14.3.1.3 General case

The intensity and reduced phase of the diffracted field as a function of the scaled transverse coordinate x/a_2 in six different planes behind the second aperture are shown in Fig. 13 for the case $\alpha = 0.4$. In Fig. 13(a) the value of the Fresnel number, $N_2 = 3.06$, was chosen such that the on-axis intensity was maximum, i.e., so that the plane is a focal plane. We see from the figure that the phase near the axis is approximately constant, so that the wave in that region is behaving like a plane wave. In Figs. 13(b) and 13(c) the Fresnel numbers are 2.5 and 2.37, respectively, and the wave is diverging in each case. In Fig. 13(d) the Fresnel number is 1.74, and the wave has changed from a diverging wave to a converging wave, i.e., it has refocused. In Fig. 13(e) the Fresnel number is 1.5, and the wave continues to converge. In Fig. 13(f) the Fresnel number is 1.05, and the plane is a focal plane. The on-axis intensity is maximum, and the phase near the axis is approximately constant. In Ref. [4] the value of α was 0.4, as it is in Fig. 13. Our results for the intensities agree well with those of Ref. [4].

14.3.2 Incident Gaussian beam

14.3.2.1 Basic equations

It follows from Eqs. (27), (28), and (10) that the complex amplitude of the field incident upon the second aperture is

$$U(\rho'', L) = U_o e^{ikL} H(\rho''/a_1), \tag{51}$$

where

$$H(w) = \frac{2\pi N_1}{u_1} e^{iN_1 \pi w^2} \left\{ e^{-iv_1^2/2u_1} - e^{iu_1/2} \left[V_0(u_1, v_1) - iV_1(u_1, v_1) \right] \right\}. \tag{52}$$

Here,

$$u_1 = 2\pi N_1 + 2i\beta^2, \quad v_1 = 2\pi N_1 w, \tag{53}$$

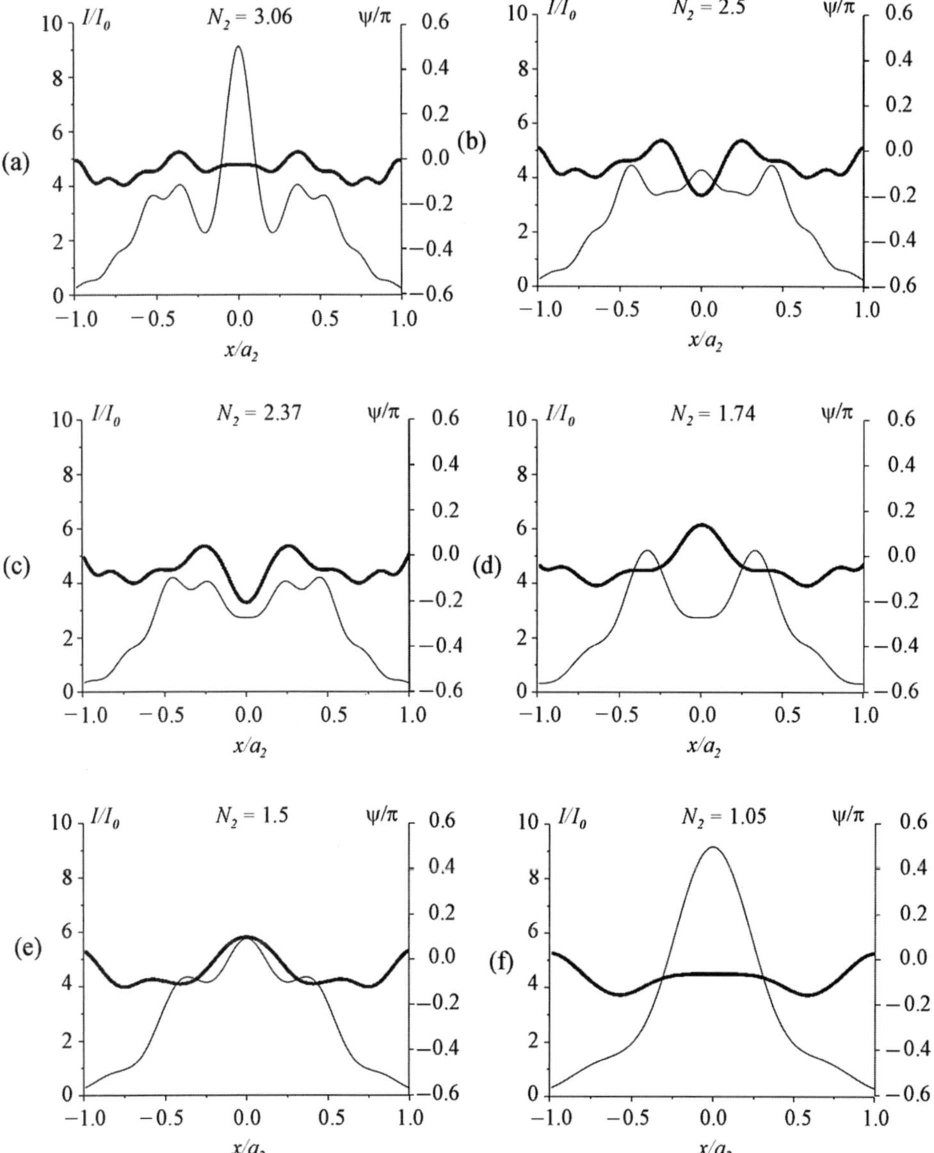

Figure 13 Plots of I/I_0 (thin line) and ψ/π (thick line) as functions of position along the x-axis for observation planes behind the second aperture with Fresnel numbers: (a) $N_2 = 3.06$, (b) $N_2 = 2.5$, (c) $N_2 = 2.37$, (d) $N_2 = 1.74$, (e) $N_2 = 1.5$, and (f) $N_2 = 1.05$.

and N_1 is the Fresnel number of the first aperture at the observation point at the center of the second aperture,

$$N_1 = a_1^2/\lambda L. \tag{54}$$

As in the previous section, let $N_1 = 1$. Equation (52) then simplifies to

$$H(w) = \frac{2\pi}{u_1}e^{i\pi w^2}\left\{e^{-iv_1^2/2u_1} - e^{iu_1/2}\left[V_0(u_1,v_1) - iV_1(u_1,v_1)\right]\right\}, \qquad (55)$$

where $u_1 = 2\pi + 2i\beta^2$, and $v_1 = 2\pi w$. In order to find the field at the observation point $P = (\rho, \theta, z)$ we substitute this incident field into Eq. (43). Upon performing the angular integration, we find that the field is independent of the angle θ and that

$$U(\rho,z) = \frac{k}{i(z-L)}U_0 e^{ikz} e^{ik\rho^2/2(z-L)}$$

$$\times \int_0^{a_2} H(\rho''/a_1)e^{ik\rho''^2/2(z-L)}J_0[k\rho\rho''/(z-L)]\rho''d\rho''. \qquad (56)$$

Let us now make the change of variable $\rho'' = \xi a_2$. After substituting this relation into the right-hand side of Eq. (56), we find that the field can be described in terms of the dimensionless variables u_2 and v_2 as

$$U(\rho,z) = -iu_2 U_0 e^{ikz} e^{iv_2^2/2u_2}\int_0^1 H(\alpha\xi)e^{iu_2\xi^2/2}J_0(v_2\xi)\xi\, d\xi, \qquad (57)$$

where

$$u_2 = 2\pi N_2, \quad v_2 = 2\pi N_2\rho/a_2. \qquad (58)$$

Here, N_2 is the Fresnel number of the second aperture at the observation point (ρ, θ, z),

$$N_2 = a_2^2/\lambda(z-L), \qquad (59)$$

and α is the ratio of the radii of the two apertures [see Eq. (48)]. By analogy with the results of the previous subsection, let us write Eq. (57) as

$$U(\rho,z) = U_0 e^{ikz} G(\rho,z), \qquad (60)$$

where

$$G(\rho,z) = -iu_2 e^{iv_2^2/2u_2}\int_0^1 H(\alpha\xi)e^{iu_2\xi^2/2}J_0(v_2\xi)\xi\, d\xi. \qquad (61)$$

It follows from Eq. (60) that the intensity and phase of the field in the lit region are given by Eqs. (29) and (30), respectively, with $G(\rho,z)$ given by Eq. (61) and the reduced phase defined as in Eq. (31). The function $G(\rho, z)$ can be evaluated by numerical integration.

14.3.2.2 On-axis intensity and phase

The on-axis intensity and phase behind the second aperture were calculated using the methods described above. Figure 14 shows plots of the on-axis intensity as a function of N_2 for: (a) an incident plane wave (which corresponds to $\beta = 0$), (b) an incident Gaussian beam with aperture-spot ratio of $\beta = 0.57$, and (c) an incident Gaussian beam with an aperture-spot ratio of $\beta = 1$. In all three cases the value of the ratio of the two aperture radii is $\alpha = 0.4$. The qualitative behavior of the three curves is very similar, but there are some differences. The peak intensity decreases as β is increased. This is to be expected, since a larger β corresponds to an amplitude distribution in the first aperture that is further from a uniform amplitude situation. Secondly, as β is increased, the curves shift toward a lower Fresnel number, i.e., toward the aperture. Finally, the minima are lower for the larger β cases.

Figure 15 shows plots of the on-axis reduced phase as a function of N_2 for $\beta = 0, 0.57$, and 1. The qualitative behavior of the three curves is very similar. However, as β increases, the curves shift toward lower values of the Fresnel number, and are shifted downward in numerical value.

14.3.2.3 General case

The intensity and reduced phase of the diffracted field as a function of the scaled transverse coordinate x/a_2 in four different planes behind the second aperture are

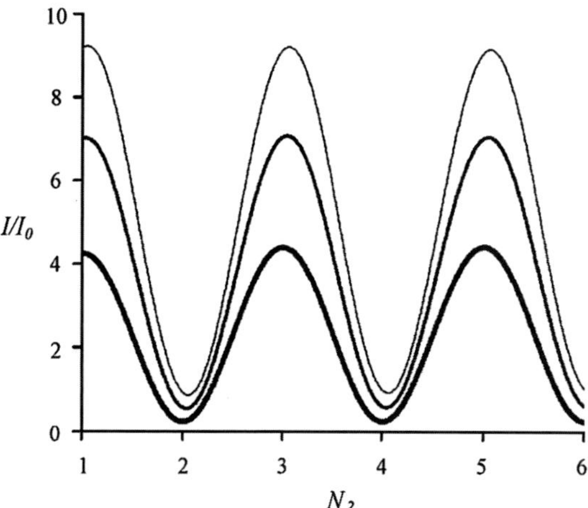

Figure 14 The on-axis intensity $I(0, z)$, in units of I_o, as a function of the Fresnel number of the second aperture, N_2, for $\beta = 0$ (thin line), $\beta = 0.57$ (medium thick line), and $\beta = 1$ (thick line), all for $\alpha = 0.4$.

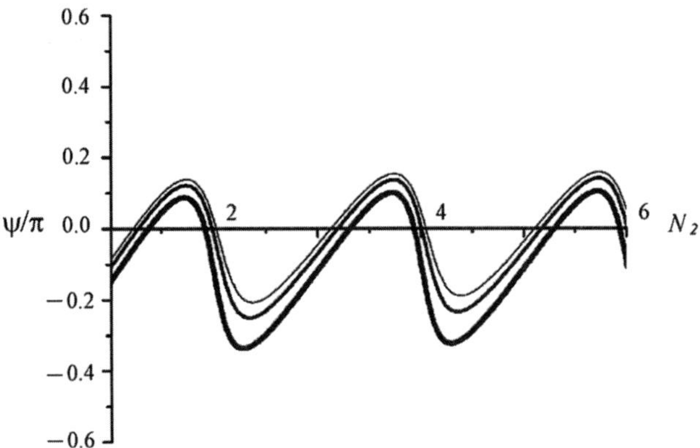

Figure 15 The on-axis reduced phase $\psi(0,z)$, in units of π, as a function of the Fresnel number of the second aperture, N_2, for $\beta = 0$ (thin line), $\beta = 0.57$ (medium thick line), and $\beta = 1$ (thick line), all for $\alpha = 0.4$.

shown in Fig. 16 for the case $\alpha = 0.4$. In Fig. 16(a) the Fresnel number is 3.04, and the plane is a focal plane. The on-axis intensity is maximum, and the phase near the axis is approximately constant. In Fig. 16(b) the Fresnel number is 2.5, and the wave is diverging. In Fig. 16(c) the Fresnel number is 1.5, and the wave has changed from a diverging wave to a converging wave. In Fig. 16(d) the Fresnel number is 1.04, and the plane is a focal plane. The on-axis intensity is maximum, and the phase near the axis is approximately constant.

Let us now compare these results to the incident plane wave results. A comparison of Figs. 16(a) and 13(a) shows that the qualitative behavior in the two cases at the $N_2 \approx 3$ focal point is very similar. The major difference is a matter of scale: the curves have similar shapes, but the maximum intensity in the Gaussian beam case is approximately $7I_o$ instead of the $9I_o$ for the plane wave case. A comparison of Figs. 16(b) and 13(b) shows that the qualitative behavior in the two cases is very similar at the location where $N_2 = 2.5$ as well. Similar results are obtained for the location where $N_2 = 1.5$ when Figs. 16(c) and 13(e) are compared. Finally, the curves for both the intensity and phase at the $N_2 \approx 1$ focal point are very similar as well [see Figs. 16(d) and 13(f)]. As with the $N_2 \approx 3$ case, the key difference is the fact that the maximum intensity in the Gaussian beam case is approximately $7I_o$ instead of $9I_o$ for the plane wave case.

14.4 Conclusions

We have investigated the intensity and phase of the diffracted field behind a circular aperture when a monochromatic plane wave is incident upon it, and when a Gaussian beam is incident upon it. We have also investigated the intensity and

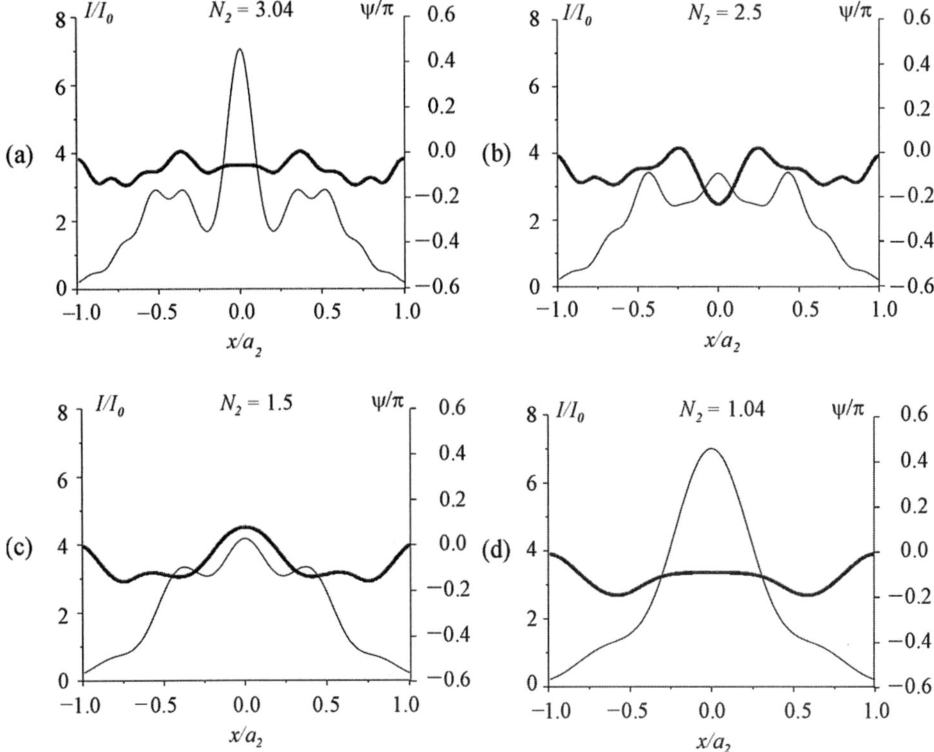

Figure 16 Plots of I/I_o (thin line) and ψ/π (thick line) as functions of position along the x-axis for observation planes after the second aperture with Fresnel numbers: (a) $N_2 = 3.04$, (b) $N_2 = 2.5$, (c) $N_2 = 1.5$, and (d) $N_2 = 1.04$.

phase of the diffracted field in a system of two circular apertures for the same incident fields.

For the single-aperture system, with a plane wave normally incident, it was shown that in the neighborhood of a focal point, the phase of the wave approaching the focal point is that of a converging wave, the phase in the focal plane is planar, and the phase of the wave exiting the focal point is that of a diverging wave. It was also shown that the wave becomes more and more divergent as the distance from the focal point is increased, until a position at which the Fresnel number becomes even is reached. At such a point the intensity of the wave is zero, and the phase of the wave is undefined, i.e., singular. It was shown that as the observation point on-axis moves away from the aperture and passes through a singular point, the nature of the wave in the neighborhood of the axis changes from that of a diverging wave to that of a converging wave, i.e., the wave refocuses.

Similar behavior was observed when a Gaussian beam was normally incident upon a circular aperture, for the case in which the waist of the Gaussian beam occurs in the plane of the aperture. It was shown that as β (the ratio of the aperture radius to the incident beam waist) increases, the focusing effect becomes weaker, i.e., the values of the intensities at the focal points decrease. This implies that the more the incident intensity deviates from a constant value (as we had in the plane wave case), the weaker the focusing effect is.

The focusing effect was also investigated for a plane wave that was normally incident upon a two-aperture system, for the case in which the separation between the two apertures is chosen such that $N_1 = 1$. It was observed, by studying the phase, that the field focuses, defocuses, and refocuses, as in the one-aperture case. We found that the effect depended crucially on α, the ratio of the radii of the two apertures. For $\alpha = 0.1$, the focusing effect was strong, with the intensities at the focal point approximately $15I_o$. This was due to the fact that, since $N_1 = 1$, the intensity and phase of the field incident upon the second aperture were both fairly constant. As α increases and the intensity incident upon the second aperture varies across it, the focusing effect becomes weaker (the peak intensities decrease) and the positions of the focal points shift slightly. Similar behavior was also observed when a Gaussian beam was normally incident upon a two-aperture system, for the case in which the separation between the two apertures is chosen such that $N_1 = 1$. As in the single-aperture Gaussian beam case, it was found that increasing β resulted in a weaker focusing effect.

14.A Derivation of Equation (7)

It is convenient to consider the real and imaginary parts of the integral on the left-hand side of Eq. (7) separately. We set

$$\int_0^1 e^{iu\xi^2/2} J_0(v\xi)\xi\,d\xi = \tfrac{1}{2}[C(u,v) + iS(u,v)], \tag{A1}$$

where

$$C(u,v) = 2\int_0^1 \cos(u\xi^2/2)J_0(v\xi)\xi\,d\xi, \tag{A2}$$

$$S(u,v) = 2\int_0^1 \sin(u\xi^2/2)J_0(v\xi)\xi\,d\xi. \tag{A3}$$

These two integrals can be expressed in terms of the Lommel functions of two variables $V_0(u,v)$ and $V_1(u,v)$,[12]

$$C(u,v) = \frac{2}{u}\left[\sin(v^2/2u) + \sin(u/2)V_0(u,v) - \cos(u/2)V_1(u,v)\right], \tag{A4}$$

$$S(u,v) = \frac{2}{u}\left[\cos(v^2/2u) - \cos(u/2)V_0(u,v) - \sin(u/2)V_1(u,v)\right]. \tag{A5}$$

Upon substituting the right-hand sides of Eqs. (A4) and (A5) in to Eq. (A1), we find

$$\int_0^1 e^{iu\xi^2/2} J_0(v\xi)\xi\,d\xi = \frac{i}{u}\left\{ e^{-iv^2/2u} - e^{iu/2}\left[V_0(u,v) - iV_1(u,v) \right] \right\}, \qquad (A6)$$

which is Eq. (7).

14.B Derivation of Equations (32) and (33)

It follows from Eq. (25) that

$$\frac{u}{2\pi N} = 1 + i\frac{\beta^2}{N\pi}. \qquad (B1)$$

Upon using Eqs. (6) and (26) we find that

$$\frac{u}{2\pi N} = 1 + i\frac{z}{\pi w_0^2/\lambda} = 1 + i\frac{z}{z_R} = \sqrt{1 + (z/z_R)^2}\,e^{i\chi} = \frac{w}{w_0}e^{i\chi}, \qquad (B2)$$

where w and χ are given by Eqs. (19) and (21), respectively. Equation (32) is the reciprocal of Eq. (B2).

It follows from Eq. (25) that

$$i\left(\frac{N\pi\rho^2}{a^2} - \frac{v^2}{2u}\right) = i\frac{N\pi\rho^2}{a^2}\left(1 - \frac{1}{1 + i\beta^2/N\pi}\right) = -\frac{\rho^2}{a^2}\frac{\beta^2}{1 + i\beta^2/N\pi}$$

$$= -\frac{\rho^2/w_0^2}{1 + iz/z_R} = -\frac{\rho^2}{w^2} + i\frac{k\rho^2}{2R}, \qquad (B3)$$

where w and R are given by Eqs. (19) and (20). Equation (B3) then gives Eq. (33).

References

1. M. De, J.W.Y. Lit, and R. Tremblay, "Multi-aperture focusing technique," *Appl. Opt.* **7**, 483–488 (1968).
2. J.W.Y. Lit and R. Tremblay, "Boundary-diffraction-wave theory of cascaded-apertures diffraction," *J. Opt. Soc. Am.* **59**, 559–567 (1969).
3. J.W.Y. Lit, R. Boulay, and R. Tremblay, "Diffraction fields of a sequence of equal radii circular apertures," *Opt. Comm.* **1**, 280–282 (1970).
4. R.R. Letfullin and T.F. George, "Optical effect of diffractive multifocal focusing of radiation on a bi-component diffraction system," *Appl. Opt.* **39**, 2545–2550 (2000).

5. R.R. Letfullin and O.A. Zayakin, "Observation of the effect of diffractive multifocal focusing of radiation," *J. Quant. Electron.* **31**, 339–342 (2001).

6. R.R. Letfullin, O.A. Zayakin, and T.F. George, "Theoretical and experimental investigations of the effect of diffractive multifocal focusing of radiation," *Appl. Opt.* **40**, 2138–2147 (2001).

7. R.R. Letfullin and T.F. George, "Diffractive multifocal focusing of Gaussian beams," *Fiber and Integrated Optics* **21**, 145–161 (2002).

8. J.W. Goodman, *Introduction to Fourier Optics*, 60, McGraw-Hill, New York (1968).

9. E. Lommel, "Theoretical and experimental investigations of diffraction phenomena at a circular aperture and obstacle," *Bayerisch. Akad. d. Wiss.* **15**, 233 (1884).

10. M.S. Soskin and M.V. Vasnetov, "Singular optics," in *Progress in Optics*, E. Wolf, Ed., **42**, 219–276, Elsevier, Amsterdam (2001).

11. J.F. Nye and M.V. Berry, "Dislocations in wave trains," *Proc. Roy. Soc. London A* **336**, 165–190 (1974).

12. M. Born and E. Wolf, *Principles of Optics*, 7th ed., Sect. 8.8, Cambridge Univ. Press, Cambridge (1999).

John T. Foley received his Ph.D. in Physics from the University of Rochester in 1978. His thesis advisor was Emil Wolf. He was post-doc at the University of Rochester for nine months, and then joined the faculty of Department of Physics and Astronomy at Mississippi State University, where he is currently Professor of Physics. He is the director of The Optics Project on the Web (WebTOP) an interactive, three-dimensional, computer graphics system for optics education. He is a Fellow of the Optical Society of America and received the 2003 George. P. Peagram Medal for excellence in teaching from the Southeastern Section of the American Physical Society.

Renat R. Letfullin earned his B.A. and M.S. in optics and spectroscopy from Samara State University (Russia) in 1984 and his Ph.D. in laser physics from Saratov State University in 1992. He is currently a Visiting Assistant Professor of Physics at Mississippi State University. Prior to 2002, he was a senior researcher in the Theoretical Department of the P.N. Lebedev Physical Institute of the Samara Branch of the Russian Academy of Sciences. He is known internationally as a theorist is in the fields of optics and kinetics of chemical pulsed lasers, with special expertise on amplifiers in two-phase active media. Dr. Letfullin has published over 80 articles and conference proceedings, including three book chapters.

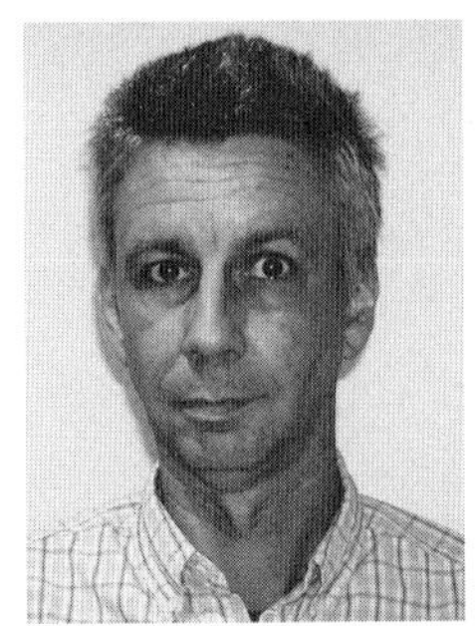

Henk F. Arnoldus obtained his Bachelor (1980) and Master (1981) degrees in Physics, both summa cum laude, from Eindhoven University of Technology, The Netherlands, with a specialization in heavy ion scattering theory. He then went to Utrecht University, The Netherlands, where he obtained his Ph.D. degree (1985) in Mathematics and Natural Sciences. From 1985 to 1988 he was postdoc with Thomas F. George in the Department of Physics at the University of New York at Buffalo. In 1988 he accepted a faculty position at Villanova University, Pennsylvania, where he stayed for six years. Currently, he is Associate Professor in the Department of Physics and Astronomy at Mississippi State University. He has published 84 papers and numerous conference proceedings, and he has edited three books. He has given 73 seminars/invited talks all over the world, including Russia, Denmark, and Korea.

CHAPTER 15

YOUNG'S INTERFERENCE EXPERIMENT: THE LONG AND SHORT OF IT

Taco D. Visser

Dedicated to Emil Wolf, in great appreciation of his friendship.

15.1 The Legacy of Thomas Young

The traditional view of Thomas Young (1773–1829) is that of an underappreciated genius. He was a child prodigy who could read fluently at the age of two and widely read the classics. By the age of 14 he was acquainted with Latin, Greek, French, Italian, Hebrew, Arabic, and Persian. Because of his many talents and wide-ranging interests, his fellow students in Cambridge nicknamed him "Phenomenon Young." After his formal education he set up a medical practice in London, but spent most of his time doing research. His initial interest was in sense perception, and he was the first to realize that the eye focuses by changing the shape of the lens. He also discovered the cause of astigmatism, and was the initiator, together with Helmholtz, of the three-color theory of perception, believing that the eye constructs its sense of color using only three receptors, for red, green and blue. In spite of all these achievements, his practice never flourished.

He also worked on deciphering the hieroglyphic text on the Rosetta Stone. He was the first to point out that the cartouches (the oval figures enclosing hieroglyphs) indicate the names of royalty. Nevertheless, all the credit for solving the mysteries of hieroglyphic writing went to the Frenchman Champollion.

Likewise, Young's seminal work on optics, which we will later discuss in more detail, was largely dismissed by his compatriots. In those days, any theory that went against the views of Newton was simply unacceptable to them. His main recognition came posthumously. As his epitaph in Westminster Abbey states, Thomas Young was *"a man alike eminent in almost every department of human learning."*

Figure 1 Thomas Young (1773–1829).

The modern-day view of Young is somewhat different. His contributions to Egyptology were not as grand as previously thought: most of his suggestions in this field were either not new at all or simply completely wrong [1]. In addition, his lack of success as a practitioner of medicine is today attributed to his total lack of bedside manner. His contributions to physics, however, are today considered to be in a class of their own. Regardless of the merits of his other activities, Thomas Young is rightfully immortalized by the experiment that now bears his name. In a lecture held for the Royal Society in 1803, he described how, by placing a card in a beam of light, it was split into two parts that resulted in a diffraction pattern that disappeared when one of the parts was obscured. In his own words [2]:

"It will not be denied by the most prejudiced that the fringes are produced by the interference of two portions of light ."

Better known today is his two-slit experiment (see Fig. 2) in which the light emanating from the slits projects a diffraction pattern of alternating dark and bright fringes [3]. The analogy between this experiment and the behavior of water waves clearly demonstrates the wavelike character of light.

Another great contribution was Young's suggestion that light vibrations, unlike sound waves, are transverse. This idea was confirmed in a series of beautiful experiments by Fresnel.

Recently, a poll was held among physicists asking them what they thought was the most beautiful physics experiment ever performed [4]. If the experiments are

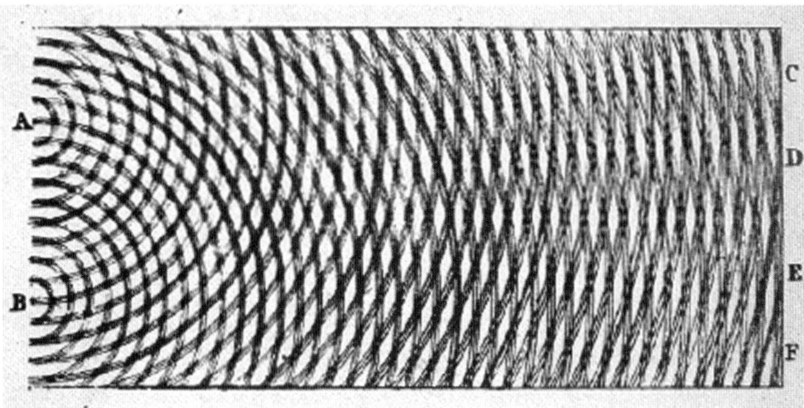

Figure 2 Young's illustration of two interfering waves, taken from Ref. [3].

ranked according to the number of times that they were cited, the result is the following:

1. Young's double-slit experiment applied to the interference of electrons by Jönsson (1961).
2. Galileo's experiment on falling bodies (ca. 1590).
3. Millikan's oil-drop experiment (1909).
4. Newton's decomposition of sunlight with a prism (1665–1666).
5. Young's light-interference experiment (1803).
6. Cavendish's torsion-bar experiment (1798).
7. Eratosthenes' measurement of the earth's circumference (ca. 250 BC).
8. Galileo's experiments with rolling balls down inclined planes (ca. 1608).
9. Rutherford's discovery of the nucleus (1911).
10. Foucault's pendulum (1851).

This result shows the historical importance of Young's experiment, both in its original form using light, and in its more recent version using electrons [5], which elegantly demonstrates the wave character of particles.

15.2 New Physics with Young's Experiment

To this very day, Young's experiment remains a source of inspiration to physicists who keep finding new ways to explore the subtleties of interference. Many examples of this can be found in a recent theme issue of the *Philosophical Transactions of the Royal Society of London*, the same society where Young lectured on his discoveries in optics, dedicated to interference phenomena [6].

Particularly in optical coherence theory, Young's experiment plays a crucial role. It was pointed out by Zernike [7] that the *visibility* of the interference fringes that are produced in such an experiment is a direct measure of the correlation properties of the field that is incident upon the pinholes. Among recent examples of the relevance of Young's experiment for coherence theory is Ref. [8] in which it was predicted that, when partially coherent fields are used, significant spectral changes may occur in the observed field. These spectral changes can in turn be used to determine the state of coherence of the incident field [9]. These examples underline the importance of Young's work for present-day research.

In the remainder of this paper we will discuss several novel effects in Young's experiment. Sections 15.3 and 15.4 are concerned with field correlations in the far zone, i.e., far away from the screen containing the pinholes. In Sect. 15.5 and Sect. 15.6 new properties of the field in the near zone, i.e., in and around the screen, are investigated.

15.3 Field Correlations in the Far Zone of Young's Experiment

Consider a plane screen $\mathcal{A}$, containing two pinholes $Q_1(\bar{\mathbf{r}}_1)$ and $Q_2(\bar{\mathbf{r}}_2)$ (see Fig. 3). The origin O of the coordinate system is taken in the screen, midway between the two pinholes, which are separated by a distance d, i.e.,

$$\bar{\mathbf{r}}_1 = (d/2, 0, 0), \tag{1}$$

$$\bar{\mathbf{r}}_2 = (-d/2, 0, 0). \tag{2}$$

We assume that the field that is incident upon the pinholes is partially coherent. A question arises naturally: *is the field in the region of superposition everywhere partially coherent?* To answer this question, we introduce a quantitative measure of the strength of the field correlations at the pair of observation points $P_1(\mathbf{r}_1), P_2(\mathbf{r}_2)$ at

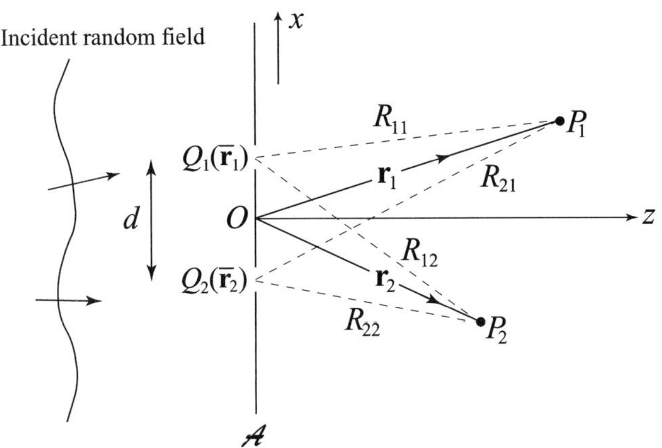

Figure 3 Illustrating the notation of Young's experiment.

frequency ω. This quantity is the *spectral degree of coherence* [10, Sect. 4.3.2], which is defined as

$$\mu(\mathbf{r}_1, \mathbf{r}_2, \omega) = \frac{W(\mathbf{r}_1, \mathbf{r}_2, \omega)}{\sqrt{S(\mathbf{r}_1, \omega)S(\mathbf{r}_2, \omega)}}, \tag{3}$$

where the *cross-spectral density* is given by the expression

$$W(\mathbf{r}_1, \mathbf{r}_2, \omega) = \langle U^*(\mathbf{r}_1, \omega)U(\mathbf{r}_2, \omega)\rangle. \tag{4}$$

In Eq. (4), $U(\mathbf{r}_i, \omega)$ represents the scalar field at position $\mathbf{r}_i$ ($i = 1, 2$), the asterisk denotes the complex conjugate, and the angular brackets denote the average taken over an ensemble of monochromatic realizations. The *spectral density* of the field at a position $\mathbf{r}$ at frequency ω is given by the diagonal element of the cross-spectral density, viz.,

$$S(\mathbf{r}, \omega) = W(\mathbf{r}, \mathbf{r}, \omega). \tag{5}$$

Let us now examine the spectral degree of coherence in the far zone of a Young's interference experiment. The field at the observation point $P_1(\mathbf{r}_1)$ is given by the sum of the field contributions of the two pinholes, i.e.,

$$U(\mathbf{r}_1, \omega) = -\frac{ikA}{2\pi}\left[U(\bar{\mathbf{r}}_1, \omega)\frac{e^{ikR_{11}}}{R_{11}} + U(\bar{\mathbf{r}}_2, \omega)\frac{e^{ikR_{21}}}{R_{21}}\right], \tag{6}$$

where R_{ij} denotes the distance Q_iP_j ($i,j = 1, 2$), $k = \omega/c$ is the wavenumber associated with frequency ω, c being the speed of light, and A is the area of each pinhole. The prefactor on the right-hand side of Eq. (6) stems from the Huygens–Fresnel principle [11, Sect. 8.8].[*] In a similar manner we find that

$$U(\mathbf{r}_2, \omega) = -\frac{ikA}{2\pi}\left[U(\bar{\mathbf{r}}_1, \omega)\frac{e^{ikR_{12}}}{R_{12}} + U(\bar{\mathbf{r}}_2, \omega)\frac{e^{ikR_{22}}}{R_{22}}\right]. \tag{7}$$

To simplify our notation, we introduce the abbreviation

$$K_{ij} = \frac{e^{ikR_{ij}}}{R_{ij}}. \tag{8}$$

We first turn our attention to pairs of points that lie in the plane $x = 0$, i.e., the plane that bisects the line joining the two pinholes and is perpendicular to the

---

[*] It is to be noted that the inclusion of obliquity factors in our analysis would not affect our results; cf. Ref. [12].

screen:

$$\mathbf{r}_1 = (0, y_1, z_1), \tag{9}$$

$$\mathbf{r}_2 = (0, y_2, z_2). \tag{10}$$

For such points we clearly have that

$$K_{11} = K_{21}, \tag{11}$$

$$K_{12} = K_{22}. \tag{12}$$

On substituting from Eqs. (6) and (7) into Eq. (4), while making use of Eqs. (11) and (12), we obtain the expression

$$W(\mathbf{r}_1, \mathbf{r}_2, \omega) = \left(\frac{kA}{2\pi}\right)^2 K_{11}^* K_{22}$$
$$\times \left\{ S_1(\omega) + S_2(\omega) + 2\left[S_1(\omega)S_2(\omega)\right]^{1/2} \Re\mu_{12}(\omega) \right\}, \tag{13}$$

where

$$S_1(\omega) = W(\bar{\mathbf{r}}_1, \bar{\mathbf{r}}_1, \omega), \tag{14}$$

$$S_2(\omega) = W(\bar{\mathbf{r}}_2, \bar{\mathbf{r}}_2, \omega), \tag{15}$$

are the spectral densities of the field at pinholes Q_1 and Q_2, respectively, $\Re$ denotes the real part, and

$$\mu_{12}(\omega) = \frac{\langle U^*(\bar{\mathbf{r}}_1, \omega) U(\bar{\mathbf{r}}_2, \omega) \rangle}{\sqrt{S_1(\omega)S_2(\omega)}} \tag{16}$$

is the spectral degree of coherence of the field at the two pinholes. Also, on substituting from Eqs. (6) and (7) into Eq. (5), while again using Eqs. (11) and (12), we obtain the formulae

$$S(\mathbf{r}_1, \omega) = |K_{11}|^2 \left\{ S_1(\omega) + S_2(\omega) + 2\left[S_1(\omega)S_2(\omega)\right]^{1/2} \Re\mu_{12}(\omega) \right\}, \tag{17}$$

$$S(\mathbf{r}_2, \omega) = |K_{22}|^2 \left\{ S_1(\omega) + S_2(\omega) + 2\left[S_1(\omega)S_2(\omega)\right]^{1/2} \Re\mu_{12}(\omega) \right\}. \tag{18}$$

Combining Eqs. (13), (17), and (18), we find for the spectral degree of coherence the expression

$$\mu(\mathbf{r}_1, \mathbf{r}_2, \omega) = \frac{K_{11}^* K_{22}}{|K_{11}||K_{22}|}, \tag{19}$$

and hence

$$|\mu(\mathbf{r}_1, \mathbf{r}_2, \omega)| = 1. \tag{20}$$

That is, *the spectral degree of coherence of the field at any two points in the plane $x = 0$ is unimodular, irrespective of the state of coherence of the field that is incident upon the two pinholes.*

We next consider pairs of points that are each other's mirror image with respect to the plane $y = 0$, i.e.,

$$\mathbf{r}_1 = (x_1, y_1, z_1), \tag{21}$$

$$\mathbf{r}_2 = (x_1, -y_1, z_1). \tag{22}$$

Obviously, for such points we have

$$K_{11} = K_{12}, \tag{23}$$

$$K_{21} = K_{22}. \tag{24}$$

On substituting from Eqs. (6) and (7) into Eq. (4), while using Eqs. (23) and (24), we obtain the expression

$$W(\mathbf{r}_1, \mathbf{r}_2, \omega) = \left(\frac{kA}{2\pi}\right)^2 \Big\{ |K_{11}|^2 S_1(\omega) + |K_{22}|^2 S_2(\omega)$$

$$+ 2[S_1(\omega)S_2(\omega)]^{1/2} \Re[K_{11}^* K_{22} \mu_{12}(\omega)] \Big\}. \tag{25}$$

On substituting from Eqs. (6) and (7) into Eq. (5), while again using Eqs. (23) and (24), we find that

$$S(\mathbf{r}_1, \omega) = S(\mathbf{r}_2, \omega) \tag{26}$$

$$= \left(\frac{kA}{2\pi}\right)^2 \Big\{ |K_{11}|^2 S_1(\omega) + |K_{22}|^2 S_2(\omega)$$

$$+ 2[S_1(\omega)S_2(\omega)]^{1/2} \Re[K_{11}^* K_{22} \mu_{12}(\omega)] \Big\}. \tag{27}$$

Substituting from Eqs. (25), (26) and (27) into definition (3) yields the result

$$\mu(\mathbf{r}_1, \mathbf{r}_2, \omega) = 1. \tag{28}$$

Stated in words, *the spectral degree of coherence of the field at two points that are each other's mirror image with respect to the plane $y = 0$, is always unity, i.e., the field at P_1*

and P_2 is fully coherent and cophaseal, irrespective of the state of coherence of the field at the two pinholes.

This result has a clear physical meaning: according to the *spectral interference law* [10, Sect. 4.3.2], if the light at P_1 and P_2 is combined (for example, by coupling the light into two fiber tips) and used in a second Young's experiment, the resulting interference pattern will have fringes with perfect visibility.

We emphasize that the surprising results expressed by Eqs. (20) and (28) also hold in the limiting case when each pinhole is illuminated by a separate laser. It is to be noted that taking obliquity factors into account does not alter the outcome of our analysis (for a proof of this, see [12]). Moreover, these results have recently been generalized to the case of partially coherent, partially polarized electromagnetic beams [13].

15.4 Phase Singularities of the Coherence Function in the Far Field

In the previous section it was demonstrated that in a Young's interference experiment with partially coherent light there exist pairs of points at which the light is fully coherent. We now turn our attention to the other extreme, and ask ourselves the question: *are there pairs of points in the region of superposition that are completely uncorrelated?* To simplify our analysis we assume that the spectral density of the field at the two pinholes is identical, i.e.,

$$S_1(\omega) = S_2(\omega). \tag{29}$$

Under this assumption the cross-spectral density at an arbitrary pair of observation points is given by the expression

$$W(\mathbf{r}_1, \mathbf{r}_2, \omega) = \left(\frac{kA}{2\pi}\right)^2 S_1(\omega) \left\{ K_{11}^* K_{12} + K_{21}^* K_{22} \right.$$

$$\left. + \mu_{12}(\omega) K_{11}^* K_{22} + \mu_{12}^*(\omega) K_{21}^* K_{12} \right\}, \tag{30}$$

where we have used Eqs. (6) and (7). In the far zone the factors K_{ij} have the approximate form

$$K_{ij} \approx \frac{\exp\left[ik(R_j - \hat{\mathbf{r}}_j \cdot \bar{\mathbf{r}}_i)\right]}{R_j}, \tag{31}$$

where $R_j = |\mathbf{r}_j|$ is the distance from the origin to the point of observation P_j, and $\hat{\mathbf{r}}_j$ is a unit vector pointing in the direction OP_j. On substituting from Eq. (31)

into Eq. (30) we find that

$$W(\mathbf{r}_1, \mathbf{r}_2, \omega) = 2\left(\frac{kA}{2\pi}\right)^2 S_1(\omega)\frac{\exp\left[ik(R_2 - R_1)\right]}{R_1 R_2}$$

$$\times \left\{\cos\left[\frac{kd}{2}(\cos\theta_1 - \cos\theta_2)\right]\right.$$

$$\left. + |\mu_{12}(\omega)|\cos\left[\frac{kd}{2}(\cos\theta_1 + \cos\theta_2) + \beta\right]\right\}, \quad (32)$$

where θ_i is the angle between $\hat{\mathbf{r}}_i$ and the positive x-axis, and β is the phase of $\mu_{12}(\omega)$. It is readily seen that Eq. (32) implies the existence of phase singularities, i.e., pairs of points at which the cross-spectral density vanishes. In particular, $W(\mathbf{r}_1, \mathbf{r}_2, \omega)$ vanishes at points for which the expression in the curly brackets vanishes (see Fig. 4). The expression in the curly brackets in Eq. (32) is independent of the distances R_1 and R_2, and in fact depends only on the directions of observation. It follows that a given zero of the cross-spectral density requires that the observation points P_1 and P_2 both lie on conical surfaces $\cos\theta_i = $ constant; a sketch of such surfaces is given in Fig. 5.

After substituting from Eq. (32) into Eq. (5), we find that the spectral density in the region of superposition can only have zero values for fully coherent fields, i.e., for fields for which $|\mu_{12}(\omega)| = 1$. Therefore, it follows from Eq. (3) that for partially coherent fields the zeros of the spectral degree of coherence $\mu(\mathbf{r}_1, \mathbf{r}_2, \omega)$ coincide with those of the cross-spectral density. The behavior of the phase of the spectral degree of coherence in the immediate vicinity of the conical zero surfaces can be readily found by noting that the expression in the curly brackets of Eq. (32) is real-valued, so that the only possible phase change of this factor on changing the

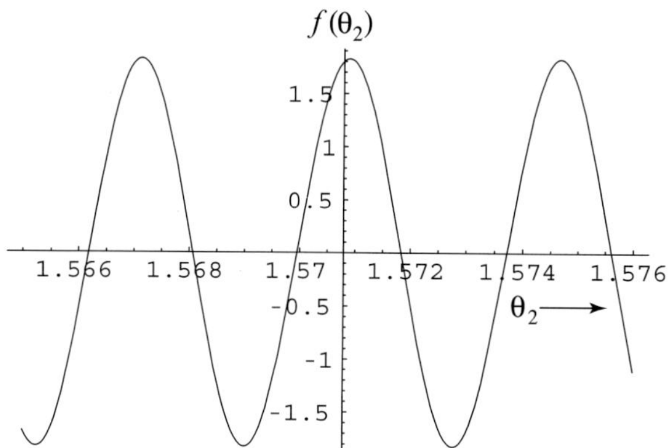

Figure 4 Some roots of the function $f(\theta_2) = \cos\left[(kd/2)(\cos\theta_1 - \cos\theta_2)\right] + |\mu_{12}(\omega)|\cos\left[(kd/2)(\cos\theta_1 + \cos\theta_2) + \beta\right]$. In this example $k = 0.333 \times 10^7$ m^{-1}, $d = 0.1$ cm, $\mu_{12}(\omega) = 0.8 + 0.3\,i$, and $\theta_1 = \pi/2$.

angles θ_1 or θ_2 is a change in sign. This corresponds to a jump in phase of $\pm\pi$, and these are the only possible singular behaviors across the singular surfaces.

The spectral degree of coherence in the region of superposition was studied numerically using Eqs. (29), (30), and (3), i.e., without the approximations that lead to Eq. (32). Let the position of the two field points $P_1(\mathbf{r}_1)$ and $P_2(\mathbf{r}_2)$ be denoted by

$$\mathbf{r}_1 = (x_1, y_1, z_1), \quad \mathbf{r}_2 = (x_2, y_2, z_2), \tag{33}$$

respectively. The phase, $\phi_\mu(\mathbf{r}_1, \mathbf{r}_2, \omega)$, of the spectral degree of coherence was calculated for the case in which x_2 is varied while y_2, z_2, and $\mathbf{r}_1$ are kept fixed. An example of its discontinuous behavior is depicted in Fig. 6. The change by an amount $\pm\pi$ of the phase across a phase singularity is clearly seen.

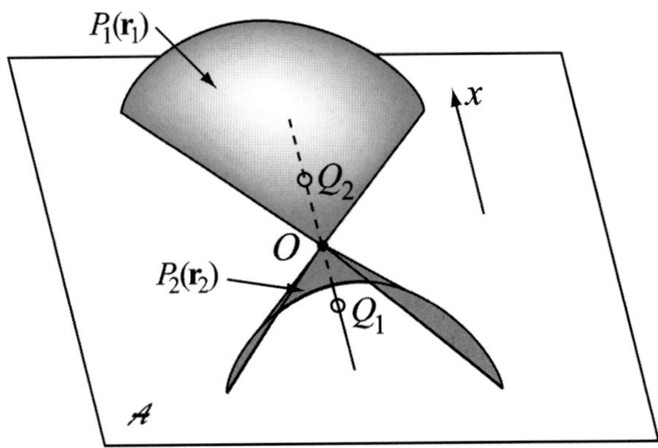

Figure 5 Schematic illustration of surfaces on which points of observation $P_1(\mathbf{r}_1)$ and $P_2(\mathbf{r}_2)$ in the far zone are located for which $W(\mathbf{r}_1, \mathbf{r}_2, \omega) = \mu(\mathbf{r}_1, \mathbf{r}_2, \omega) = 0$.

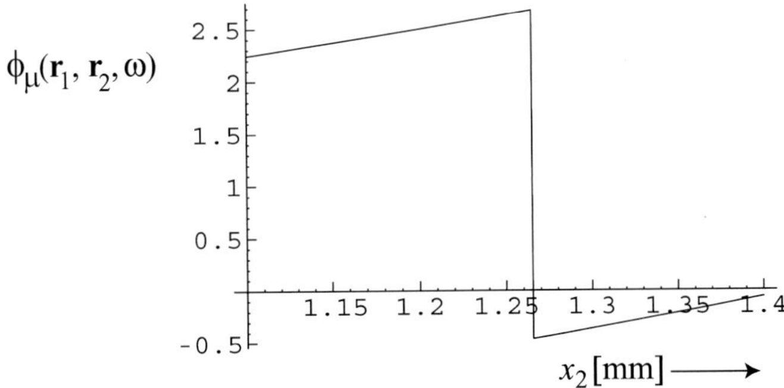

Figure 6 Illustrating the discontinuous behavior of the phase ϕ_μ of the spectral degree of coherence across a phase singularity. In this example $\mathbf{r}_1$, y_2, and z_2 are kept fixed while x_2 is varied. Here $k = 0.333 \times 10^7$ m^{-1}, $d = 0.1$ cm, $\mu_{12}(\omega) = 0.8 + 0.3\,i$, $\mathbf{r}_1 = (0, 0, 1.5)$ m, $y_2 = 0.9$ mm, and $z_2 = 1.5$ m.

We have also studied the phase of the spectral degree of coherence in a plane parallel to the screen that contains the apertures. An example is shown in Fig. 7. The vertical line indicates the location of a phase singularity, i.e., a set of points $P_2(\mathbf{r}_2)$ for which $\mu(\mathbf{r}_1, \mathbf{r}_2, \omega) = 0$, and hence the phase of the spectral degree of coherence is singular. Four pairs of contour lines show the discontinuity of the phase ϕ_μ across the phase singularity. In all four cases the phase undergoes a jump equal to $\pm\pi$, in agreement with the asymptotic behavior predicted by Eq. (32).

We note that the phase singularities of the spectral degree of coherence can easily be observed. Detecting the phase change requires interfering the light from the vicinity of the pair of points P_1 and P_2. This can be done by coupling the light from these points into another Young's interference experiment (for example, by using optical fibers) and observing the behavior of the spectral interference fringes produced in this secondary experiment as the point P_2 is moved across the phase singularity (see Fig. 8). In Fig. 9 the fringe patterns shown would be observed in this secondary experiment for a selection of points P_1, P_2. The point $P_1(\mathbf{r}_1)$ was chosen as in Fig. 7, and the point P_2 was taken along a line of constant phase at several points in the vicinity of the phase singularity; the choices of P_2 (A, B, C, D, and E) are illustrated in Fig. 7. It can be seen in Fig. 9 that the π phase change results in the minima of the secondary fringe pattern becoming maxima, and vice versa, in accordance with the spectral interference law [10, Sect. 4.3.2].

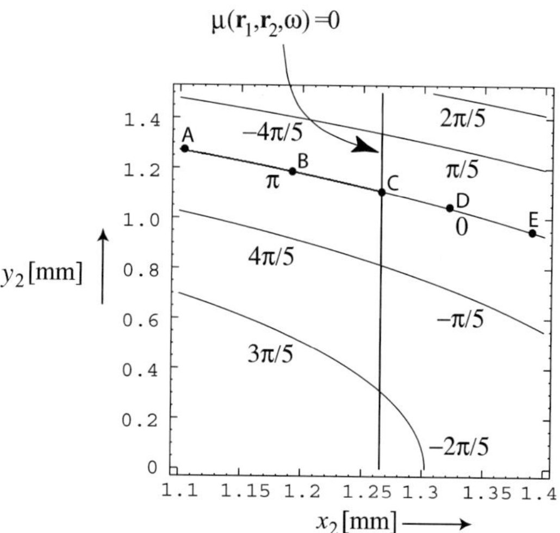

Figure 7 Contours of equal phase of the spectral degree of coherence $\mu(\mathbf{r}_1, \mathbf{r}_2, \omega)$ near a phase singularity (the vertical line) in a plane parallel to the screen. In this example $k = 0.333 \times 10^7$ m^{-1}, $d = 0.1$ cm, $\mu_{12}(\omega) = 0.8 + 0.3\mathrm{i}$, $\mathbf{r}_1 = (0, 0, 1.5)$ m, and $z_2 = 1.5$ m (after [14]).

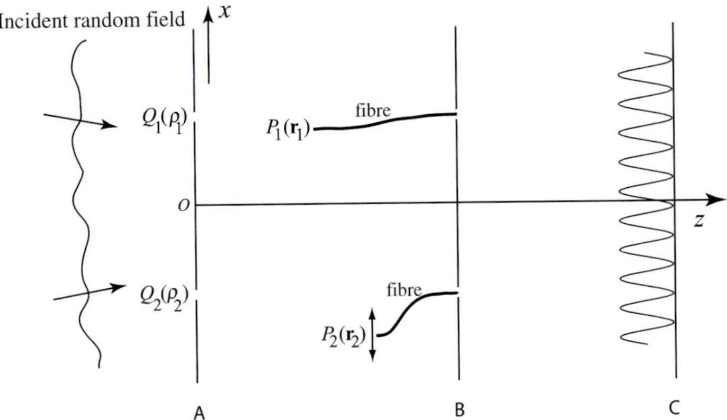

Figure 8 Combining the light from the two observation points P_1 and P_2 in a secondary Young's interference experiment using fibers. The end points of the fibers act as point sources situated in a second screen **B**. The visibility of the resulting interference fringes on a third screen **C** changes as the point P_1 is kept fixed while P_2 is scanned across a phase singularity of $\mu(\mathbf{r}_1, \mathbf{r}_2, \omega)$.

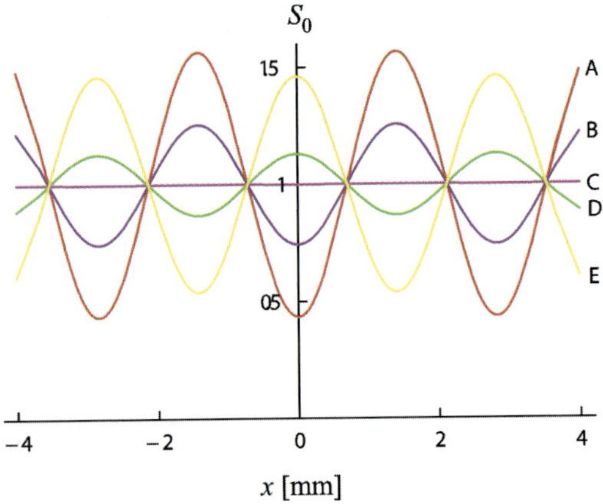

Figure 9 Illustrating the spectral interference pattern formed along the x-direction by combining the light from the two observation points P_1 and P_2 in a secondary Young's interference experiment. The observation plane was taken to be at $z = 1.5\,\mathrm{m}$, and the spacing of the pinholes taken to be $d = 0.1\,\mathrm{cm}$. The positions **A**, **B**, **C**, **D**, and **E** of the points P_2 are illustrated in Fig. 7. S_0 is the spectral intensity normalized to the value of the spectral intensity on the curve **C**.

15.5 Phase Singularities of the Poynting Vector near the Screen

In most discussions of Young's interference experiment the precise nature of the screen is conveniently omitted. In this section we investigate the field in the vicinity of subwavelength apertures in a realistic screen, i.e., a screen with a finite thickness and a finite conductivity. Not only does our analysis reveal new effects such as the creation and annihilation of phase singularities of the Poynting vector, but it also shows that in certain cases the two apertures can strongly influence each other through surface plasmons that are generated on the screen. The latter will be discussed in Sect. 15.6.

We first study a single, infinitely long slit in a thin metal plate. The slit runs in the y-direction. The incident field, taken to have time dependence $\exp(-i\omega t)$, propagates in the positive z-direction, perpendicular to the plate. In this case, a scalar approach, as was used in the previous sections, does not suffice, so rigorous electromagnetic theory must be used to obtain the field. Specifically, the following integral equation for the electric field [15] has to be solved:

$$\hat{E}_i(x,z) = \hat{E}_i^{(\mathrm{inc})}(x,z) - i\omega\Delta\varepsilon \int_{\mathcal{D}} \hat{G}_{ij}^E(x,z;x',z')\hat{E}_j(x',z')\,dx'\,dz', \qquad (34)$$

where $\Delta\varepsilon = \varepsilon_0 - \varepsilon_{\mathrm{plate}}$ is the difference between the vacuum permittivity and the permittivity of the metal plate, $\hat{\mathbf{G}}^E$ is the electric Green's tensor pertaining to the plate without the slit, and $\hat{E}_i^{(\mathrm{inc})}$ is the ith component ($i = x,y,z$) of the incident field, i.e., the field that is present when there is no slit in the plate. The integration is over the domain $\mathcal{D}$ of the slit. For points which lie within the slit, Eq. (34) is a Fredholm equation of the second kind for $\hat{\mathbf{E}}$, which can be solved numerically by the collocation method with piecewise-constant basis functions. The electric field at observation points outside the domain of the slit is then calculated by substituting this solution back into Eq. (34). With the electric field determined everywhere, the magnetic field $\hat{\mathbf{H}}$ follows directly from Maxwell's equations.

We are interested in the singular optics behavior of the real-valued, two-dimensional time-averaged Poynting vector field,

$$\mathbf{S}(x,z) = \frac{1}{2}\Re\left\{\hat{\mathbf{E}}(x,z) \times \hat{\mathbf{H}}^*(x,z)\right\}. \qquad (35)$$

The phase ϕ_S of the Poynting vector is given by the pair of relations

$$\sin\phi_S(x,z) \equiv \frac{S_z(x,z)}{|\mathbf{S}(x,z)|},$$

$$\cos\phi_S(x,z) \equiv \frac{S_x(x,z)}{|\mathbf{S}(x,z)|}. \qquad (36)$$

It follows from these equations that $\phi_S(x,z)$ is singular at those points where $\mathbf{S}(x,z) = 0$.

An example of the power flow field (i.e., the time-averaged Poynting vector) near a narrow slit in a thin silver plate is shown in Fig. 10. In this example the incident field is taken to be TE polarized (i.e., the $\hat{\mathbf{E}}$ field is parallel to the slit). It is seen that the field exhibits several phase singularities, namely vortices and saddles. In addition, the aperture is seen to have a funnel-like effect on the field, corre-

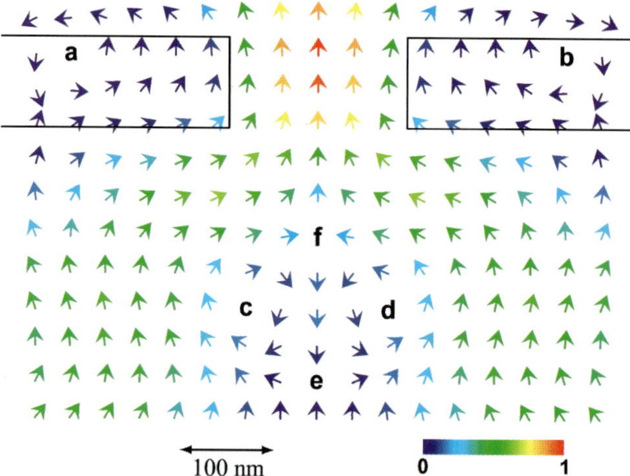

Figure 10 Behavior of the time-averaged Poynting vector near a 200-nm-wide slit in a 100-nm-thick silver plate with refractive index $n = 0.05 + i2.87$. The incident light (coming from below) has a wavelength $\lambda = 500$ nm. The left-handed vortices (a and d) and the right-handed optical vortices (b and c) each have a topological charge of $+1$, whereas the topological charge of the saddle points (e and f) is -1. The color coding indicates the modulus of the (normalized) Poynting vector (see legend).

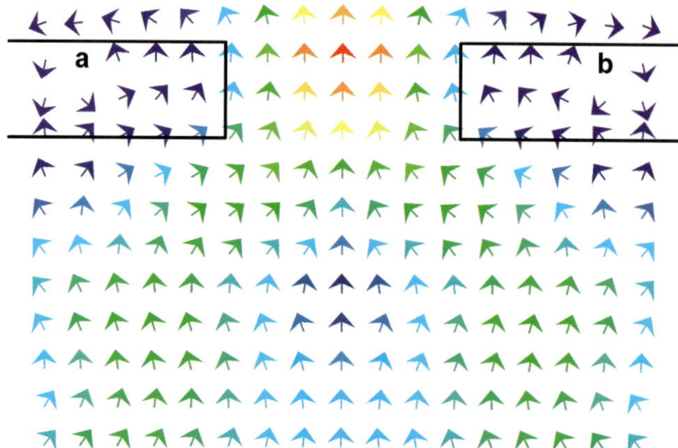

Figure 11 Behavior of the time-averaged Poynting vector when the slit width is increased to 250 nm. The two vortices a and b remain, whereas the four phase singularities c, d, e, and f that were visible in Fig. 10 have annihilated each other.

sponding to an enhanced light transmission [16]. When the slit width is increased in a continuous manner, the four singularities below the slit (c, d, e, and f) move together and eventually annihilate each other. In this process topological charge is conserved. As can be seen from Fig. 11, the annihilation results in a smoother power flow field, corresponding to a greater power transmission. Further examples of such creation and annihilation events are given in [17]. The relation between such events and the onset of guided modes within the slit is discussed in [18].

15.6 Surface Plasmons on the Screen and the Light Transmission Process

When the incident field is TM polarized (i.e., when the $\hat{\mathbf{E}}$ field is perpendicular to the slit), surface plasmons may be generated [19]. These are electromagnetic fields that travel along the interface between a metal and a dielectric. The field component normal to the interface decays exponentially. Because the surface plasmon decay length on the interface (i.e., the propagation distance over which the amplitude of the fields decreases by a factor $1/e$) is much larger than the skin depth of the metal, it is possible for plasmons that are generated at one slit in the screen to travel to another slit. When a plasmon exchanges momentum with the screen, for example at a slit, it can be converted into a propagating light field. There are conflicting reports in the literature on whether the generation of plasmons actually helps or frustrates the light transmission process [20,21]. In this section we demonstrate some unexpected consequences of plasmon excitation for the light transmission through two narrow slits.

The two-slit configuration that we analyzed is depicted in Fig. 12. A metal plate of thickness d with two parallel slits is illuminated by a plane, monochromatic, TM polarized wave that is normally incident upon it. The slits each have a width w and are separated by a distance b. The plate is surrounded by vacuum. Let us write the complex-valued relative permittivity of the metal plate as

$$\epsilon_m = \epsilon'_m + i\epsilon''_m, \qquad (\epsilon'_m, \epsilon''_m \in \mathfrak{R}) \tag{37}$$

and the complex-valued wavenumber of the plasmon along the interface as

$$k_x = k'_x + ik''_x, \qquad (k'_x, k''_x \in \mathfrak{R}). \tag{38}$$

One can then derive the *plasmon dispersion relation* [19]

$$k'_x = \frac{\omega}{c}\left(\frac{\epsilon'_m}{\epsilon'_m + 1}\right)^{1/2}, \tag{39}$$

with c the speed of light in vacuum. Using the integral equation formalism described in Sect. 15.5 we have analyzed the light transmission process for two parallel narrow slits. First, for TM polarized incident light the amplitude of the field along a cross section was calculated some distance away from the slits. In Fig. 13 the exponential fall off of E_z, the electric field component normal to the two in-

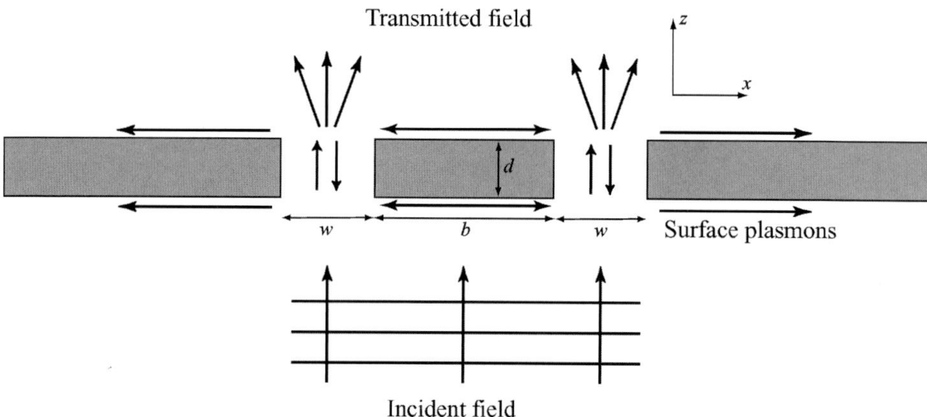

Figure 12 A two-slit configuration. Surface plasmons can be generated on both sides of the plate. When a plasmon that is excited at one slit reaches the other slit, it can be reflected or converted into a propagation wave field.

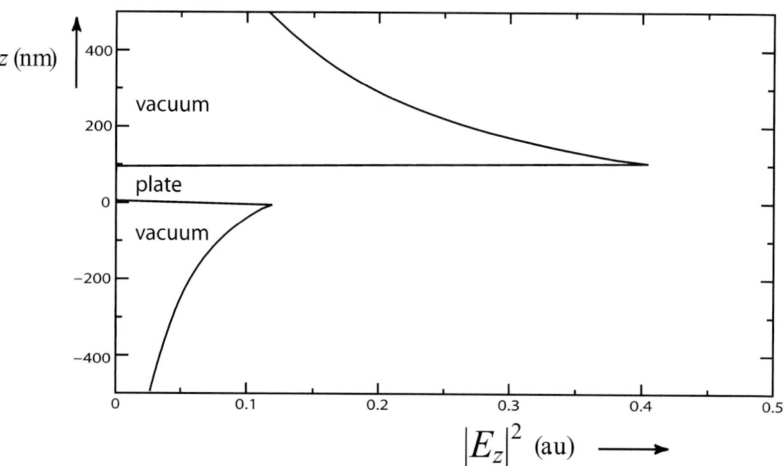

Figure 13 The transverse field profile $|E_z|^2$ (in arbitrary units), showing the exponential fall off of the field amplitudes normal to the interfaces that is typical of a surface plasmon. In this example the cross section is taken at a distance of two wavelengths from the nearest slit. The two slits are both 30-nm wide. The silver plate has a thickness of 100 nm, and its permittivity $\epsilon_m = (0.05 + i2.87)^2$. The slit spacing is 450 nm. The incident TM field, traveling in the positive z-direction, has a wavelength $\lambda = 500$ nm.

terfaces, that is characteristic for surface plasmons can be seen. We conclude that plasmons are indeed generated. (Note that for TE polarized light the amplitude of E_z is zero everywhere.)

In Fig. 14 the light transmission of a two-slit system is shown as a function of the separation distance b between the two slits (cf. [22]). For TE polarized light, no plasmons are generated and the spacing hardly affects the field transmission. For TM polarized light, however, a very strong modulation of the light transmission can be seen. The modulation period coincides exactly with the plasmon wavelength $\lambda_{\mathrm{sp}} = 2\pi/k'_x$, with k'_x given by Eq. (39). Clearly, the plasmons that are generated at each of the two slits and travel toward the other can interfere with each other either constructively or destructively. In the former case they give rise to enhanced transmission (i.e., a transmission greater than one); in the latter case they cause frustrated transmission (i.e., a transmission smaller than one).

We conclude that for TM polarized light, the light transmission through two narrow parallel slits is dominated by the effect of surface plasmons. In contrast to claims in the literature (Refs. [21] and [22]), they can either give rise to an enhanced transmission or to a frustrated transmission. The effect of this *surface plasmon-induced transmission* depends on the spacing between the two slits. Experimental verification of this prediction has already taken place.

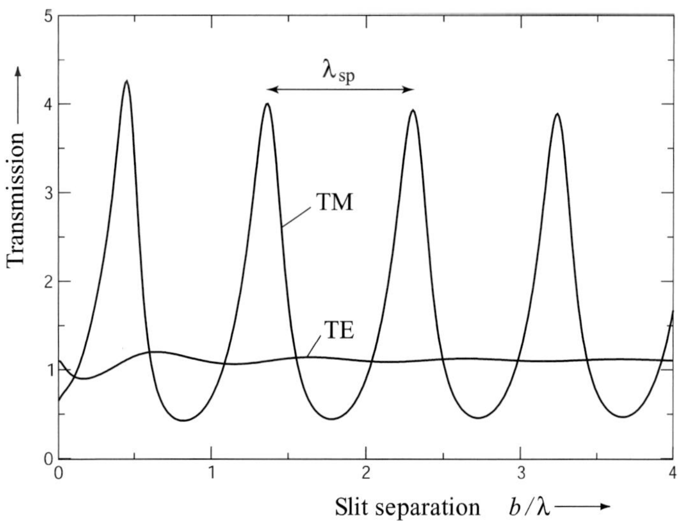

Figure 14 The light transmission for a two-slit system as a function of the separation distance b between the two slits. For TE polarized light no plasmons are generated, in contrast to the TM case. The transmission is normalized to the intensity that is incident onto the two slits according to geometrical optics.

15.7 Conclusions

We have predicted several new effects in Young's interference experiment. When the experiment is carried out using partially coherent light, there exist pairs of points in the region of superposition where the field is fully coherent, irrespective of the state of coherence of the field that is incident upon the two pinholes. Also, there are pairs of points that are completely incoherent. These points correspond to phase singularities of the spectral degree of coherence. This quantity undergoes a π phase jump across the singular surfaces. It was discussed how these singularities can be observed using a secondary Young's experiment.

In the vicinity of the screen that contains the apertures, the time-averaged Poynting vector exhibits a rich variety of phase singularities. The nature and location of these singularities depend on the thickness and conductivity of the screen. When a parameter of the configuration, for example, the width of the slits is varied in a continuous way, annihilation or creation of these singularities may occur. The total topological charge is conserved in such events.

We have analyzed the role of surface plasmons that can be generated on metallic screens. It was found that, due to their long propagation distance, they may lead to "cross talk" between the two apertures. Depending on the distance between the apertures, the plasmons can give rise to a light transmission that is either enhanced or frustrated. Experiments that validate these predictions have already taken place.

References

1. M. Pope, *The Story of Decipherment: From Egyptian Hieroglyphs to Maya Script*, Thames and Hudson, London (1999).
2. T. Young, "The Bakerian Lecture. Experiments and calculations relative to physical optics," *Phil. Trans. R. Soc. Lond.* **94**, 1–16 (1804).
3. T. Young, *A Course of Lectures on Natural Philosophy and the Mechanical Arts*, 2 Vols., Johnson, London (1807).
4. R.P. Crease, *Physics World*, September (2002). See also: R. Crease, *The Prism and the Pendulum: The Ten Most Beautiful Experiments in Science*, Random House, New York (2003).
5. C. Jönsson, "Elektroneninterferenzen an mehreren künstlich hergestellten Feinspalten," *Zeitschrift für Physik* **161**, 454–474 (1961). An English translation was published by D. Brandt and S. Hirschi, in the *American Journal of Physics* **42**, 4–11 (1974).
6. Theme issue, Interference: 200 years after Thomas Young's discoveries, *Phil. Trans. R. Soc. Lond.* A **360**, 805–1069 (2002).
7. F. Zernike, "The concept of degree of coherence and its applications to optical problems," *Physica* **5**, 785–795 (1938).

8. D.F.V. James and E. Wolf, "Spectral changes produced in Young's interference experiment," *Opt. Commun.* **81**, 150–154 (1991).

9. D.F.V. James and E. Wolf, "Some new aspects of Young's interference experiment," *Phys. Lett. A* **157**, 6–10 (1991).

10. L. Mandel and E. Wolf, *Optical Coherence and Quantum Optics*, Cambridge University Press, Cambridge (1995).

11. M. Born and E. Wolf, *Principles of Optics*, 7th (expanded) ed., Cambridge University Press, Cambridge (1999).

12. H.F. Schouten, T.D. Visser, and E. Wolf, "New effects in Young's interference experiment with partially coherent light," *Opt. Letters* **28**, 1182–1184 (2003).

13. G.S. Agarwal, A. Dogariu, T.D. Visser, and E. Wolf, "Generation of complete coherence in Young's interference experiment with random electromagnetic beams" (submitted).

14. H.F. Schouten, G. Gbur, T.D. Visser, and E. Wolf, "Phase singularities of the coherence functions in Young's interference pattern," *Opt. Letters* **28**, 968–970 (2003).

15. T.D. Visser, H. Blok, and D. Lenstra, "Theory of polarization-dependent amplification in a slab waveguide with anisotropic gain and losses," *IEEE J. Quant. Elect.* **35**, 240–249 (1999).

16. T.W. Ebbesen et al., "Extraordinary optical transmission through sub-wavelength hole arrays," *Nature* **391**, 667–669 (1998).

17. H.F. Schouten, G. Gbur, T.D. Visser, D. Lenstra, and H. Blok, "Creation and annihilation of phase singularities near a sub-wavelength slit," *Optics Express* **11**, 371–380 (2003).

18. H.F. Schouten, T.D. Visser, D. Lenstra, and H. Blok, "Light transmission through a subwavelength slit: Waveguiding and optical vortices," *Phys. Rev. E* **67**, 036608 (2003).

19. H. Raether, *Surface Plasmons on Smooth and Rough Surfaces and on Gratings*, Springer, Berlin (1988).

20. J.A. Porto, F.J. Garcia-Vidal, and J.B. Pendry, "Transmission resonances on metallic gratings with very narrow slits," *Phys. Rev. Lett.* **83**, 2845–2848 (1999).

21. Q. Cao and P. Lalanne, "Negative role of surface plasmons in the transmission of metallic gratings with very narrow slits," *Phys. Rev. Lett.* **88**, 057403 (2002).

22. H.F. Schouten, T.D. Visser, G. Gbur, D. Lenstra, and H. Blok, "Surface plasmon-induced enhanced and frustrated light transmission through two narrow slits" (submitted).

Taco D. Visser received his Ph.D. at the University of Amsterdam. He works at the Department of Physics and Astronomy of the Free University, also in Amsterdam. His research interests are diffraction, near-field optics and scattering theory. He has collaborated with Emil Wolf since 1994.

ᎧCHAPTER 16Ꭵ

QUALITATIVE DESCRIPTION OF THE WOLF EFFECT AND DIFFERENCES BETWEEN THE DOPPLER AND THE WOLF SHIFTS

Valerian I. Tatarskii

16.1 Introduction

In the papers of Emil Wolf [1,2] a statement that at first seems paradoxical but becomes evident after simple analysis, was formulated. The statement is that the spectrum of a source depends on its spatial coherence properties. This phenomenon was later called the "Wolf effect."

In this paper we present a simple physical picture explaining the influence of spatial coherence on the spectrum, and on the basis of this explanation we analyze the differences between the Doppler and the Wolf shifts. For the simplest illustration of the role of spatial coherence, in this paper we compare two limiting cases: the completely spatially coherent and completely spatially incoherent cases. This paper is not a review, but rather a qualitative physical picture of the Wolf effect, based on the ideas in Ref. [3]. The models considered in this paper are rather simple, perhaps even oversimplified, and do not pretend to be a comprehensive description of the subject, which can be found in the review paper [5] and in the book [4].

In Sect. 16.2 we consider the Young experiment, but starting with Sect. 16.3 we analyze the case of radiation, which is produced by the plane circular disk. We consider the cases of completely spatially coherent sources, and the cases of wide, narrow, and discrete spectral lines. Sect. 16.4 is devoted to a qualitative explanation of increasing spatial coherence in the process of propagation. In Sect. 16.5 the differences between the Doppler shift and the Wolf shift are discussed.

We do not discuss here dynamic scattering (see [5]) related to the time-dependent fluctuations of dielectric permittivity. The reason for this restriction is that dynamic scattering includes the real Doppler effect.[*]

16.2 The Relation between Interference and the Wolf Effect

The temporal autocorrelation function of the electric field E in the point $\mathbf{r}$

$$B(\mathbf{r},\tau) = \langle E(\mathbf{r},t)E^*(\mathbf{r},t+\tau)\rangle; \quad B^*(\mathbf{r},-\tau) = B(\mathbf{r},\tau) \tag{1}$$

determines both the intensity I and the spectrum W of the field:

$$I(\mathbf{r}) = B(\mathbf{r},0), \tag{2}$$

$$W(\mathbf{r},\omega) = \frac{1}{2\pi}\int_{-\infty}^{\infty} B(\mathbf{r},\tau)\exp(-i\omega\tau)\,d\tau. \tag{3}$$

The inverse Fourier transform determines $B(\mathbf{r},\tau)$ in terms of $W(\mathbf{r},\omega)$:

$$B(\mathbf{r},\tau) = \int_{-\infty}^{\infty} W(\mathbf{r},\omega)\exp(i\omega\tau)\,d\omega. \tag{4}$$

Let us consider the Young experiment: the field created by two point sources, located in different points (Fig. 1).

The field in some point $\mathbf{r}$ is the sum of two contributions, caused by these two sources:

$$E(\mathbf{r},t) = E_1(\mathbf{r},t) + E_2(\mathbf{r},t). \tag{5}$$

This relation follows from the linearity of electrodynamics.

---

[*] For example, the moving with the speed V sphere of radius R, having dielectric permittivity $\varepsilon > 1$, may be treated as a time-dependent ε field, described by the function $\varepsilon(x,y,z,t) = 1 + (\varepsilon - 1)\theta(R - \sqrt{(x-Vt)^2 + y^2 + z^2})$. Here, $\theta(u) = 1$ for $u > 0$ and $\theta(u) = 0$ for $u < 0$. The frequency shift in the scattered electromagnetic field from this object may be treated either as the real Doppler effect, or as dynamic scattering. Therefore, in the case of dynamic scattering the real Doppler effect may contribute to the frequency shift as well as the Wolf shift.

It is now traditional in radar meteorology to measure wind speed using the Doppler shift, which appears in the radar signal returns scattered by atmospheric turbulence [6].

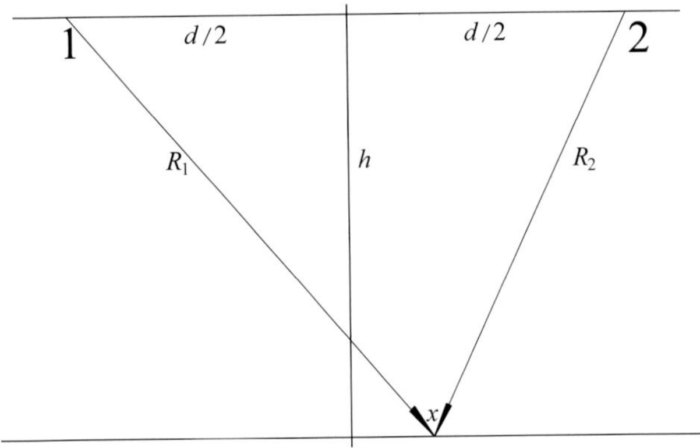

Figure 1 Interference experiment.

For the autocorrelation function of the total field we obtain

$$B(\mathbf{r},\tau) = \left\langle \left[E_1(\mathbf{r},t) + E_2(\mathbf{r},t)\right]\left[E_1^*(\mathbf{r},t+\tau) + E_2^*(\mathbf{r},t+\tau)\right]\right\rangle$$
$$= B_{11}(\mathbf{r},\tau) + B_{22}(\mathbf{r},\tau) + B_{12}(\mathbf{r},\tau) + B_{21}(\mathbf{r},\tau). \tag{6}$$

The term B_{11} presents the autocorrelation function of source 1 in the absence of source 2, and similarly for B_{22}. The terms B_{12} and B_{21} appear only if the correlation in two sources exists. Because these sources are located in different points, such a correlation is called *spatial* correlation.

If we set $\tau = 0$ in Eq. (6), we obtain the intensity $I(\mathbf{r})$:

$$I(\mathbf{r}) = I_{11}(\mathbf{r}) + I_{22}(\mathbf{r}) + I_{12}(\mathbf{r}) + I_{21}(\mathbf{r}). \tag{7}$$

If we consider the Fourier transform of $B(\mathbf{r},\tau)$ with respect to τ, we obtain the spectrum

$$W(\mathbf{r},\omega) = \frac{1}{2\pi}\int_{-\infty}^{\infty}[B_{11}(\mathbf{r},\tau) + B_{22}(\mathbf{r},\tau) + B_{12}(\mathbf{r},\tau) + B_{21}(\mathbf{r},\tau)]\exp(-i\omega\tau)d\tau$$
$$= W_{11}(\mathbf{r},\omega) + W_{22}(\mathbf{r},\omega) + W_{12}(\mathbf{r},\omega) + W_{21}(\mathbf{r},\omega). \tag{8}$$

The term $[I_{12}(\mathbf{r}) + I_{21}(\mathbf{r})]$ describes the interference fringes in the Young experiment. The corresponding term $[W_{12}(\mathbf{r},\omega) + W_{21}(\mathbf{r},\omega)]$ describes the change of spectrum with respect to the background spectrum $W_{11}(\mathbf{r},\omega) + W_{22}(\mathbf{r},\omega)$, i.e., the Wolf effect. If we set $\tau = 0$ in Eq. (4), we come to the important relationship between these two phenomena:

$$I_{12}(\mathbf{r}) + I_{21}(\mathbf{r}) = \int_{-\infty}^{\infty} [W_{12}(\mathbf{r}, \omega) + W_{21}(\mathbf{r}, \omega)]\, d\omega. \quad (9)$$

The left hand side of Eq. (9) is due to T. Young, the right hand side is due to E. Wolf.

We reach the important conclusion: the appearance of the interference fringes and changes in the spectrum are closely related; both appear together as the result of the spatial coherence of sources. If there is no Wolf effect, i.e., $W_{12}(\mathbf{r}, \omega) + W_{21}(\mathbf{r}, \omega) = 0$, there should be no interference, i.e., $I_{12}(\mathbf{r}) + I_{21}(\mathbf{r}) = 0$. If one of these phenomena can be observed, another also must appear (this statement is correct even in the case of monochromatic radiation, when the position of the spectral line does not change; see the example below). The visibility of interference fringes may serve as an estimation of the relative contribution of the Wolf effect to the spectrum change.

The first phenomenon (interference fringes) has been well known since 1801 (Thomas Young), but the second was theoretically discovered by Emil Wolf in 1986.

16.2.1 Example

In this subsection we will show that even in the case of a very narrow (quasi-monochromatic) spectrum, the Wolf effect exists in the Young experiment and formula (9) is correct. At first glance it seems that for a quasi-monochromatic radiation, the interference fringes in the Young experiment exist, but there is no frequency change, i.e., no Wolf effect. We will show here that in this case the Wolf effect exists and formula (9) remains correct, but the Wolf effect manifests itself not in the frequency shift, but in the coordinate dependence of the intensity of the spectral line.

We will consider the simple example of two completely spatially coherent sources in the Young experiment. Because the mean field must be zero, $\langle E \rangle = 0$, we must introduce randomness in the temporal behavior of E.

If the distance between pinholes is d, the distance between the plane of pinholes and the plane of observation is h, and the displacement of the point of observation from the middlepoint is x (see Fig. 1), then the distances from the point of observation to the pinholes are equal to

$$R_{1,2} = \sqrt{\left(\frac{d}{2} \pm x\right)^2 + h^2} = R \pm \frac{xd}{2R} + \cdots, \quad R = \sqrt{\left(\frac{d}{2}\right)^2 + h^2}. \tag{10}$$

The field created by source 1 is given by the formula

$$\begin{aligned}
F_1(t) &= \frac{1}{R_1} \exp\left[i\omega_0\left(t - \frac{R_1}{c}\right) - i\varphi\left(t - \frac{R_1}{c}\right)\right] \\
&\approx \frac{1}{R} \exp\left[i\omega_0 t - ik_0 R - \frac{ik_0 xd}{2R} - i\varphi\left(t - \frac{R_1}{c}\right)\right],
\end{aligned} \tag{11}$$

where $k_0 = \omega_0/c$. Here, $\varphi(t)$ is a random phase. We assume that φ has a Gaussian multivariate probability distribution, $\langle\varphi\rangle = 0$, and $\langle\varphi^2\rangle \gg 1$. For any random value u having a Gaussian probability distribution with $\langle u\rangle = 0$, the following formula is correct:

$$\langle\exp(-iu)\rangle = \exp\left(-\frac{\langle u^2\rangle}{2}\right). \tag{12}$$

Using Eq. (12) and taking into account that $\langle\varphi^2\rangle \gg 1$ we obtain

$$\begin{aligned}
\langle F_1(t)\rangle &= \frac{\exp\left[i\omega_0\left(t - \frac{R_1}{c}\right)\right]}{R_1} \left\langle -i\varphi\left(t - \frac{R_1}{c}\right)\right\rangle \\
&= \frac{\exp\left[i\omega_0\left(t - \frac{R_1}{c}\right)\right]}{R_1} \exp\left(-\frac{\langle\varphi^2\rangle}{2}\right) \approx 0.
\end{aligned}$$

Similarly, $\langle F_2(t)\rangle \approx 0$.

The difference between two Gaussian values also is a Gaussian random value. Thus, $u = \varphi(t_1) - \varphi(t_2)$ is Gaussian and its variance

$$\langle u^2\rangle = \langle[\varphi(t_1) - \varphi(t_2)]^2\rangle = D_\varphi(t_1 - t_2) = D_\varphi(t_2 - t_1) \tag{13}$$

is the temporal structure function of the phase fluctuations, which may be expressed in terms of the autocorrelation function of phase fluctuations.[*]

Similarly, the field created by source 2 is given by the formula

$$F_2(t) = \frac{1}{R_2} \exp\left[i\omega_0\left(t - \frac{R_2}{c}\right) - i\varphi\left(t - \frac{R_2}{c}\right)\right]$$

$$\approx \frac{1}{R} \exp\left[i\omega_0 t - ik_0 R + \frac{ik_0 x d}{2R} - i\varphi\left(t - \frac{R_2}{c}\right)\right]. \tag{14}$$

The correlation matrix for fields is defined by the formula

$$B_{ik}(\tau) \equiv \langle F_i(t + \tau)F_k^*(t)\rangle, \tag{15}$$

averaging over φ fluctuations. Using Eq. (11) and Eq. (14) we obtain:

$$B_{11}(\tau) = \frac{\exp[i\omega_0\tau]}{R^2}\left\langle \exp\left\{i\left[\varphi\left(t - \frac{R_1}{c}\right) - \varphi\left(t + \tau - \frac{R_1}{c}\right)\right]\right\}\right\rangle$$

$$B_{22}(\tau) = \frac{\exp[i\omega_0\tau]}{R^2}\left\langle \exp\left\{i\left[\varphi\left(t - \frac{R_2}{c}\right) - \varphi\left(t + \tau - \frac{R_2}{c}\right)\right]\right\}\right\rangle$$

$$B_{12}(\tau) = \frac{1}{R^2} \exp\left[i\omega_0\tau - \frac{ik_0 x d}{R}\right]\left\langle \exp\left\{i\left[\varphi\left(t + \tau - \frac{R_2}{c}\right) - \varphi\left(t - \frac{R_1}{c}\right)\right]\right\}\right\rangle$$

$$B_{21}(\tau) = \frac{1}{R^2} \exp\left[i\omega_0\tau + \frac{ik_0 x d}{R}\right]\left\langle \exp\left\{i\left[\varphi\left(t - \frac{R_1}{c}\right) - \varphi\left(t + \tau - \frac{R_2}{c}\right)\right]\right\}\right\rangle. \tag{16}$$

Each difference of two Gaussian values $\varphi(t_1) - \varphi(t_2) = u$ is a Gaussian random value with zero mean value. Using formula (12) and approximate formulas $R_{1,2} = R \pm xd/2R$, we obtain the following expressions for B_{ik}:

$$B_{11}(\tau) = B_{22}(\tau) = \frac{1}{R^2} \exp\left[i\omega_0\tau - \frac{1}{2}D_\varphi(\tau)\right]$$

$$B_{12}(\tau, x) = \frac{1}{R^2} \exp\left[i\omega_0\tau - \frac{ik_0 x d}{R}\right] \exp\left[-\frac{1}{2}D_\varphi\left(\tau + \frac{xd}{Rc}\right)\right] \tag{17}$$

---

[*] We use the terminology of the theory of stationary random functions, where the quantity $\langle[f(t_1) - f(t_2)]^2\rangle = D(t_1 - t_2)$ is called as a "structure function." This function is closely related to the autocorrelation function $\langle f(t_1)f(t_2)\rangle = B(t_1 - t_2)$ by the relation $D(t_1 - t_2) = 2B(0) - 2B(t_1 - t_2)$. Even if $\langle f^2(t)\rangle = B(0) \to \infty$, the difference $2B(0) - 2B(t_1 - t_2)$ for small $|t_1 - t_2|$ remains finite.

$$B_{21}(\tau,x) = \frac{1}{R^2}\exp\left(i\omega_0\tau + \frac{ik_0xd}{R}\right)\exp\left[-\frac{1}{2}D_\varphi\left(\tau + \frac{xd}{Rc}\right)\right] = B_{12}^*(-\tau,x).$$

In the following we use the diffusive model for the structure function of the phase fluctuations

$$D_\varphi(\tau) = 2\Omega|\tau|. \tag{18}$$

Such a model leads to a Lorentzian shape of the spectrum.

Another useful model is

$$D_\varphi(\tau) = \Omega^2\tau^2,$$

which corresponds to a random frequency $\omega = \omega_0 + \Delta\omega$, $\langle\Delta\omega\rangle = 0$, $\langle(\Delta\omega)^2\rangle = \Omega^2$, leads to a Gaussian shape of the spectrum.

For the spectrum corresponding to Eq. (18), we obtain after evaluation of the standard integrals:

$$W_{11}(\omega) = W_{22}(\omega) = \frac{1}{2\pi R^2}\int_{-\infty}^{\infty}\exp(-i\omega\tau)\exp(i\omega_0\tau - \Omega|\tau|)\,d\tau$$

$$= \frac{\Omega}{\pi R^2\left[\Omega^2 + (\omega_0 - \omega)^2\right]}$$

$$W_{12}(\omega,x) = \frac{\Omega}{\pi R^2\left[\Omega^2 + (\omega_0 - \omega)^2\right]}\exp\left[-\frac{ik_0xd}{R} - i(\omega_0 - \omega)\frac{xd}{Rc}\right] \tag{19}$$

$$W_{12}(\omega,x) + W_{21}(\omega,x) = 2\,\mathrm{Re}\,W_{12}(\omega,x)$$

$$= \frac{2\Omega}{\pi R^2\left[\Omega^2 + (\omega_0 - \omega)^2\right]}\cos\left[\frac{k_0xd}{R} + (k_0 - k)\frac{xd}{R}\right].$$

Here, $k = \omega/c$ and $k_0 = \omega_0/c$.

The values $W_{11}(\omega)$ and $W_{22}(\omega)$ present the original spectra, which would appear in the absence of interference. The term $W_{12}(\omega,x) + W_{21}(\omega,x)$ describes the Wolf effect.

The entire spectrum in the point x is given by the formula

$$W(\omega,x) = W_{11}(\omega) + W_{22}(\omega) + W_{12}(\omega,x) + W_{21}(\omega,x)$$

$$= \frac{2\Omega}{\pi R^2\left[\Omega^2 + (\omega_0 - \omega)^2\right]}\left\{1 + \cos\left[(2k_0 - k)\frac{xd}{R}\right]\right\}. \tag{20}$$

In the point $x = 0$, the entire spectrum is given by the formula

$$W(\omega,x) = \frac{4\Omega}{\pi R^2 \left[\Omega^2 + (\omega_0 - \omega)^2\right]} = 2\left[W_{11}(\omega) + W_{22}(\omega)\right].$$

The spectrum in some shifted point is shown in Fig. 2.

Figure 3 shows the x-dependencies of several spectral components $W(\omega_i,x)$ for different ω_j.

For a very small Ω, the linewidth becomes very small and the Wolf shift can be observed only in a rather narrow range of wavenumbers. The example of the Wolf shift for such a situation ($\Omega/ck_0 = 10^{-3}$) is presented in Fig. 4 for $x = 0$ (red) and

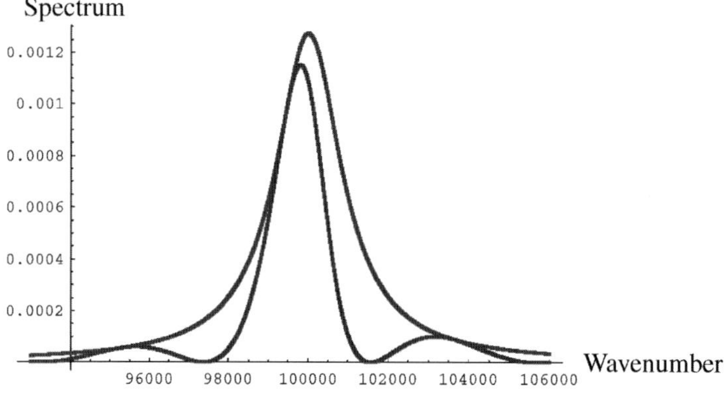

Figure 2 Two spectra corresponding to $\Omega = 10^3$ sec^{-1}, $R = 1$ cm, $k_0 = 10^5$ cm^{-1}, $d = 0.01$ cm, $x = 0$ (blue), and $x = 0.15$ cm (red).

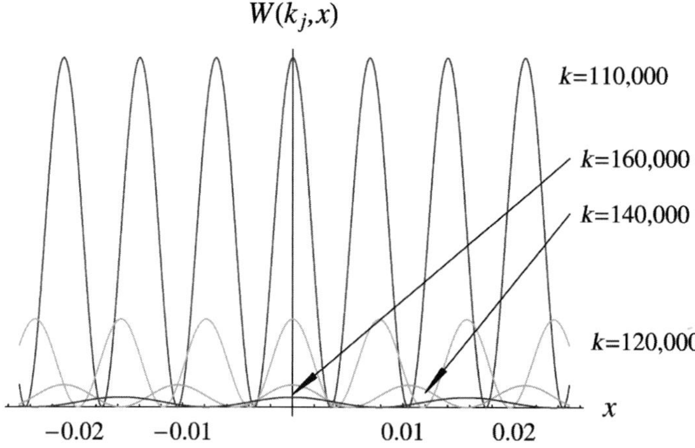

Figure 3 The functions $W(k_i,x)$ for $k = 110,000$ cm^{-1} (red), $k = 120,000$ cm^{-1} (green), $k = 140,000$ cm^{-1} (lightblue), and $k = 160,000$ cm^{-1} (violet). Other parameters: $\Omega = 10^3$ sec^{-1}, $R = 1$ cm, $k_0 = 10^5$ cm^{-1}, $d = 0.01$ cm.

$x = 0.003$ cm (blue). Because $W(k_{\max}, x = 0.003) \approx W(k_{\max}, x = 0)/200$, the normalized values $W(k,x)/W(k_{\max},x)$ are presented.

In the case of a very small Ω, the following approximate equality is true:

$$\frac{\Omega}{\pi[\Omega^2 + (\omega - \omega_0)^2]} \equiv \delta_\Omega(\omega - \omega_0) \approx \delta(\omega - \omega_0). \tag{21}$$

This relation means that $\delta_\Omega(0) = 1/(\pi\Omega) \to \infty$ for $\Omega \to 0$, and the integral of $\delta_\Omega(\omega - \omega_0)$ over all ω is equal to 1. Therefore, for $\Omega \to 0$ using the identity $\delta(\omega - \omega_0)f(\omega) = \delta(\omega - \omega_0)f(\omega_0)$, we may present Eq. (20) in the form

$$\lim_{\Omega \to 0} W(\omega,x) = \frac{2}{R^2}\delta(\omega - \omega_0)\left\{1 + \cos\left[(2k_0 - k)\frac{xd}{R}\right]\right\}$$
$$= \frac{2}{R^2}\delta(\omega - \omega_0)\left[1 + \cos\left(k_0\frac{xd}{R}\right)\right]. \tag{22}$$

In this case, the Wolf effect manifests itself not in the form of a frequency shift, but in the form of the dependency of the intensity of the spectral line on the coordinate. Figure 4 illustrates this fact: the shift in the maximum of the spectrum is very small ($\Delta k/k_0 \approx 2 \times 10^{-5}$, while $\Omega/(ck_0) = 10^{-3}$), but the change of spectrum intensity is about 200.

Thus, in the case of monochromatic radiation, the Wolf effect also exists in the Young experiment in the form of the coordinate dependence of the intensity of the spectral line.

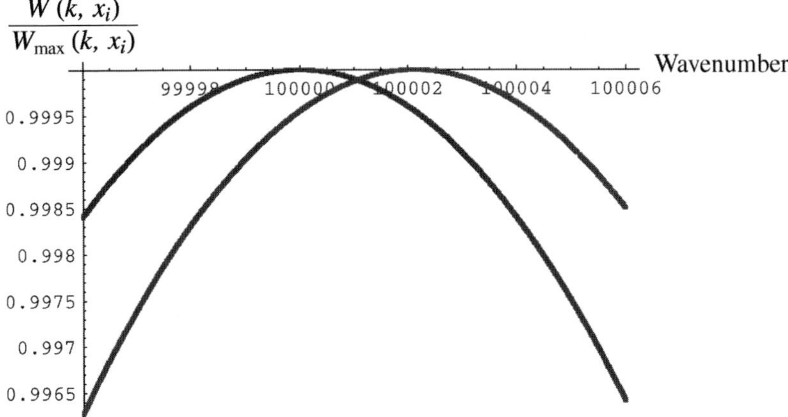

Figure 4 The normalized shapes of thin spectral lines ($\Delta\omega/\omega_0 = 0.001$) in the Young experiment for $x = 0$ (red) and $x = 0.003$ cm (blue).

16.3 Physics of the Spectrum Changes in Radiation Problems

The spectrum of a source, $W(\omega)$, describes the energy distribution between different frequencies. The spectrum may depend on the direction a priori, but for a completely spatially incoherent source, the Lambert cosine law asserts that $W(\omega)$ is proportional to the apparent area of the source, which is independent of ω. Thus, for spatially incoherent sources, the spectrum $W(\omega)$ is independent of the direction of observation. If the ratio $W(\omega_1)/W(\omega_2) = N$ for some direction, it remains the same N for any other directions.

A totally different picture appears for completely spatially coherent sources. The typical spatially coherent source is an antenna. The radiation pattern of a uniformly fed round disk of radius a and wave number $k = \omega/c$ in the direction determined by the angle θ is given by the well-known formula

$$F(k, \theta) = P_0(k) \left[\frac{2J_1(ka \sin \theta)}{ka \sin \theta} \right]^2. \tag{23}$$

Here, $P_0(k) = F(k, 0)$ is the intensity of radiation at the axes of the beam. The function $F(k, \theta)$ for two different k and two different P_0 is presented in Fig. 5.

In this example, if one measures intensities of radiation in the direction of the main lobe, the result would be that the higher frequency (narrower beam width) is more intense than the lower one. The corresponding measurement in the direction $\theta > 0.058$ (for this example) would give the opposite result: the lower frequency is more intense than the higher one. An example with several frequencies is shown in Fig. 6. Thus, the spectrum of radiation for a completely spatially coherent source depends on the direction.

In the following we consider four different cases: continuous spectrum and discrete spectrum, and coherent and partially coherent cases.

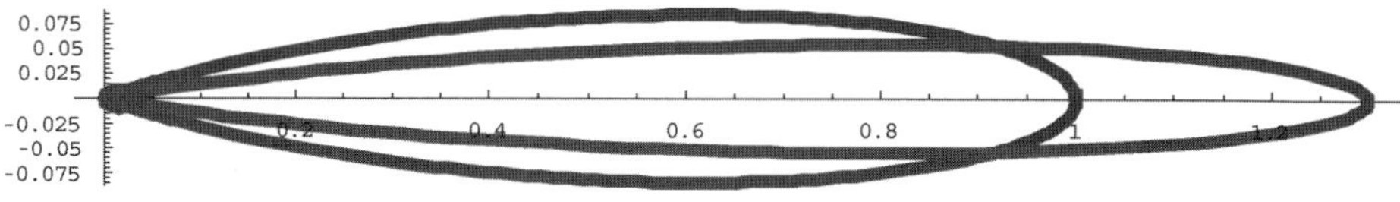

Figure 5 The radiation patterns of a uniformly fed round disk at two different frequencies, having different intensities P_0. The intensity of the higher frequency radiation for $\theta = 0$ is larger than the intensity of the lower frequency radiation, but for $\theta > 0.058$, the situation is reversed.

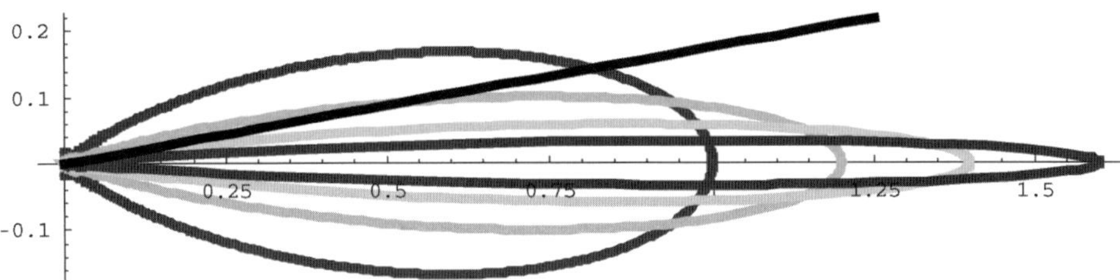

Figure 6 The radiation patterns of a uniformly fed round disk for several different frequencies, having different intensities P_0. The intensity of radiation of the higher frequency for $\theta = 0$ is greater than the intensities of radiation of all lower frequencies, but for direction (shown by the black line) the intensities are ordered in the inverse order.

16.3.1 Continuous spectrum, coherent source

If the spectrum of radiation at the axis of the source (the axis directions may be related to the axis of symmetry, but in general its direction is arbitrary) is described by the expression $P_0(k)$, and the radiation pattern of the source with respect to the chosen axis is described by the expression $Q(\theta,k)$, then the spectrum of radiation in the fixed direction θ is determined by the formula

$$W(k,\theta) = P_0(k)Q(k,\theta). \tag{24}$$

We consider two different models of source. In the first example,

$$Q(k,\theta) = \left[\frac{2J_1(ka\sin\theta)}{ka\sin\theta}\right]^2 \tag{25}$$

corresponds to the radiative pattern of a uniformly fed round disk of radius a, and $P_0(k)$ is described by the Planck formula

$$P_{0,Pl}(k) = \frac{C(T)k^3}{\exp\left(\dfrac{k}{T}\right) - 1}, \tag{26}$$

where T is measured in the units of $hc/2\pi K_B = 0.229$ deg/cm, and the normalization constant $C(T)$ is chosen such that the maximum value $P_{0,Pl}(2.82T) = 1$.

The plots of $W(k,\theta)$ for $ka = 10$ and several values of θ are shown in Fig. 7. The black (Planck) curve corresponds to $\theta = 0$, and three red curves correspond to $\theta = 0.0005$ rad (the upper red line), $\theta = 0.001$ rad (the middle red line), and $\theta = 0.0025$ rad (the lowest red line).

The red shift that appears corresponds to the picture shown in Fig. 6.

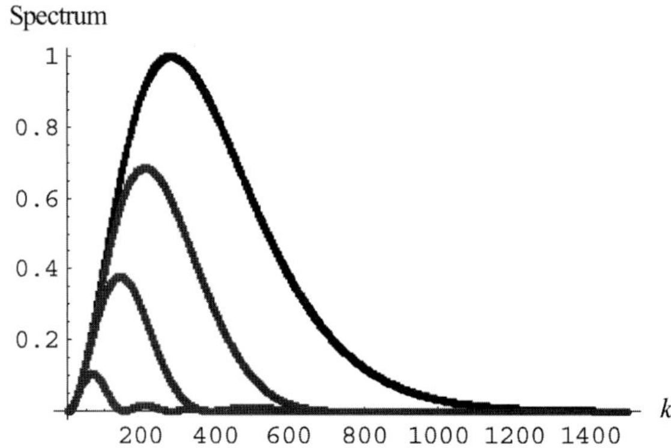

Figure 7 Spectra of uniformly fed disk for different θ. For $\theta = 0$, the Planck spectrum was chosen. For $\theta > 0$, the red shift appears.

We should emphasize that while discussing the shift of the spectrum, we always mean the shift with respect to some reference spectrum. In the examples considered above, we always consider the spectrum of radiation along the axis of symmetry as the reference spectrum. This choice is natural, but not unique. In principle, we may consider the spectrum along some direction $\theta_0 > 0$ as a reference spectrum (for instance, one of the red curves in Fig. 7). In this case, for $\theta > \theta_0$ we obtain the red shift, but for $\theta < \theta_0$, the shift would be blue.

It is possible to construct a more sophisticated model of the source so that the shift would be blue with respect to the reference spectrum of radiation along the axis of symmetry. Such a source must have the minimum of radiation pattern along the axis.

To construct such a source, we consider an antenna having a nonuniform distribution of current $j(r)$, shown in Fig. 8. The negative current corresponds to the phase shift π.

Analytically, this curve is described by the formula

$$j(r) = \exp\left[-\frac{(r-r_0)^2}{2a^2}\right] \cos\left(\frac{r}{L}\right). \tag{27}$$

The corresponding radiation pattern is determined by the Fourier-Bessel transform

$$Q(k,\theta) = \left[\int_0^\infty J_0(kr\sin\theta)\exp\left[-\frac{(r-r_0)^2}{2a^2}\right]\cos\left(\frac{r}{L}\right)r\,dr\right]^2. \tag{28}$$

For $\theta = 0$ the function (28) is independent of k; thus all frequencies have the same intensity. The plot presented in Fig. 8 corresponds to $ka = 10, kr_0 = 20$, and

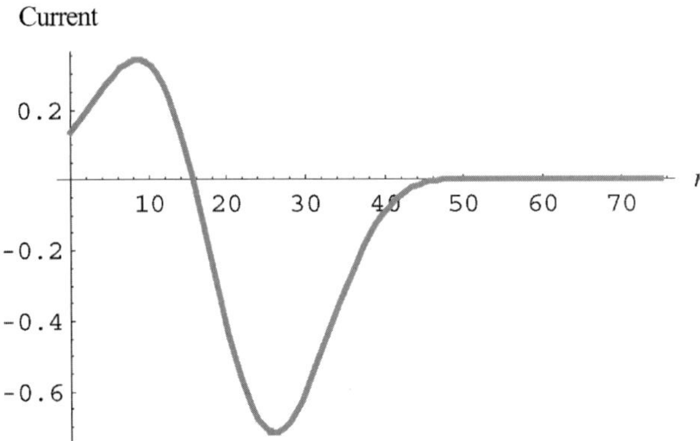

Figure 8 Example of the distribution of current for an antenna having a minimum of radiative pattern along the axis.

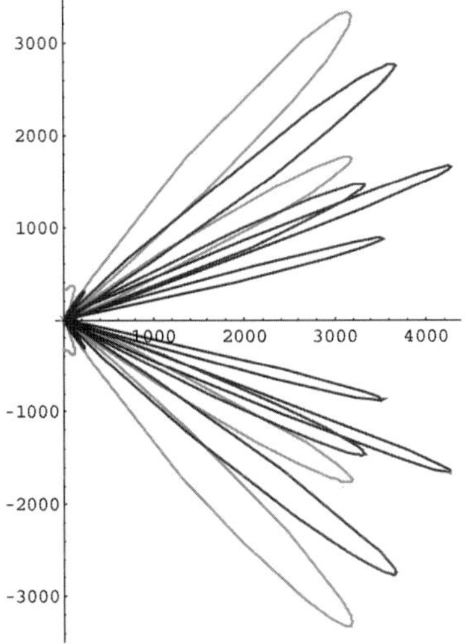

Figure 9 Example of radiative pattern corresponding to the current distribution of Fig. 8.

$kL = 3$. The examples of corresponding radiation patterns obtained numerically for different k are shown in Figs. 9 and 10.

In Fig. 9, the green line corresponds to $k = 0.5$, the blue line to $k = 0.6$, and the red line to $k = 1$. For the type of current considered in this example, the direction $\theta = 0$ corresponds to the minimum of radiative pattern.

To better illustrate the situation, in Fig. 10 the function $Q(k, \theta)$ is presented in Cartesian coordinates. The red line in Fig. 10 corresponds to $k = 0.5$ and the blue line to $k = 0.7$. If we consider, for example, the fixed value $\theta = 0.35$, the blue line is above the red one. This means that the radiation at higher frequency in this direction is more intense than at lower frequency (blue shift). For some other directions, for example $\theta = 0.8$, we observe the red shift.

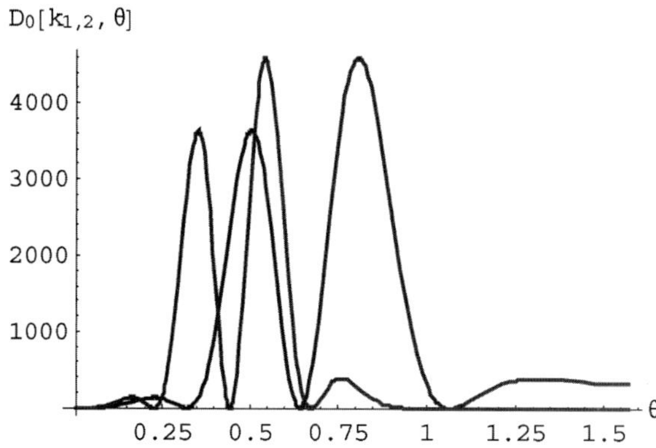

Figure 10 Radiative patterns for two frequencies for the current distribution of Fig. 8.

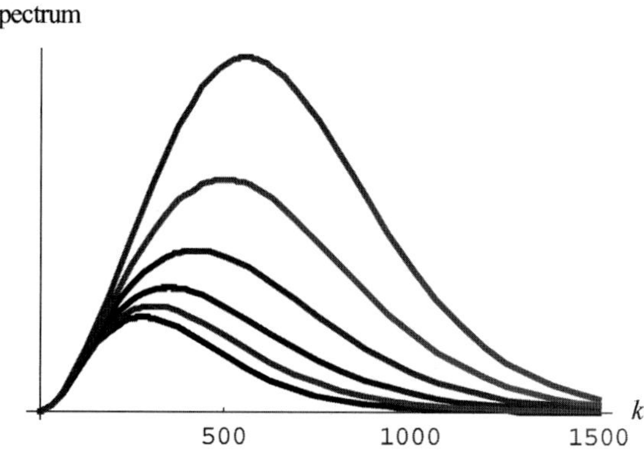

Figure 11 Example of blue shift for the current distribution of Fig. 8.

Using Eq. (24), we can find the spectrum in some arbitrary direction (see Fig. 11). We take the Planck distribution in $\theta = 0$ direction (black), and obtain the spectra in $\theta = 10^{-6}$ direction (the closest to the black curve), $\theta = 1.5 \times 10^{-6}$, $\theta = 2.5 \times 10^{-6}$, $\theta = 5 \times 10^{-6}$, and $\theta = 7.5 \times 10^{-6}$ direction (the upper curve).

We observe the blue shift for all these spectra. The intensities of all the spectra are larger than for $\theta = 0$ because the chosen reference spectrum corresponds to the minimum of radiative pattern. Physically, this means that if we observe the Planck spectrum in the $\theta = 0$ direction for a source having for all frequencies the distribution of currents shown in Fig. 8, we will observe the more intense and blue-shifted radiation in some other directions.

16.3.2 Continuous spectrum, partially coherent source

In the previous subsection we considered the case of completely spatially coherent sources. Now we analyze the case of partially spatially coherent sources. The example we consider here does not describe the general case of partially coherent sources, but corresponds to a rather simple special situation. We assume that the source of radiation [current $j(\mathbf{r})$] consists of two parts: completely spatially coherent component $A(k)\exp(i\varphi)$ with random phase φ; and spatially δ-correlated component (spatial noise) $N(k,\mathbf{r})$, which is statistically independent of φ. Thus,[*]

$$j(\mathbf{r}) = A(k)\exp(i\varphi) + N(k,\mathbf{r}), \tag{29}$$

where

$$\langle N(k,\mathbf{r}')N(k,\mathbf{r}'')\rangle = \mathfrak{N}(k)\delta(\mathbf{r}' - \mathbf{r}'') \tag{30}$$

and

$$\langle j(\mathbf{r}')j^*(\mathbf{r}'')\rangle = |A(k)|^2 + \mathfrak{N}(k)\delta(\mathbf{r}' - \mathbf{r}''). \tag{31}$$

If we consider the radiation of this current in the far zone using the Fraunhofer diffraction approximation, it is possible to obtain the following formula for the intensity of radiation $\langle E(\mathbf{r})E^*(\mathbf{r})\rangle$ of a disk of radius a:

$$\langle E(\mathbf{r})E^*(\mathbf{r})\rangle = \frac{|A(k)|^2 a^4}{16r^2}\left[\frac{2J_1(ka\sin\theta)}{ka\sin\theta}\right]^2 + \frac{\mathfrak{N}(k)a^2}{16\pi r^2}. \tag{32}$$

Here, r is the distance and θ is the angle between the axis of symmetry and the direction from the center of disk to the point of observation.[†]

Therefore, we add to the right-hand side of Eq. (24) an additional term, describing a completely spatially incoherent source of radiation. According to Eq. (32), the radiation pattern of such a source is independent of θ. As the result, we obtain

$$W(k,\theta) = P_0(k)Q(k,\theta) + P_1(k). \tag{33}$$

---

[*] Of course, any realistic function $\langle j(\mathbf{r}')j(\mathbf{r}'')\rangle$ must have some finite correlation radius $L < \infty$ for the first term, and some nonzero correlation radius $l > 0$ for the second term. If L is much larger and l is much smaller than all other important scales, it is possible to use approximation (31).

[†] A little longer calculation leads to the formula for the coherence function that corresponds to the same type of source:

$$\langle E(\mathbf{r}_1)E^*(\mathbf{r}_2)\rangle$$

$$= a^4|A(k)|^2\frac{\exp[ik(r_1-r_2)]}{16r_1 r_2}\left[\frac{2J_1(ka\sin\theta_1)}{ka\sin\theta_1}\right]\left[\frac{2J_1(ka\sin\theta_2)}{ka\sin\theta_2}\right]$$

$$+ \frac{\mathfrak{N}(k)a^2\exp[ik(r_1-r_2)]}{16\pi r_1 r_2}\left[\frac{2J_1(ka\sqrt{\sin^2\theta_1 + \sin^2\theta_2 - 2\sin\theta_1\sin\theta_2\cos(\alpha_1-\alpha_2)})}{ka\sqrt{\sin^2\theta_1 + \sin^2\theta_2 - 2\sin\theta_1\sin\theta_2\cos(\alpha_1-\alpha_2)}}\right].$$

The appearance of the factor $2J_1(ka\sin\theta_{\mathrm{eff}})/ka\sin\theta_{\mathrm{eff}}$ in this formula corresponds to that in the van Cittert-Zernike theorem [3,6].

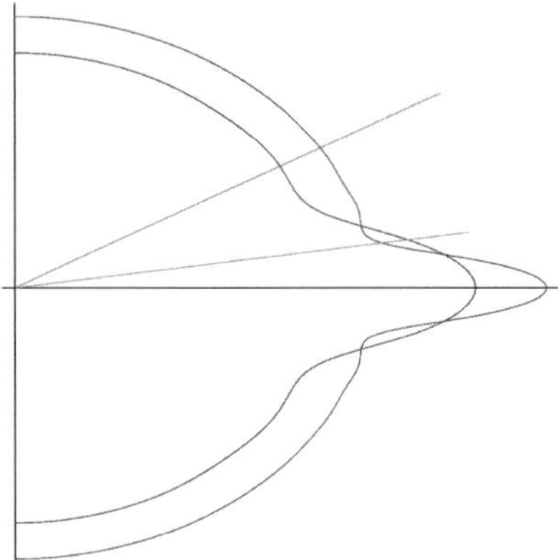

Figure 12 Radiative pattern for the combination of spatially coherent and incoherent sources.

The ratio P_0/P_1 describes the degree of coherence. For example, if $P_1(k) = \gamma P_0(k)$, we obtain

$$W(k,\theta) = P_0(k)[Q(k,\theta) + \gamma\cos\theta]. \tag{34}$$

The example of two radiative patterns, which correspond to the uniform current distribution across the disk [Eq. (25)], the Planck spectrum [Eq. (26)] with $T = 100$, and parameters $k_1 = 180$ (red), $k_2 = 350$ (blue), $a = 0.05$, and $\gamma = 2$, is shown in Fig. 12.

For small θ, the coherence is significant (bumps in radiation patterns near $\theta = 0$), but for large θ, radiation patterns became circular, corresponding to the incoherent component of radiation.

In this example, the highest frequency is more intense for $\theta = 0$. Then, for the angle θ corresponding to the orange straight line, the red shift appears. Finally, for the angle θ corresponding to the green straight line, the red shift disappears.

The spectra, corresponding to several values of θ, $ka = 10$, and $\gamma = 0.2$, are presented in Fig. 13.

In general, there are two maxima at positions k_{01} and k_{02} in the spectra. When θ increases, the position of k_{01} decreases (red shift). When θ approaches the region of incoherent radiative pattern (circle in Fig. 12), the second maximum at highest k appears, and this second maximum becomes dominant when θ is inside the incoherent radiative pattern. The position of k_{02} tends to the position of the maximum of the Planck function while θ increases. Thus, the red shift disappears in this region.

The blue shift situation [Eq. (28) and Figs. 9 and 10] is more complicated. In this case, the instability with respect to small changes of θ appears. The blue and the red shifts may replace each other while θ varies.

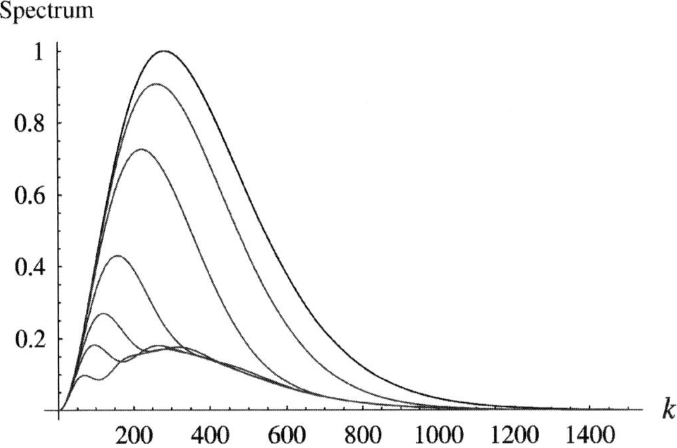

Figure 13 Spectra for partially coherent source. The red shift appears as θ increases, but for large θ, corresponding to the incoherent part of the radiative pattern, the new maximum appears at the position of the maximum of the Planck curve, which becomes dominant while θ increases, and the red shift disappears.

16.3.3 Discrete spectrum, coherent source

In the case of a discrete spectrum, the function $P_0(k)$ entered in Eq. (23) has the form

$$P_0(k) = \sum_i P_i \delta(k - k_i). \tag{35}$$

An example of this spectrum is shown in Fig. 14 [here, $P_i = P_{0,Pl}(k_i)$].

If we consider the radiation in some direction θ, the previous formula (24) is valid. Using this formula and Eq. (35) we obtain the following Fig. 15.

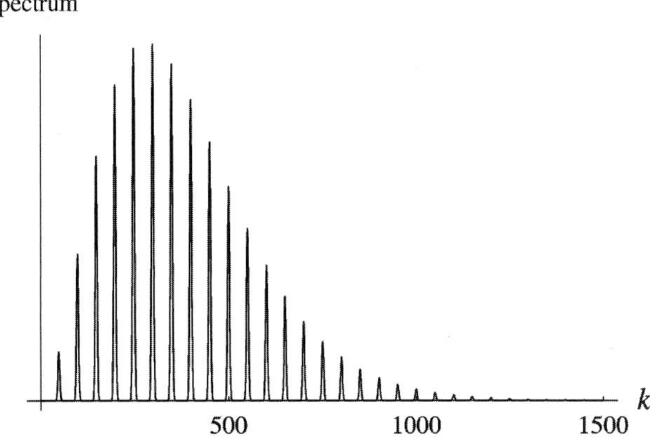

Figure 14 Discrete spectrum with envelope corresponding to the Planck curve.

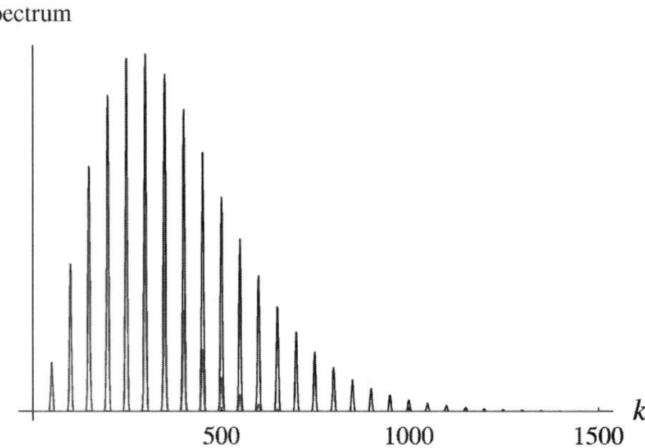

Figure 15 The Wolf red shift appears for the envelope of the discrete spectrum, but each line remains at its original position.

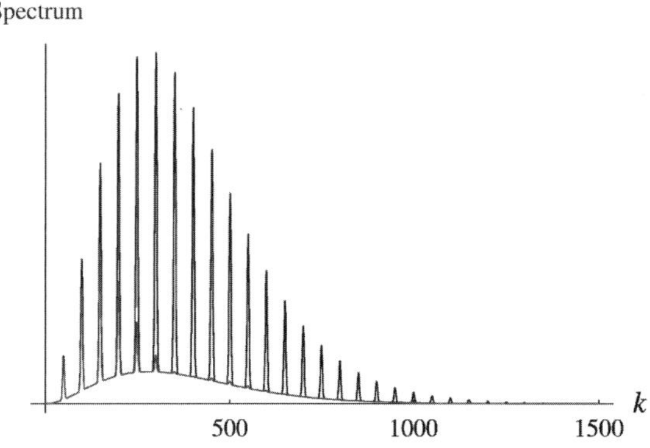

Figure 16 The Wolf red shift for the partially coherent discrete spectrum.

The important feature of the Wolf effect for the discrete spectrum is that each line remains at its original position. Only the magnitudes of each line change. In contrast to the continuous case, only the maximum of the envelope of the discrete spectrum shifts.

In the presence of an incoherent continuous component, the picture looks as in the Fig. 16.

The value $\gamma = 0.1$ in this example.

16.4 Increasing Spatial Coherence in a Process of Propagation

Usually all natural sources of radiation are incoherent. Thus, the question arises: how do coherent or partially coherent sources appear? First, we must emphasize the difference between temporal and spatial coherence. For example, the completely temporally incoherent point source, located in the point r_0, produces a completely spatially coherent field at the sphere $|\mathbf{r} - \mathbf{r}_0| = R_1$. Indeed, the phases of the field in any point of this sphere at the same time are equal; i.e., we have complete spatial coherence. At the surface of another sphere $|\mathbf{r} - \mathbf{r}_0| = R_2$ we obtain another spatially coherent field, but the fields at these two spheres will be incoherent with respect to each other.

To analyze the appearance of a spatial coherence of a field, radiated by a spatially incoherent source, we first consider two incoherent sources, located in the points $r_1 = (0, a/2)$ and $r_2 = (0, -a/2)$ (see Fig. 17).

At the large distance $L \gg a$, so that the angle $\theta = a/L \ll 1$, the two wavefronts are intersected in the point $(0, L)$. Some indefinite phase ψ of the resulting field appears in the point $(0, L)$. If the observation point is shifted along one of a phase front, the additional optical path $L\varphi\theta$ and the additional phase shift $kL\varphi\theta$ appears, so that the phase difference in the shifted point is equal to $kL\varphi\theta$.

We may consider the fields in the initial and shifted points as spatially coherent, if the phase difference $kL\varphi\theta < 2\pi$. Thus, the angle of coherence φ is equal to

$$\varphi = \frac{\lambda}{L\theta} = \frac{\lambda}{a} \tag{36}$$

and the coherence radius

$$\rho = L\varphi = \frac{\lambda}{\theta} = L\frac{\lambda}{a}. \tag{37}$$

Formulas (36) and (37) qualitatively express the van Cittert-Zernike theorem [7,8]. The coherence radius $L\lambda/a$ of a completely spatially incoherent source

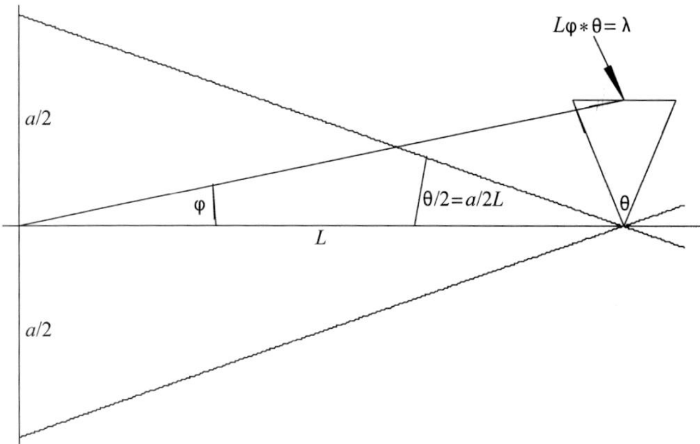

Figure 17 Derivation of the dependence of the coherence radius on distance for spatially incoherent sources.

coincides with the beam-width $(\lambda/a)L$ of the completely spatially coherent source. For a source of incoherent radiation of a scale a, a formula similar to Eq. (37) will be valid.

It follows from Eq. (37) that the coherence radius is proportional to the distance L from the source of incoherent radiation. For large L, the value ρ also may become large. If we consider some scatterer having a linear scale D and illuminated by a spatially incoherent source from large distance L, the coherence radius ρ of the incident wave may be large in comparison with D. In such case, the scatterer will be equivalent to the partially spatially coherent source of the secondary waves. This mechanism may explain the appearance in nature of partially coherent sources of radiation.

16.5 Differences in the Doppler and the Wolf Shifts

1. According to Eq. (24), the Wolf effect may be described by the multiplication of the initial spectrum $P_0(k)$ by some factor $Q(k,\theta)$, depending on θ and k. This means that if $P_0(k) = 0$ in some region of k, the modified spectrum also is zero in this region. The Wolf effect cannot shift the spectrum outside the initial region, where this spectrum was concentrated. The additional factor $Q(k,\theta)$ may only change the shape of the spectrum and the positions of maxima or minima of the original spectrum inside the original domain.

There is no similar restriction for the Doppler shift. The Doppler shift may be larger than the width of the initial spectrum. The change due to the Doppler effect spectrum $W_D(\omega)$ has the form

$$W_D(\omega) = W_0 \left(\frac{1+\beta}{\sqrt{1-\beta^2}}\omega \right), \tag{38}$$

where $\beta = v/c$, v is the radial velocity, $\beta > 0$ corresponds to the red shift. The ω-axis of spectrum is compressed by the factor $K = (1+\beta)/\sqrt{1-\beta^2}$, which may be very large if $\beta \to 1$.

Three spectra are presented in Fig. 18. The Planck spectrum (black) and the Doppler shifted spectra corresponding to $\beta = 0.8$ (orange, $K = 3$) and $\beta = 0.99$ (red, $K = 14$). The red-shifted spectra look different than in the Fig. 7, because the position of maximum is located far away from the region of the original spectrum.

2. If we consider a relatively narrow spectral line (in Fig. 19 we present the Gaussian line having $\delta\omega/\omega = 0.01$) and compare the Doppler and the Wolf shifts, we again see that the Wolf shift $\Delta\omega$ does not exceed the width of the line $\delta\omega$,

$$\Delta\omega \lesssim \delta\omega. \tag{39}$$

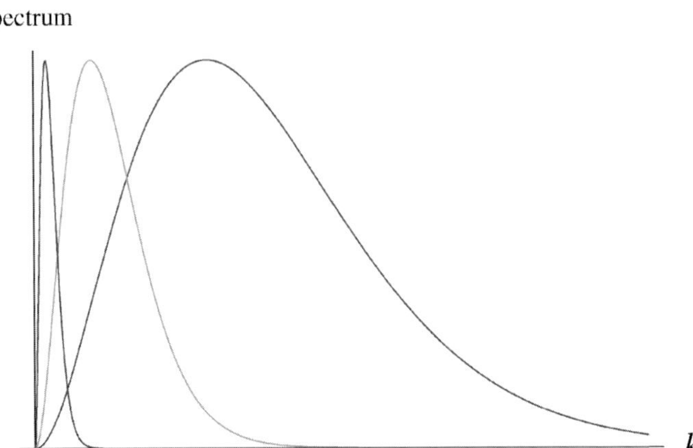

Figure 18 The Planck spectrum (black), the spectrum corresponding to $\beta = 0.8$ (orange), and the spectrum corresponding to $\beta = 0.99$ (red).

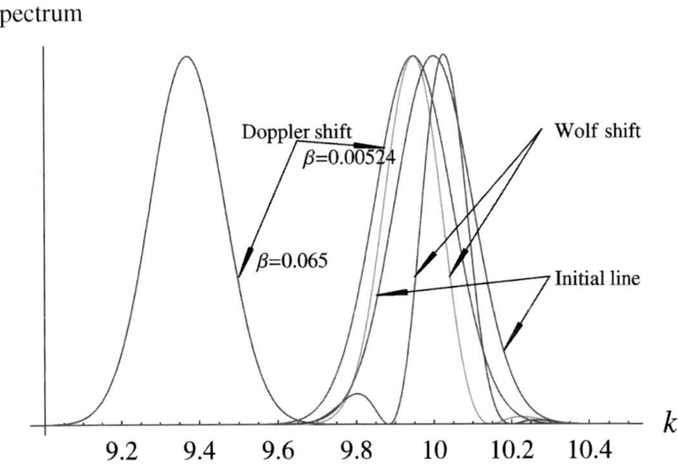

Figure 19 The original Gaussian line (black), the red and blue Wolf shifted lines (orange and blue; they correspond to different radii of the sources), and the Doppler shifted lines for $\beta = 0.065$ (red) and $\beta = 0.00245$ (magenta). The value $\beta = 0.00245$ was chosen such that the maxima of orange and magenta curves have the same position.

Thus, for narrow lines the Wolf shift is also small. At the same time, the Doppler shift may be much larger than $\delta\omega$ (the line, corresponding to $\beta = 0.065$ in Fig. 19).

If in the case of narrow lines we compare the Wolf shift and the Doppler shift for small β, it is difficult to distinguish between them (compare the Doppler red shifted line for $\beta = 0.00524$ and the Wolf shifted lines in Fig. 19). The value $\beta = 0.00524$ was chosen such that the maxima of Doppler and Wolf red-shifted curves have the same position.

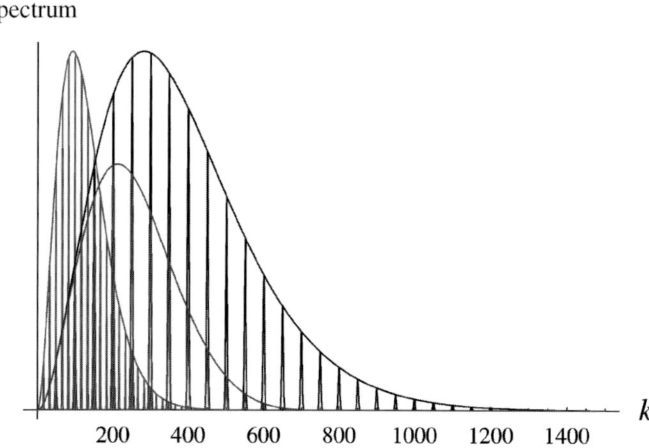

Figure 20 The black lines present the discrete spectrum, having the Planck distribution as an envelope. The red lines represent the red Wolf shifted lines and their envelope. The magenta lines represent the Doppler shifted initial spectrum, corresponding to $\beta = 0.8$ and their envelope.

3. Now we consider the discrete spectrum. The discrete spectrum consists of such narrow lines that it is impossible to observe the Wolf shift inside a single line. In Fig. 20, three spectra are presented. The black lines present the discrete spectrum, having the Planck distribution as an envelope. The red lines present the red Wolf shift. Only the maximum of the envelope is shifted; all lines remain on their original positions. The intensities of the Wolf shifted lines also change. The magenta lines present the Doppler shifted spectrum, corresponding to $\beta = 0.8$. Not only the envelope but each line is shifted, and also the distance between lines is reduced in $K = (1 + \beta)/\sqrt{1 - \beta^2}$ times. In this case, the difference between the Doppler shift and the Wolf shift is also evident.

16.6 Conclusions

(1) The interference fringes in the Young experiment may exist only if the Wolf effect exists [see Eq. (9)].
(2) The Wolf effect is caused by the dependence of the radiative pattern on frequency for spatially coherent (or partially coherent) sources. For spatially incoherent sources, the radiative pattern is independent of frequency and the Wolf effect does not appear.
(3) In contrast to the Doppler shift, the possible Wolf shift is always in the same region in which the original spectrum is concentrated. Thus, the Wolf shift may be large for wide spectra, and it is small for narrow spectral lines. The Doppler shift may have an arbitrary value independent of the width of spectrum.

(4) For the Wolf shift, the change of position of maximum is accompanied by a change of intensity of the spectrum.

(5) For discrete spectra, the main difference between the Doppler and Wolf shifts is that the positions of spectral lines change in the case of the Doppler shift, but then do not change in the case of the Wolf shift (only the position of the maximum of the envelope may change for the Wolf shift). The distance between the spectral lines does not change in the case of the Wolf shift, but it does change in the case of Doppler shift.

(6) For narrow spectral lines it is difficult to distinguish the Wolf shift from the small Doppler shift (see Fig. 18).

Acknowledgments

Professor Emil Wolf discussed this paper with me and made many comments. It is my pleasure to thank him for his attention. I am grateful to Zel Technologies for support of this work.

References

1. E. Wolf, *Phys. Rev. Lett.* **56**, 1370 (1986).
2. E. Wolf, *Nature* **326**, 363 (1987).
3. V.I. Tatarskii, *Pure Appl. Opt.* **7**, 953 (1998).
4. L. Mandel and E. Wolf, *Optical Coherence and Quantum Optics*, Sect. 5.8, Cambridge University Press, Cambridge (1995).
5. E. Wolf and D.F.V. James, *Rep. Prog. Phys.* **59**, 771 (1996).
6. R.J. Doviak and D.S. Zrnic, *Doppler Radar and Weather Observations*, 1st ed., Academic Press, San Diego (1984) and 2nd ed. (1993).
7. P.H. van Cittert, *Physica* **1**, 201 (1934).
8. F. Zernike, *Physica* **5**, 785 (1938).

Valerian I. Tatarskii received his M.S. degree in physics and mathematics from Moscow State University in 1952, his Ph.D. degree in physics and mathematics from the Acoustical Institute USSR Academy of Sciences in 1957, and his Doctor of Sciences in physics and mathematics from Gorky State University in 1964. From 1953–1956 he was a Junior Scientist with the Geophysical Institute, USSR Academy of Sciences, and from 1956–1959, he was a Junior Scientist with the Institute of Atmospheric Physics,

USSR Academy of Sciences. From 1959–1978 he worked as a Senior Scientist with the Institute of Atmospheric Physics, USSR Academy of Sciences. He was Head of the Laboratory with the Institute of Atmospheric Physics, USSR Academy of Sciences from 1978–1990. In 1990 he becomes Head of the Department with the Lebedev Physical Institute, USSR Academy of Sciences. From 1991–2001 he was a Senior Research Scientist at the University of Colorado/CIRES & NOAA/ERL. From 2001 to the present he has been Senior Scientist at Zel Technology & NOAA/ETL. Dr. Tatarskii has published more than 130 titles in wave propagation in random and turbulent media, theory of turbulence, quantum and statistical optics, and mathematics. He authored *Wave Propagation in a Turbulent Media* (1959 in Russian, 1961 and 1967 in English), *The Effect of the Turbulent Atmosphere on Wave Propagation* (1965 in Russian, 1971 revised edition in English), and coauthored with S.M. Rytov and Yu. A. Kravtsov, *Principles of Statistical Radiophysics* (1978 in Russian, 1989 in English). Dr. Tatarskii is a Corresponding member of the Russian (former USSR) Academy of Sciences (1976), Fellow of the Optical Society of America, Member of the USA National Academy of Engineering (1994), Member of New York Academy of Sciences (1994), Fellow of the Institute of Physics (1999), Associate Editor of the international Institute of Physics journal *Waves in Random Media* (1991–1998), Member of the Editorial Board of the journal *Soviet Physics*, Uspekhi (1985–1996), Member of the Editorial Board of the *Journal of Electromagnetic Waves and Applications*, and a consultant of the *Great Russian Encyclopedia* (formerly the *Great Soviet Encyclopedia*). He was the USSR State Prize winner (1990) and winner of the Max Born Award of the Optical Society of America in 1994. He was a chair of two International meetings in Wave Propagation in Random Media (Tallinn, 1989, and Seattle, 1992).

❧CHAPTER 17❦

THE SIGNIFICANCE OF PHASE AND INFORMATION

Michael A. Fiddy and H. John Caulfield

17.1 Introduction

Many years ago, a paper by Emil Wolf inspired one of us (MAF) to look more closely at the analytic properties of propagating and scattered optical fields. In 1962, Wolf discussed how Michelson's interferometer could provide information about the energy distribution in the spectrum of a light beam from measurements of the visibility of interference fringes [1]. A simple relationship exists between the visibility function $V(\tau)$ and the modulus of the complex degree of coherence, $\gamma(\tau)$. The methods of Hanbury Brown, and Twiss [1] showed that $|\gamma(\tau)|$ could be determined in various ways, and the important question arose as to how one might recover spectral information, $g(\nu)$ from $|\gamma(\tau)|$. Since $V(\tau)$ is proportional to the Fourier transform of $g(\nu)$, one might think that only the autocorrelation function of $g(\nu)$ might be found, without explicitly knowing the phase of $\gamma(\tau)$. How does one determine the missing phase? The deep insight used to address this question was to take into account the analytic properties from which a complete recovery of $g(\nu)$ became possible. It is the underlying analytic properties of the functions that we so routinely employ in modeling optical processes that can provide a deeper understanding of relationships between field parameters and information; this is what we address here.

The complex degree of coherence and the spectral energy density are related by a Fourier transform, namely

$$\gamma(\tau) = \int_0^\infty g(\nu)\exp(-2\pi i \nu \tau)\,d\nu. \tag{1}$$

PROC. PHYS. SOC., 1962, VOL. 80

Is a Complete Determination of the Energy Spectrum
of Light Possible from Measurements of
the Degree of Coherence?

By E. WOLF

Department of Physics and Astronomy, University of Rochester,
Rochester, New York

MS. received 30th April 1962

PROC. PHYS. SOC., 1962, VOL. 80

Temporal Coherence of Black Body Radiation

By Y. KANO AND E. WOLF

Department of Physics and Astronomy, University of Rochester,
Rochester, New York

MS. received 30th April 1962

Groundbreaking papers on the significance of phase references. [Reprinted with permission from Institute of Physics Publishing, Copyright (1962).]

If the light is quasi-monochromatic, the visibility of the interference fringes is given by [2]

$$V(\tau) = A|\gamma(\tau)|, \tag{2}$$

where $A = 2(I_1 I_2)^{1/2}/(I_1 + I_2)$ and I_1 and I_2 are the (time-averaged) intensities of the two partial beams.

The physically and mathematically important constraint that results in $\gamma(\tau)$ being analytic is that $g(\nu)$ is a *causal* function. This imparts constraints on the real and imaginary parts of $\gamma(\tau)$. The well known theorem due to Titchmarsh [3] proves that (1) causality in the frequency domain; (2) $\gamma(\tau)$ being analytic and regular in the lower half of the complex plane (lhp); and (3) the real and imaginary parts of $\gamma(\tau)$ being related by a Hilbert transform, are equivalent statements. The Hilbert transform relationship derives from taking the real and imaginary parts of a Cauchy integral over a contour, including the real axis and a semicircle in the lhp whose radius tends to infinity. Invoking Jordan's lemma [3] allows one to write, on the real axis

$$\mathrm{Im}\{\gamma(\tau)\} = P \int_{-\infty}^{\infty} \frac{\mathrm{Re}\{\gamma(\tau')\}}{\tau' - \tau} d\tau'. \tag{3}$$

This leads to many powerful dispersion relationships for Fourier transforms of causal functions. An extension to these dispersion relationships, when considering the real and imaginary parts of $\log\gamma(\tau)$, is possible, provided that this function satisfies Titchmarsh's theorem. This occurs if $\gamma(\tau)$ has no zeros in the lhp, a condition known as the minimum phase condition. When this condition is satisfied, it follows that a Hilbert transform can be written between the real and imaginary parts of $\log\gamma(\tau)$, allowing the phase of $\gamma(\tau)$ to be calculated from its magnitude. Kano and Wolf [1] proved that for the specific case of *blackbody radiation*, $\gamma(\tau)$ has a zero-free lhp and no zeros on the real axis either. This was based on the analytical expression for the real coherence tensor of blackbody radiation first given by Bourret [4] and corrected by Kano and Wolf. They showed that

$$\gamma(\tau) = 90\zeta(4, 1 + i\tau)/\pi^4, \tag{4}$$

where ζ is the generalized Riemann-zeta function, which has no real zeros and a zero-free lhp. This is one of what might be many asymmetric spectral profiles that can be uniquely determined from knowledge of the absolute value of the complex degree of coherence. Wolf hypothesized that the existence of zeros in the lhp would have a physical significance, and this remains a tantalizing proposition.

What is so fascinating about this is that inherent mathematical properties of the functions we use in our physical models allow this inverse problem to be solved from just one data set, namely real positive measurements of the visibility function. One is led to wonder what distinguishes the class of problems for which inversion is possible from magnitude-only data from those for which it appears that phase information must somehow be independently measured. Numerous interferometric techniques exist that can provide phase information, but they are not always convenient to implement. Can we preprocess our data or alter our experiments in such a way that only real positive measurable quantities are required to extract the information we seek without recourse to interferometry? The question is essentially the title of this paper, exploring the significance of phase and what information it carries.

17.2 Analyticity and Phase

Exploiting the analytic properties of the Fourier transforms of causal functions was studied for many years in the context of phase retrieval [4–8]. It became clear that in one-dimensional problems, the presence or absence of zeros on the lhp determined the degree of ambiguity one might expect in attempting the recovery of phase from magnitude data, see, e.g., Ref. [9]. Indeed, it became almost a sport to create ambiguities based on this deeper understanding of phase ambiguities, by "complex zero-flipping." Such a flip, or complex conjugation of a zero's coordinates, left the Fourier magnitude unchanged but altered the Fourier phase (see Fig. 1). Extending this understanding to two- or higher-dimensional problems has remained elusive, however. This is because of the fundamental difference between

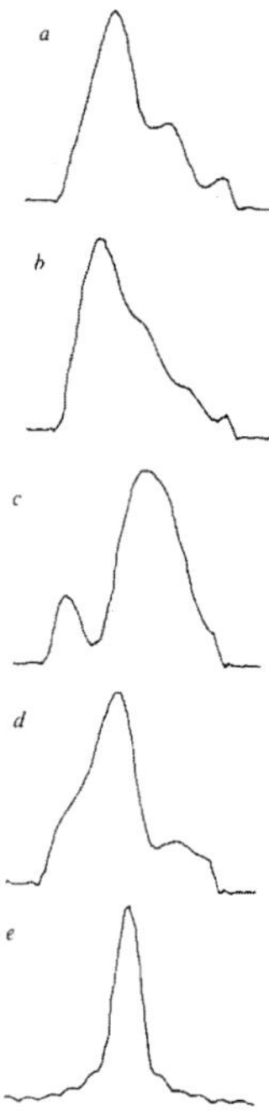

Figure 1 All four different real possitive objects shown in a, b, c and d produce the same far field intensity, shown in e. The four different objects have been generated by reflecting or 'flipping' the zeros of $F(z)$ about the x-axis, as described in the text. [Reprinted with permission from M.A. Fiddy and A.H. Greenaway, "Object reconstruction from intensity data," *Nature* **276**, 421 (1978). Copyright (1978) by Macmillan Magazines Ltd.]

the properties of analytic functions of one complex variable, which, like polynomials, can be factored into a product of their roots or zeros; i.e., the fundamental theorem of algebra applies. In two-, three- or higher-dimensional problems, this is not the case.

Conspiring to make things more complicated were two other landmark papers in the 1980s. One was by Oppenheim and Lim arguing that phase is far more important than magnitude information in imaging, i.e., two-dimensional problems [10] (see Fig. 2). The other was by Hayes and McClellan [11] reminding the less mathematical amongst us that in two- and higher-dimensional problems, the Fourier transforms of causal functions, or bandlimited functions for that matter, were not generally factorizable. Indeed, while the zero set of one-dimensional functions has dimension zero, i.e., they occur at isolated points, in n-dimensions the zero set has a volume of $2n - 2$ dimensions [12] and there is no analogous tool such as the Weierstrass or Hadamard product through which to represent the function by its zero locations. Also, zeros can never be isolated points in analytic functions of two or more complex variables.

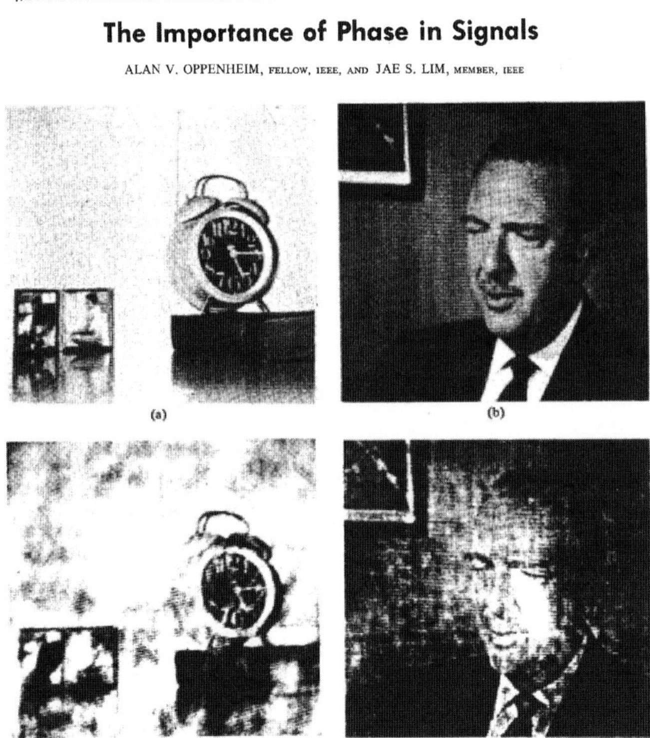

Figure 2 (a) Original image A. (b) Original image B. (c) Image synthesized from the Fourier transform phase of image A and the magnitude of image B. (d) Image synthesized from the Fourier transform magnitude of image A and the phase of image B. [Reprinted with permission from A.V. Oppenheim and J.S. Lim, "The importance of phase in signals," *Proc. IEEE* **70**, 192–198 (1981). Copyright (1981) by IEEE.]

For the last 25 years or so, phase retrieval problems have come and gone, in the sense that practical methods have been developed to recover phase information from magnitude data, but they have not relied explicitly on the underlying analytic properties of the wave. These properties may have implicitly been used to justify algorithmic success (e.g., by stating that a nonfactorizable function should have a unique phase, since zero-flipping cannot occur), but there remains to be found a satisfying analytical relationship between magnitude and phase. Dr. Wolf initiated a curiosity in this subject that has not faltered, and deeper insights can be expected by understanding the role of analyticity in optics. It should be admitted that in inverse problems, the reality of noisy sampled data sometimes appears to make the explicit use of analyticity, e.g., for superresolution, somewhat impractical. Nevertheless, its existence as a ghost in the mathematical framework that we employ should not be dismissed.

17.3 On the Absence of Magic

When information is lost in a measurement process, it is truly and irreversibly lost; it is absent from the data. What we mean by information restoration is solving the inverse problem; i.e., determining what information was there before the information-destroying operation was performed. Absent magic, the missing information must come from somewhere other than the data themselves, or have been explicitly encoded in the measured data a priori. In all the cases discussed here and also in our other paper in this volume: "Thinking backward: holography and the inverse problem," the information is injected into a computer in the form of assumptions about the problem and the measurements. If those assumptions hold, the required restoration will be accurate, while the converse will also be true.

17.4 What is Phase?

Dr. Wolf recently pointed out that, when discussing measurements of phase, one is usually assuming that the field is monochromatic [13]. He makes the important point that the question as to the true meaning of the phase of an optical field is not asked because it is typical to assume that fields are monochromatic. In practice, every field has finite bandwidth and its phase fluctuates rapidly and randomly. However, if a field is spatially completely coherent at each frequency, one can associate a monochromatic field of that frequency that represents a statistically averaged behavior of the fluctuating field. The phase of this associated monochromatic field is proportional to the phase of the spectral degree of coherence of the original field. Wolf explains the distinction to be made between coherence and monochromaticity. The ability of two beams to produce interference fringes when superimposed is a measure of the coherence between the two fields. The point is well taken, however, that two fields represented by $V_1(t) = a\exp[i\phi_1(t)]$ and $V_2(t) = a\exp[i\phi_2(t)]$ will have an interference term given by $2a^2\langle\cos[\phi_2(t) - \phi_1(t)]\rangle$ where $\langle\rangle$ denotes ensemble average. For monochromatic fields for which $\phi_j(t) = \alpha_j + \omega t$, where α_j are constants, the interference term is proportional to $\cos(\alpha_2 - \alpha_1)$, which is nonzero unless $\alpha_2 - \alpha_1 = n\pi/2$. However, even for nonmonochromatic fields, interference is possible, if, for example, $\phi_2(t) = \phi_1(t) + \beta$, i.e., if the phases differ from each other by a constant not equal to $n\pi/2$. Thus even rapid fluctuations in time of these phases does not necessarily mean that they are not mutually coherent; monochromaticity is not a necessary condition for interference. Thus, although an optical field is never strictly monochromatic, it may be completely spatially coherent, not only at a single frequency but at all frequencies in its spectrum [13]. Indeed, the original off-axis holograms of Leith and Upatnieks exploited that phenomenon before they ever used laser holography.

For a steady state optical field, i.e., a statistically stationary field, its cross-spectral density is given by

$$W(r_1, r_2, \omega) = \int_{-\infty}^{\infty} \langle V^*(r_1, t) V(r_2, t + \tau) \rangle \exp(i\omega\tau) d\tau. \tag{5}$$

It is known that if an optical field is spatially completely coherent at frequency ω and for all points r_1, r_2 in a domain D then W necessarily factorizes into [14]

$$W(r_1, r_2, \omega) = U^*(r_1, \omega) U(r_2, \omega), \tag{6}$$

where U satisfies the Helmholz equation throughout D; this relationship implies that the field consists of a single mode. This $U(r, \omega)$ can be identified with the space-dependent part of a monochromatic field $V(r, t) = U(r, \omega)\exp(i\omega\tau)$, which confirms that we can associate with any field that is spatially coherent at a frequency ω, a monochromatic field of the same frequency that yields the cross-spectral density of the field. It is this deterministic field $U = |U(r, \omega)|\exp[i\phi(r, \omega)]$ whose amplitude is associated with the square root of the spectral density and whose phase is associated with the spectral degree of coherence. The phase function is therefore the same for both the averaged field variable U of a spatially coherent field and its spectral degree of coherence.

17.5 Can We Do Without Phase?

This question is being posed only in the context of solving an inverse problem, i.e., to recover signal or image information from measurements of its scattered field or its Fourier magnitude. Wolf has recently demonstrated that, indeed, one can [15]. The approach is inspired by diffraction tomography methods based on the Green's function phase-retrieval technique proposed by Teague [16]. For weakly scattering objects, intensity measurements taken on several planes have been shown to be sufficient for recovery of the object distribution, especially when the scattering object has a well-localized Fourier spectrum. Teague's original approach relied on determining the phase from intensity measurements by solving a 2D Poisson equation, which in turn required a paraxial approximation to hold and an absence of vortices. Earlier innovative work by Carney, Wolf, and Agarwal [17] showed how measurements based on a generalization of the optical (cross-section) theorem could also allow object reconstruction without phase information. The total power extinguished from the incident field as a result of scattering and absorption specifies the imaginary part of the scattered field, or more precisely, the imaginary part of the scattering amplitude. By making measurements of the extinguished power with two incident plane waves propagating in different directions, and with different relative phases, one can extract the absorptive part of the scattering object.

It is rather tempting, especially in the field of Fourier optics, to assume that two parameters are always necessary in order to fully describe a propagating or diffracted field, its magnitude, and its phase, ϕ. We write for a two-dimensional problem, that

$$F(x,y) = |F(x,y)|\exp(\phi(x,y)) = \iint f(p,q)\exp(-ik[\,px + qy\,])\,dp\,dq, \qquad (7)$$

where k is the wavenumber, $2\pi/\lambda$. It would seem that only if the phase ϕ is a constant can its importance be ignored when retrieving the function f. Symmetries imposed in the object domain, for example replicating $f(p,q)$ in such a way that a Hermitian function existed would ensure that $F(x,y)$ is real, but not necessarily real and positive. The function in the (p,q) domain needs to be a positive definite function for F to be identical to an intensity $|F|^2$. Holography exploits this, since reconstructing a hologram can be regarded as generating the Fourier transform, optically, of the transmittance of the hologram, which is $|H|^2 = |R + F|^2$, where R is a reference wave and F is the complex FT of the object to be recovered. Provided R is an off-axis plane wave, a reconstruction of f convolved with a δ-function can be spatially distinct from the autocorrelation function of f.

The magnitude of F, $|F|$, is also real and positive but is not typically a bandlimited function. Nevertheless, there are conditions under which it can provide a reconstruction of f directly, without the need to compute or measure a phase function. This can be understood by considering the bandlimited intensity data, $|G|^2$, written as $|1 + F|^2$, which can allow $|G|$ to be approximated by $\{(1 + F)(1 + F)^*\}^{1/2} = \{1 + 2Re\{F\} + |F|^2\}^{1/2}$, which can be increasingly well approximated by $1 + Re\{F\}$ as $|F| \ll 1$.

We note that these examples suffer from the same kind of problems associated with the method proposed over 10 years ago by Devaney [18] in which backpropagation of intensity data could sometimes spatially separate object and autocorrelation function information, in much the same way that Gabor envisioned the first hologram might work even for an in-line rather than off-axis hologram. Separation of the desired object information from (often severely low-pass filtered and therefore broadened) background information becomes the challenging problem to solve.

17.6 The Role of Reference Points

Holography, as discussed above, provides a mechanism through which a magnitude-only function, specifically the real transmittance of an amplitude hologram, represents a function that can encode information about $f(p,q)$ from a function of the form $|F(x,y)|^2$. The role of a reference wave, or reference point in the object domain, is the key.

The role of reference points in assisting with phase retrieval has been discussed many times in the past, exploiting for example Rouche's theorem to enforce the minimum phase condition [19], and to ensure irreducibility of a field thereby ensuring a unique phase by exploiting Eisenstein's criterion [20].

In a recent paper by Lohmann et al. [21], the relative importance of Fourier amplitude and phase was discussed. The widespread belief that phase is more important was questioned through a series of elegant examples. They showed that swapping Fourier phases or amplitudes in spectra with those from different objects, did indeed confirm the earlier work of Oppenheim and Lim [10], but only up to a point, when the object is real and positive. In Ref. [21], examples are shown of different centrosymmetric objects with a reference point located at the origin, whose magnitude equaled the total energy of the object. These were then reconstructed from their Fourier amplitude only, or Fourier phase only, with the phase of the reference point either zero of $\pi/2$. It turned out that even for non-symmetric objects possessing complex values, whether the Fourier magnitude or the Fourier phase regenerated the object (and its twin), depended on whether the phase of the reference point was zero or $\pi/2$, respectively (see Fig. 3).

It is surprising that such a dramatic effect can arise from changing the phase of only one pixel in the object domain; it brings into question whether it is fair to say phase is more important than magnitude; this is illustrated in Figs. 4 and 5.

In Fig. 6 the Fourier transform of the logarithm of the Fourier intensity is shown for increasing values of the amplitude of the reference point. As is evident,

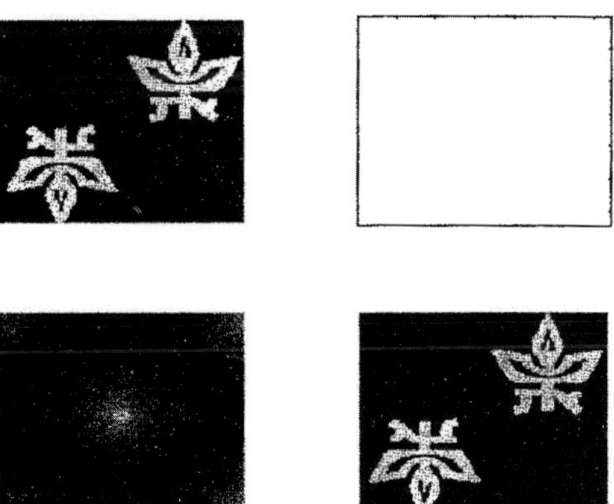

Figure 3 Reconstructions from an asymmetric object with totally random phase: peak phase zero ampilitude-only reconstruction (upper left), peak phase zero phase-only reconstruction (upper right), peak phase 90-deg amplitude-only reconstruction (lower left), peak phase 90-deg phase-only reconstruction (lower right). [Reprinted with permission from A.W. Lohmann, D. Mendlovic, and G. Shabtay, "Significance of phase and amplitude in the Fourier domain," *J. Opt. Soc. Amer.* **20**, 753 (2003). Copyright (2003) by the Optical Society of America.]

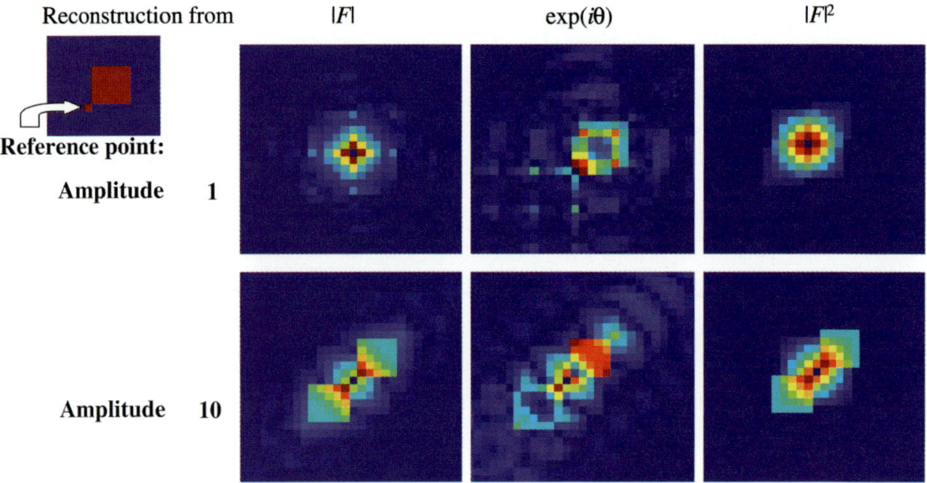

Figure 4 This figure illustrates reconstructions from Fourier magnitude, Fourier phase, and Fourier intensity of an object consisting of a simple square, but with a reference point inserted at one corner. Two amplitudes for the reference point are given, 1 and 10. As expected, the larger the amplitude of the reference point, the more prominent is the reconstruction of the square, even from Fourier magnitude in the case of an amplitude of 10. The phase reconstructions are not that good.

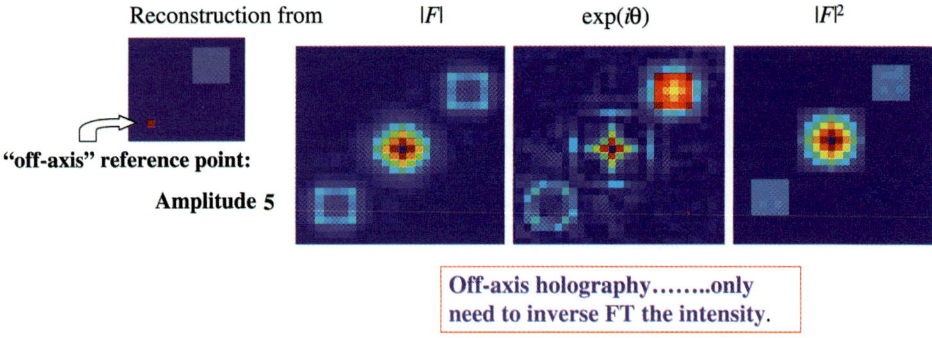

Figure 5 The figure above illustrates the case for an intermediary value of the amplitude of the reference point of 5. Also, the location of the reference point is now removed from the square by a distance greater than the diagonal of the square. This "off-axis" reference point ensures that the inverse Fourier transform of the intensity alone reveals the object.

the reconstruction of the square becomes increasingly clear as the reference point's amplitude increases. An interpretation of this is that as that amplitude increases, the Fourier transform of the square plus the reference point can be thought of as approximating the exponential of the Fourier transform of the square alone. If this exponential exists and its Fourier transform has a causal compact support, then it follows that the *logarithm* of the Fourier transform of the square plus the refer-

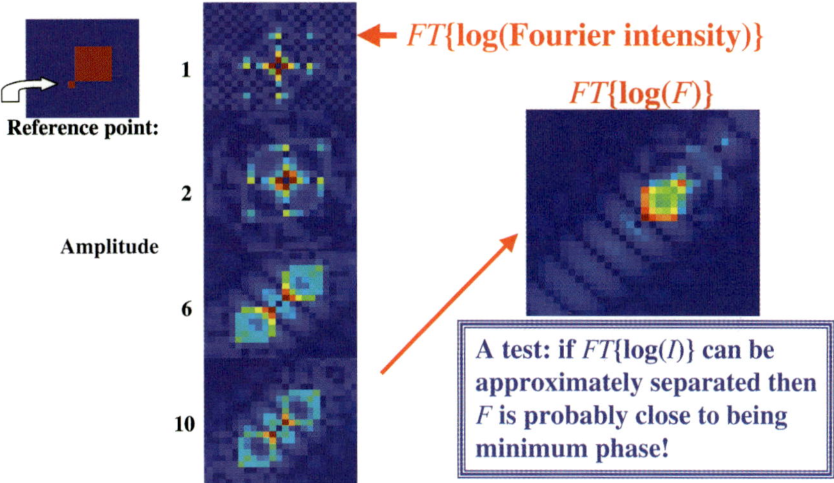

Figure 6 Trivial case: with a strong reference wave added to F, i.e., $1 + F$ then $\sim \exp(F)$ for $|F| \ll 1$ and $FT\{1 + F\} = \delta + f$; thus we expect to recover f when a (strong) reference point is present somewhere in the object domain.

ence point is also analytic. The latter is the condition for the 2D minimum phase condition to be satisfied, which allows one to recovery of the phase of F from its magnitude using a logarithmic Hilbert transform. It is therefore not surprising that one can recover the object from Fourier intensity data when this somewhat trivial condition has been satisfied. It does suggest a test, however, for whether phase can be computed from magnitude in this way, because if the Fourier transform of the logarithm of the Fourier intensity appears to separate into two objects as shown, then the Fourier transform of that object is probably close to being a minimum phase function.

Millane and Hsiao [22] recently commented on this work, arguing that in the examples above, the phase functions used were not independent of the true phase function. They meant by this that the central reference point was so dominating that the Fourier phase of interest is a small perturbation on this known background. They concluded that phase dominance does indeed appear to be a general phenomenon. If this is the case, then it explains why trying to reconstruct objects with poor or missing phase information is so difficult. But it still begs the question as to why this is so, and whether some kind of general preprocessing of the data could not allow straightforward object reconstruction from the Fourier magnitude. With a suitable reference point this appears to be true, but when thinking about whether a reference point is or is not part of the object and whether it is necessary to know of its existence a priori seem to be moot points. An intensity distribution, given the clarification of the meaning of phase by Wolf, should be an interferogram, and the phase of the object should be encoded in a way that the object structure can be extracted.

17.7 Phase and Information

Hinted at and even expressly addressed in what precedes this paragraph is the possibility that phase may carry more information than amplitude in an optical wavefront. But caution should be recommended in reaching a universal conclusion on this point. A single wavefront propagating through space or transformed by diffraction-limited optics should conserve information. That is, the transformation is unitary. Gabor captured this idea in terms of conservation of logons—joint spatial-angular (phase) information. But the way information is distributed within phase and amplitude may not be constant. Consider a uniform beam (having very little spatial amplitude information) converging toward a point (and thus having substantial phase information). At the focal plane, there is much spatial information telling us where the beam was focused but little valuable phase information, as the amplitude is negligible at most points. At least in terms of practical and valuable information, the distribution of that information between amplitude and phase may change considerably under unitary transformation. A good Bayesian, however, might suggest that we ought to measure and use all of the available information, even the amplitude and phase information on low-amplitude points if we seek the most accurate determination of the focal point. Those are issues too remote from our central theme to warrant further discussion here. This paragraph is offered not as proof of any thesis but as a cautionary note about the danger of making a sweeping conclusion that phase carries more information than amplitude.

17.8 Conclusions

Given Wolf's more precise definition of what is meant by the phase of an optical field, it is tempting to state that the importance of phase in inverse Fourier problems should be interpreted in a manner consistent with this. In linear problems, one can interpret the far-field scattering pattern as the linear combination of spherical wavelets from each point in the plane emerging from the object. It is the relative phase of these wavelets that, when they combine, generate the scattered field. The intensity of this field is an interferogram when employing light with a high degree of spatial coherence, and these relative phases are in principle embodied in the measured intensity. The challenge lies in how to extract the information. Inversion of intensity data provides only an autocorrelation function, which, as we have described above, may only yield useful information about the object under specific circumstances. If any part of the object is known a priori, or a conscious decision to introduce a reference point or interpret that there is a reference point present, then one has a handle on how to "read out" the Fourier phase and magnitude in much the same way as one would replay a hologram. The reference point provides a constraint against which the remainder of the information in the measured data can be interpreted. This is not to say that determining the object is straightforward, only that, in principle, it becomes possible. Iterative methods, criteria to enforce a minimum phase condition, or, recursively, recovering f from its autocorrelation function, all appear to have this need for a reference point of some kind. The more prominent the reference point the better. X-ray diffraction techniques based on the heavy-atom method exploit the same idea. This is a kind of holography, a means to record patterns that allow a solution of the inverse problem given the proper information to inject into the solver. The difficulties come when nothing is known a priori of this kind. When information is lost and we have no a priori information of any kind about the problem and/or the measurement system; the situation is hopeless. For all other situations, this is where the problems get interesting.

References

1. E. Wolf, "Is a complete determination of the energy spectrum of light possible from measurements of the degree of coherence?" *Proc. Phys. Soc.* **80**, 1269 (1962); Y. Kano and E. Wolf, "Temporal coherence of black body radiation," *Proc. Phys. Soc.* **80**, 1273 (1962); R. Hanbury Brown and R.Q. Twiss, *Nature*, **178**, 27 (1956) and *Nature* **178**, 1046 (1956).
2. M. Born and E. Wolf, *Principles of Optics*, Pergamon Press, Oxford (1959).
3. E.C. Titchmarsh, *Introduction to the Theory of Fourier Integrals*, 2nd ed., Oxford University Press, Oxford (1948).
4. R.C. Bourret, *Nuovo Cim.* **18**, 347 (1960).
5. E.L. O'Neil and A. Walther, "The question of phase retrieval in optics," *Opt. Act.* **10**, 33 (1963).
6. D. Kohler and L. Mandel, "Source reconstruction from the modulus of the correlation function," *J. Opt. Soc. Amer.* **63**, 126–134 (1973).
7. R.E. Burge, M.A. Fiddy, A.H. Greenaway, and G. Ross, "The phase problem," *Proc. Roy. Soc. Lond. A* **350**, 191–212 (1976).
8. M.A. Fiddy, "The role of analyticity in image recovery," in *Image Recovery: Theory and Applications*, H. Stark, Ed., 499–529, Academic Press, FL (1987).
9. M.A. Fiddy and A.H. Greenaway, "Object reconstruction from intensity data," *Nature* **276**, 421 (1978).
10. A.V. Oppenheim and J.S. Lim, "The importance of phase in signals," *Proc. IEEE* **69**, 529–536 (1981).
11. M.H. Hayes and J.H. McClellan, "Reducible polynomials in more than one variable," *Proc. IEEE* **70**, 192–198 (1982).

12. L.I. Ronkin, *Entire Functions in Several Complex Variables, 3rd ed.*, A.G. Vitushkin, Ed., 30, Springer, Berlin (1985).

13. E. Wolf, "Significance and measurability of the phase of a spatially coherent optical field," *Opt. Lett.* **28**, 5 (2003).

14. L. Mandel and E. Wolf, *Optical Coherence and Quantum Optics*, Cambridge University Press, Cambridge (1995).

15. G. Gbur and E. Wolf, "Hybrid diffraction tomography without phase information," *J. Opt. Soc. Amer. A* **19**, 2194 (2002).

16. M.R. Teague, "Deterministic phase retrieval: A Green's function solution," *J. Opt. Soc. Amer.* **73**, 1434 (1983).

17. P. Scott Carney, E. Wolf, and G.S. Agarwal, "Diffraction tomography using power extinction measurements," *J. Opt. Soc. Amer. A* **16**, 2643 (1999).

18. A.J. Devaney, "Diffraction tomographic reconstruction from intensity data," *IEEE Trans. Image Process.* **1**, 221 (1992).

19. R.E. Burge, M.A. Fiddy, A.H. Greenaway, and G. Ross, "The application of dispersion relations (Hilbert transforms) to phase retrieval," *J. Phys. D* **7**, L65–68 (1974).

20. M.A. Fiddy, B.J. Brames, and J.C. Dainty, "Enforcing irreducibility for phase retrieval in two dimensions," *Opt. Lett.* **8**, 96–98 (1983).

21. A.W. Lohmann, D. Mendlovic, and G. Shabtay, "Significance of phase and amplitude in the Fourier domain," *J. Opt. Soc. Amer. A* **14**, 2901 (1997).

22. R.P. Millane and W.H. Hsiao, "On apparent counterexamples to phase dominance," *J. Opt. Soc. Amer. A.* **20**, 753 (2003).

23. M.A. Fiddy and U. Shahid, "Minimum phase and zero distributions in 2D signals," *SPIE Proc.* Vol. 5202, 201–208, Bellingham, WA (2003).

24. D. Dudegeon and R. Mersereau, *Multidimensional Digital Signal Processing*, Prentice-Hall, NJ (1978).

Prof. Wolf and Michael Fiddy at the Second Annual Symposium held by University of North Carolina at Charlotte in November 2003.

Michael Fiddy received his Ph.D. in Physics from the University of London in 1977, and was a post-doc in the Department of Electronic and Electrical Engineering at University College London before becoming a tenured faculty member at Queen Elizabeth College/Kings College, London University in 1979. Between 1982 and 1987, he held visiting professor positions at the Institute of Optics Rochester and the Catholic University of America in Washington, DC. Dr. Fiddy moved to the University of Massachusetts Lowell in 1987, where he was Electrical and Computer Engineering Department Head from 1994 until 2001. In January 2002, Dr. Fiddy moved to UNC Charlotte as the founding director of the newly created Center for Optoelectronics and Optical Communications. He was the topical editor for signal and image processing for the journal of the Optical Society of America from 1994 until 2001 and has been the Editor in Chief of the journal *Waves in Random Media* since 1996. He has chaired a number of conferences in his field, and is a fellow of the Optical Society of America, the Institute of Physics, and the International Society for Optical Engineering (SPIE). His research interests are in inverse problems and optical information processing.

LOCAL INTERFACE TECHNIQUES IN WAVE-OPTICAL ENGINEERING

Frank Wyrowski and Jari Turunen

18.1 Introduction

Optical engineering, which addresses modeling, design, fabrication, and testing of optical systems, is often associated with imaging systems and the design of lenses. Thus, classical optical engineering is largely based on a ray-optical representation of electromagnetic fields and the modeling of their propagation by geometrical optics, i.e., ray tracing. This geometrical theory of field propagation is obtained from general electromagnetic wave theory at the short-wavelength limit, as illustrated in Chapter 3 of Ref. [1]. It serves well the propagation of smoothly modulated fields that are are typically of concern in the modeling of imaging systems and illumination systems. As soon as the truncation of fields by apertures (see Sect. 8.8 in Ref. [1] and Refs. [2,3]) or other high-frequency field modulations become important in an application, wave-optical propagation techniques must be applied. Strong modulations may be inherent in the field we wish to obtain; consider for instance the generation of a top-hat profile laser beam. Such modulations may be introduced by strong aberrations (see Chapter 9, Ref. [1]), or by the use of micro-structured elements [4]. Independently of the reason, geometrical optics modeling starts to fail.

Often the use of geometrical optics for modeling the propagation of fields is directly associated with a ray-bundle representation of the field, which the optical designer can employ to evaluate the quality of an imaging system by investigating spot diagrams in or near the image plane. The representation of fields by rays is indeed a basic characteristic for modeling in conventional optical engineering. If the geometrical optics propagation model starts to fail, we need to give up the ray-optical field representation as well. It should be noted, however, that a wave-optical field representation in combination with a geometrical optics propagation model can give very reasonable results, as we will see in Sect. 18.5.

In modern optical engineering, systems with increasingly complex nonimaging functionality are of concern. The innovative potential of optics and photonics lies particularly in the capability to generate tailored electromagnetic radiation. At this stage we arrive at a domain called photon management. Because of the complex functionality, one is typically not allowed to use a ray-optical field representation. Instead, the evaluation of the optical system requires access to the entire electromagnetic field information. As an example, consider the calculation of the M^2 parameter of a Gaussian beam, which requires the evaluation of the complex amplitude of the beam at the observation plane. Thus we may conclude that optical engineering for photon management must rely on a wave-optical field model instead of the ray-bundle representation. Moreover, electromagnetic radiation that is best adapted to a given application is not smoothly modulated a priori, which implies that often microstructured elements must be included in a system. Thus, in substantial parts of the system a geometrical optics propagation model may not be suitable. We arrive at the conclusion that a wave-optical generalization of optical engineering is required for photon management, which is itself one pillar of the innovative potential of photonics. We refer to the systematic inclusion of wave optics in optical engineering as wave-optical engineering (WOE) [5].

It would be logical to apply a fully rigorous electromagnetic modeling of the propagation of electromagnetic fields through optical systems in WOE. However, an exact solution of Maxwell's equations is at best a time consuming but often an unmanagable numerical task. Therefore it is important to develop approximate propagation models suitable for wave-optical engineering. In this chapter we describe how the knowledge of elementary rigorous solutions of the propagation problem can be used to develop more general propagation techniques (to be called local interface techniques) suitable for the analysis and design of photon management systems.

This chapter is organized as follows. We begin in Sect. 18.2 with a specification of the propagation problem of concern. Then, in Sect. 18.3 we describe the use and limitations of rigorous electromagnetic techniques to solve the problem. Section 18.4 provides the basic ideas behind local interface approximations. In Sect. 18.5 and 18.6 we describe two classes of local approximations in some detail. Finally, comments on the presented techniques are provided in Sect. 18.7 and some conclusions are drawn in Sect. 18.8.

18.2 Problem Statement

The principle configuration of a general optical system is illustrated in Fig. 1. An electromagnetic field enters the system from the left and propagates through a sequence of an arbitrary number of sections of homogeneous media separated by

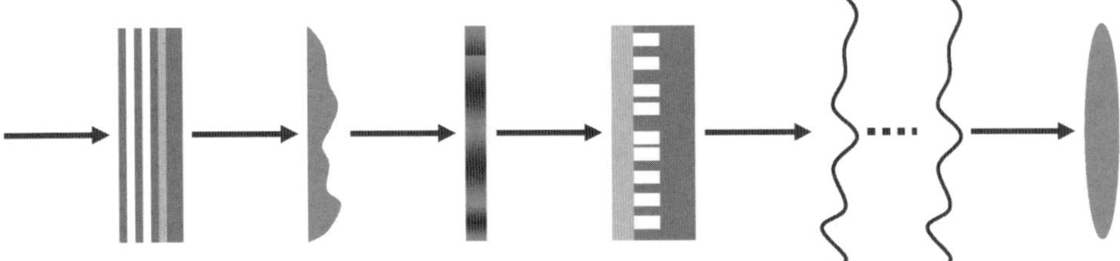

Figure 1 Illustration of an optical system consisting of a sequence of interfaces of arbitrary form between sections of homogeneous dielectric media.

optical interfaces. The interfaces may be spherical or aspherical refractive or reflective surfaces, microstructured surface profiles, refractive-index-modulated layers, or combinations of these. The incident field is assumed to be known, and our basic goal in modeling the system is to predict the field in the output region after it has passed through the whole system. Physically, the solution of this problem requires a proper mathematical field model and its propagation through homogeneous media as well as through interfaces. In wave-optical engineering, an electromagnetic field model is mandatory. Depending on the source, different models are appropriate. The harmonic field model is useful not only for monomode laser sources, but also as a basis for partially coherent radiation in the space-frequency domain (see Sect. 18.7). Using a harmonic field model, the electromagnetic field in any homogeneous region is completely represented by the complex amplitudes of the x- and y-components of the electric field in an arbitrary z-plane in the region, that is, by $E_x(x,y,z_0)$ and $E_y(x,y,z_0)$ if we use the plane $z = z_0$ as a reference. The two complex amplitudes can be formally combined by introducing the harmonic field operand $\mathbf{f}(x,y,z_0) = [E_x(x,y,z_0), E_y(x,y,z_0)]$. Mathematically, the propagation of the field thus means the application of propagation operators, in the form of integrals or algorithms, on the harmonic field operand. In homogeneous media the propagation operator can be expressed in terms of the angular spectrum of plane waves integral, or the Fresnel integral in paraxial approximation [17]. The propagation through interfaces states a boundary value problem in terms of Maxwell's equations. Here we focus on the propagation of the harmonic field $\mathbf{f}_0$ through a single interface, which states the fundamental problem to be solved in order to propagate a field through the entire system. Figure 2 illustrates this fundamental problem. We wish to determine $\mathbf{f}_1(x,y,z_1)$ from the knowledge of $\mathbf{f}_0(x,y,z_0)$, the shape of the interface, and the refractive indices n_0 and n_1 (the transmission geometry is considered here, but the reflection mode of operation can be treated analogously). It is remarkable that this basic problem possesses a rigorous solution for only a few special interface geometries, as we see next.

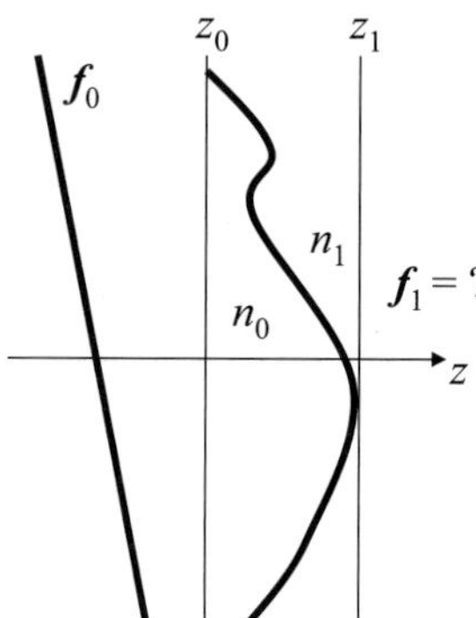

Figure 2 A plane wave f_0 propagates through an interface between two homogeneous dielectrics with refractive index n_0 and n_1. The interface is of general shape and therefore a rigorous solution of the propagation problem does not exist. The virtual planes in z_0 and z_1 encompass the interface in a way that minimizes the distance $z_1 - z_0$.

18.3 Rigorous Solutions of the Propagation Problem

In rigorous theory one seeks exact solutions of Maxwell's equations in the presence of the electromagnetic boundary conditions defined by the interface. Thus we have a scattering problem: fields in regions separated by the interface are matched to each other. Rigorous solutions of this problem exist only for some special geometries that are of interest in WOE. The most important of such geometries are plane interfaces (also film stacks) and gratings. In the case of plane interfaces, just two plane waves are created because of the symmetry, and Fresnel's equations are obtained. In the case of gratings, Bloch's theorem leads to an infinite but countable number of plane waves in both regions, and matching is possible by truncation. This leads to the rigorous electromagnetic theory of gratings [6–10], which provides answers to many questions in WOE. Some of these are discussed in Ref. [9]. However, this approach has certain severe practical limitations, which we will discuss below. First of all, rigorous theory is applicable mostly to single interfaces, not to entire systems such as that illustrated in Fig. 1.

The most generally applicable rigorous technique is the Fourier modal method (FMM) [4,10], but in some cases (especially for continuous surface profiles) coordinate-transformation methods [11,12] are more natural and efficient. All these methods are directly applicable only to periodic structures, i.e., gratings. If a grating is illuminated by a plane wave, it produces a discrete set of diffracted plane waves with the propagation directions determined by the grating equations. The reflected and transmitted fields are thus expressed as plane-wave superpositions (Rayleigh expansions) with unknown complex amplitudes associated with individual plane waves.

In FMM the complex amplitudes of the diffraction orders, and thereby also the diffraction efficiencies of different orders, are obtained by matching the Rayleigh expansions with discrete waveguide mode expansions of the field within the grating. The numerical procedure in FMM can be divided into two main parts: first, the modal propagation constants and wave forms are determined by solving a matrix eigenvalue problem; then the matching problem that leads to a set of simultaneous equations is solved. Unfortunately both numerical problems (but in particular the eigenvalue problem) scale badly when the grating period d is increased. This is because the number of propagating orders increases with d, and so does the size of the matrices that must be treated in the numerical solution. For one-dimensionally periodic gratings, the size of the boundary value matrix increases with $\sim d^2$, and for biperiodic gratings the increase is $\sim d^4$. With such scaling it is obvious that no foreseeable development in computer technology will permit rigorous analysis of macroscopic diffractive structures. At present the practical upper limit is $d \approx 1000\lambda$ for one-dimensionally periodic gratings and $d \approx 30\lambda$ for two-dimensionally periodic gratings. It should be noted that these limits are only indicative; if the grating structure is simple, much larger periods can be used, but for metallic gratings the limits can be substantially lower, especially if the conductivity is high.

As already mentioned, the most efficient rigorous methods apply to periodic structures only. There are ways to use these methods also for nonperiodic structures, but then the computational complexity is increased even further. One way is to model the incident field as a superposition of plane waves that illuminate only a part of a single grating period. In this case the diffraction problem has to be solved independently for each incident plane-wave component and the results are superimposed coherently. Another way is to embed the nonperiodic structure as a part of a periodic structure such that the modulated part forms only a small fraction of the grating period. If rigorous grating theory is now applied to the periodic structure and the output field is truncated such that information about the periodicity is lost, the truncated field can be propagated further without a significant error. These techniques can be used to model, e.g., the transmission of subwavelength holes in a metallic screen [13], but they are definitely not suitable for analyzing large nonperiodic structures.

18.4 Concept of Local Interface Techniques

The considerations presented above clearly show that rigorous solution of the propagation problem through interfaces is possible for planar and laterally periodic interfaces, and for microscopic nonperiodic structures. Thus there is a compelling need to develop approximate, yet sufficiently accurate methods for the analysis of light reflection and transmission by optical interfaces.

The basic concept behind local interface techniques is to divide a general propagation problem into a set of simpler problems that can be solved rigorously—the results are then combined to obtain an approximate, but sufficiently accurate, solution of the original problem. We apply two basic concepts to divide the general problem into manageable parts: (1) decomposition of the incident field into laterally truncated fragments, which illuminate approximately only elementary fractions of the interface, and (2) decomposition of the response of the interaction of the incident field with the interface into approximately independent interactions with local interface features. We refer to the the first technique as local elementary interface approximation (LEIA) and to the second as local independent response approximation (LIRA). In LEIA, the most important elementary interfaces are planar, periodic, and spherical interfaces. The strong response of abrupt profile transitions in the interfaces is a particular motivation to develop LIRA.

18.5 Local Elementary Interface Approximations

As discussed above, we know rigorous solutions of the propagation of harmonic fields through planar and periodic interfaces between two homogeneous dielectrics. How can we use this knowledge to derive propagation methods through more general interfaces? We will discuss the answer to this question using the rigorous solution for the plane interface. We wish to emphasize that we only present qualitative arguments of this method to propagate fields through interfaces. More quantitative discussions are found in Refs. [14–16].

Fresnel's equations, together with Snell's law of refraction and the law of reflection, constitute the rigorous solution to propagate a plane wave through a plane interface separating dielectric media with refractive indices n_0 and n_1. This rigorous solution can be extended to a general incident harmonic field using the angular spectrum representation [17]: an appropriate number of plane waves is propagated individually through the interface and the resulting fields are superimposed afterward. The left-hand side of Fig. 3 displays the amplitude of a Gaussian beam after internal reflection at a plane interface using this method. The simulation predicts the Goos-Haenchen shift, though no evanescent waves were used in the simulation. In case of a local plane wave approximation, which means that a local instead of a global decomposition of the incident harmonic field into plane waves is applied, the Goos-Haenchen shift is not predicted (see the right-hand side of Fig. 3). This underlines the approximate nature of the local plane wave approximation even for smooth harmonic fields. Comparison of the figures also shows that the critical angle leads to a truncation of the beam for the local plane-wave model.

In what follows the propagation of a general harmonic field through a plane interface is considered to be solved, and therefore to be available for wave-optical engineering. Next we turn to the situation illustrated in Fig. 2. We restrict our discussion to a plane incident field for the sake of simplicity. Moreover, we use a 1D formalism (however, the generalization to 2D is straightforward). A plane wave $\mathbf{f}_0 = [E_x(x, z_0), E_y(x, z_0)]$ is to be propagated through the interface, and the resulting transmitted field $\mathbf{f}_1 = [E_x(x, z_1), E_y(x, z_1)]$ is of concern. Again, the reflection mode of operation can be investigated analogously. As illustrated in Fig. 4, the interface may be interpreted as being piecewise planar provided that it does not contain abrupt transitions. In other words, we may identify an interval $[x_I - \Delta_-(x_I), x_I + \Delta_+(x_I)]$ around any point x_I, and consider the interface planar within this interval. The extent of this interval, that is $\Delta(x_I) = \Delta_-(x_I) + \Delta_+(x_I)$,

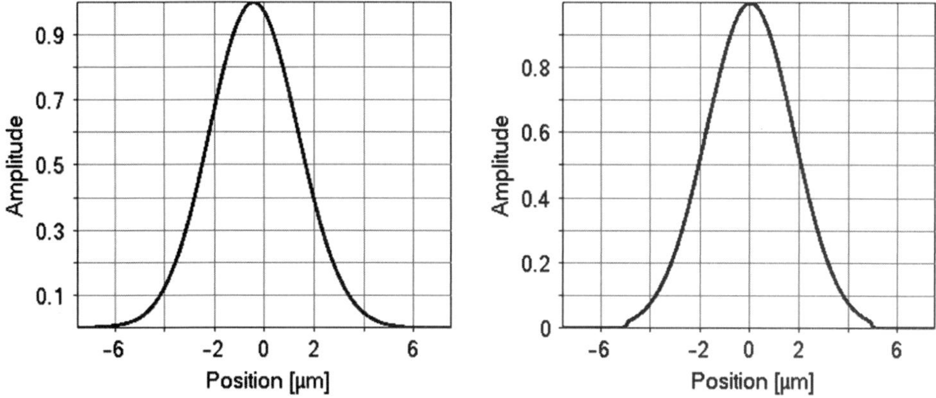

Figure 3 The internal reflection of a Gaussian beam at a plane interface was simulated by two different approaches, namely the global decomposition of the Gaussian beam into plane waves (left) and the local plane wave approximation (right). The latter method is typical in thinking and practice for most optical engineers and laser physicists. The parameters are: TM-polarization, $\lambda = 632.8$ nm, beam waist radius 2.5 μm, angle of incidence 50 degrees, $n_0 = 1.5$, and $n_1 = 1.0$. The resulting critical angle is 41.8 degrees.

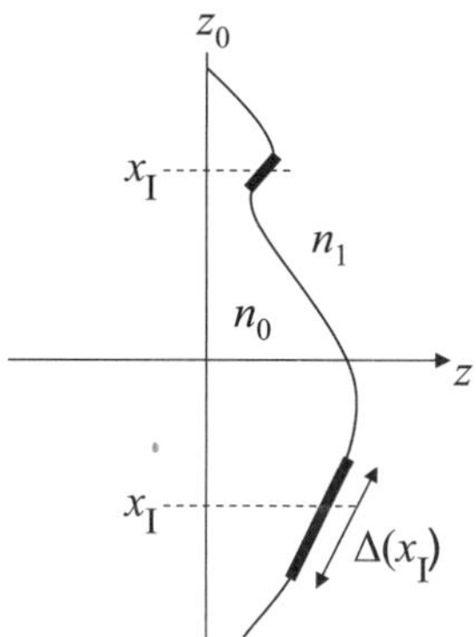

Figure 4 Locally a smooth interface may be considered as planar: the extent of the plane local substitute depends on the local curvature of the interface at x_I.

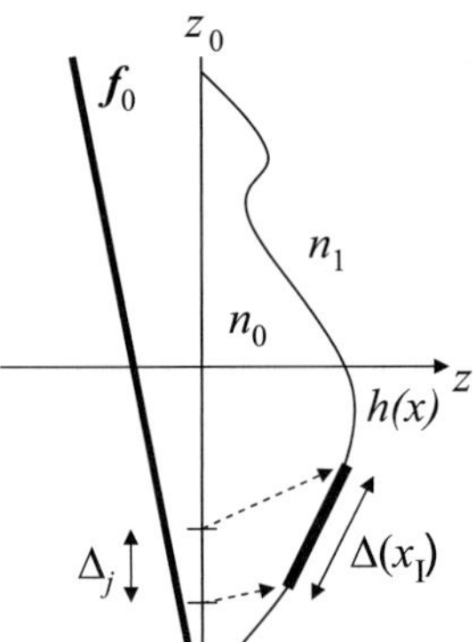

Figure 5 If we propagate a part of the plane field toward the interface $h(x)$, it is often possible to laterally restrict the major part of the propagated field to a local plane interface. Because of diffraction it is never possible to do that rigorously. However, the smaller the distance between z_0 and the interface $h(x_I)$, the smaller the diffraction effects are for a given section size Δ_j. The section size itself must be adapted to the extent $\Delta(x_I)$ of the local plane interface. Thus, the relation between $h(x_I)$, $\Delta(x_I)$, and Δ_j determines the accuracy of the approximation.

can be evaluated by basic mathematical means at any point x_I. In case the incident field $\mathbf{f}_0$ would be restricted to an approximately plane interface fraction around x_I, the propagation problem would be solved in an approximation, which is as good as the approximation of the local planarity of the interface. Other approximations are not involved. However, the incident field $\mathbf{f}_0$ passes the whole interface; thus, it is not possible to take advantage of the local planarity of the interface directly. Fortunately, it is possible to decompose the incident field in such a way that often allows the local application of the plane interface solution of the propagation problem. To this end we use the simple decomposition[*]

$$\mathbf{f}_0(x,z_0) = \sum_j \mathbf{f}_0^{(j)}(x,z_0) = \sum_j \left[\mathbf{f}_0(x,z_0)\mathrm{rect}\left(\frac{x-x_j}{\Delta_j}\right) \right] \tag{1}$$

in the plane $z = z_0$, with $\mathrm{rect}(x/\Delta) = 1$ for $x \in (x - \Delta/2, x + \Delta/2)$ and $\mathrm{rect}(x/\Delta) = 0$ elsewhere. Equation (1) does not contain any approximation but expresses a lateral decomposition of $\mathbf{f}_0$ into fractions $\mathbf{f}_0^{(j)}$. By a suitable choice of Δ_j one can try to restrict the size of the propagated $\mathbf{f}_0^{(j)}$ to a plane part of the interface, as illustrated in Fig. 5. Because of diffraction, that cannot be done rigorously, but often can be done to a good approximation, as long as the modulation of the interface is smooth enough on the wavelength scale.

---

[*] A similar decomposition is also possible on the interface itself, leading to other versions of local plane interface approximation methods.

Decomposition (1), in combination with the interpretation of the interface as being locally planar, constitutes the basic ingredients of the local plane interface approximation (LPIA). If we use the same arguments for a locally linear grating instead of a locally planar interface, we obtain the local linear grating approximation (LLGA). In general, we refer to LEIA if we decompose a general propagation problem through interfaces into various elementary ones by utilizing a field decomposition of form (1) in combination with a local interpretation of the interface as an elementary one, for which the propagation problem is solved.

Obviously it is not straightforward to determine the parameters of decomposition (1) in such a way that all field fractions $\mathbf{f}_0^{(j)}$ propagate through local plane interfaces only. If we propagate $\mathbf{f}_0^{(j)}$ to the interface, diffraction increases the size of the field and typically it is larger than allowed by the corresponding $\Delta(x_{\mathrm{I}})$. That prevents us from choosing intervals Δ_j small on the wavelength scale. On the other hand, the larger we choose Δ_j, the more results we are likely not able to restrict the propagated $\mathbf{f}_0^{(j)}$ enough in order to illuminate a planar fraction of the interface. That is, we have the same problem as we have for too small Δ_j. This basic argumentation shows that LPIA is only applicable if the modulation of the interface is smooth on the wavelength scale. A more quantitative discussion of the limitations of general LPIA, and also a method to determine an optimum decomposition for a given interface, is subject to further research. In what follows we simplify the discussion by an additional approximation.

Let us neglect wave-optical effects for the short-distance propagation of $\mathbf{f}_0^{(j)}$ from the plane z_0 to the interface; that is, we propagate $\mathbf{f}_0^{(j)}$ by geometrical optics to the interface. Then, we obtain a direct relationship between the size Δ_j and the extent of the field $\mathbf{f}_0^{(j)}$ on the interface as illustrated in Fig. 6, which may be directly restricted to a plane section of the interface. Thus, locally we reduce the problem to the propagation of a plane field through a plane interface. Also the resulting transmitted plane field fragment is propagated to the plane z_1 by geometrical optics, and we arrive at the situation illustrated in Fig. 6. By a suitable choice of intervals in the plane $z = z_0$ we reduce the propagation of the plane field through the general interface to propagations of plane field sections through plane interface sections. We refer to this technique as geometrical optics LPIA. Because the geometrical optics approximation is connected to an electromagnetic field representation $\mathbf{f} = [E_x(x,z), E_y(x,z)]$, this method is suitable for WOE as long as the approximation delivers the accuracy required in an application. Because we neglected diffraction while propagating $\mathbf{f}_0^{(j)}$ from z_0 to z_1, the distance $z_1 - z_0$ should be small enough to provide sufficient accuracy in practice. A more detailed discussion of geometrical optics LPIA can be found in Refs. [14] and [15]. A further refinement that approximates the interface by local spherical fractions instead of planar ones has been suggested [16]. In this method, Coddington's equations for

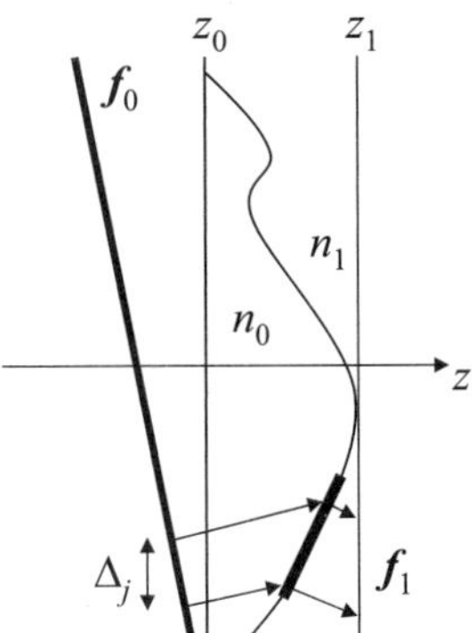

Figure 6 Neglecting diffraction in the region between z_0 and z_1 provides a simple model to propagate a section of the incident plane field through the interface. It is always possible to choose the section Δ_j small enough to ensure that the plane field passes a local plane interface only. Fresnel's formulas and Snell's law of refraction can then be applied directly to obtain the transmitted plane field, which is propagated to the plane z_1.

spherical fields are applied instead of Snell's law for plane waves. Mathematically this technique can be understood as a higher-order Taylor expansion of the interface. Thus it typically allows a lower number of sections in decomposition (1).

It is worth mentioning that the combination of geometrical optics LPIA with a ray-bundle representation of the incident electromagnetic field leads to ray tracing through the interface; it reduces to the basic technique in conventional optical engineering.

Before presenting some numerical results, an important special case of the geometrical optics LPIA should be discussed. Let us assume as a further approximation a paraxial situation: all characteristic angles with respect to the z-axis are small. That includes a plane incident wave with a **k**-vector almost parallel to the z-axis and an interface with small local slopes orthogonal to the z-axis, that is, a thin smooth interface. Then, we may neglect the change of direction of $\mathbf{f}_0^{(j)}$ when passing the interface. Moreover, Fresnel's equations provide identical losses in all sections. The situation is illustrated in Fig. 7. Obviously, the propagation through the interface reduces to the inclusion of the optical path from z_0 to z_1 through any x_I in the phase of $\mathbf{f}_0^{(j)}$. This technique is well known as the thin element approximation (TEA) and is often also called paraxial or scalar approximation in grating theory [18]. It is a scalar technique because the E_x and E_y channels become decoupled and identical. It is a paraxial technique because refraction at the interface is neglected. TEA is the most popular technique in paraxial diffractive optics, even though it is the lowest approximation level of LPIA [19]. It is based on geometrical optics and the paraxial approximation. However, since it was originally connected to a wave-optical field representation, it is commonly understood as a propagation technique

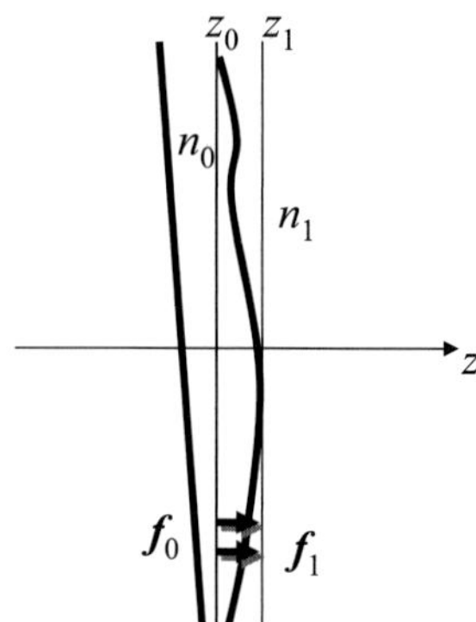

Figure 7 In a paraxial approximation the refraction of the plane field fragment at the interface may be neglected. Moreover, Fresnel's equations provide identical contributions for all plane field fragments. Then the propagation from z_0 to z_1 causes a change of the phase of the field fragment only. This phase change is proportional to the optical path calculated along a ray that passes straight through the interface from z_0 to z_1. This method is called thin element approximation (TEA). It constitutes the lowest approximation level of LPIA.

well suited in wave optics [18]. From a propagation-model point of view, the more general geometrical optics LPIA is more accurate than TEA. But in combination with a ray-bundle representation, it leads to the ray-tracing technique, which is typically understood as less accurate than wave-optical modeling with TEA. This makes it very clear that it is of fundamental importance to distinguish between the modeling of a field itself and its propagation. Moreover, the possible use of geometrical optics LPIA in WOE emphasizes that the electromagnetic field representation is mandatory for modeling in wave-optical engineering, while physical optics field propagation techniques are not always necessary. In what follows we present some numerical results using geometrical optics LPIA.

To compare geometrical optics LPIA with a rigorous reference, we start with the simulation of the propagation through a sinusoidal grating and investigate the diffraction efficiency of the first order for increasing period d. Figure 8 shows the results: the solid line describes the efficiency curve calculated using the rigorous Fourier modal method (see Sect. 18.3). The dashed line is the result of geometrical optics LPIA. For periods down to about five wavelengths it predicts the efficiency very well. Between four and two wavelengths the tendencies are correct, and for smaller periods geometrical optics LPIA fails, because the local curvature of the sinusoidal surface profile becomes too small to allow the use of LPIA. The dotted line represents the result obtained by TEA, which does not predict any dependence of the efficiency on the grating period. This example shows that geometrical optics LPIA constitutes a powerful extension of methods to propagate electromagnetic fields through interfaces. Figure 9 shows analogous results for a blazed (triangular-

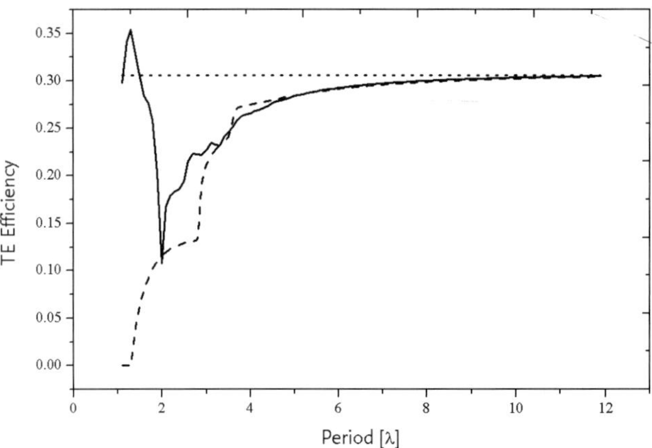

Figure 8 TE mode efficiencies in first order of sinusoidal gratings (refractive index 1.5 embedded in vacuum) of different periods. Solid line: FMM. Dashed line: geometrical optics LPIA. Dotted line: TEA.

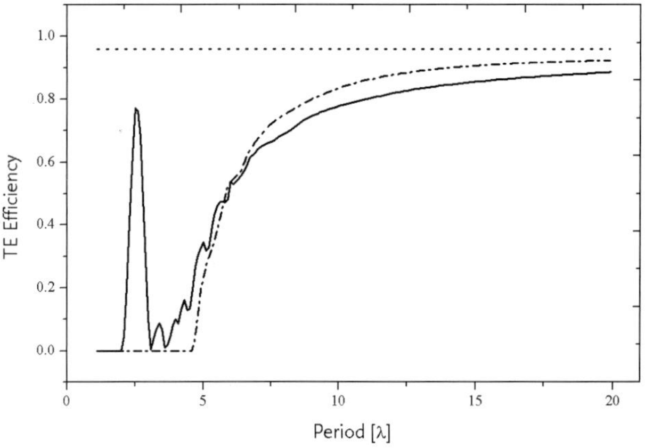

Figure 9 TE mode efficiencies in first order of blazed gratings (refractive index 1.457 embedded in vacuum) of different periods. Solid line: FMM. Dashed line: geometrical optics LPIA. Dotted line: TEA.

profile) grating. Because of the abrupt transitions in the profile of a blazed grating, geometrical optics LPIA does not predict the accurate values even for rather large periods. However, the values given by geometrical optics LPIA are much more accurate than those given by TEA. Moreover, for a period of about five wavelengths, internal reflection occurs and geometrical optics as discussed in this section does not predict any transmitted light, that is, the efficiency falls to zero. However, rigorous theory shows significant efficiency for smaller periods. It has been shown that this effect is due to multiple reflection of the internally reflected light and can be qualitatively predicted also by LPIA if this multiple reflection is included in the model [20]. Both grating simulation experiments demonstrate the capability of the geometrical optics LPIA to model the propagation of harmonic fields through interfaces with high accuracy, as long as the interface does not possess local curvatures strong enough to prevent a plane interface approximation. However, the blazed-grating example shows that if critical locations such as abrupt transitions appear only sporadically, the results are still reasonable.

Though the simulations for gratings are interesting, for a better understanding of the limitations of geometrical optics LPIA, its particular strength lies in the propagation through nonperiodic interfaces. Then grating theory is only applicable for very small structure details and we must rely on analysis by suitable approximate techniques. Often the thin-lens approximation is used in wave-optical simulations. However, we know already that this technique is restricted to paraxial situations. As an example, we consider the propagation of a laser beam through a spherical lens. The parameters are: a focal length of 100 mm; a lens diameter of 1.25 mm; a refractive index of 1.457; and the waist of an ideal Gaussian beam of 300 μm. We like to calculate the M^2 beam parameter in the focal plane of the lens. Therefore, as is typical in WOE, we need access to a wave-optical representation of the electromagnetic field in the form of its complex amplitude, and thus we must propagate the complex amplitude through the lens system. In the paraxial case the resulting [2] factor is equal to 1 for TEA and 1.05 if we use geometrical optics LPIA. This underlines the fact that TEA is a suitable model in the paraxial case. Moreover, in the paraxial case a spherical lens performs well, as we know. Next we change the parameters as follows: a focal length of 5 mm; a lens diameter of 3 mm; and the waist of the ideal Gaussian beam of 500 μm. Obviously, that describes a highly nonparaxial situation, but we still apply a spherical lens. Now we obtain $M^2 = 1$ for TEA and 2.74 if we use geometrical optics LPIA. TEA fails completely and geometrical optics LPIA predicts the effect of the spherical aberrations due to the use of a spherical lens in a nonparaxial system. Figure 10 depicts the spherical aberration directly after the lens, as calculated by geometrical optics LPIA. Because of it geometrical optics, LPIA also leads to the expected shift of the focal distance from 5 mm (TEA) to 4.7 mm.

The examples demonstrate that geometrical optics LPIA is suitable as an approximate propagation technique for gratings and other microstructured interfaces, as long as the appearance of details close to the wavelength do not become dominant, and for wave-optical modeling of the propagation through macroscopic smooth surfaces. Thus, it builds a bridge from diffractive optics to the modeling of systems, which are of major concern in more conventional optical engineering. Because of their proven functionality, local plane interface approximations seem to be powerful modeling techniques in the wave-optical generalization of optical engineering. Though the geometrical optics LPIA as discussed here works well, it still represents a low approximation level of the more general LPIA concept. Still to be

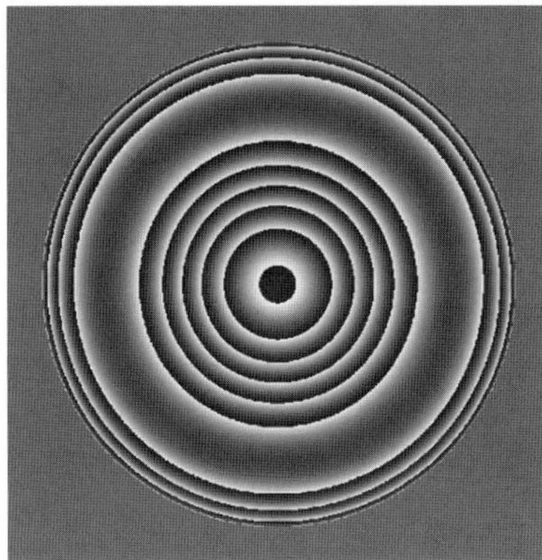

Figure 10 Geometrical optics LPIA is used to determine the spherical aberration, which occurs when a Gaussian beam propagates through a spherical lens with high numerical aperture. The parameters are: focal length, 5 mm; lens diameter, 3 mm; refractive index of lens, 1.457; and waist of ideal Gaussian beam, 500 μm.

investigated is an efficient technique not applying the geometrical optics approximation, thus leading to a truly wave-optical version of LPIA to be formulated. Moreover, the inclusion and role of multiple reflection and refraction should be considered. Several ways to combine the fragments $\mathbf{f}_1^{(j)}$ behind the interface exist and need a more thorough understanding.

So far we have discussed the local elementary interface model considering a plane interface as the elementary one. A very similar technique can be based on local spherical interfaces [16]. The decomposition (1) is also the base for other LEIA approaches, such as the use of a local linear grating as an elementary interface. Also in the resulting local linear grating approximation (LLGA) we apply locally the rigorous solution of the propagation of a plane wave through the elementary interface (in this case a linear grating) and combine the resulting field fragments together in a suitable manner. Some examples of LLGA are presented in Refs. [20] and [21].

Next we turn to an approximation, the strength of which lies in the handling of small features in interfaces. In the case of LEIA, we reduce our general propagation problem by the assumption of laterally independent, elementary propagation channels, which is expressed by Eq. (1). The technique to be discussed does not laterally decompose the incident field, but assumes an independence of the response of the interaction of the incident field with laterally separated structure details.

18.6 Local Independent Response Approximations

The fact that sharp transitions of the physical properties of the diffracting object have substantial effects in the diffraction pattern is well known, and well-established theories based on this fact have been formulated [24]. Classically, this Maggi-Rubinowich or Miyamoto-Wolf boundary diffraction method deals with apertures in opaque screens. The diffracted field behind the aperture is presented as a superposition of a "geometrical wave" and a perturbationlike contribution generated at the abrupt boundary of the aperture.

However, transitions of the profile height in a purely dielectric interface also cause substantial perturbations in the transmitted and reflected fields. Recently these have been considered quantitatively and a method that may be seen as an extension of the classical boundary diffraction technique has been introduced. The method is capable of dealing with transitions of any type [25]. Again the diffracted field is expressed as a superposition of a geometrical wave, given by TEA, and perturbations caused by the transitions. The perturbations are evaluated by applying rigorous diffraction theory.

Figure 11 illustrates rigorously calculated phase and amplitude profiles produced at the plane $z = z_1$ by abrupt vertical transitions of different heights on an otherwise planar interface. The dotted lines give the predictions of TEA, according to which the amplitude remains constant and an abrupt phase shift occurs at the interface, its magnitude being proportional to the surface step height. The solid lines represent rigorously calculated results. Only the propagating part of the field is considered and TE polarization (the electric field points in the y-directions) is considered. In all cases, both the phase and the magnitude of $E_y(x)$ contain rapid spatial oscillations that are damped as one moves laterally away from the transition.

Rather similar though quantitatively different results are obtained in TM polarization (electric field perpendicular to the y-direction). We stress that corresponding results can be obtained for transitions of arbitrary shape; for example slanted, undercut, or smoothly shaped transitions, as well as narrow grooves, or trenches can be analyzed. Different angles of incidence can also be considered.

The results presented in Fig. 2 are valid if the transition is an isolated feature. However, they can also be applied to more general structures, provided that the neighboring transitions are not too close to each other (local independent response approximation, LIRA). Then any general structure consisting of an arbitrary number of individual features can be modeled using a superposition of the transmission function provided by TEA with local corrections [26]. Therefore, LIRA turns out to be a well-suited approach for the propagation of the electromagnetic field represented by $\mathbf{f}_0 = [E_x(x, z_0), E_y(x, z_0)]$ through an interface that consists mainly of transition-type features.

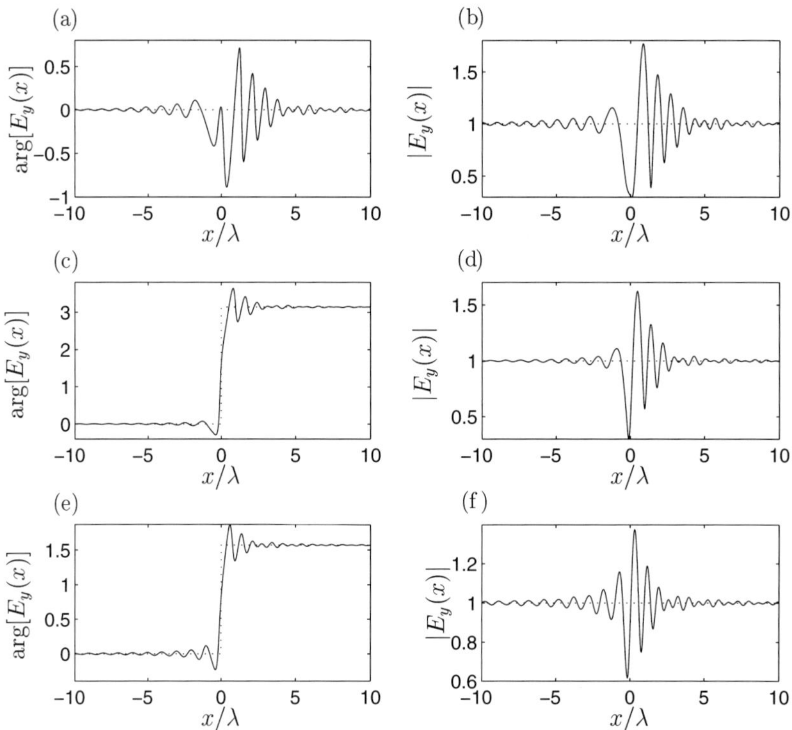

Figure 11 Phase (left) and amplitude (right) of the *y*-component of the propagating part of the electric field immediately behind abrupt vertical surface-relief transitions of different heights, corresponding to phase delays of (a, b) 2π radians, (c, d) π radians, and (e, f) $\pi/2$ radians in TEA (dotted lines). (Courtesy of T. Vallius.)

Mathematically, assuming a *y*-invariant structure and TE polarization, we have

$$E_y(x, z_1) = E_y^{\mathrm{TEA}}(x, z_1) + \sum_{j=1}^{J} P_j(x - x_j, z_1), \tag{2}$$

where E_y^{TEA} is the field given by TEA, J is the number of transitions centered at $x = x_j$. The perturbation terms are of the form

$$P_j(x, z_j) = \begin{cases} E_{yj}^{\mathrm{RIG}}(x, z_1) - E_{yj}^{\mathrm{TEA}}(x, z_1), & \text{if } |x| < \Delta \\ 0 & \text{otherwise} \end{cases}. \tag{3}$$

Here E_{yj}^{TEA} is the result given by rigorous theory for the transition and Δ is a truncation parameter, which can typically chosen to be approximately 10 wavelengths.

It is important to note that the perturbation terms can be precalculated for all different transition shapes found in the structure, and an archive of them can be formed. As a result, the application of Eq. (2) is a simple numerical task, and the computation time does not depend on, e.g., the grating period as in rigorous theory. Moreover, and very important, the method applies equally well to periodic and nonperiodic structures. The high numerical efficiency also facilitates efficient design of complicated profiles in the nonparaxial domain [27].

Figure 12 illustrates LIRA analysis of a $1 \rightarrow 8$ array illuminator. The phases and amplitudes of the propagating part of the field $E_y(x, z_1)$ are plotted here using both LIRA and rigorous diffraction theory, along with the diffraction efficiencies of some central orders. It is seen that, although there are some differences in the transmitted phases and (especially) amplitudes of $E_y(x, z_1)$ predicted by the two methods, the far-field diffraction patterns are remarkably similar, which demonstrates the power of LIRA as a numerically efficient simulation technique in wave-optical engineering.

The approach presented above can also be extended to three-dimensionally modulated structures, which again need not be two-dimensionally periodic [28,29]. In this case one must pay attention to the orientation of the transitions

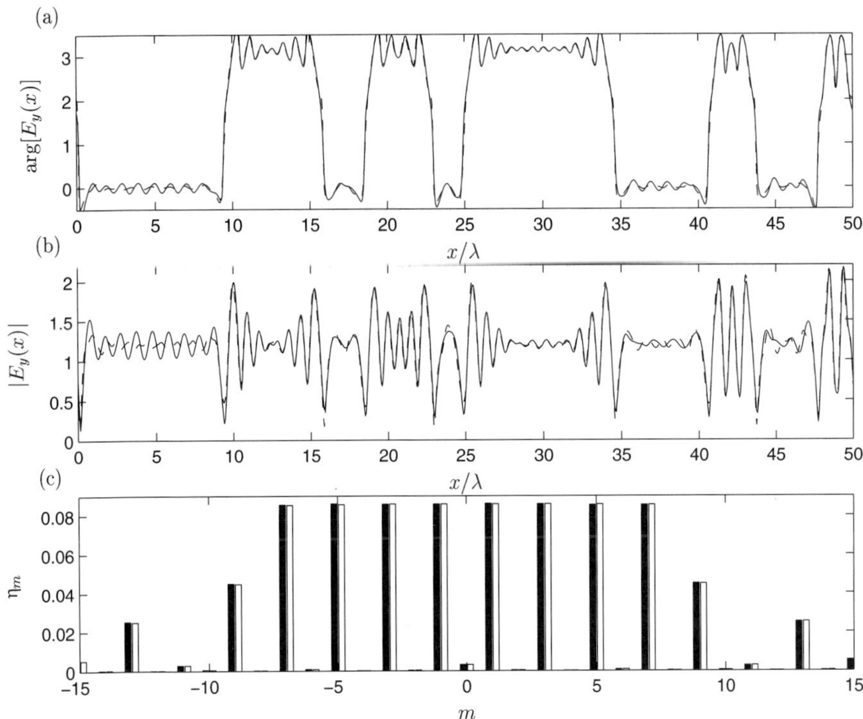

Figure 12 (a) Phase and (b) amplitude of the y-component of the propagating part of the electric field immediately behind a binary $1 \rightarrow 8$ beam-splitter grating with a period $d = 50\lambda$ (solid lines: LIRA, dashed lines: rigorous theory). (c) The diffraction efficiencies of some central orders (solid bars: rigorous theory, empty bars: LIRA). (Courtesy of T. Vallius.)

with respect to the polarization state of the incident field to determine the transformation $\mathbf{f}_0 \to \mathbf{f}_1$.

In general LIRA has been found to be a good approximation when the neighboring features are at least $\sim$1–2 wavelengths apart, but this depends somewhat on the feature shape.

18.7 Extensions to General Fields

Thus far we have considered fully spatially and temporally coherent harmonic fields only. The LEIA and LIRA techniques can be extended to partially coherent light in a rather straightforward manner, although the computational complexity is of course increased.

Let us first consider essentially monochromatic but spatially partially coherent incident fields. Such a field can be represented as a superposition of a partially correlated set of plane waves, in the same manner as an arbitrary fully spatially coherent field can be represented as a superposition of fully correlated plane waves that propagate in different directions [17]. The transmission of each such plane wave is analyzed separately and the results are superimposed according to the original correlations between the incident plane waves.

In the case of spatially coherent but temporally partially coherent beamlike fields, as well as pulses, the analysis is best performed via the frequency domain (the space-frequency domain field representation). The response is analyzed separately for each spectral component of the incident field and the results are then superimposed to get the temporal response by Fourier-transform techniques. In this process the superposition depends on whether coherent pulses, stationary fields, or nonstationary fields between these two extremes are of concern. In the case of conventional pulses, different frequency components are fully correlated, but in the case of stationary fields, they are completely uncorrelated.

18.8 Conclusions

In this paper we have described a class of approximate methods to analyze wave propagation in almost arbitrarily complex optical systems. These methods are based on local application of rigorous diffraction theory to connect the fields at planes on the input and output sides of the interface. Propagation from the output plane of one interface to the input plane of the next one can be treated by geometrical optics or by wave theory, depending on the field. Combination of LEIA and LIRA appears attractive and is a subject of further research. We believe that the methods presented here will form the backbone of much of the wave-optical engineering modeling in the future.

Acknowledgments

We wish to thank Thomas Paul from LightTrans GmbH and Tuomas Vallius from University of Joensuu for providing us with many of the figures.

References

1. M. Born and E. Wolf, *Principles of Optics*, 6th ed., Pergamon Press, Oxford (1980).
2. J.J. Stamnes, *Waves in Focal Regions*, Adam Hilger, Bristol (1986).
3. Y. Li and E. Wolf, "Three-dimensional intensity distributions near the focus in systems of different Fresnel numbers," *J. Opt. Soc. Am. A* **1**, 801 (1984).
4. J. Turunen and F. Wyrowski, Eds., *Diffractive Optics for Industrial and Commercial Applications*, Akademie Verlag, Berlin (1997).
5. F. Wyrowski and J. Turunen, "Wave-optical engineering," in *International Trends in Applied Optics*, A.H. Guenther, Ed., SPIE Press, Bellingham, WA (2002).
6. R. Petit, Ed., *Electromagnetic Theory of Gratings*, Springer-Verlag, Berlin (1980).
7. T.K. Gaylord and M.G. Moharam, "Analysis and applications of optical diffraction by gratings," *Proc. IEEE* **73**, 894–937 (1985).
8. J. Turunen, "Diffraction theory of microrelief gratings," in *Micro-optics: Elemenents, Systems and Applications*, H.P. Herzig, Ed., Taylor & Francis, London (1997).
9. J. Turunen, M. Kuittinen, and F. Wyrowski, "Diffractive optics: Electromagnetic approach," in *Progress in Optics* Vol. 40, E. Wolf, Ed., Elsevier, Amsterdam (2000).
10. L. Li, "Mathematical reflections on the Fourier modal method in grating theory," in *Mathematical Modelling in Optical Science*, G. Bao, L. Cowsar, and W. Masters, Eds., SIAM, Philadelphia (2001).
11. J. Chandezon, D. Maystre, and G. Raoult, "A new theoretical method for diffraction gratings and its numerical application," *J. Opt.* **11**, 235–241 (1980).
12. L. Li, J. Chandezon, G. Granet, and J.P. Plumey, "Rigorous and efficient grating-analysis method made easy for optical engineers," *Appl. Opt.* **38**, 304–313 (1999).
13. T. Vallius, J. Turunen, M. Mansuripur, and S. Honkanen, "Transmission through single subwavelength holes and effects of surface plasmons," *Journal of the Optical Society of America A* **21**, 456–463 (2004).
14. A. Pfeil, F. Wyrowski, A. Drauschke, and H. Aagedal, "Analysis of optical elements with the local plane-interface approximation," *Appl. Opt.* **39**, 3304–3313 (2000).

15. A. Pfeil and F. Wyrowski, "Analysis and design of optical elements with the local plane-interface approximation," *Proc. SPIE* Vol. 4436, 68–79, Bellingham, WA (2001).

16. H. Lajunen, J. Tervo, J. Turunen, T. Vallius, and F. Wyrowski, "Simulation of light propagation by local spherical interface approximation," *Appl. Opt.* **34**, 6804–6810 (2003).

17. L. Mandel and E. Wolf, *Coherence and Quantum Optics*, Cambridge University Press, Cambridge (1995).

18. J.W. Goodmann, *Introduction to Fourier Optics*, McGraw-Hill, New York (1968).

19. H. Aagedal, F. Wyrowski, and M. Schmid, "Paraxial beam splitting and shaping," in *Diffractive Optics for Industrial and Commercial Applications*, J. Turunen and F. Wyrowski, Eds., 165–188, Akademie Verlag, Berlin (1997).

20. E. Noponen, J. Turunen, and A. Vasara, "Electromagnetic theory and design of diffractive-lens arrays," *J. Opt. Soc. Am. A* **10**, 434–443 (1993).

21. E. Noponen and J. Turunen, "Binary high-frequency-carrier diffractive optical elements: Electromagnetic theory," *J. Opt. Soc. Am. A* **11**, 1097–1109 (1994).

22. K. Blomstedt, E. Noponen, and J. Turunen, "Surface-profile optimization of diffractive imaging lenses," *J. Opt. Soc. Am. A* **18**, 521–525 (2001).

23. N. Sergienko, J.J. Stamnes, V. Kettunen, M. Kuittinen, J. Turunen, P. Vahimaa, and A.T. Friberg, "Comparison of electromagnetic and scalar methods for evaluation of diffractive lenses," *J. Mod. Opt.* **46**, 65–82 (1999).

24. A. Rubinowicz, "The Miyamoto–Wolf diffraction wave," *Progress in Optics* Vol. 4, E. Wolf, Ed., North-Holland, Amsterdam (1965).

25. V. Kettunen, M. Kuittinen, and J. Turunen, "Effects of abrupt surface-profile transitions in non-paraxial diffractive optics," *J. Opt. Soc. Am. A* **18**, 1257–1260 (2001).

26. T. Vallius, V. Kettunen, M. Kuittinen, and J. Turunen, "Step-discontinuity approach for non-paraxial diffractive optics," *J. Mod. Opt.* **48**, 1869–1879 (2001).

27. T. Vallius, K. Jefimovs, V. Kettunen, M. Kuittinen, P. Laakkonen, and J. Turunen, "Design of non-paraxial array illuminators by step-transition perturbation approach," *J. Mod. Opt.* **48**, 1869–1879 (2001).

28. T. Vallius, V. Kettunen, M. Kuittinen, and J. Turunen, "Step-transition perturbation approach for pixel-structured non-paraxial diffractive elements," *J. Opt. Soc. Am. A* **19**, 1129–1135 (2002).

29. T. Vallius, "*Advanced numerical diffractive optics*," Dissertation 33, Chapter 8, Department of Physics, University of Joensuu, Joensuu, Finland (2003).

An Honor and a Pleasure

We became aware of Professor Wolf's unparalleled contributions to theoretical optics as soon as we started our own scientific careers in the mid-1980s, learning to know him as the father of modern coherence theory, among his other outstanding achievements. It has long been a source of amazement for both of us how anybody can produce a flow of groundbreaking publications for half a century and still be as productive as ever.

In the late 1980s we both had the honor to meet Professor Wolf personally for the first time, and were truly impressed by his character. We had great pleasure to invite Professor Wolf as the Guest of Honor to the European Optical Society Topical Meetings on Diffractive Optics in 1995 (Prague) and 1999 (Jena). One of us (JT) greatly enjoyed having him as the plenary speaker and the Guest of Honor in the 2003 major event of the International Commission for Optics, the ICO Topical Meeting on Polarization Optics, and showing his "castle" to Professor Wolf and his lovely wife, Marlies. Since Professor Wolf refused to follow his wife into the underground prison cell/torture chamber of the castle on the guided tour, it must be concluded that he has no plans to retire from scientific work!

It is a great pleasure for us to notice in our own recent activities that Professor Wolf's pioneering theoretical work on coherence theory is rapidly becoming recognized as belonging to the basic toolbox of a modern wave-optical engineer. Modeling light sources as being fully coherent or incoherent is simply no longer adequate in a large number of industrial applications.

Professor Wolf's unique discoveries and ideas will influence the development of optics of this century in ways that are yet to be seen.

Frank Wyrowski received his doctoral degree in applied physics from University of Essen, Germany, in 1988. In 1992 he received his Dr. habil. at the same university. He then joined the Philips Company in Eindhoven, the Netherlands, where he worked in the field of laser materials processing with special emphasis on the use of diffractive optics. In 1994 he became head of the Department for *Holography and Diffractive Optics* at Berlin Institute of Optics GmbH, Germany. In 1996 he was appointed Professor of Technical Physics at the Friedrich-Schiller-University of Jena, Germany. In 1998 he founded the company LightTrans GmbH. Prof. Wyrowski has published approximately 100 peer-reviewed articles in the fields of diffractive optics, holography, information processing, halftoning, and optical engineering. He is a co-editor (with J. Turunen) of the book *Diffractive Optics for Industrial and Commercial Applications*. In 2004 he became a fellow of SPIE—The International

Society for Optical Engineering. Currently he is, among other things, a member of the Board of Editors of the *Journal of Modern Optics*, a member of the Board of Directors of SPIE, and a consultant at his company LightTrans GmbH.

Jari Turunen received his doctoral degree in technical physics from Helsinki University of Technology, Finland, in 1990. He then joined Heriot-Watt University in Edinburgh, Scotland, where he developed diffractive elements for optical interconnects for three years. In 1994 he was appointed Professor of Optoelectronics at the University of Joensuu, Finland, and in 2000 he was nominated Professor of Physics at the same university. Professor Turunen has published approximately 150 peer-reviewed articles in the fields of diffractive optics, optical coherence theory, integrated optics, acousto-optics, and surface-acoustic-wave technology. He is a co-editor (with F. Wyrowski) of the book *Diffractive Optics for Industrial and Commercial Applications* and was the recipient of the Väisälä Foundation Physics Prize in 2000, awarded by the Finnish Academy of Sciences and Letters. His main hobby is castle design and construction. Currently he is, among other things, a topical editor of *JOSA A*, responsible for the topical area diffraction and gratings.

❧ CHAPTER 19 ❧

BACKWARD THINKING: HOLOGRAPHY AND THE INVERSE PROBLEM

H. John Caulfield and Michael A. Fiddy

19.1 Introduction: Inverse Problems

The dominant paradigm of holography is best expressed through communication theoretic concepts. Ultimately, this is permitted by placing the information of interest on a spatial carrier wave. The information recorded in a 2D plane relates to a wave emanating from a 3D scattering structure. This fact has tied holography to inverse problems at frequencies for which phase information is tough to record otherwise. Emil Wolf was the first to suggest using holography to solve inverse scattering problems [1] and Hank Carter confirmed the usefulness of his analysis [2]. The inverse problem and its relation to holography has been largely neglected for the past three decades. This is true for both quantitative imaging of scattering obstacles and for the synthesis of scattering structures. We argue here that there are strong reasons present now, that were not present 30 years ago, to reinvigorate this connection to inverse problems. We must extend it well beyond the weak scattering case that Wolf was able to solve analytically. Strong scattering problems will always be ill-posed, meaning that solutions may not exist, or may be ambiguous or highly unstable, so exact solutions are precluded. However, good solutions can be obtained in situations where an optical readout appears to give poor results.

How does one solve inverse problems? As you read this page, you are aware of only a small fraction of the effort your brain must make to accomplish this. Your eye forms an image on your retina where it is detected with a very nonuniformly spaced set of detectors of various sizes. But you never see that detected pattern of pixels. You do not see the 2D pattern at all. Instead you see a moving, colored, smooth 3D world. You see your brain's solution to the inverse problem. Like most inverse problems, perception from vision is an ill-posed problem. That is, there

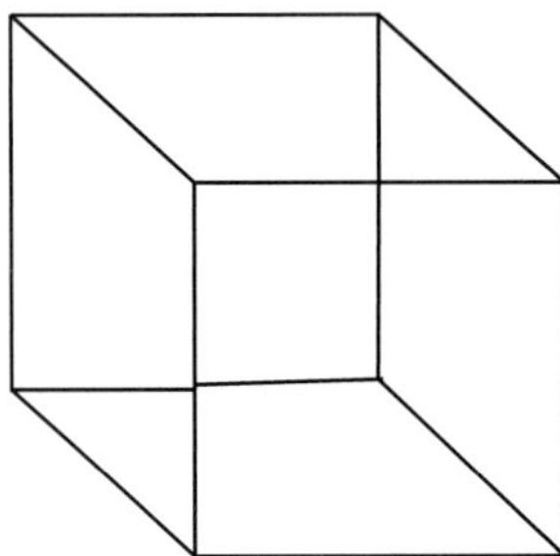

Figure 1 Desperate to show you a 3D scene, your mind has trouble with this singular (ambiguous) drawing. Which vertex is closest to you? Are you sure? Can you see both interpretations simultaneously? Can you force yourself to see this as only a flat drawing?

are infinitely many 3D scenes that could have caused the detected pattern on the retina. The brain must persuade you that one of those is really there, but which one?

Figure 1 illustrates the problem very clearly. What should you see? What do you see? You may not see the truth—a collection of squares and straight lines in a plane. You are most likely to see it as a 3D wire diagram. But it could equally well be either of two such figures. Your brain protects you from ambiguity by making definite decisions. In cases like this one, you see only one interpretation at a time, but what you see switches back and forth between the two 3D interpretations. It is extremely hard to avoid seeing the 3D figure and simply see it as flat.

One would expect that there has to be some irreversible loss of information in the projection of a 3D scene into a 2D image. Some great scientists apparently believe that holography overcomes that problem. They join the vast majority of nonholographers who believe holograms do magic. Of course they do not. The only way to overcome that loss is to inject information into the inversion step, e.g., by computation. If the information we inject is correct, our inverse solver is likely to be accurate. If the information we inject is wrong, all bets are off. Your vision processor offers some lovely illustrations of the errors resulting from incorrect assumptions. One of its assumptions is that light travels in a straight line from the object to the eye. This allows us to see virtual images where no "real" image exists. But it also causes us to see sticks bend as they enter the water and people to appear on the other side of the mirror from us. Your visual processor knows nothing of Snell's law or of reflection.

So now you know the following things about inverse problems:

(1) Brains do them quickly and well most of the time.
(2) The problem is ill posed or singular. There is not enough information to get an unambiguous answer.

(3) The only way to solve the singular inverse problem is to inject information directly or through constraints (i.e., using prior knowledge about the propagation of waves, the anticipated shape of objects, the type of noise, etc., referred to as regularization).

(4) If we make the correct guess about the injected information, our solutions are often fairly good. But, if our assumptions are wrong, our solutions may be nonsense. There really was no one on the other side of the mirror when you looked into it this morning.

These are quite general remarks that apply in all fields. They are mathematical laws, not physical ones.

19.2 2D Holograms of 3D Scenes and the Holographic Principle

Much intellectual mischief has come from not understanding what is discussed in this section. It is tempting to believe that holograms are magic, i.e., that they can produce more information than they record; however there is no getting around two facts.

First, there is more spatial information in a 3D volume than there is on a 2D surface surrounding that volume. The minimum meaningful area is about a wavelength squared and the minimum meaningful volume is about a wavelength cubed. Doubling the radius of a sphere increases the surface area by 4 and increases the volume by 8. If you reject magic, you realize that there is a conceptual problem here somewhere. Holograms routinely produce optically perfect copies of some 3D scenes. Reconciling those observations is critical to the understanding of holography and the inverse problem.

Consider current black hole theory. For convenience, consider a spherically symmetric black hole whose event horizon is a sphere. We do not know much about what is inside a black hole, but we can reason about the entropy change when it absorbs new material. It is provable and universally accepted that the information (negentropy) content of a black hole is essentially equal to the number of minimum meaningful areas (Planck length squared) that is required to tile the event horizon sphere. The 3D internal structure has no more information than is contained on the 2D surface. Magic of some sort has occurred. This turns out to be profoundly important and extremely surprising in cosmology. Not understanding either how holograms work or the details of the solution to the inverse problem in holography, cosmologists call this the *Holographic Principle*.

The hologram analogy is clear since the amount of spatial information in a 2D hologram cannot exceed the area divided by the minimum meaningful area (a wavelength squared), yet it produces lovely 3D images. Apparently magic is

possible after all. The Holographic Principle holds that the universe is like a hologram, in that our seemingly 3D universe could be virtual, reconstructed from fields
and constrained by laws, that are confined to a distant, vast surface. Support for this
concept comes from studies of black holes and arguments that the maximum entropy or information content of any region of space is defined not by its volume but
by its surface area. How much information can be contained in a region of space
is derived by considering a spherical distribution of matter contained by a surface
of area A. For a black hole having an area of A Planck areas, one can show that it
has $A/4$ units of entropy [3]. Since entropy is not supposed to decrease, then the
original distribution of matter confined by a surface A also had no more than $A/4$
units of entropy of information.

Recall that a hologram records the means by which a wavefront that can produce a 3D image is carved out of a very specific reference beam—a concept independently called the *Michelangelo Principle*[*] by Caulfield and Yuri Denisyuk (see
Ref. [4]). The hologram, the reference beam, and the wave equation conspire to
solve the inverse problem jointly. The hologram encrypts the wavefront. The reference beam decrypts it. The wave equation works out the consequences of the
wavefront. One of the pieces of information that must be injected to solve the inverse problem, therefore, is the detailed specification of the reference beam. Get
it wrong, and the 3D object inferred by our inverse solver will be wrong. We will
return to this point subsequently when we discuss ultrasonic holograms. Note that
the usual way we use a hologram to produce a 3D image is simply a specialized
optical processor designed for that purpose. Optical computing is the heart of conventional holography.

Other assumptions are also being made in conventional holography. They are
usually but not universally correct. The most obvious assumption made in conventional hologram viewing is that the index of refraction of the medium in which
the object is located is uniform. Even if the object is transparent (a situation we
will encounter in ultrasonic holography), we still make that implicit assumption.
Sometimes that assumption is quite wrong. The object and surrounding medium
may contain index variations and hence refraction that can render the solution to
inverse problem invalid. In other cases, this assumption is so drastically wrong that
our inverse solver gives pure nonsense. Emmett Leith dealt with that case very
cleverly several decades ago [4]. Consider the situation sketched in Fig. 2. The
light from the object passes through a diffuser en route to the hologram plane. An
optical solver of the inverse problem in this case does not produce a recognizable
image of the object as suggested in Fig. 3. Leith figured out, however, that the ex-
tra information needed to recover the image information could be inserted into the
optical computation by reinserting the diffuser in its proper location, as suggested

---

[*] Recall that when asked how he could carve such a spectacular statue, Michelangelo replied that
he simply chipped away the parts of the marble that did not belong to the statue.

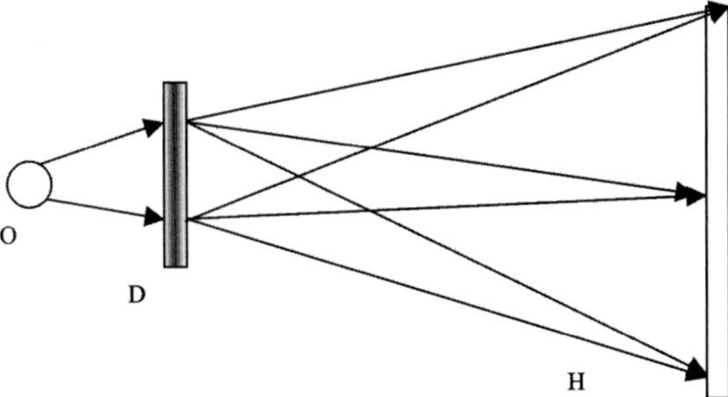

Figure 2 An object (O) hidden behind a diffuser (D) can produce a hologram (H) that does not yield to the simplest inverse problem solver.

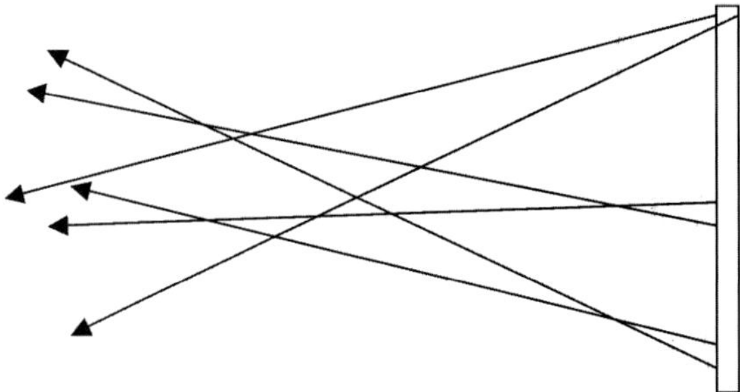

Figure 3 The usual inverse problem solver makes the same assumption your brain makes—that light travels in a straight line from the object to the detection (hologram) plane. Because that assumption did not apply to the situation in Fig. 2, the actual object shape is not recovered.

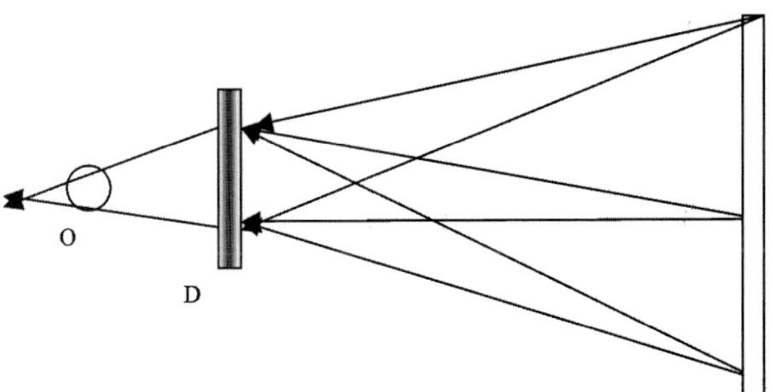

Figure 4 To recover the object information, we need to insert the missing information (in this case, it is inserted physically into the optical processor).

in Fig. 4. This technique has been called optical phase conjugation. It recovers the object information rather well, but imperfectly—as pointed out by Wolf [5] in what was once a controversial but is now universally acknowledged observation.

Thus, there are clearly cases in which the Holographic Principle does not apply. But, there are also cases where it does apply, and Wolf was the first to point this out. Under appropriately restricted circumstances (far field, weak scattering, etc.) the 3D Fourier transform of the object and hence the object itself can be recovered from the 2D recorded wavefront. Well, that is almost the Holographic Principle, if you allow for the fact that the hologram does not record the wavefront but *allows* it to be reconstructed by insertion of a priori information about the recording reference beam. Likewise, it is possible to reconstruct an optically accurate image of a 3D opaque object from that wavefront. The critical thing to notice is the phrase "optically accurate." Our colleague Joseph Shamir has devoted considerable attention to the reconstruction of 3D images from a 2D wavefront, and our treatment is essentially equivalent to his [6]. It is rearranged somewhat to show how the "magic" of recovering a 3D image from a 2D wavefront can occur.

An open problem, however, is when we wish to reconstruct information about a penetrable scattering 3D object. Here, we can recover the wavefront at the surface of the object but inverting the multiply scattered waves to infer structural information about the object is hard. This is clearly an ill-posed problem since without prior knowledge, many objects could be consistent with such data. In conventional holography (whether the inverse problem is solved by an optical computer or an electronic computer), the mechanism enforcing a no-magic dictum is scattering and diffraction. It is not that diffraction must be used in image reconstruction (although it normally is). Rather, it is that scattering and diffraction were involved in the formation of the hologram, and diffraction limitations were there from recording onward.

Let us consider a sphere of radius $r \gg \lambda$, where λ is the wavelength. It contains $M = 4\pi r^2/\lambda^2$ spatially resolvable pixels. The sphere contains $N = 4/3\pi r^3/\lambda^3$ optically definable voxels. The critical number is the ratio $N/M = r/\lambda \gg 1$. By hypothesis, the number of resolvable elements in the recovered 3D image is about M, not N. Only for an $F/1$ system is the recoverable voxel of volume λ^3. For all other image voxels, the volume is greater than λ^3. Roughly, for F-number K, the voxel is an ellipsoid of radius $K\lambda$ and length $K^2\lambda$. Roughly $V/\lambda^3 = K^3$. As most image voxels will have $K \gg 1$, it is not at all surprising that a 3D image with voxels larger than the minimum can be recorded on a 2D surface using minimum sized pixels.

Not being experts in general relativity, we cannot assert that the Holographic Principle for black holes can be resolved in the same way; but it is not unreasonable. It would amount to asserting that space itself is stretched inside a black hole so that the minimum meaningful voxel volume becomes much greater than the cube of the Planck constant.

19.3 Paradigms of Holography

Perhaps only Claude Shannon had more influence on communication theory than Dennis Gabor, the inventor of holography. The two great reinventors of holography, Yury Denisyuk and Emmett Leith, independently employed that interpretation [4]. The world sends a message over a noisy channel—a recording medium. That medium has a channel capacity

$$C = N\log(1 + \text{SNR}), \tag{1}$$

where N is its space-bandwidth product ($A\Omega/\lambda^2$) and SNR is the achievable single pixel signal-to-noise ratio. That is, it can support almost perfect recording of an image with information content C, if that message is ideally encoded. Holography (particularly diffuse light holography of a distant object) is regarded as an almost perfect encryption. The decryption has the wonderful property of being performable by optical computing. The decryption "key" is the re-enactment of the recording situation (geometry, wavelength, etc.). Duplicate those, and good optical decryption is assured.

There are, however, many other paradigms and explanations of holography. Profs. Lohmann, Leith, Denisyuk, and we have long lists of valid interpretations, as different people require different explanations before they can understand holography easily or they arrive at its elegant reality from different perspectives. Examples include regarding the readout of a hologram as recovery of an autocorrelation function, or information storage in a hologram being possible only because a reference wave ensures minimum phaselike properties [7].

One of the interpretations is due almost entirely to our honoree, Emil Wolf. That is the view that holography allows the recording of a complex wavefront coming (ultimately) from an object, and the object details can be recovered from that wavefront by solving the inverse problem—finding out what object had to have been there to have caused the observed wavefront. In general, a computer must be used. Although Wolf's theory [1] and Carter's test of his theory [2] worked brilliantly, the significance of holography in inverse problems has been largely dormant within holography ever since, as mentioned earlier. We note that using holograms to store and retrieve the degree of coherence of a partially coherent field was addressed by Wolf et al. in 1999 [8]. Also, somewhat related to this, the inverse scattering problem for strongly scattering media, namely determining the spectral density of refractive index fluctuations in isotropic random media, was solved using measurements of the degree of coherence of the incident and scattered field [9].

We argue here that so much has changed in the last 30 years that a reexamination of the optical decryption of a holographically stored wavefront deserves to be revisited. In simple problems, holography provides a solution to the inverse problem that is indistinguishable from, and exactly what one would expect from, the communication theory approach. In hard problems, specifically those for which there is not a simple relationship between the scattering properties of an object and the wavefront recorded holographically, an inverse problem solution may be the only way to recover useful object information.

19.4 The Inverse Scattering Problem and Experimental Data

The inverse problem is a favorite among mathematicians because it is very challenging and good solutions are elusive except for relatively trivial problems of weakly scattering penetrable obstacles (the first Born approximation) or those for which permittivity fluctuations are slow on the scale of the wavelength being employed (the Rytov approximation) [10,11]. Methods are often problem specific and hard to implement numerically. Amazon.com gave 20 book titles when searched for "inverse problems." Google gave over one million web sites in response to the same search goal. There is a vital journal devoted to what its title suggests "Inverse Problems."[*]

A connection between an inverse problem and holography is described here. Building on classical optical concepts and the Fourier-based models that are so pervasive in treatments of communication theory and holography, one can write

$$F(x,y) = \iint f(u,v)\exp\left[-ik(ux + vy)\right] du\, dv, \qquad (2)$$

where F is the (far) field scattered from a weakly scattering obstacle f and k is the wavenumber. The wavefront F is complex, requiring both amplitude and phase information to be described fully. An assumption leading to this integral is that the field emerging from the scattering object is well approximated by the product of the incident field (conveniently chosen to be a plane wave) and the complex transmittance of the obstacle. Herein lies a fundamental problem, namely that except for relatively large-scale features (large on the scale of a wavelength) in a relatively thin object (i.e., thin in the direction of propagation on the scale of the wavelength), this relationship does not hold. It is a remarkable fact that it serves us so well under a wide range of experimental situations, when on closer inspection, one might argue from a theoretical standpoint that it should not! By interfering F with a tilted plane wave, P, a square-law recording medium captures the amplitude and phase of F in a real modulation of the recording medium, allowing recovery of f by optical computation of the Fourier transform of $|F + P|^2$, i.e., by scattering from the hologram, provided its features satisfy those same properties listed above.

---

[*] (http://www.iop.org/EJ/S/1/NAL004900/journal/0266-5611)

A more exact statement of the properties we demand of f, is that on passing through the scattering object, f, only weak, or more exactly, single scattering takes place. It is this condition that allows one to represent the exit field as a product of the incident field with a transmittance. When this is not the case, and in principle it hardly ever is, then the integral equation of scattering indicates that the recorded scattered field is actually given by

$$F(x,y) = \iint f(u,v)F_T(u,v)\exp\left[-ik(ux + vy)\right] du\, dv, \tag{3}$$

where one can write $F_T = \Psi/\Psi_0$, $\Psi_0(u,v,k)$ is the incident wave, and $\Psi(u,v,k)$ is the total field within f [10]. Clearly, when $\Psi = \Psi_0$ we satisfy the first Born approximation, and Eq. (3) reduces to Eq. (2).

Intuitively the nature of Eq. (3) is not surprising, since when multiple scattering occurs in an obstacle, emerging wavefront modifications arise with potential ambiguity as to their origin. This is precisely the difficulty in solving the inverse problem. In some communication problems, signals are severely corrupted by multipath and noise but a model for the channel can often be developed, allowing a deconvolution or optimization problem to be solved, which permits the restoration of the information of interest. This is not so straightforward in 2D and 3D (i.e., imaging) problems because of the inherent nonlinearity of the equation above, as Ψ the total field depends on f.

Holography provides a possible solution to this challenging inverse problem, like no other. Consider a scattering obstacle f that varies only slightly over a range of wavelengths $\Delta\lambda$. Knowing $\Delta\lambda$, one can make a succession of holograms in this wavelength range, and reconstruct a set of "secondary source" functions, $f(u,v)F_T(u,v,k)$, from each one. Since f is approximately constant through this ensemble, and the set of $F_T(u,v,k)$ are essentially random functions, a dynamic sweeping of the optical frequency over $\Delta\lambda$ should reveal the fixed features of the object f.

When a hologram is made, the field emerging from the obstacle is scattered through 4π steradians, but only the part falling on the recording medium is easily recoverable. If one were to locate a recording medium around the entire scatterer, then for a given incident field direction given by $\mathbf{r}_0$, data on $f(u,v)F_T(u,v,k)$ lie on the Ewald circle (or sphere in 3D) indicated below in the uv domain, or k-space as it is often called. This has been elegantly described by Emil Wolf in both his original paper and several since [11,12].

In Fig. 5(a) we indicate the data measured from a single backscatter or monostatic experiment and in Fig. 5(b) how changing the wavelength of the incident wave varies the locus of data points in k-space. Changing the direction of the incident wave and changing the incident wavelength allow a multiplexing of information about $f(u,v)F_T(u,v,k)$ to be intercepted and recorded holographically, but

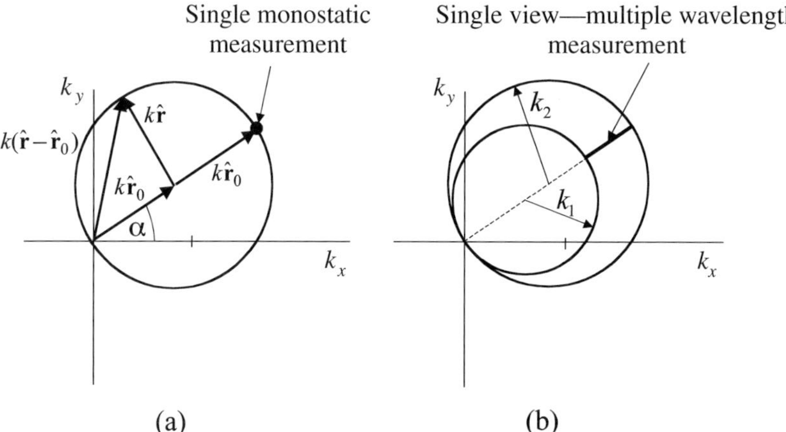

Figure 5 The Ewald sphere analysis.

this still represents only a limited covering of k-space, given the finite physical size of the hologram. This inevitably limits the information one can hope to retrieve about f and contributes to the ill-posed nature of this kind of inverse problem as mentioned earlier. This is precisely the case even when the first Born approximation is valid, but frequently when dealing with a very weak scatterer, sufficient information is recovered about f that this low-pass filtering can be overlooked. Also, if optical decryption is employed, the eye-brain processor can often be relied upon to incorporate prior knowledge of f and thereby appear to recreate a better reconstruction than it perhaps would be if a crude measure like a mean square error were the criterion for judgment.

19.5 Resolution, Phase, and Evanescent Waves

When Wolf pointed out, for example, that evanescent waves are not recorded (in the vast majority of holograms, anyway), it caused a great uproar in the optics community that puzzled Wolf and us. Many years later, we realized why. Part of the mystique of holography is that it records the whole of the information and leads to optically perfect copies of the object. The consequences of Fig. 5 should dispel that. A sponsor of our work on ultrasonic holography was shocked to learn that optical reconstruction of the wavefront (even if it were done at the proper far infrared wavelength to match the recording ultrasound wavelength) would produce only a restricted image of the object, because of the data truncation and noise. The fact that many holographic reconstructions look so good, is in part because when viewing by eye, as mentioned above, we are quite forgiving about noise and our brain employs a lot of prior knowledge in interpreting what we see. We appear to produce a fairly accurate image of a penetrable object that exists in a uniform index of refraction environment. That object, however, is not the physical object, as we explained above. Even using diffuse illumination, i.e., a large number of incident wavefront directions, the associated spatial multiplexing that occurs is still limited by the extent of k-space captured by the hologram, which in turn dictates the resolution achievable in the reconstruction. It is this limited resolution, or more exactly the point spread function associated with the geometry of the hologram, that can make the separation of the twin images encoded by the hologram so difficult, even when numerically rather than optically decrypting. Also, it is well known [13], that over a finite range of incident field directions, classes of weakly scattering structures exist that are nonscattering scatterers, and hence would not contribute to the hologram. This is a concrete example of the lack of uniqueness associated with inverse scattering problems.

We also note here that Wolf and Gbur have shown recently that the scattered field in the far field, F, or the scattering amplitude, can be calculated in a stable fashion from the complex amplitude measured over a plane at an arbitrary distance from the scattering object [14] without the need to include evanescent waves. Another important observation is that within the limits of the Rytov approximation, object recovery is possible when a hologram cannot be made by using two scattered field intensities at two closely spaced planes beyond the scatterer [15]. A comparison between the first Born and Rytov approximation can be found in Fiddy [10], and Cairns and Wolf [16].

19.6 Broadening the Concept of Holography

There are several reasons to believe that what Prof. Wolf started will become a major part of a broadened definition of holography. We list some of those reasons below (in no particular order).

(1) *Moore's Law* has worked ever since Wolf's papers and it shows no signs of failing. Computers are much faster, smaller, and cheaper than when he first suggested solving the inverse problem using holograms. Moore's Law looks good until at least 2010.

(2) *Special computer hardware*, such as DSP (digital signal processor) chips and Xyron's ZOTS (zero overhead transaction processor)[*] have sped up computations even more.

(3) *Algorithms* have improved dramatically. The critical Fourier transform operation has moved from the super fast FFT to the even faster FFTW.[†]

(4) *Digital image processing* is far advanced since Wolf's paper, providing us with tools that improve the resulting image.

---

[*] *(http://www.xyronsemi.com/xnews.shtml)*

[†] *(http://www.fftw.org)*

(5) *3D displays* unimaginable then are on the market now, allowing a dramatic viewing of the computed image.

(6) *Image compression* has improved tremendously, allowing us to send even complex 3D images anywhere we wish at quite low bandwidths.

(7) *Optical processing* has continued to evolve and the technology has become more sophisticated, allowing all-optical processing of information to be more easily realized.

The problems that can benefit from a deeper look at the relationship between holography and a wider class of inverse problems are becoming clearer and we examine them next.

19.7 Holography in Other Fields: Ultrasound

Invented soon after Gabor's work by Hungarian Pal Greguss [17], ultrasonic holography has never quite lived up to the hopes we all had for it. Ultrasonic holography would be vitally important if it could yield beautiful 3D images as light does. Here are some of the reasons.

(1) Ultrasound penetrates metal, flesh, plastics, etc., so it can look inside things that are opaque to the visible part of the electromagnetic spectrum.

(2) Unlike other penetrating means, such as x rays, ultrasound is not dangerous at the insonification levels needed.

(3) Ultrasound measures properties of soft tissues such as impedance and velocity that may be far more valuable in medical diagnosis than, say, x-ray opacity.

Unfortunately, ultrasonic holography has been dominated by the wrong paradigm, namely optical reconstruction. As a result, it does not fulfill its promise. The problem should be clear to any holographer who has read thus far—optical reconstruction of a wavefront from a hologram formed in ultrasound has a huge ratio of recording-to-reconstructing wavelengths (by a factor of about 1000). This leads to a demagnification factor of about 1000 in an optically decrypted visible image. That problem is avoided in the lateral directions by using a lens to form an image plane hologram. But the image is "squashed" in depth by about 1000. Depth information is gained by viewing holograms of images taken at multiple depths through the object (achieved by moving the object relative to the system or moving the system with respect to the object). A human observer of such a sequence of 2D slices (tomograms) in real time can imagine what the 3D object might look like, but the hologram produces no useful 3D image. One is reminded of the pioneering work of Hildebrandt, who developed a medical imaging system based on the interference on a water surface of scattered ultrasound from a submerged human body

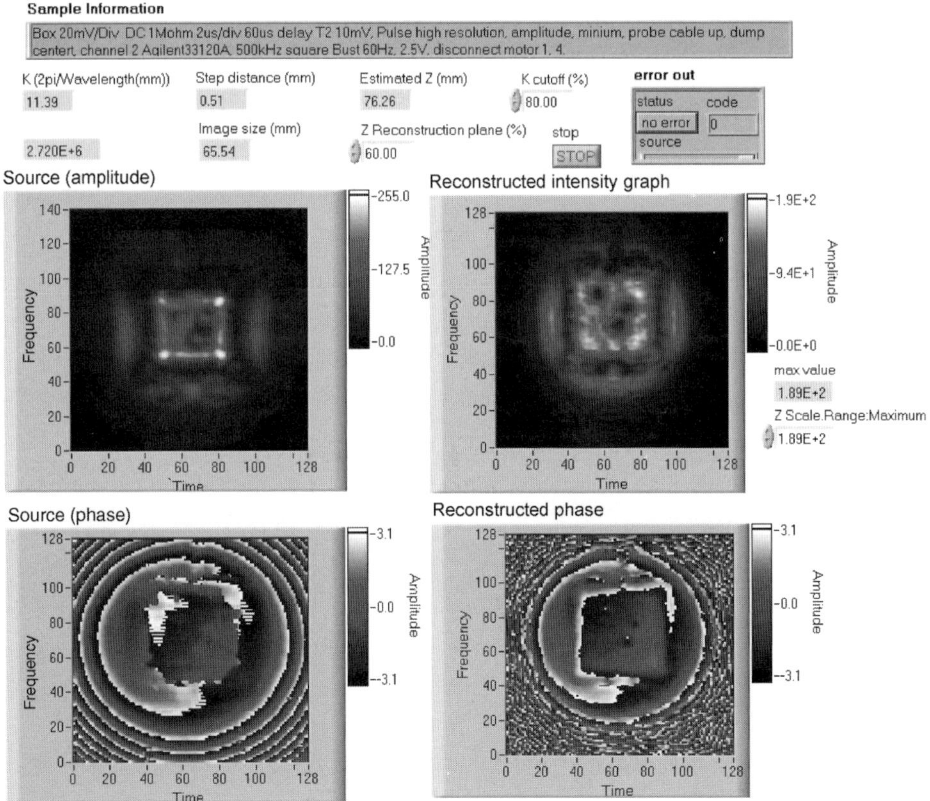

Figure 6 This figure shows a second plane recovered from the same hologram—something quite new to ultrasonic holography. Indeed any plane can be viewed using a single hologram, just as is done in optical holography.

with a reference wave. Light scattering, even at grazing incidence off the ripple pattern, produced severely compressed optical images. Much simpler ultrasonic cameras based on A-scans or B-scans are available, so why bother with ultrasonic holograms?

The obvious way to avoid that problem and get beautiful 3D images from a hologram is to apply a numerical inverse problem solution at the same (virtual) wavelength as used in recording. In the computer, half-millimeter waves are as visible as half-micron waves. Actually, computer images are better in that they are easier to manipulate, store, transmit, compress, search, etc., than optical images. Moreover, once holograms have been made at several wavelengths, one can in principle reconstruct an ensemble of images of $f(u,v)F_T(u,v,k)$ and extract the common function, the desired image, f. It is assumed that the scattering parameter of interest, such as impedance, is not frequency dependent over the wavelength range used. With ultrasound, tissue is highly multiply scattering and both longitudinal and shear waves add to the complexity of the inverse problem, the tissue requiring a tensor description. Nevertheless, 3D images of good quality should be amenable to extraction from such data.

Several years ago, one of us (HJC) began work on this topic at the Karmanos Cancer Institute in Detroit. Lacking a suitable detector array, we went to discrete arrays of transducers that are spaced much greater than Nyquist in both directions. We placed those arrays on "paddles" that could be used to enclose the regions of the breast where cancers were most likely to occur. Then we transmitted ultrasound from various transducers one at a time and received the transmitted sound on all transducers on the other paddle. We then reversed the roles of the two paddles. The resulting data set was then subjected to an inverse problem solver injecting all of the a priori data available, in compliance with what any Bayesian would expect. The resulting tomograms were remarkable (Fig. 7).

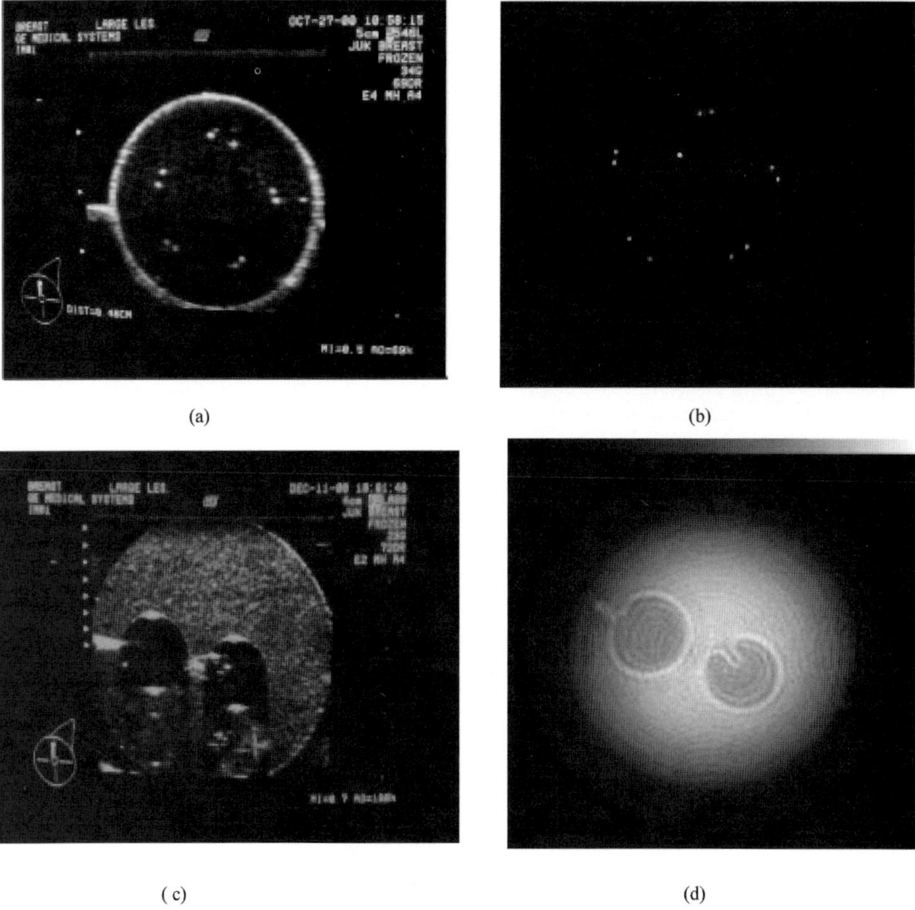

Figure 7 On a microcalcification phantom, a medical B-scan gave the image (a), while reflection inverse solution gave (b). On a cyst phantom, medical B-scan gave (c), while transmission inverse methods gave (d).

Since then, we have been engaged in building a suitable transducer array to allow direct recording of the amplitude and phase of the sound wave at any instant. The inverse problem for this hologram will be easier to solve and perhaps lead to a medically interesting system.

The potential of using holography to solve strongly scattering inverse problems with microwave radiation or the recently popular THz waves, for seeing through walls and dielectric media follows the same logic as for ultrasound.

19.8 Electron Holograms

Gabor invented holography as a means to obtain good electron images. An electron hologram is an interference pattern between the scattered object wave, F, transmitted through the object and a reference wave passing (ideally unchanged) through the field-free vacuum and the object itself. For a weakly scattering situation, the recorded hologram is $|F + P|^2 = |F + \Psi_0|^2$. In other words, the reference wave is assumed to be the incident wave. For a strongly scattering obstacle this is not going to be the case. Since field emission transmission electron microscopes are commercially available, electron holography has become quite feasible. They are used to study electromagnetic microfields such as the magnetic fields of ferromagnetic nanowires or of thin ferromagnetic films. Unlike the wavelength-difference problem described above for ultrasound, Gabor seems to have envisioned optical reconstruction with its attendant huge magnification. This is extremely appealing and much work has been done to try to introduce a scattering structure that would produce an off-axis reference beam without the associated electron charging and perturbations that go with it. With or without an off-axis reference wave, the appeal of an electron hologram, optically decrypted, is enormous. The hologram wavefronts can always be numerically reconstructed to give the phase and amplitude variations of the transmitted electron wave with respect to an unperturbed reference wave. The application of inverse solutions to electron beam holograms appears to have been pioneered by Lichte [19–24].

19.9 Conclusions

In this paper we have built on the pioneering work for our honoree, Emil Wolf, in connecting holography to solving inverse problems. We have examined how holography remains a powerful tool for a much broader class of strongly scattering and severely ill-posed problems than might have been considered to date. Recovery of weakly scattering obstacles or thin masks or slowly modulating media or surfaces, permits a straightforward recovery of object information from the inversion of the information about the wavefront stored in a hologram. Essentially, this information can be backpropagated into the object domain either numerically or by optical computing/decryption. The function recovered in the object domain is not simply related to the object itself for more strongly scattering structures, and the solution of this all-pervasive inverse problem has made little progress over the last 30 years. Methods relying on iterative Born techniques, modified gradient methods or higher-order Born and Rytov approximations have proved numerically challenging and still somewhat limited in their range of applicability. The inevitable problem of limited noisy data from which to work, adds to the ill-posed nature of all inverse problems, and the finite size of a hologram and its associated noise impact holographic inversions also. However, the fact remains that a hologram conveniently captures relevant wavefront information, which then forms boundary conditions, so to speak, on how that captured wave, albeit of limited extent and quality, propagates to and from the three-dimensional object domain. Varying the incident wave direction and wavelength allows considerable additional information about the obstacle to be gathered, and in compliance with Bayes' two laws that were introduced here, offers numerical or optical computational opportunities to solve this inverse problem. Ensembles of images registered and overlaid in 3D will build up an image of f, while the field term, F_T, rather like a bipolar noise term, averages out. Moreover, when recording holograms using waves other than electromagnetic waves, the same remarkable possibilities exist. Indeed, when recording holograms using wavelengths shorter than visible radiation, dynamic reconstructions employing visible wavelength sweeping may provide visually stunning 3D images of strong scatterers, on a time frame with which numerical electronic computing, as opposed to analog optical computation, can never compete.

By thinking backwards, with respect to the information content of a hologram, Prof. Wolf's ideas have, and will continue to move us forward!

References

1. E. Wolf, "Three-dimensional structure determination of semi-transparent objects from holographic data," *Opt. Comm.* **1**, 153–156 (1969).
2. W.H. Carter, "Computational reconstruction of scattering objects from holograms," *J. Opt. Soc. Amer.* **60**, 306–314 (1970).
3. J. Bekenstein, "Information in the holographic universe," *Scientific American*, August, 59–65 (2003).
4. H.J. Caulfield, "Holography: A tribute to Yuri Denisyuk and Emmet Leith," in *SPIE Proc.* Vol. 4737, Bellingham, WA (2002).
5. G.S. Agarwal, A.T. Friberg, and E. Wolf, "Scattering theory of distortion correction by phase conjugation," *J. Opt. Soc. Amer.* **73**, 529–538 (1983); see also N. Taket and M.A. Fiddy, "Theory of distortion correction by phase conjugation for strongly scattering media," *J. Opt. Soc. Amer. A* **4**, 112–117 (1987).

6. H.J. Caulfield, J.W. Goodman, and J. Shamir, "Progress in holography: Special issue," *J. Opt. Soc. Amer. A* **9**, 1138 (1992).

7. M.A. Fiddy and H.J. Caulfield, "The significance of phase and information," in this volume.

8. E. Wolf, T. Shirai, G. Agarwal, and L. Mandel, "Storage and retrieval of correlation functions of partially coherent fields," *Opt. Lett.* **24**, 367–369 (1999), based on earlier work by M. Lurie *J. Opt. Soc. Amer.* **56**, 1369 (1966), and **58**, 614 (1968).

9. S.A. Ponomarenko and E. Wolf, "Solution to the inverse scattering problem for strongly fluctuating media using partially coherent light," *Opt. Lett.* **27**, 1770–1772 (2002).

10. M.A. Fiddy, "Inversion of optical scattered field data," *J. Phys. D* **19**, 301–317 (1986).

11. E. Wolf, "Principles and development of diffraction tomography," in *Trends in Optics*, A. Consortini, Ed., Academic Press (1996).

12. R.P. Porter, "Generalized holography with application to inverse scattering and inverse source problems," in *Progress in Optics*, Vol. 27, E. Wolf, Ed., 315–397, North-Holland, Amsterdam (1989).

13. E. Wolf and T. Habashy, "Invisible bodies and uniqueness of the inverse scattering problem," *J. Mod. Optics* **40**, 785–792 (1993).

14. E. Wolf and G. Gbur, "Determination of the scattering amplitude and of the extinction cross-section from measurements at arbitrary distances from the scatterer," *Phys. Lett. A* **302**, 225–228 (2002).

15. G. Gbur and E. Wolf, "Diffraction tomography without phase information," *Opt. Lett.* **27**, 1890–1892 (2002).

16. B. Cairns and E. Wolf, "Comparison of the Born and the Rytov approximations for scattering on quasi-homogeneous media," *Opt. Comm.* **74**, 284–289 (1990).

17. P. Greguss, *Ultrasonic Imaging*, Focal Press, Elsevier Science, Amsterdam (1969).

18. B.P. Hildebrand and B.B. Brenden, "An introduction to acoustical holography," *Plenum Press*, NY (1974).

19. H. Lichte, "Electron holography approaching atomic resolution," *Ultramicroscopy* **20**, 293–304 (1986).

20. H. Lichte, "Electron image plane off-axis electron holography of atomic structures," *Advances in Optical and Electron Microscopy* **12**, 25–91 (1991).

21. H. Lichte, "Holography—just another method of image processing?" *Scanning Microscopy Suppl.* **6**, 433–440 (1992).

22. H. Lichte, E. Völkl, and K. Scheerschmidt, "Electron holography: II First steps of high resolution electron holography into materials science," *Ultramicroscopy* **47**, 231–240 (1992).

23. A. Orchowski and H. Lichte, "High resolution electron holography of real structures at the example of a $\sum = 13$ grain boundary in gold," *Ultramicroscopy* **69**, 199–209 (1996).

24. W.D. Rau and H. Lichte, "Electron holography surmounts resolution limit of electron microscopy," *Phys. Rev. Letters* **74**, 399–402 (1995).

An Emil Wolf Anecdote

I can now publicly tell the story of how, without even knowing it, Emil saved my job over 35 years ago when he was already the most prominent opticist in the world and I was a neophyte researcher at Texas Instruments Incorporated (TI). What has changed to allow this is the publication by Sean Johnston of the story of George Stroke and the history of holography.[*] That article chronicles some of Stroke's efforts to claim credit for other people's work in this field.

I had just published something called "Local reference beam holography." It was clever work and, as I later found, invented in parallel by my friend Tom Cathay. The day after my paper was published, the Director of Central Research at TI called me into his office for an important conference.

He announced gravely that a prominent professor had accused me of stealing his work.

"Don't tell me. Let me guess," I said. Is that professor's name Stroke?"

Being convinced now that the charges against me were true, he said, "Yes, you must know of his work in the area of your paper."

"No, I know of no work of his in this field, but I do know of him," I replied. "He likes to claim other people's work. Let me show you what the most prominent opticist in the world said."

I then brought him a copy of Emil's review of George's newly issued book on holography in the *Journal of the Optical Society of America*. It said in part: "This sort of referencing, which presents unsupported claims and not real references at all evokes a question of ethics. The reader should reflect on the scientific climate that would be created if other authors were to adopt such a way of referencing."

My job was saved.

While others praise Emil for his brilliance and the fact that he is a nice human being, I want to praise him for the courage to say the truth even if it seemed risky.[†]

---

[*] S.E. Johnston, "Telling tales: George Stroke and the historiography of holography," *History and Technology* **20**, 29–51 (2004).

[†] Mary Warga, the late Executive Director of the Optical Society of America, searched long and hard for someone of such distinction and integrity to write that review.

H. John Caulfield holds research appointments at Fisk University and Alabama A&M University, is CTO of half a dozen small companies, and consults for numerous other companies and nonprofit institutions. Widely published (author, editor, or coauthor of 12 books, 36 book chapters, more than 200 refereed journal papers, and many popular articles along with inventor or coinventor on 27 U.S. patents), he has been widely honored, including these honors from SPIE: Fellow, President's Award, Governors' Award, and the Dennis Gabor Medal. He also edited the SPIE journal *Optical Engineering*.

CHAPTER 20

SEVERAL CONTROVERSIAL TOPICS IN CONTEMPORARY OPTICS: DISPERSIVE PULSE DYNAMICS AND THE QUESTION OF SUPERLUMINAL PULSE VELOCITIES

Kurt E. Oughstun

20.1 Historical Development of Dispersive Wave Theory

Early considerations of the wave theory of light represented the optical wavefield as a coherent superposition of monochromatic scalar wave disturbances. Dispersive wave propagation was first considered in this manner by Sir William R. Hamilton [1] in 1839, when the concept of group velocity was first introduced. In that paper, Hamilton compared the phase and group velocities of light, stating that [1] *"the velocity with which such vibration spreads into those portions of the vibratory medium which were previously undisturbed, is in general different from the velocity of a passage of a given phase from one particle to another within that portion of the medium which is already fully agitated; since we have velocity of transmission of phase $= s/k$, but velocity of propagation of vibratory motion $= ds/dk$,"* where s denotes the angular frequency and k the wavenumber of the disturbance in Hamilton's notation. Subsequent to this definition, Stokes [2] posed the concept of group velocity as a "Smith's Prize examination" question in 1876. Lord Rayleigh [3] then mistakenly attributed the original definition of the group velocity to Stokes, stating that *"when a group of waves advances into still water, the velocity of the group is less than that of the individual waves of which it is composed; the waves appear to advance through the group, dying away as they approach its anterior limit. This phenomenon was, I believe, first explained by Stokes, who regarded the group as formed by the superposition of two infinite trains of* *waves, of equal amplitudes and of nearly equal wavelengths, advancing in the same direction.*" Rayleigh [4] then applied these results to explain the difference between the phase and group velocities of light with respect to their observability, arguing that "*Unless we can deal with phases, a simple train of waves presents no mark by which its parts can be identified. The introduction of such a mark necessarily involves a departure from the original simplicity of a single train, and we have to consider how in accordance with Fourier's theorem the new state of things is to be represented. The only case in which we can expect a simple result is when the mark is of such a character that it leaves a considerable number of consecutive waves still sensibly of the given harmonic type, though the wavelength and amplitude may vary within moderate limits at points whose distance amounts to a very large multiple of* λ *... From this we see that ... the deviations from the simple harmonic type travel with the velocity dn/dk and not with the velocity n/k,*" where n denotes the angular frequency and k the wavenumber in Rayleigh's notation.

The distinction between the signal and group velocities originated in the early research by Voigt [5,6] and Ehrenfest [7] on elementary dispersive waves, and by Laue [8], who first considered the problem of dispersive wave propagation in a region of anomalous dispersion where the absorption is both large and strongly dependent upon the frequency. Subsequently, the distinction between the front and signal velocities was considered by Sommerfeld [9,10], who showed that no signal could travel faster than the vacuum speed of light c and that the signal front progressed with the velocity c in a dispersive medium, as well as by Brillouin [11,12] who provided a detailed description of the signal evolution in a Lorentz model dielectric. In his 1907 paper, Sommerfeld [9] stated that (as translated by Brillouin [12]): "*It can be proven that the signal velocity is exactly equal to c, if we assume the observer to be equipped with a detector of infinite sensitivity, and this is true for normal or anomalous dispersion, for isotropic or anisotropic medium, that may or may not contain conduction electrons. The signal velocity has absolutely nothing to do with the phase velocity. There is nothing, in this problem, in the way of Relativity theory.*"

The "signal velocity" referred to here by Sommerfeld has since become known as the front velocity, the signal velocity being described by Brillouin in terms of the moment of transition from the forerunner evolution to the signal evolution in the dynamical field evolution due to an initial Heaviside step function modulated signal. Brillouin's asymptotic analysis, based upon the then newly developed method of steepest descent due to Debye [13], provided the first detailed description of the frequency dispersion of the signal velocity in a single resonance Lorentz model dielectric. Based upon this seminal analysis, Brillouin concluded that [11,12] "*The signal velocity does not differ from the group velocity, except in the region of anomalous dispersion. There the group velocity becomes greater than the velocity in vacuum if the reciprocal c/U < 1; it even becomes negative ... Naturally, the group velocity has a meaning only so long as it agrees with the signal velocity. The negative parts of the group velocity have no physical meaning ... The signal velocity is always less than or at most equal to* *the velocity of light in vacuum.*" This research then established the asymptotic theory of pulse propagation in dispersive, absorptive media. An essential feature of this approach is its adherence to relativistic causality through careful treatment of the dispersive properties of both the real and imaginary parts of the complex index of refraction.

At approximately the same time, Havelock [14,15] completed his research on wave propagation in dispersive media based upon Kelvin's stationary phase method [16]. It appears that Havelock was the first to employ the Taylor series expansion of the wavenumber (κ in Havelock's notation) about a given wavenumber value κ_0 that the spectrum of the wave group is clustered about, referring to this approach as the group method. In addition, Havelock [15] stated that "*The range of integration is supposed to be small and the amplitude, phase, and velocity of the members of the group are assumed to be continuous, slowly varying, functions of* κ." This research then established the group velocity method for dispersive wave propagation. Since the method of stationary phase [17] requires that the wavenumber be real valued, this method cannot properly treat causally dispersive, attenuative media. Furthermore, notice that Havelock's group velocity method is a significant departure from Kelvin's stationary phase method with regard to the wavenumber value κ_0 about which the Taylor series expansion is taken. In Kelvin's method, κ_0 is the stationary phase point of the wavenumber κ, while in Havelock's method κ_0 describes the wavenumber value about which the wave group spectrum is peaked. This apparently subtle change in the value of κ_0 results in significant consequences for the accuracy of the resulting group velocity description.

There were then two different approaches to the problem of dispersive pulse propagation: the asymptotic approach (based upon Debye's method [12] of steepest descent) that provided a proper accounting of causality but was considered to be mathematically unwieldy without any simple, physical interpretation, and Havelock's group velocity approximation (based upon Havelock's reformulation of Kelvin's asymptotic method of stationary phase) that violates causality but possesses a simple, physically appealing interpretation. It is interesting to note that both methods are based upon an asymptotic expansion technique but with two very different approaches: the method of stationary phase that relies on coherent interference, and the method of steepest descent that relies on attenuation.

The asymptotic approach was revisited in 1930 by Baerwald [18], who reconsidered Brillouin's description of the signal velocity in causally dispersive systems, and also in 1941 by Stratton [19], who reformulated the problem in terms of the Laplace transform and derived an alternate contour integral representation of the propagated signal. Stratton appears to have first referred to the forerunners described by Sommerfeld and Brillouin as precursors. The first experimental measurement of the signal velocity was attempted by Shiren [20] in 1962, using pulsed microwave ultrasonic waves within a narrow absorption band. His experimental results were *"found to lie within theoretical limits established by calculations of Brillouin and Baerwald."* However, a more detailed analysis of these experimental results by Weber and Trizna [21] indicated that the velocity measured by Shiren was in reality that for the first precursor and not the signal. Subsequent research by Handelsman and Bleistein [22] in 1969 provided a uniform asymptotic description of the arrival and initial evolution of the signal front. The first experimental measurements of the precursor fields originally described by Sommerfeld and Brillouin were then published by Pleshko and Palócz [23]; it is apparent that they were the first to refer to the first and second precursors as the Sommerfeld and Brillouin precursors, respectively. Although their experiments were conducted in the microwave domain on waveguiding structures with dispersion characteristics that are similar to those described by a single resonance Lorentz model dielectric, the results established the physical propriety of the asymptotic approach.

The group velocity approximation was also refined and extended during this same time period, most notably by Eckart [24], who considered the close relationship between the method of stationary phase and Hamilton-Jacobi ray theory in dispersive but nonabsorptive media. This opened a new avenue of research into complex rays and dispersion surfaces [25–29] with direct application to pulse propagation in spatially inhomogeneous, lossy dispersive media, but only when the material attenuation is nondispersive. The equivalence between the group velocity and the energy transport velocity in loss-free media was then established [30–32], thereby providing a physical basis for the group velocity in lossless systems. In addition, the quasi-monochromatic or slowly-varying envelope approximation was precisely formulated by Born and Wolf [33] in the context of partial coherence theory. This completed the mathematical and physical basis for the group velocity approximation, which was then generalized [34] and extended [35,36] to any order of dispersion.

The description of the velocity of energy transport through a causally dispersive medium, originally considered by Brillouin, was reinvestigated by Schulz-DuBois [37] in 1969 and finally by Loudon [38], who in 1970 provided a correct description of the energy velocity in a single-resonance Lorentz model dielectric. This description showed that the energy velocity and group velocity are different in the region of anomalous dispersion in causally dispersive dielectrics. Based upon this critical result, Sherman and Oughstun [39,40] then presented a complete physical description of dispersive pulse dynamics in a causally dispersive medium in terms of the energy velocity and attenuation of time-harmonic waves. This description reduces to the approximate group velocity description in the limit as the material loss goes to zero. Based upon an extension [41] of Loudon's energy velocity to a multiple-resonance Lorentz model dielectric, this energy velocity description may then be directly extended to this more general physical situation.

Finally, the signal velocity in a dispersive medium was considered once again by Trizna and Weber [42] in 1982 in a numerical study, concluding that *"one cannot realistically separate signal from precursor."* A precise definition and asymptotic description of the signal velocity was then given by Oughstun and Sherman [43] in 1988 in connection with the modern asymptotic theory of dispersive pulse propagation [43–46]. The physical propriety and related observability of this signal velocity definition was then demonstrated [46,47] through a numerical experiment, thereby completing the physical interpretation of the asymptotic description.

Recently published research [48,49] has identified the space-time domain within which the group velocity approximation is valid. The group velocity description of dispersive pulse propagation is based on both the slowly varying envelope approximation and the Taylor series approximation of the complex wavenumber about some characteristic angular frequency ω_c of the initial pulse at which the pulse spectrum is peaked. The quasi-monochromatic or slow varying envelope approximation, precisely formulated by Born and Wolf in the context of partial coherence theory, is a hybrid time and frequency domain representation [50] in which the temporal field behavior is separated into the product of a temporally slow varying envelope function and an exponential phase term whose angular frequency is centered about ω_c. The envelope function is assumed to be slow varying on the time scale $\Delta t_c \sim 1/\omega_c$, which is equivalent [51] to the assumption that its spectral bandwidth $\Delta\omega$ is sufficiently narrow that the inequality $\Delta\omega/\omega_c \ll 1$ is satisfied. The frequency dependence of the wavenumber may then be approximated by the first few terms of its Taylor series expansion about the characteristic pulse frequency ω_c with the assumption [35,36,50] that improved accuracy can always be obtained through the inclusion of higher-order terms; this assumption has been proven incorrect [48,49], optimal results being obtained using either the quadratic or the cubic dispersion approximation of the wavenumber.

Because of the slowly varying envelope approximation, together with the neglect of the frequency dispersion of the material attenuation, the group velocity approximation is invalid in the ultrashort pulse regime in a causally dispersive material or system, its accuracy decreasing as the propagation distance Δz increases. This is in contrast with the modern asymptotic description whose accuracy increases in the sense of Poincaré as the propagation distance increases. There is then a critical propagation distance $z_c > 0$ such that the group velocity description using either the quadratic or cubic dispersion approximation provides an accurate description of the pulse dynamics when $0 \le \Delta z < z_c$ (the accuracy increasing as $\Delta z \to 0$), while the modern asymptotic theory provides an accurate description when $\Delta z > z_c$ (the accuracy increasing as $\Delta z \to \infty$). This critical distance z_c depends upon both the dispersive material and the input pulse characteristics, including the pulse shape, temporal width, and characteristic angular frequency ω_c. For example, $z_c = \infty$ for the trivial case of vacuum for all pulse shapes, whereas $z_c \sim z_d$ for an ultrashort, ultrawideband pulse in a causally dispersive dielectric with e^{-1} penetration depth z_d at the characteristic frequency ω_c of the input pulse.

In spite of these results, the group velocity approximation remains central to the description of ultrashort pulse dynamics in both linear and nonlinear optics with little regard to its domain of validity. This is seemingly supported by the apparent agreement between experimental measurements and the results predicted by the group velocity approximation. Herein lies the central controversy considered in this paper. Related to this is the controversy regarding the possibility of superluminal pulse velocities, since the group velocity can assume any value between $-\infty$ and $+\infty$ in a region of anomalous dispersion.

20.2 Integral Representation of the Propagated Pulse and Causality

The propagated plane wave, pulsed optical field $A(z,t)$ that results from the initial pulse $A(z_0,t) = f(t)$ at the plane $z = z_0$ is given by the *Fourier-Laplace integral representation* [46]

$$A(z,t) = \frac{1}{2\pi} \int_C \tilde{f}(\omega) e^{i[\tilde{k}(\omega)\Delta z - \omega t]} d\omega \tag{1}$$

for all $\Delta z \geq 0$. Here $\tilde{f}(\omega)$ is the temporal angular frequency spectrum of the initial pulse function $f(t) = A(z_0,t)$, C denotes the contour of integration $\omega = \omega' + ia$, where $\omega' = \Re\{\omega\}$ ranges from negative to positive infinity, and a is a constant greater than the abscissa of absolute convergence for $f(t)$. The spectrum $\tilde{A}(z,\omega)$ of the optical field $A(z,t)$ satisfies the *Helmholtz equation*

$$[\nabla^2 + \tilde{k}^2(\omega)]\tilde{A}(z,\omega) = 0, \tag{2}$$

where

$$\tilde{k}(\omega) \equiv \beta(\omega) + i\alpha(\omega) = \frac{\omega}{c} n(\omega) \tag{3}$$

is the complex wavenumber of the plane wave field with *propagation factor* $\beta(\omega) = \Re\{\tilde{k}(\omega)\}$ and *attenuation coefficient* $\alpha(\omega) = \Im\{\tilde{k}(\omega)\}$, and where $n(\omega) = \sqrt{\varepsilon(\omega)}$ denotes the complex index of refraction of the dispersive dielectric with relative dielectric permittivity $\varepsilon(\omega)$ and relative magnetic permeability $\mu = 1$. Here $\Re\{*\}$ denotes the real part and $\Im\{*\}$ the imaginary part of the quantity $*$ appearing within the brackets. Causality [52,53] requires that the real and imaginary parts of the dielectric permittivity $\varepsilon(\omega) = \varepsilon'(\omega) + i\varepsilon''(\omega)$ satisfy the *Kramers-Kronig relations* (or *Plemelj formulae*)

$$\varepsilon'(\omega) = \frac{1}{\pi}P\int_{-\infty}^{\infty}\frac{\varepsilon''(\zeta)}{\zeta-\omega}d\zeta, \quad \varepsilon''(\omega) = -\frac{1}{\pi}P\int_{-\infty}^{\infty}\frac{\varepsilon'(\zeta)}{\zeta-\omega}d\zeta, \tag{4}$$

where the symbol P indicates that the principal part of the indicated integration is to be taken. The frequency dependence of the real part of the dielectric permittivity is then seen to imply the frequency dependence of the imaginary part, and vice versa. The consequences of this intimate interrelationship have far-reaching implications in the analysis of linear dispersive pulse propagation phenomena. As stated by Bohren and Huffman [54], these consequences of the Kramers-Kronig relations "*are almost trivial, but it is disturbing how often they are blithely ignored.*"

For the asymptotic theory of dispersive pulse propagation [12,46], the integral representation given in Eq. (1) is expressed as

$$A(z,t) = \frac{1}{2\pi}\int_C \tilde{f}(\omega)e^{(\Delta z/c)\phi(\omega,\theta)}d\omega \tag{5}$$

with *complex phase function*

$$\phi(\omega,\theta) \equiv i\frac{c}{\Delta z}[\tilde{k}(\omega)\Delta z - \omega t] = i\omega[n(\omega)-\theta] \tag{6}$$

and nondimensional *space-time parameter* $\theta \equiv ct/\Delta z$. The fact that this exact integral representation of the propagated optical wavefield satisfies relativistic causality is expressed by the following theorem (originally proved by Sommerfeld [9,10] for a Heaviside unit step function modulated signal in a single resonance Lorentz model dielectric and later extended [43,46] to an arbitrary plane wave pulse in a general causally dispersive medium):

> If $f(t) = 0$ for all $t < 0$ and if $\Re\{i\omega[n(\omega)-\theta]\} \to -\infty$ as $|\omega| \to \infty$ with $\omega'' > 0$ for all $\theta < 1$, then $A(z,t) = 0$ for all $\Delta z > 0$ when $\theta < 1$.

This precise statement of the luminal arrival of the signal front then proves that any information that may be present in the signal will follow at some later space-time point with $\theta > 1$.

20.3 Havelock's Classical Group Velocity Approximation

The group velocity approximation is a hybrid time and frequency domain representation [50] in which the temporal pulse behavior is separated into the product of a slowly varying envelope function and an exponential phase term whose angular frequency is centered about some fixed characteristic frequency of the initial pulse. Consider then the specific form of the initial pulse at the plane $z = z_0$ that is given by $f(t) = u(t)\sin(\omega_c t + \psi)$ with envelope $u(t)$ and constant carrier frequency ω_c. The propagated plane wave pulse is then given by the Fourier-Laplace integral representation [46]

$$A(z,t) = \frac{1}{2\pi}\Re\left\{ ie^{-i\psi}\int_C \tilde{u}(\omega - \omega_c)e^{i[\tilde{k}(\omega)\Delta z - \omega t]}d\omega \right\} \tag{7}$$

for all $\Delta z \geq 0$, where $\psi = 0, \pi/2$ for either a cosine or sine wave carrier, respectively. Here $\tilde{u}(\omega)$ is the temporal angular frequency spectrum of the initial pulse envelope function $u(t)$. In the slowly varying envelope approximation, the envelope function $u(t)$ is assumed to be slowly varying on the time scale $\Delta t_c \sim 1/\omega_c$, which is equivalent [51] to the quasi-monochromatic approximation that the spectral bandwidth $\Delta\omega$ of $\tilde{u}(\omega)$ is sufficiently narrow that the inequality $\Delta\omega/\omega_c \ll 1$ is satisfied. The complex wavenumber $\tilde{k}(\omega)$ is then expanded in a Taylor series about the carrier frequency ω_c with the assumption [34–36,50] that this series may be truncated after a few terms with some undefined error. It is typically assumed that the attenuation coefficient $\alpha(\omega) = \Im\{\tilde{k}(\omega)\}$ is sufficiently small that its frequency dispersion is entirely negligible in comparison to that for the propagation factor $\beta(\omega) = \Re\{\tilde{k}(\omega)\}$, so that $\alpha(\omega) \approx \alpha(\omega_c)$; this is entirely compatible with the stationary phase foundation [14–16] of the group velocity description, which requires that the wavenumber be real-valued since this frequency-independent attenuation factor may then be taken outside of the integration. In addition, the propagation factor $\beta(\omega)$ is typically represented by the *quadratic dispersion approximation*

$$\beta(\omega) \approx \beta(\omega_c) + \beta^{(1)}(\omega_c)(\omega - \omega_c) + \frac{1}{2!}\beta^{(2)}(\omega_c)(\omega - \omega_c)^2, \tag{8}$$

where $\beta^{(j)}(\omega) \equiv \partial^j\beta(\omega)/\partial\omega^j$. The coefficient $\beta^{(1)}(\omega_c)$ is the inverse of the *group velocity* evaluated at the carrier frequency, while the coefficient $\beta^{(2)}(\omega_c)$ describes the so-called *group velocity dispersion* [50]. With this substitution, Eq. (7) becomes

$$A(z,t) \approx \frac{e^{-\alpha(\omega_c)\Delta z}}{[2\pi\beta^{(2)}(\omega_c)\Delta z]^{1/2}}\Re\left\{ e^{i[\beta(\omega_c)\Delta z - \omega_c t - \psi - 3\pi/4]} \right.$$

$$\left. \int_{-\infty}^{\infty} u(t')\exp\left[-i\frac{(\beta'(\omega_c)\Delta z + t' - t)^2}{2\beta^{(2)}(\omega_c)\Delta z} \right]dt' \right\}, \tag{9}$$

and the pulse phase propagates through the dispersive medium at the phase velocity

$$v_p(\omega) \equiv \frac{\omega}{\beta(\omega)},$$

(10)

while the pulse envelope propagates through the dispersive medium at the group velocity

$$v_g(\omega) \equiv \frac{1}{\partial\beta(\omega)/\partial\omega}$$

(11)

evaluated at the input pulse carrier frequency ω_c. The propagated pulse shape is then seen to be proportional to the Fresnel transform of the initial pulse envelope shape in this approximation. In particular, the propagated pulse structure is seen to depend on the value of the time scale parameter [34] $T_F \equiv \sqrt{2\pi\beta^{(2)}(\omega_c)\Delta z}$, which relies on the value of the group velocity dispersion. If the initial pulse width T and propagation distance Δz are such that the inequality $T \gg T_F$ is satisfied, then the scale of variation of $u(t)$ is much larger than T_F and the expression (9) for the propagated pulse may be approximated as

$$A(z,t) \approx -u(t - \beta^{(1)}(\omega_c)\Delta z)e^{-\alpha(\omega_c)\Delta z}\sin(\beta(\omega_c)\Delta z - \omega_c t - \psi).$$

(12)

The pulse is then seen to propagate undistorted in shape with an overall amplitude decay when $T \gg T_F$, corresponding to the geometric optics approximation in the near zone of an aperture. At the opposite extreme when $T \ll T_F$ the contribution from the quadratic phase terms in the exponential of the integrand in Eq. (9) is negligible in comparison to the linear phase term, and the propagated pulse becomes proportional to the Fourier transform of the initial pulse envelope function, corresponding to the Fraunhofer or far zone in the analogous diffraction problem. Broader input pulses then propagate undistorted for larger distances than do shorter input pulses in a given dispersive material.

It has long been argued [35,36,50] that improved accuracy in the group velocity approximation can always be obtained through the inclusion of higher-order terms in the Taylor series approximation of the propagation factor. In particular, the cubic dispersion approximation

$$\beta(\omega) \approx \beta(\omega_c) + \beta^{(1)}(\omega_c)(\omega - \omega_c) + \frac{1}{2!}\beta^{(2)}(\omega_c)(\omega - \omega_c)^2$$

$$+ \frac{1}{3!}\beta^{(3)}(\omega_c)(\omega - \omega_c)^3$$

(13)

is also widely used [55] because the cubic term introduces a small degree of asymmetry into the propagated pulse. However, it has recently been established that *"With the exception of a small neighborhood about some characteristic frequency of the initial pulse, the inclusion of higher-order terms in the Taylor series approximation of the complex wavenumber in a causally dispersive, attenuative medium beyond the quadratic approximation is practically meaningless from both the physical and the mathematical points of view* [49]." In particular, optimal results in the global sense are obtained with either the quadratic dispersion approximation given in Eq. (8) or with the cubic dispersion approximation given in Eq. (13); neither is really better than the other, simplicity favoring the quadratic dispersion approximation, while asymmetric pulse distortion requires that the cubic dispersion approximation be employed.

20.4 The Modern Asymptotic Theory of Dispersive Optical Pulse Propagation

The form of the contour integral appearing in Eq. (5) is most appropriate for asymptotic analysis as the propagation distance Δz becomes large on some suitable physical scale that is a characteristic of both the dispersive material and the initial pulse structure. This asymptotic description [9–12,43–46] is obtained from a mathematically well-founded expansion about several variable saddle points of the complex phase function $\phi(\omega, \theta)$ defined in Eq. (6) that provide the exponentially dominant contributions to the integral representation (5) of the propagated field. A complete understanding of these saddle-point dynamics, together with the manner in which they interact with the initial pulse spectrum, provides a detailed, accurate description of the entire dynamical evolution of the propagated pulse in the dispersive, absorptive medium for all $\Delta z > z_c$, the accuracy of this approximation increasing in the sense of Poincaré [17] as the propagation distance increases above some critical propagation distance $z_c > 0$.

The set of saddle points of the complex phase function $\phi(\omega, \theta) = i\omega \times [n(\omega) - \theta]$ is determined by the condition that $\phi(\omega, \theta)$ be stationary at a saddle point, in which case $\phi'(\omega, \theta) = 0$, where the prime denotes differentiation with respect to ω, so that

$$n(\omega) + \omega n'(\omega) = 0. \tag{14}$$

The solutions of this saddle-point equation then give the desired saddle-point locations in the complex ω-plane as a function of the space-time parameter $\theta = ct/\Delta z$. The saddle points will then evolve with time at any fixed propagation distance Δz. Because of the general symmetry relations [43,46] $n(-\omega) = n^*(\omega^*)$ and $\phi(-\omega, \theta) = \phi^*(\omega^*, \theta)$ that are satisfied by a causal medium, if $\omega_j(\theta)$ is saddle point of $\phi(\omega, \theta)$, then so is $-\omega_j^*(\theta)$.

An appreciation of the physical significance of the saddle points can be obtained from the relation $(\Delta z/c)\phi(\omega,\theta) = i[\tilde{k}(\omega)\Delta z - \omega t]$ for the complex phase function. Upon differentiating this expression with respect to ω, one obtains $(\Delta z/c)\phi'(\omega,\theta) = i\{[\partial\tilde{k}(\omega)/\partial\omega]\Delta z - t\}$. Since $\phi'(\omega,\theta) = 0$ at each saddle point $\omega_j(\theta)$ of ϕ, then

$$\frac{\Delta z}{t} = \frac{1}{(\partial\tilde{k}(\omega)/\partial\omega)_{\omega=\omega_j}} = \tilde{v}_g(\omega_j) \tag{15}$$

and the *complex group velocity* is real-valued at the saddle points.

With the saddle-point locations known for $\theta \geq 1$, the asymptotic analysis then proceeds by expressing the integral representation given in Eq. (5) in terms of an integral $I(z,\theta)$ with the same integrand but with a new contour of integration $P(\theta)$ to which the original contour C may be deformed [43–46]. By Cauchy's residue theorem, the integral representation (5) of $A(z,t)$ and the contour integral $I(z,\theta)$ are related by

$$A(z,t) = I(z,\theta) - \Re\{2\pi i\Lambda(\theta)\}, \tag{16}$$

where

$$\Lambda(\theta) = \sum_p \operatorname*{Re\,s}_{\omega=\omega_p}\left\{\frac{1}{2\pi}\tilde{f}(\omega)\exp[(\Delta z/c)\phi(\omega,\theta)]\right\} \tag{17}$$

is the sum of the residues of the poles that were crossed in the deformation from C to $P(\theta)$, and where

$$I(z,\theta) = \frac{1}{2\pi}\int_{P(\theta)}\tilde{f}(\omega)\exp[(\Delta z/c)\phi(\omega,\theta)]d\omega. \tag{18}$$

For the asymptotic evaluation of the contour integral $I(z,\theta)$ as $\Delta z \to \infty$, the path $P(\theta)$ is taken as a union of Olver-type [55] paths [43–46] with respect to a subset of the set of saddle points of $\phi(\omega,\theta)$ subject to the condition that $P(\theta)$ evolves continuously with θ for all $\theta \geq 1$. Not all saddle points in this set may be appropriate in the asymptotic description because the Olver-type paths with respect to them may not be deformable to the original contour C owing, for example, to the presence of the branch cuts of $\phi(\omega,\theta)$. Such saddle points are said to be inaccessible; otherwise they are said to be accessible. The dominant accessible saddle point (or points) refers to the saddle point (or points) that has the largest value of $\Re\{\phi(\omega,\theta)\}$ at it, and hence, has the least exponential attenuation associated with it. By comparison, Brillouin's interpretation of this asymptotic method required that the contour of integration C be deformed so that it lay along the entire path of steepest descent through the accessible saddle points of the complex phase function. Olver's theorem proved this requirement unnecessary, with important consequences regarding the physical significance of whether or not a particular pole singularity is crossed in deforming the contour C to $P(\theta)$.

If $\omega_j(\theta)$ and $-\omega_j^*(\theta)$ are the dominant accessible first-order saddle points at a particular value of θ, and if they are isolated from each other as well as from all other saddle points of $\phi(\omega, \theta)$ at that value of θ, then the asymptotic approximation of $I(z, \theta)$ as $\Delta z \to \infty$ is obtained from Olver's theorem as

$$I(z, \theta) \sim \Re \left\{ \left[-\frac{c}{2\pi\Delta z \phi^{(2)}(\omega_j, \theta)} \right]^{1/2} \tilde{f}(\omega_j) \exp\left[(\Delta z/c)\phi(\omega_j, \theta) \right] \right.$$

$$\left. + \left[-\frac{c}{2\pi\Delta z \phi^{(2)}(-\omega_j^*, \theta)} \right]^{1/2} \tilde{f}(\omega_j^*) \exp\left[(\Delta z/c)\phi(\omega_j^*, \theta) \right] \right\}. \quad (19)$$

The dynamical evolution of the saddle points then provides a nearly complete description of the dynamical evolution of the transient field behavior associated with dispersive pulse propagation.

The residue contribution to $A(z, t)$ is nonzero only if $\tilde{f}(\omega)$, or $\tilde{u}(\omega - \omega_c)$, has poles. Consider the case of the pulse envelope modulated carrier wave given in Eq. (7) in which case Eq. (17) becomes

$$\Lambda(\theta) = \sum_p \operatorname*{Re}_{\omega = \omega_p} s \left\{ \frac{1}{2\pi} i \exp(-i\psi) \tilde{u}(\omega - \omega_c) \exp\left[(\Delta z/c)\phi(\omega, \theta) \right] \right\}. \quad (20)$$

If the envelope function $u(t)$ of the initial field $A(z_0, t)$ at the plane $z = z_0$ is bounded for all time t, then $\tilde{u}(\omega - \omega_c)$ can have poles only if $u(t)$ does not tend to zero too fast as $t \to \infty$. Hence, the implication of a nonzero residue contribution is that the field $A(z, t)$ oscillates with angular frequency ω_c for positive times t at the plane $z = z_0$ and will tend to do so at larger values of $z > z_0$ for sufficiently large t. As a result, this contribution to the asymptotic behavior of the propagated field describes the steady-state behavior of the signal. The arrival of this signal contribution is determined by the dynamics of the dominant saddle point that becomes exponentially negligible in comparison to the pole contribution. A detailed knowledge of the saddle-point dynamics for a given dispersive material is then seen to be a critical ingredient for a detailed description of dispersive pulse propagation in that material, not just for the transient field behavior described by Eq. (19) but also for the steady-state behavior described in Eq. (20).

In a multiple resonance Lorentz model [56] dielectric the complex index of refraction is described by [46]

$$n(\omega) = \left(1 - \sum_{j=0}^{N} \frac{b_{2j}^2}{\omega^2 - \omega_{2j}^2 + 2i\delta_{2j}\omega}\right)^{1/2}, \qquad (21)$$

where ω_{2j} is the undamped resonance frequency, b_{2j} is the plasma frequency, and δ_{2j} the phenomenological damping constant for the $(2j)^{\text{th}}$ resonance line of the dielectric material. This causal model [52] provides an accurate description of both the normal and anomalous dispersion phenomena observed in homogeneous, isotropic, locally linear optical materials. The regions of anomalous dispersion approximately extend over each frequency domain $(\omega_{2j}, \omega_{2j+1})$, where $\omega_{2j+1} \equiv \sqrt{\omega_{2j}^2 + b_{2j}^2}$. In this case the saddle point Eq. (14) has at least two sets of saddle points that are symmetrically situated about the imaginary axis. One pair of saddle points (the distant saddle points [11,12,43,46]) evolve in the high-frequency region $|\omega| \geq \omega_{2N+1}$ of the complex ω-plane above the uppermost absorption band of the material, while another pair of saddle points (the near saddle points [11,12,43,46]) evolve in the low-frequency region $|\omega| \leq \omega_0$ of the complex ω-plane below the lowermost absorption band of the dielectric. If the dielectric material is described by multiple resonance lines, then additional middle saddle points will appear in the region $|\omega| < \omega_{2N}$ below the uppermost absorption of the dielectric. The asymptotic description of the propagated pulse may then be expressed either in the form

$$A(z,t) \sim A_S(z,t) + A_m(z,t) + A_B(z,t) + A_c(z,t) \qquad (22)$$

as $\Delta z \to \infty$, or by an expression that is a superposition of expressions of the form given in Eq. (22).

Here $A_S(z,t)$ denotes the contribution from the distant saddle points with nonuniform asymptotic approximation given by Eq. (19) for $\theta > 1$ and is referred to as the first or Sommerfeld precursor. This nonuniform approximation breaks down at $\theta = 1$, when the distant saddle points are at infinity. The uniform asymptotic description [22,45,46] of the Sommerfeld precursor, uniformly valid in the space-time parameter $\theta = ct/\Delta z$ for all $\theta \geq 1$, must then be used in place of Eq. (19) for the initial pulse evolution. The instantaneous angular oscillation frequency of the Sommerfeld precursor is approximately given by the real part of the distant saddle point location [43,46] in the right half of the complex ω-plane. The Sommerfeld precursor then describes the signal front that arrives at $\theta = 1$ with infinite oscillation frequency (but zero amplitude for a finite energy input pulse) and consequently propagates at the speed of light c in vacuum. As

θ increases away from unity, the amplitude of the Sommerfeld precursor rapidly increases to a maximum value and then decreases monotonically for all larger θ, while the instantaneous oscillation frequency monotonically decreases from infinity and approaches the real frequency value at the upper end of the uppermost absorption band as $\theta \to \infty$. The Sommerfeld precursor is then seen to be a characteristic of the high frequency response of the dispersive material.

The field component $A_B(z,t)$ appearing in Eq. (22) denotes the contribution from the near saddle points with nonuniform asymptotic approximation given by Eq. (19) with just the single upper near saddle point contribution over the space-time domain $1 < \theta < \theta_1$ and with both near saddle point contributions over the space-time domain $\theta > \theta_1$. The critical value [11,12,43–46] θ_1 denotes the space-time point at which the two first-order near saddle points coalesce into a single second-order saddle point, at which point the nonuniform expansion breaks down. This contribution to the total field evolution is referred to as the first or Brillouin precursor. The uniform asymptotic description [45,46,57] of the Brillouin precursor, uniformly valid in the space-time parameter $\theta = ct/\Delta z$ for all $\theta > 1$, must then be used in place of Eq. (19) to describe the pulse evolution about the space-time point $\theta = \theta_1 \approx \theta_0$, where $\theta_0 = n(0)$ describes the space-time point at which the upper near saddle point crosses the origin. The instantaneous oscillation frequency of the Brillouin precursor is approximately given by the real part of the near saddle point location [43,46] in the right half of the complex ω-plane. The amplitude of the Brillouin precursor is found [43–46] to rapidly increase to its peak value at $\theta \cong \theta_0$, where the field attenuation vanishes, and then to decrease monotonically for larger space-time values, while the instantaneous oscillation frequency monotonically increases from its zero value at the space-time point $\theta = \theta_1$ and approaches the real frequency value $\omega_0' \cong \sqrt{\omega_0^2 - \delta_0^2}$ that characterizes the lower end of the lower absorption band as $\theta \to \infty$. The Brillouin precursor is then seen to be a characteristic of the low frequency response of the dispersive material. Finally, notice that the peak amplitude point in the Brillouin precursor experiences zero exponential attenuation, decreasing only algebraically as $(\Delta z)^{-1/2}$ and travelling at the velocity $v_B = c/\theta_0 = c/n(0)$.

The field component $A_m(z,t)$ appearing in Eq. (22) describes the middle precursor [44] that is due to any of the middle saddle points that may become asymptotically dominant over the distant and near saddle points for a finite space-time interval that typically occurs between the Sommerfeld and Brillouin precursor evolutions. The asymptotic description of this middle precursor field is given by a superposition of terms of the form given in Eq. (19), each term arising from a separate pair of middle saddle points that are introduced by each additional resonance line beyond that described by the single-resonance Lorentz model. Not all of these middle saddle points will contribute to the asymptotic field behavior, however, so that the middle precursor may or may not be present in the dynamical field evolution; a necessary condition [44] for the appearance of the middle precursor is given in terms of the energy velocity in the dispersive dielectric [38,41]. The middle precursors (if present) are typically a characteristic of the intermediate frequency response of the dispersive medium below the upper absorption band of the material.

The final contribution appearing in Eq. (22) is the pole contribution $A_c(z,t)$. As described in connection with Eq. (20), this contribution, when present, describes the steady-state behavior of the signal. A canonical problem of considerable historical interest [5–12,18–21,42–46] in this regard is provided by the Heaviside step-function signal $f(t) = u(t)\sin(\omega_c t)$ with fixed carrier frequency $\omega_c > 0$ and with envelope function given by the Heaviside unit step-function [$u(t) = 0$ for $t < 0$, and $u(t) = 1$ for $t > 0$]. The proper solution of this problem then entails a careful description of the signal arrival [43–46] and provides a detailed description of the signal velocity [43,44,46,47] in a Lorentz model dielectric.

20.5 Accuracy of the Group Velocity Description of Ultrashort Pulse Dynamics

The accuracy of the group velocity approximation of ultrashort pulse dispersion is now considered in order to establish the space-time domain over which this approximate description is valid. A double resonance Lorentz model of a fluoride-type glass with infrared ($\omega_0 = 1.74 \times 10^{14} r/s$, $b_0 = 1.22 \times 10^{14} r/s$, $\delta_0 = 4.96 \times 10^{13} r/s$) and near-ultraviolet ($\omega_2 = 9.145 \times 10^{15} r/s$, $b_2 = 6.72 \times 10^{15} r/s$, $\delta_2 = 1.434 \times 10^{15} r/s$) resonance lines is considered with complex index of refraction given by Eq. (21) with $N = 2$. The angular frequency dispersion of the real and imaginary parts of the complex wavenumber $\tilde{k}(\omega) = (\omega/c)n(\omega)$ for this double-resonance Lorentz model dielectric is illustrated in Fig. 1. The upper and lower solid curves in each part of the figure describe the exact frequency dependence of $\beta(\omega) \equiv \Re\{\tilde{k}(\omega)\}$ and $\alpha(\omega) \equiv \Im\{\tilde{k}(\omega)\}$, respectively, while the dashed curves describe the cubic dispersion approximation when (a) $\omega_c = \omega_{min} = 1.615 \times 10^{15} r/s$, and (b) $\omega_c = 0.87\,\omega_2 = 8.0 \times 10^{15} r/s$. The cubic dispersion approximation is seen to provide a reasonably accurate estimate of the local frequency dispersion of the propagation factor $\beta(\omega)$ about the carrier frequency within the passband, where the dispersion is normal when $\omega_c = \omega_{min}$; but the accuracy of this approximation is seen to decrease [48,49] as ω_c is shifted toward either absorption band, where the dispersion becomes anomalous. The inclusion of higher-order terms in the Taylor series approximation of the complex wavenumber only serves to further decrease its accuracy in a global sense [48,49]. The cubic dispersion approximation of the attenuation coefficient is not as accurate as that for the propagation factor, making the necessity of the approximation $\alpha(\omega) \approx \alpha(\omega_c)$ used in the group velocity description all the more important.

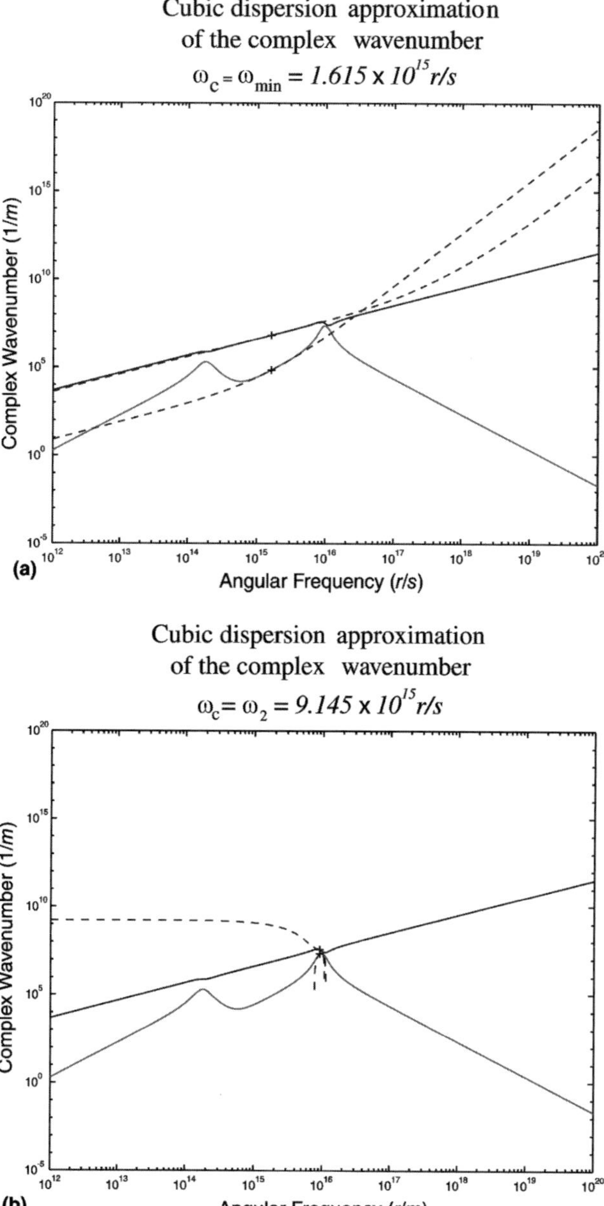

Figure 1 Angular frequency dependence of the real (upper solid curves) and imaginary (lower solid curves) parts of the complex wavenumber for a double-resonance Lorentz model of a fluoride-type glass with infrared and near-ultraviolet resonance lines. (a) The dashed curves describe the cubic dispersion approximation about the minimum dispersion point in the passband between the absorption bands; (b) they describe that approximation about the upper resonance frequency.

Because of its central importance in ultrashort optical pulse technology, a unit amplitude Gaussian envelope pulse $f(t) = u(t)\sin(\omega_c t + \pi/2)$ is considered, where

$$u(t) = \exp(-t^2/T^2), \qquad (23)$$

with initial full pulse width $2T > 0$ measured at the $\exp(-1)$ amplitude points. Because the Fourier spectrum of such a Gaussian envelope function is an entire function of complex ω, its asymptotic representation is comprised solely of precursor fields, so that [58–60]

$$A(z,t) \sim A_S(z,t) + A_m(z,t) + A_B(z,t), \qquad (24)$$

as $\Delta z \to \infty$, where $A_c(z,t) = 0$. The propagated pulse structure at any fixed propagation distance $\Delta z > 0$ is numerically determined from the Fourier integral representation given in Eq. (1), using the fast Fourier transform (FFT) algorithm. A comparison between the propagated pulse structure using the exact dispersion relation for the complex wavenumber $\tilde{k}(\omega) = (\omega/c)n(\omega)$ with $n(\omega)$ given by Eq. (21) and that obtained using the cubic dispersion approximation given in Eq. (13) then reveals the accuracy of the group velocity approximation as a function of both the propagation distance Δz and the initial temporal pulse width $2T$. The computed error between these two results then yields the space-time domain over which the group velocity approximation is valid.

Because the approximate group velocity pulse travels with a velocity that is different from the actual pulse velocity, the first step in this numerical comparison is to shift the approximate pulse to the position of the actual pulse such that the peak amplitude points are coincident. Of course, this readily observed difference in the pulse velocities is an obvious error in the group velocity description. An example of this procedure is presented in Fig. 2 for a $2T = 4.12fs$ Gaussian envelope pulse with carrier angular frequency $\omega_c = \omega_2 = 9.145 \times 10^{15} r/s$ at the upper undamped resonance frequency of the double-resonance Lorentz model dielectric, where the group velocity is superluminal and negative. Part (a) of the figure illustrates the approximate group velocity and actual pulses at one absorption depth $\Delta z = z_d$ in the medium, where $z_d = \alpha^{-1}(\omega_c)$; the actual pulse is clearly not traveling at a rate given by the classical group velocity $v_g(\omega_c) = 1/\beta'(\omega_c)$. Part (b) of the figure illustrates the actual and approximate pulses when the approximate group velocity pulse (dashed curve) has been temporally shifted such that its peak amplitude point is coincident with that for the actual pulse (solid curve). The error between these two pulses is then computed in two different ways. The first error measure (error 1) is given by the integral of the square of the difference between the two aligned pulses. This error then measures both shape and energy differences. The second error measure (error 2) is obtained by first renormalizing both pulses

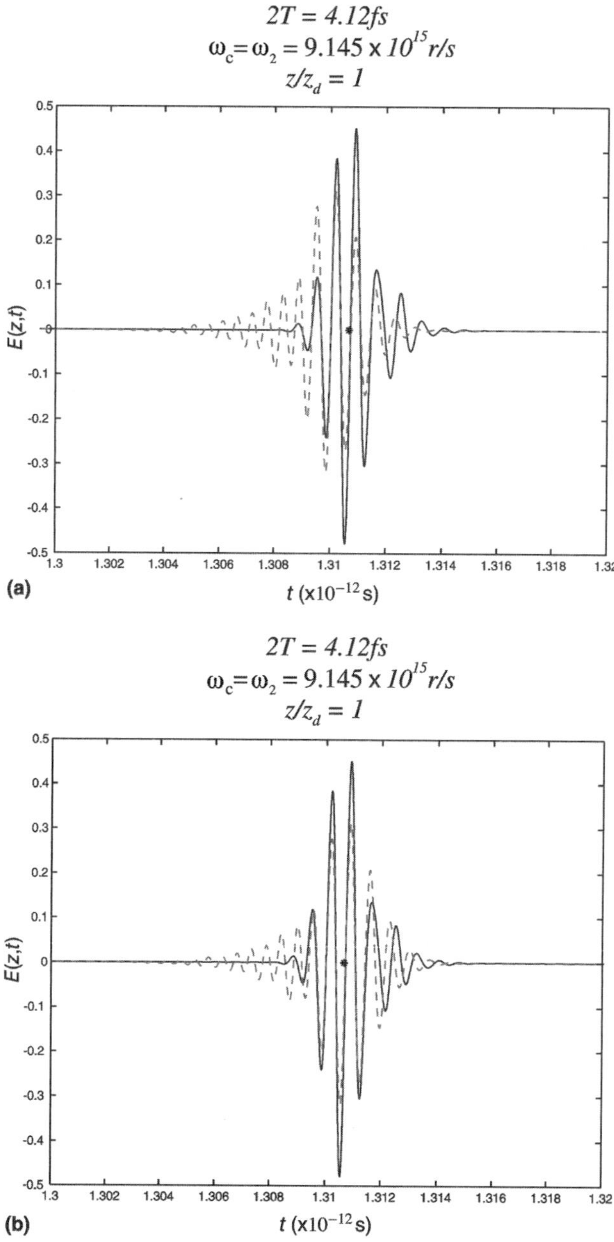

Figure 2 Approximate group velocity (dashed curves) and exact (solid curves) pulses at one absorption depth due to the same input Gaussian envelope pulse with carrier frequency at the upper resonance frequency of a double-resonance Lorentz model dielectric. (a) The effect of the velocity difference between the exact and approximate pulses is shown. (b) The approximate pulse has been shifted in time in order that their peak amplitude points are coincident.

by the square root of their respective pulse energies and then taking the integral of the square of their differences. This error then measures the shape difference between the two aligned pulses at a given fixed propagation distance.

The numerical results for these two error measures are depicted in Fig. 3 as a function of the relative propagation distance $\Delta z/z_d$, when the input pulse carrier angular frequency ω_c is set equal to the angular frequency ω_{min} at the minimum dispersion point in the passband between the two absorption bands of the double-resonance Lorentz model dielectric. The results are presented for four different values of the input Gaussian envelope pulse width $2T$. The results clearly show that, as expected, the error decreases with increasing initial pulse width at any fixed value of the propagation distance. The first error measure (error 1) is seen in Fig. 2(a) and initially increases with increasing propagation distance, reaching a peak value at approximately one absorption depth ($\Delta z/z_d \sim 1$), and then decreasing to zero as the relative propagation distance $\Delta z/z_d$ increases above unity. This behavior is due to the fact that the group velocity approximate pulse amplitude decays exponentially with penetration distance at a faster rate than does the actual ultrashort pulse so that the observed monotonic decrease in error with increasing propagation distance $\Delta z/z_d > 1$ primarily describes the slow, nonexponential amplitude decay with propagation distance of the actual pulse in the lossy dielectric. This difference between the two pulse amplitudes is eliminated in the second error measure (error 2), which describes the relative structural error between the group velocity approximate and exact pulses, as illustrated in Fig. 3(b). This structural-error measure increases with increasing propagation distance, the rate of increase decreasing with increasing initial pulse width. For the shortest initial pulse considered ($2T = 7.78fs$), the error exceeds 10% when $\Delta z/z_d \geq 0.15$, while for the longest initial pulse considered ($2T = 77.8fs$), the error exceeds 10% when $\Delta z/z_d > 1.53$.

The error typically increases for both error measures when the input pulse carrier frequency is shifted toward either the lower or the upper absorption band of the double-resonance Lorentz model dielectric. Numerical results for these two error measures are presented in Fig. 4 as a function of the relative propagation distance $\Delta z/z_d$ when $\omega_c = \omega_2 = 9.145 \times 10^{15} r/s$. The results are presented for four different values of the input Gaussian envelope pulse width $2T$, each value corresponding to the same number of oscillations between the $\exp(-1)$ amplitude points as presented in Fig. 3 at the minimum dispersion point ($N_{osc} = 1, 3, 5, 10$). The error again decreases with increasing initial pulse width at any fixed value of the propagation distance. The first error measure (error 1) is seen in Fig. 3(a) of the figure to initially increase with increasing propagation distance, reaching a peak value at approximately one absorption depth for the two largest initial pulse width cases ($2T = 6.87fs$ and $2T = 13.7fs$), and then decreasing to zero as the relative propagation distance increases above unity. In the three-oscillation pulse case

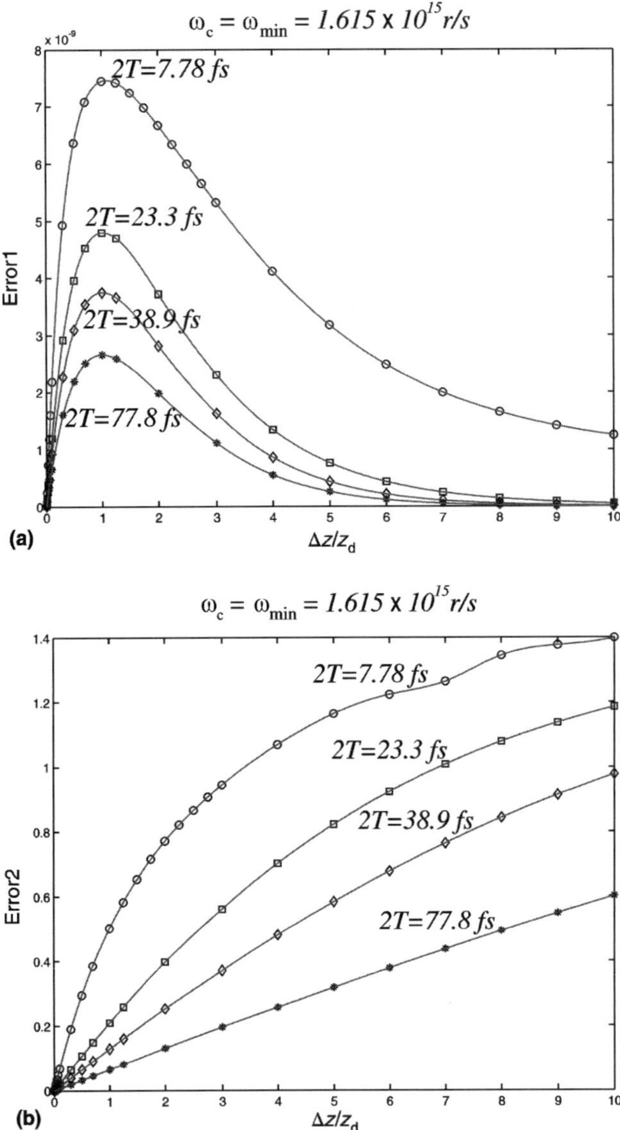

Figure 3 Error resulting form the group velocity description of the propagated Gaussian envelope pulse with a cubic dispersion approximation of the propagation factor $\beta(\omega)$ in a double-resonance Lorentz model dielectric as function of the relative penetration depth $\Delta z/z_d$ when the input pulse carrier frequency ω_c is equal to the angular frequency ω_{min} at the minimum dispersion point in the passband between the two absorption bands for different values of the input pulse width $2T$.

$(2T = 4.12fs)$ this peak in the first error measure occurs at $\Delta z/z_d \approx 0.4$, while for the single-oscillation pulse case $(2T = 1.37fs)$ this peak in the first error measure occurs at $\Delta z/z_d \approx 0.2$, with a secondary peak appearing at $\Delta z/z_d \approx 1$. The

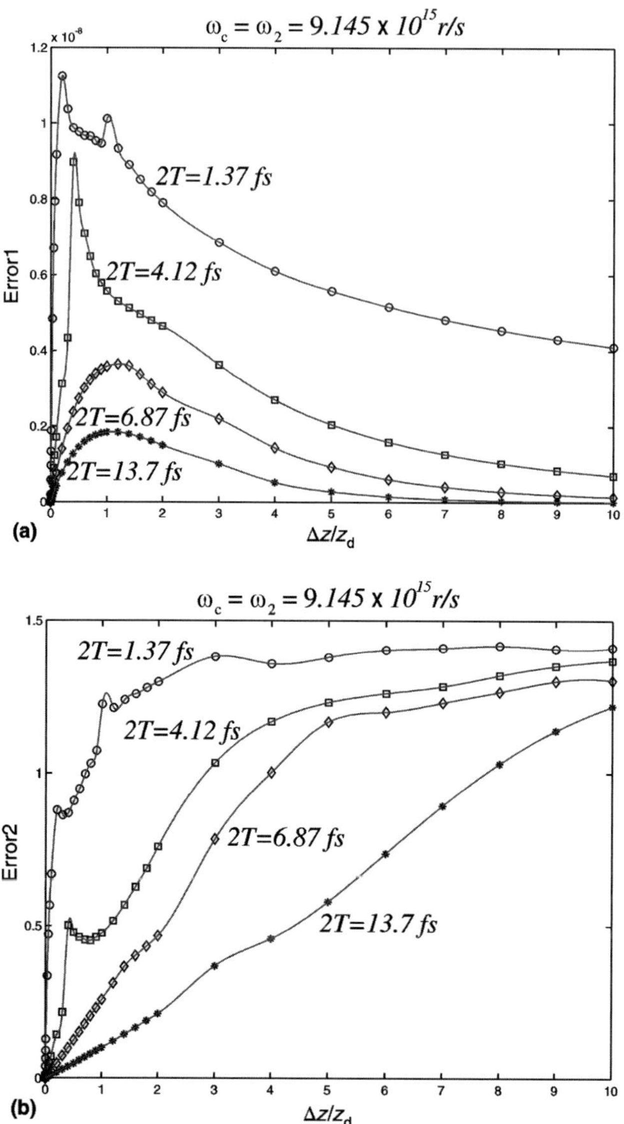

Figure 4 Error resulting from the group velocity description of the propagated Gaussian envelope pulse with a cubic dispersion approximation of the propagation factor $\beta(\omega)$ in a double-resonance Lorentz model dielectric as a function of the relative penetration depth $\Delta z/z_d$ when the input pulse carrier frequency ω_c is equal to the upper resonance frequency ω_2 of the dielectric for different values of the input pulse width $2T$.

structural error measure (error 2) is seen in Fig. 4(b) to increase with increasing propagation distance for both the five- and ten-oscillation input pulse cases, while it exhibits a more complicated behavior for the one- and three-oscillation input pulse cases. For the shortest initial pulse considered here ($2T = 1.37fs$) the second error measure exceeds 10% when $\Delta z/z_d \geq 0.0052$; while for the longest initial pulse considered ($2T = 13.7fs$), this error measure exceeds 10% when $\Delta z/z_d \geq 0.99$.

20.6 The Question of Superluminal Pulse Velocities

A number of velocity measures have been introduced for the purpose of describing the rate at which some particular feature of a pulse travels through a dispersive material. The most important of these are the phase [3], group [1–4,30–32], energy [38,41], signal [5,7–12,18,43–47], and centroid [61–65] velocities. Each of these velocity measures in free space equals the velocity of light c in vacuum, but they are in general different in a causally dispersive material such as that described by the Lorentz model.

The phase velocity $v_p(\omega) \equiv \omega/\beta(\omega)$ describes the rate at which the cophasal surfaces propagate through the dispersive medium [3] [cf. Eqs. (9) and (10)]. Since the phase of a spatially coherent optical field can only be measured indirectly [66], this velocity measure does not have any separate, measurable physical meaning in spite of the fact that it plays a central role in the mathematical description of pulse dispersion, as described by Eq. (1). In particular, the phase velocity of a pulse is superluminal [i.e., $v_p(\omega_c) > c$] when the input pulse carrier frequency ω_c is above the uppermost absorption band of a Lorentz model dielectric, as illustrated in Fig. 5(a) for a single-resonance Lorentz model dielectric when $\omega_c > \omega_1$. For an ultrashort pulse with above-resonance carrier frequency whose temporal energy centroid is moving subluminally, the phase velocity is then seen to describe the motion of a space-time point where there is negligible pulse energy.

The group velocity $v_g(\omega) \equiv (\partial\beta(\omega)/\partial\omega)^{-1}$ describes the rate at which the envelope of a group of waves travels through the dispersive medium [1–4] [cf. Eqs. (9) and (11)]. As described by Rayleigh [3], a group of waves is defined as moving beats following each other in a regular pattern as, for example, that obtained from the coherent superposition of two monochromatic waves with slightly different amplitudes and frequencies. Although the group velocity does indeed describe the beat velocity of such an infinite wave group, its extension to the description of the velocity of an ultrashort pulse in a causally dispersive medium is invalid. This is readily evident in Fig. 2(a), where the group velocity approximate pulse is seen to be moving at a faster rate than is the actual pulse. This example illustrates the extreme dispersion case when $\omega_c = \omega_2$. The group velocity value $v_g(\omega_c)$ is then seen to be a poor measure of the actual pulse velocity when the material loss is not negligible, although it can describe the initial pulse evolution when the material loss is near the minimum dispersion point, where the material loss is minimal. However, even in that minimum dispersion situation, the pulse velocity will diverge from the group velocity as the pulse evolution progresses deeper into the dispersive material.

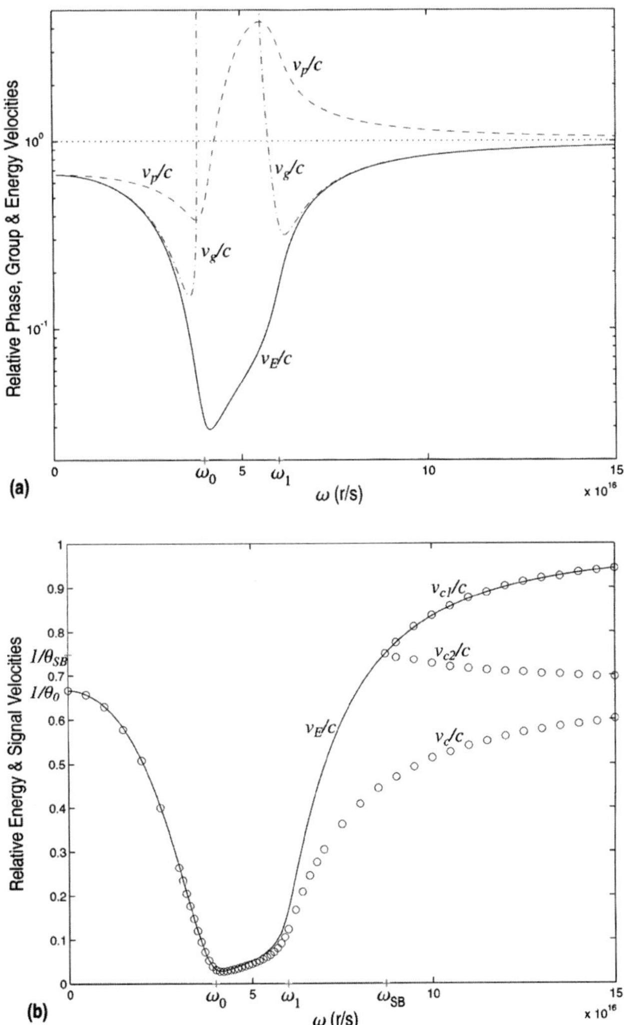

Figure 5 Frequency dependence of (a) the relative phase velocity (dashed curve), group velocity (dash-dot curve), and energy velocity (solid curve); and (b) the relative energy velocity (solid curve) and signal velocity (open circles) in a single-resonance Lorentz model dielectric.

The energy transport velocity $v_E(\omega_c) \equiv \langle \mathbf{S} \cdot \hat{\mathbf{Z}} \rangle / \langle U \rangle$ for a monochromatic plane wave field with angular frequency ω_c traveling in the positive z-direction, defined [38] as the ratio of the time-average Poynting vector $\langle \mathbf{S} \rangle = (c/4\pi)\langle \mathbf{E} \times \mathbf{H} \rangle = (c/8\pi)n_r(\omega_c)|\mathbf{E}|^2$ to the total time-average electromagnetic energy density $\langle U \rangle = \langle U_{\mathrm{em}} \rangle + \langle U_{\mathrm{med}} \rangle$ in the coupled medium-field system, describes the time-average rate at which electromagnetic energy is transported through the dispersive medium. For a multiple resonance Lorentz model dielectric, the energy transport velocity is given by [41]

$$v_E(\omega_c) = \cfrac{c}{n_r(\omega_c) + \cfrac{1}{n_r(\omega_c)} \sum_j \cfrac{b_j^2 \omega_c}{(\omega_c^2 - \omega_j^2)^2 + 4\delta_j^2 \omega_c^2}}. \tag{25}$$

Notice that this expression yields relativistically causal results, where $0 < v_E(\omega_c) \leq c$ for all $\omega_c \in [0, \infty]$. In regions of normal dispersion, the energy velocity reduces to the corresponding expression for the group velocity provided that the material loss at the carrier angular frequency ω_c is not too large, while in regions of anomalous dispersion the two results can be significantly different, particularly when the group velocity assumes either superluminal or negative values.

The signal velocity [5–12] describes the rate at which the signal arrival due to the pole contribution $A_c(z,t)$ first appears in the asymptotic description given in Eq. (22). This contribution, when present, describes the steady-state behavior of the signal. The modern asymptotic theory [43–46] has shown that the energy velocity for a monochromatic wave forms an upper envelope for the signal velocity in a Lorentz model dielectric, as illustrated in Fig. 5(b). Based upon this connection between the energy velocity of a monochromatic signal and the signal velocity of a Heaviside step-function signal, a new physical description [39,40,46] of dispersive pulse dynamics in Lorentz model dielectrics has been developed in terms of the energy transport velocity and attenuation of monochromatic waves in the dispersive, attenuative material. This physical description reduces to the group velocity description in the limit of zero material loss (i.e., as $\delta_j \rightarrow 0$ for each resonance line). Most important, this energy velocity description provides an accurately detailed description of the precursor fields in dispersive pulse dynamics when the pulse dispersion is in the mature dispersion regime (typically when $\Delta z > z_d$).

A comparison of the relative phase, group, and energy velocities is given in Fig. 6(a) for the double-resonance Lorentz model dielectric whose complex wavenumber dependence on angular frequency is illustrated in Fig. 1, and in Fig. 6(b) when each of the phenomenological damping constants for the double-resonance Lorentz model dielectric considered in Fig. 1 has each been reduced by a factor of 10 (i.e., when $\delta_j \rightarrow \delta_j/10$ for $j = 0, 2$). Notice that the group velocity becomes negative in the upper absorption band, but not in the lower absorption band in part (a) of the figure; however, when the damping constants are reduced, as in part (b) of the figure, the group velocity then becomes negative in both absorption bands. Notice also that the group velocity becomes nearly identical with the energy velocity throughout the normal dispersion regions when the damping constants are decreased, as in Fig. 6(b).

For the numerical Gaussian pulse example illustrated in Fig. 2, the peak amplitude point of the actual pulse is moving with the average velocity $v_{\text{peak}} \cong 0.91c$

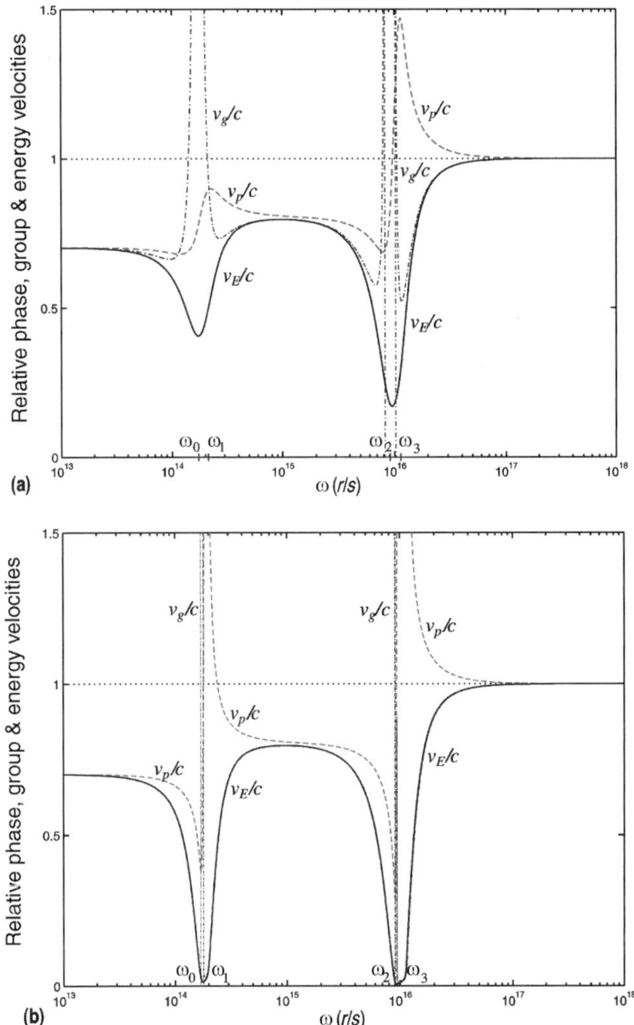

Figure 6 Frequency dependence of the relative phase velocity (dashed curves), group velocity (dash-dot curves), and energy velocity (solid curves) for (a) the double-resonance Lorentz model dielectric considered in Fig. 1; and (b) the same double-resonance Lorentz model dielectric when each of the phenomenological damping constants have been reduced by a factor of 10.

and the peak amplitude point of the group velocity approximate pulse is moving with the average velocity $v_{\text{peak}}^{\text{app}} \cong -0.30c$, while the phase velocity is given by $v_p(\omega_c) \cong 0.82c$, the group velocity is given by $v_g(\omega_c) \cong -0.41c$, and the energy velocity is given by $v_E(\omega_c) \cong 0.18c$ at the input pulse carrier angular frequency $\omega_c = \omega_2 = 9.145 \times 10^{15} r/s$. The peak amplitude values reported here are "time-of-flight" values that result from numerical measurements of the initial and final pulse positions. This average peak amplitude velocity measure changes as the propagation distance into the dispersive dielectric increases, as does the instantaneous peak amplitude velocity. The instantaneous peak amplitude velocity of an ultrashort Gaussian pulse has been shown [59,60] to evolve with increasing propagation distance along the group velocity curve toward the energy velocity curve as the instantaneous oscillation frequency at the peak amplitude point shifts away from the region of anomalous dispersion and into the normal dispersion region either above or below that absorption band. It is then not surprising that the numerical velocity values given above are all significantly different as each describes a different feature of the pulse that may only be valid either in the limit of vanishingly small propagation distance (the group velocity), or else in the large propagation distance asymptotic limit (the energy velocity).

It is clear that a more physically meaningful pulse velocity measure needs to be considered in order to accurately describe the complicated pulse evolution that occurs in ultrashort dispersive pulse dynamics. One possible measure is given by the pulse centrovelocity [61]

$$v_{CE} \equiv \left| \nabla \left(\int_{-\infty}^{\infty} t E^2(\mathbf{r},t)\,dt \Big/ \int_{-\infty}^{\infty} E^2(\mathbf{r},t)\,dt \right) \right|^{-1}, \qquad (26)$$

which describes the temporal center of gravity of the pulse intensity. A more appropriate velocity measure would track the temporal centroid of the Poynting vector of the pulse. This pulse centroid velocity of the Poynting vector was first introduced by Lisak [62] in 1976. Recent descriptions [63–65] of its properties have established its efficacy in describing the evolution of the pulse velocity with propagation distance in a Lorentz model dielectric. The instantaneous centroid velocity of the pulse Poynting vector is defined as [65]

$$v_{CI} = \lim_{\Delta z \to 0} \left(\Delta z / \Delta \langle t \rangle \right) \qquad (27)$$

for a plane wave pulse propagating in the positive z-direction through the dispersive medium, where $\Delta z = z_2 - z_1$, $\Delta \langle t \rangle = \langle t_2 \rangle - \langle t_1 \rangle$ with

$$\langle t_j \rangle \equiv \int_{-\infty}^{\infty} t S_z(z_j,t)\,dt \Big/ \int_{-\infty}^{\infty} S_z(z_j,t)\,dt, \qquad (28)$$

where $S_z = \hat{\mathbf{z}} \cdot \mathbf{S}$ denotes the z-component of the Poynting vector.

The evolution of the instantaneous centroid velocity of the pulse Poynting vector with the relative propagation distance $\Delta z / z_d$ for Gaussian pulse propagation in a double-resonance Lorentz model dielectric is illustrated in Fig. 7 for several values of the input pulse width. In part (a) of the figure, the input pulse carrier frequency is set at the minimum dispersion point $\omega_{\min}$ in the passband between

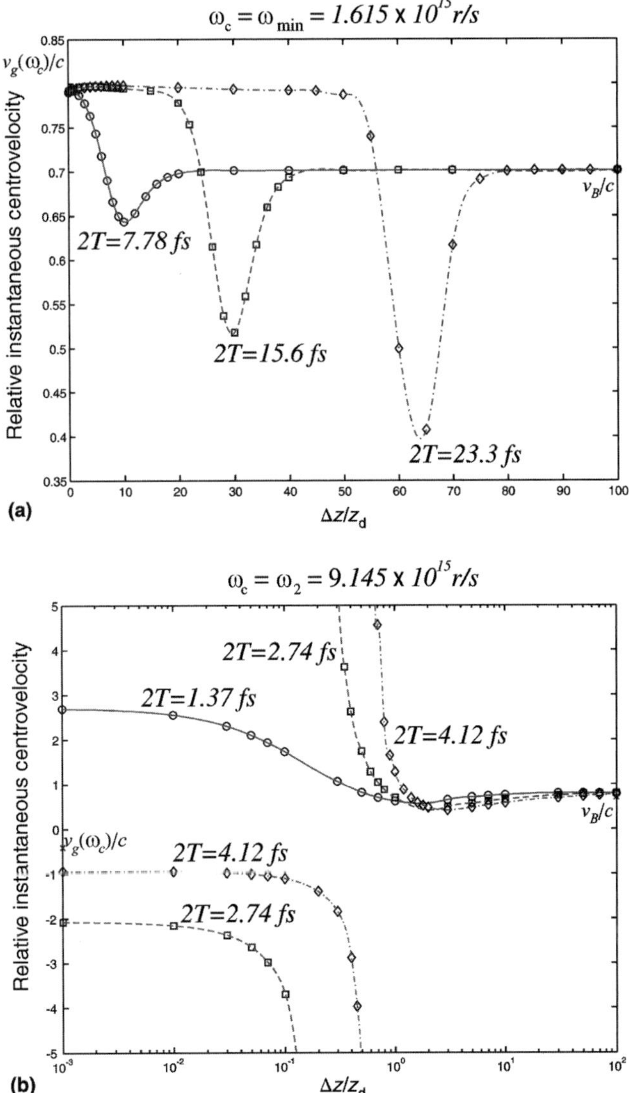

Figure 7 Evolution of the relative instantaneous centroid velocity of the pulse Poynting vector with relative propagation distance for Gaussian pulse propagation in the double-resonance Lorentz model dielectric whose frequency dispersion is presented in Fig. 1. In part (a) the input pulse carrier frequency is at the minimum dispersion point in the passband between the two absorption bands, and in part (b) the input pulse carrier frequency is at the upper resonance frequency.

the two absorption bands, where the material dispersion is normal. For each input pulse width case considered in Fig. 7(a), the instantaneous centrovelocity is approximately given by the classical group velocity $v_g(\omega_c) \cong 0.7920c$ evaluated at the input carrier frequency ω_c in the limit of vanishingly small propagation distance. This classical group velocity limit is actually obtained at a sufficiently small propagation distance in the limit as the initial pulse width increases and the initial pulse spectrum narrows about the carrier frequency. In the opposite limit as $\Delta z/z_d \to \infty$ the centroid velocity is found to approach the velocity $v_B = c/n(0)$ at which the peak amplitude point in the Brillouin precursor travels through the dispersive material. The transition of the pulse centroid velocity between these two limits is marked by a narrow dip in velocity as the pulse evolves from the approximate group velocity behavior at small propagation distances to its asymptotic behavior for large propagation distances. The minimum point in this dip occurs at a propagation distance whose value increases with increasing initial pulse width, while the minimum value decreases with increasing initial pulse width. At the leading edge of each dip when the centrovelocity rapidly decreases, the pulse begins to separate into a pair of middle and Brillouin precursors and the temporal pulse centroid is found to occur at a space-time point between these two pulse components where the pulse energy is minimal. Finally, notice that the centrovelocity for a Gaussian pulse is subluminal and nonnegative (i.e., $0 \leq v_{CI} < c$) for all propagation distances at this input carrier frequency.

In Fig. 7(b) the input pulse carrier frequency is set equal to the upper resonance frequency ω_2 of the double-resonance Lorentz model dielectric where the dispersion is anomalous. The classical group velocity limit $v_g(\omega_c) \cong -0.4076c$ is again approached at a sufficiently small propagation distance in the limit as the initial pulse width increases and the initial pulse spectrum narrows about the carrier frequency ω_c, but at a much slower rate than that obtained at the minimum dispersion point. However, notice that if the initial pulse width is sufficiently small (as it is for the 1.37*fs* pulse case), then the initial pulse spectrum is extremely ultrawideband such that the classical group velocity limit is not obtained as $\Delta z \to 0$ and the centrovelocity remains positive for all propagation distances $\Delta z \geq 0$. In the opposite limit as $\Delta z/z_d \to \infty$ the centroid velocity is again found to approach the velocity $v_B = c/n(0)$ at which the peak amplitude point in the Brillouin precursor travels through the dispersive material. For a sufficiently long initial pulse width, the transition of the ultrawideband pulse centroid velocity between these two limits is marked by a rapid decrease in centrovelocity to $-\infty$ and then from $+\infty$ to subluminal values before approaching the asymptotic limit $v_B = c/n(0)$ set by the peak amplitude point of the Brillouin precursor. The discontinuous jump in the centrovelocity from $-\infty$ to $+\infty$ is found to occur at a relative propagation distance whose value increases with increasing pulse width, provided that the initial pulse spectrum is ultrawideband.

In spite of the fact that the instantaneous centrovelocity of the pulse Poynting vector can take on both negative and superluminal values for sufficiently small relative propagation distances, the pulse itself is found to only undergo a slight change in shape. There is no superluminal movement of the pulse when the instantaneous pulse centrovelocity is superluminal, nor is there any retrogression in position when the pulse centrovelocity is negative. Sommerfeld's theorem firmly establishes that electromagnetic field energy cannot move forward of any space-time point in the pulse at a superluminal rate. Finally, notice that this seemingly nonphysical behavior only occurs in the immature dispersion regime ($0 \leq \Delta z < z_c$), where the group velocity approximation applies in the limit as $\Delta z \to 0$.

20.7 Conclusions

The analysis and numerical results presented in this paper have established the following results:

(1) The group velocity approximation is valid only in the immature dispersion regime $0 \leq \Delta z < z_c$, its accuracy increasing as $\Delta z \to 0$. The asymptotic description is valid in the mature dispersion regime $\Delta z > z_c$, its accuracy increasing in the sense of Poincaré [17] as $\Delta z \to \infty$. The critical distance z_c depends upon the input pulse type and initial pulse length, as well as upon the input pulse carrier frequency for a given dispersive material.

(2) The instantaneous centroid velocity of the pulse Poynting vector is a convenient, albeit sometimes misleading, measure of the pulse evolution in a dispersive medium for input Gaussian envelope pulses. Although this velocity measure can take on both negative and superluminal values for relative propagation distances in the immature dispersion regime, these nonphysical values mark the initial transition into the mature dispersion regime where the asymptotic description applies.

(3) Superluminal values of the instantaneous centroid velocity of the pulse Poynting vector are ephemeral and do not describe any real, observable motion of the pulse.

(4) Negative values of the instantaneous centroid velocity of the pulse Poynting vector are also short-lived and do not describe any retrogression in position of the pulse. In each case, the energy transport velocity for each monochromatic component present in the initial pulse spectrum travels with the relativistically causal velocity given in Eq. (25) so that, taken together with the attenuation, the propagated pulse may always be constructed [39,40,46] in a causal manner from the initial Gaussian pulse. There can then be no superluminal motion of the pulse, nor can there be any pulse retrogression.

Acknowledgment

The research presented in this paper has been supported, in part, by AFOSR Grant 49620-01-0306 and by NSF Grant PHY99-07949.

References

1. W.R. Hamilton, "Researches respecting vibration, connected with the theory of light," *Proc. Royal Irish Academy* **1**, 341–349 (1839).

2. G.G. Stokes, Smith's Prize examination question no. 11 (February 2, 1876); reprinted in G.G. Stokes, *Mathematical and Physical Papers*, Vol. 5, 362, Cambridge University Press, Cambridge (1905).

3. Lord Rayleigh, "On progressive waves," *Proc. London Math. Soc.* **9**, 21–26 (1877).

4. Lord Rayleigh, "On the velocity of light," *Nature* **24**, 52–55 (1881).

5. W. Voigt, "Ueber die Aenderung der Schwingungsform des Lichtes beim Fortschreiten in einem dispergirenden oder absorbirenden Mittel," *Ann. Physik und Chemie* (Leipzig) **68**, 598–603 (1899).

6. W. Voigt, "Weiteres zur Aenderung der Schwingungsform des Lichtes beim Fortschreiten in einem dispergirenden und absorbirenden Mittel," *Ann. Physik* (Leipzig) **4**, 209–214 (1901).

7. P. Ehrenfest, "Mißt der Aberrationswinkel in Fall einer Dispersion des Äthers die Wellengeschwindigkeit?" *Ann. Physik* (Leipzig) **33**, 1571 (1910).

8. A. Laue, "Die Fortpflanzung der Strahlung in dispergierenden und absorbierenden Medien," *Ann. Physik* **18**, 523 (1905).

9. A. Sommerfeld, "Ein Einwand gegen die Relativtheorie der Elektrodynamik und seine Beseitigung," *Physik. Z.* **8**, 841 (1907).

10. A. Sommerfeld, "Über die Fortpflanzung des Lichtes in disperdierenden Medien," *Ann. Physik* (Leipzig) **44**, 177–202 (1914); see also *Weber Festschrift*, 338 (1912).

11. L. Brillouin, "Über die Fortpflanzung des Licht in disperdierenden Medien," *Ann. Physik* (Leipzig) **44**, 203–240 (1914).

12. L. Brillouin, *Wave Propagation and Group Velocity*, Academic Press (1960).

13. P. Debye, "Näherungsformeln für die Zylinderfunktionen für grosse Werte des Arguments nd unbeschränkt veranderliche Werte des Index," *Mathematische Ann.* **67**, 535–558 (1909).

14. T.H. Havelock, "The propagation of groups of waves in dispersive media," *Proc. Roy. Soc. A* **81**, 398 (1908).

15. T.H. Havelock, *The Propagation of Disturbances in Dispersive Media*, Cambridge University Press, Cambridge (1914).

16. Lord Kelvin, "On the waves produced by a single impulse in water of any depth, or in a dispersive medium," *Proc. Roy. Soc.* **42**, 80 (1887).

17. E.T. Copson, *Asymptotic Expansions*, Cambridge University Press, Cambridge (1965).

18. H. Baerwald, "Über die Fortpflanzung von Signalen in dispergierenden Systemen," *Ann. Physik* **7**, 731–760 (1930).

19. J.A. Stratton, *Electromagnetic Theory*, §5.18, McGraw-Hill (1941).

20. N.S. Shiren, "Measurement of signal velocity in a region of resonant absorption by ultrasonic paramagnetic resonance," *Phys. Rev.* **128**, 2103–2112 (1962).

21. T.A. Weber and D.B. Trizna, "Wave propagation in a dispersive and emissive medium," *Phys. Rev.* **144**, 277–282 (1966).

22. R.A. Handelsman and N. Bleistein, "Uniform asymptotic expansions of integrals that arise in the analysis of precursors," *Arch. Rational Mech. Analysis* **35**, 267–283 (1969).

23. P. Pleshko and I. Palócz, "Experimental observation of Sommerfeld and Brillouin precursors in the microwave domain," *Phys. Rev. Lett.* **22**, 1201–1204 (1969).

24. C. Eckart, "The approximate solution of one-dimensional wave equations," *Rev. Modern Physics* **20**, 399–417 (1948).

25. R.M. Lewis, "Asymptotic theory of wave propagation," *Arch. Rational Mech. Analysis* **20**, 191–250 (1065).

26. R.M. Lewis, "Asymptotic theory of transients," in *Electromagnetic Wave Theory*, J. Brown, Ed., 845–869, Pergamon Press, Oxford (1967).

27. L.B. Felsen, "Rays, dispersion surfaces, and their uses for radiation and diffraction problems," *SIAM Review* **12**, 424–448 (1970).

28. K.A. Connor and L.B. Felsen, "Complex-space-time rays and their application to pulse propagation in lossy dispersive media," *Proc. IEEE* **62**, 1586–1598 (1974).

29. D. Censor, "Fermat's principle and real space-time rays in absorbing media," *J. Phys. A* **10**, 1781–1790 (1977).

30. M.A. Biot, "General theorems on the equivalence of group velocity and energy transport," *Phys. Rev.* **105**, 1129–1137 (1957).

31. G.B. Whitham, "Group velocity and energy propagation for three-dimensional waves," *Comm. Pure and Applied Math.* **14**, 675–691 (1961).

32. M.J. Lighthill, "Group velocity," *J. Inst. Maths. Applies.* **1**, 1–28 (1964).

33. M. Born and E. Wolf, *Principles of Optics*, Chapter 10, Pergamon Press, Oxford (1959).

34. J. Jones, "On the propagation of a pulse through a dispersive medium," *Am. J. Phys.* **42**, 43–46 (1974).

35. D.G. Anderson and J.I.H. Askne, "Wave packets in strongly dispersive media," *Proc. IEEE* **62**, 1518–1523 (1974).

36. D. Anderson, J. Askne, and M. Lisak, "Wave packets in an absorptive and strongly dispersive medium," *Phys. Rev. A* **12**, 1546–1552 (1975).

37. E.O. Schulz-DuBois, "Energy transport velocity of electromagnetic propagation in dispersive media," *Proc. IEEE* **57**, 1748–1757 (1969).

38. R. Loudon, "The propagation of electromagnetic energy through an absorbing dielectric," *J. Phys. A* **3**, 233–245 (1970).

39. G.C. Sherman and K.E. Oughstun, "Description of pulse dynamics in Lorentz media in terms of the energy velocity and attenuation of time-harmonic waves," *Phys. Rev. Lett.* **47**, 1451–1454 (1981).

40. G.C. Sherman and K.E. Oughstun, "Energy velocity description of pulse propagation in absorbing, dispersive dielectrics," *J. Opt. Soc. Am. B* **12**, 229–247 (1995).

41. K.E. Oughstun and S. Shen, "The velocity of energy transport for a time-harmonic field in a multiple resonance Lorentz medium," *J. Opt. Soc. Am. B* **5**, 2395–2398 (1988).

42. D. Trizna and T. Weber, "Brillouin revisited: Signal velocity definition for pulse propagation in a medium with resonant anomalous dispersion," *Radio Sci.* **17**, 1169–1180 (1982).

43. K.E. Oughstun and G.C. Sherman, "Propagation of electromagnetic pulses in a linear dispersive medium with absorption (the Lorentz medium)," *J. Opt. Soc. Am. B* **5**, 817–849 (1988).

44. S. Shen and K.E. Oughstun, "Dispersive pulse propagation in a double resonance Lorentz medium," *J. Opt. Soc. Am. B* **6**, 948–963 (1989).

45. K.E. Oughstun and G.C. Sherman, "Uniform asymptotic description of electromagnetic pulse propagation in a linear dispersive medium with absorption (the Lorentz medium)," *J. Opt. Soc. Am. A* **6**, 1394–1420 (1989).

46. K.E. Oughstun and G.C. Sherman, *Electromagnetic Pulse Propagation in Causal Dielectrics*, Springer-Verlag, Berlin (1994).

47. K.E. Oughstun, P. Wyns, and D.P. Foty, "Numerical determination of the signal velocity in dispersive pulse propagation," *J. Opt. Soc. Am. A* **6**, 1430–1440 (1989).

48. K.E. Oughstun and H. Xiao, "Failure of the quasimonochromatic approximation for ultrashort pulse propagation in a dispersive, attenuative medium," *Phys. Rev. Lett.* **78**, 642–645 (1997).

49. H. Xiao and K.E. Oughstun, "Failure of the group-velocity description for ultrawideband pulse propagation in a causally dispersive, absorptive dielectric," *J. Opt. Soc. Am. B* **16**, 1773–1785 (1999).

50. P.N. Butcher and D. Cotter, *The Elements of Nonlinear Optics*, Chapter 2, Cambridge University Press, Cambridge (1990).

51. L. Mandel and E. Wolf, *Optical Coherence and Quantum Optics*, Chapter 3, Cambridge University Press, Cambridge (1995).

52. H.M. Nussenzveig, *Causality and Dispersion Relations*, Chapter 1, Academic Press (1972).

53. K.E. Peiponen, E.M. Vartiainen, and T. Asakura, "Dispersion relations and phase retrieval in optical spectroscopy," in *Progress in Optics*, E. Wolf, Ed., Vol. 37, 57–94, Elsevier Press, Amsterdam (1997).

54. C.F. Bohren and D.R. Huffman, *Absorption and Scattering of Light by Small Particles*, Chapter 8, Wiley & Sons, New York (1983).

55. F.W.J. Olver, "Why steepest descents?" *SIAM Review* **12**, 228–247 (1970).

56. H.A. Lorentz, *The Theory of Electrons*, Dover Publications, New York (1952); L. Rosenfeld, *The Theory of Electrons*, Dover Publications, New York (1965).

57. J.A. Solhaug, K.E. Oughstun, J.J. Stamnes, and P.D. Smith, "Uniform asymptotic description of the Brillouin precursor in a Lorentz model dielectric," *J. Eur. Opt. Soc. A, Pure and Applied Optics* **7**, 575–602 (1998).

58. C.M. Balictsis and K.E. Oughstun, "Uniform asymptotic description of ultrashort Gaussian pulse propagation in a causal, dispersive dielectric," *Phys. Rev. E* **47**, 3645–3669 (1993).

59. K.E. Oughstun and C.M. Balictsis, "Gaussian pulse propagation in a dispersive, absorbing dielectric," *Phys. Rev. Lett.* **77**, 2210–2213 (1996).

60. C.M. Balictsis and K.E. Oughstun, "Generalized asymptotic description of the propagated field dynamics in gaussian pulse propagation in a linear, causally dispersive dielectric," *Phys. Rev. E* **55**, 1910–1921 (1997).

61. R. Smith, "The velocities of light," *Am. J. Phys.* **38**, 978–984 (1970).

62. M. Lisak, "Energy expressions and energy velocity for wave packets in an absorptive and dispersive medium," *J. Phys. A: Math. Gen.* **9**, 1145–1158 (1976).

63. J. Peatross, S.A. Glasgow, and M. Ware, "Average energy flow of optical pulses in dispersive media," *Phys. Rev. Lett.* **84**, 2370–2373 (2000).

64. K.E. Oughstun and N.A. Cartwright, "Dispersive pulse dynamics and associated pulse velocity measures," *J. Opt. A: Pure and Appl. Optics* **4**, S125–S134 (2002).

65. N.A. Cartwright and K.E. Oughstun, "On the pulse centroid velocity of the Poynting vector," *J. Opt. Soc. Am. A* **21**, 439–450 (2004).

66. E. Wolf, "Significance and measurability of the phase of a spatially coherent optical field," *Opt. Lett.* **28**, 5–6 (2003).

Ari Friberg, Prof. Wolf, and Kurt Oughstun (right) at SPIE Conference AM100: Tribute to Emil Wolf: Engineering Legacy of Physical Optics.

Kurt E. Oughstun is professor of electrical engineering and mathematics, and computer science. His areas of expertise are electromagnetic and optical field theory, wave propagation phenomena, and applied mathematics.

CHAPTER 21

TOTAL INTERNAL REFLECTION TOMOGRAPHY FOR THREE-DIMENSIONAL SUBWAVELENGTH IMAGING

David G. Fischer and P. Scott Carney

21.1 Introduction

Near-field imaging has gained a great deal of exposure in recent years for its ability to resolve subwavelength structure in optically thin media [1–9]. It has many variants, including total internal reflection microscopy (TIRM) [5–7], photon scanning tunneling microscopy (PSTM) [9,10], and near-field scanning optical microscopy (NSOM) [1–4], but common to all is the use of evanescent waves for illumination and/or detection. In many instances, image interpretation is difficult, owing to the complex interaction between the incident field and the sample, as well as between the scattered field and the near-field probe.

These difficulties are exacerbated when near-field techniques are applied to relatively thick samples. In addition to the problem of reconstructing a three-dimensional function of position (the dielectric susceptibility) from two-dimensional data sets (measurements of the scattered field in various planes), a thick object may exhibit strong scattering, with the consequence that the scattered field is a nonlinear function of the susceptibility. Even when the scattering is weak, the detected field may not be simply related to the subwavelength structure of the object, as it is, for example, in the case of diffraction from a 2D object [11,12].

In this chapter, we will discuss a new form of near-field imaging that makes use of TIRM measurements to produce computed reconstructions of the susceptibility of the sample. This method provides tomographic views and subwavelength resolution. Since the system is free from the moving (and often ill-characterized) probe present in PSTM and NSOM, the analysis of the problem is greatly simplified. Indeed the experiment is well modeled as a half-space problem and an exact solution for the Green's function (absent the sample) is well known. The linearized inverse scattering problem may then be solved in a computationally efficient and stable manner. In Sect. 21.2 we review the fundamentals of diffraction tomography and observe the emergence of the classical resolution limits. In Sect. 21.3 we examine the properties of near-field evanescent waves and the role they play in achieving super-resolution in a variety of near-field methods. In Sect. 21.4 we describe the basic TIRM measurement scheme and its extension to total internal reflection tomography (TIRT). In Sect. 21.5 we address the structure of the TIRT data and the development of fast, stable reconstruction algorithms, followed by numerical simulations in Sect. 21.6. Finally, in Sect. 21.7 we describe the instrument currently under construction at NASA to implement this modality.

21.2 Conventional Imaging

Conventional optical imaging systems are limited by diffraction to a resolution of approximately half the illuminating wavelength [13,14]. This so-called Abbe-Rayleigh resolution limit [15,16] is not a fundamental one, but is a consequence of the measurement scheme. In particular, the Abbe-Rayleigh limit arises only when the evanescent field in the near zone of the scatterer is inaccessible. While this was the case for more than a century after the theoretical predictions of Abbe and Rayleigh, the so-called near field can now be practically measured, as we will see later. Let us first review the conventional case.

Consider a scattering experiment in which a monochromatic field is incident on a localized dielectric medium with complex susceptibility $\eta(\mathbf{r})$ (see Fig. 1). For simplicity, we ignore the effects of polarization and consider the case of a scalar field $U(\mathbf{r})$ that obeys the reduced wave equation

$$\nabla^2 U(\mathbf{r}) + k^2 U(\mathbf{r}) = -4\pi k^2 \eta(\mathbf{r}) U(\mathbf{r}),\tag{1}$$

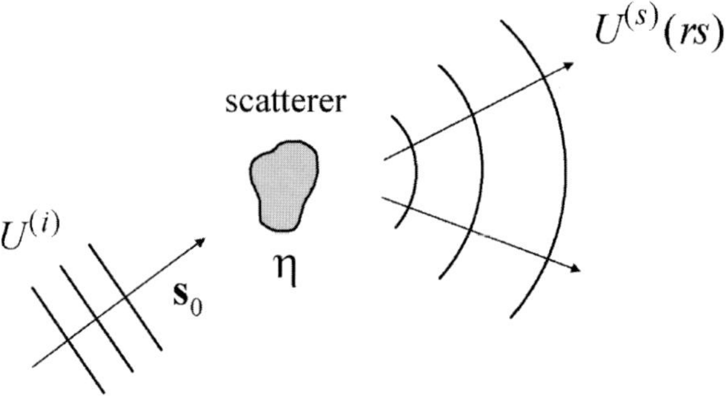

Figure 1 Illustrating the scattering geometry.

where k is the free-space wavenumber. Following standard procedures, we find that $U(\mathbf{r})$ satisfies the integral equation [13]

$$U(\mathbf{r}) = U^{(i)}(\mathbf{r}) + k^2 \int d^3r' G(\mathbf{r}, \mathbf{r}') U(\mathbf{r}') \eta(\mathbf{r}'), \tag{2}$$

where the outgoing Green's function $G(\mathbf{r}, \mathbf{r}')$ is given by

$$G(\mathbf{r}, \mathbf{r}') = \frac{\exp(ik|\mathbf{r} - \mathbf{r}'|)}{|\mathbf{r} - \mathbf{r}'|}, \tag{3}$$

and $U^{(i)}(\mathbf{r})$ is the incident field. We restrict ourselves to the weak-scattering approximation (also known as the first Born approximation [13]), which is particularly suited to the study of subwavelength structures. Accordingly, the scattered field $U^{(s)}(\mathbf{r}) = U(\mathbf{r}) - U^{(i)}(\mathbf{r})$ may be calculated perturbatively to lowest order in η with the result

$$U^{(s)}(\mathbf{r}) = k^2 \int d^3r' G(\mathbf{r}, \mathbf{r}') U^{(i)}(\mathbf{r}') \eta(\mathbf{r}'). \tag{4}$$

We take the incident field to be a unit amplitude plane wave traveling in the direction of the unit vector $\mathbf{s}_0$, i.e., $U^{(i)}(\mathbf{r}) = \exp(ik\mathbf{s}_0 \cdot \mathbf{r})$. Utilizing the asymptotic form of the outgoing Green's function given by

$$\frac{\exp(ik|\mathbf{r} - \mathbf{r}'|)}{|\mathbf{r} - \mathbf{r}'|} \sim \frac{\exp(ikr)}{r} \exp(-ik\mathbf{s} \cdot \mathbf{r}'), \tag{5}$$

as $kr \to \infty$ with the unit vector $\mathbf{s}$ kept fixed, we find that the scattered field in the far zone has the form

$$U^{(s)}(r\mathbf{s}) \sim \frac{\exp(ikr)}{r} a(\mathbf{s}, \mathbf{s}_0), \tag{6}$$

where the scattering amplitude is given by the expression [13]

$$a(\mathbf{s}, \mathbf{s}_0) = k^2 (2\pi)^3 \tilde{\eta}\big[k(\mathbf{s} - \mathbf{s}_0)\big], \tag{7}$$

and

$$\tilde{\eta}(\mathbf{K}) = \frac{1}{(2\pi)^3} \int d^3r\, \eta(\mathbf{r}) \exp(-i\mathbf{K} \cdot \mathbf{r}) \tag{8}$$

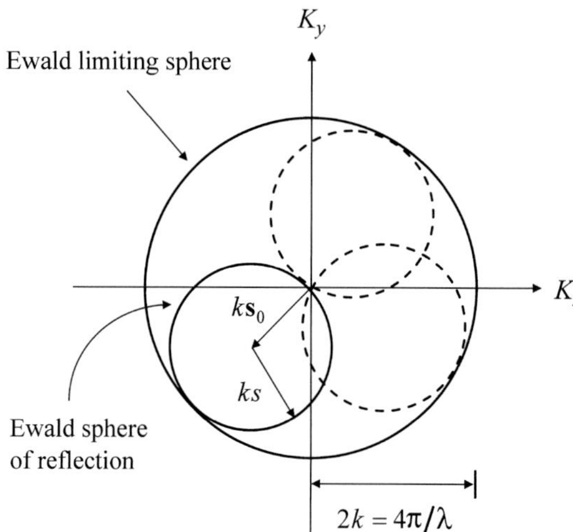

Figure 2 Illustrating the Ewald sphere of reflection and the Ewald-limiting sphere.

is the three-dimensional spatial Fourier transform of the dielectric susceptibility. This is the fundamental result of diffraction tomography and was first derived by Prof. Wolf in 1969 [14].

Equation (7) illustrates that, for a weakly scattering medium, there is a one-to-one mapping between the scattering amplitude for real $\mathbf{s}, \mathbf{s}_0$ and the low spatial frequency components of the dielectric susceptibility. Specifically, for a fixed direction of incidence $\mathbf{s}_0$, the scattering amplitude is mapped onto the surface of a sphere of radius k centered at $-k\,\mathbf{s}_0$ in the 3D Fourier space of the dielectric susceptibility (Fig. 2).

As $\mathbf{s}_0$ is varied, those surfaces fill a sphere of radius $2k$ centered at the origin (known as the Ewald limiting sphere [13,17]). Consequently, one may obtain a low-pass filtered estimate of the susceptibility, namely

$$[\eta(\mathbf{r})]_{\mathrm{LP}} = \int_{|\mathbf{K}|\leq 2k} d^3K \tilde{\eta}(\mathbf{K})\exp(i\mathbf{K}\cdot\mathbf{r}). \tag{9}$$

Furthermore, this estimate is unique since, due to the analyticity of $\tilde{\eta}(\mathbf{K})$,[*] the low-spatial frequency components of the dielectric susceptibility can, in principle, be analytically continued to the exterior of the Ewald limiting sphere [18]. In practice, however, techniques based on analytic continuation, such as band-limited extrapolation, are known to be unstable in the presence of measurement noise [19–21].

---

[*] It follows from the 3D version of the Plancherel-Polya theorem that, since the domain of localization of the medium is finite, the 3D spatial Fourier transform of the dielectric susceptibility is the boundary value on the real axes K_x, K_y, K_z of an entire analytic function of three complex variables.

21.3 Evanescent Wave Illumination

We have seen that the low-spatial frequency components of the susceptibility may be determined by illuminating a sample with homogeneous waves and measuring the scattered far field. To determine some subset of the Fourier components that lie outside the Ewald limiting sphere and consequently improve resolution, the illuminating field (or the measured field) must contain non-negligible evanescent field components.

Consider a general plane-wave form for the incident field:

$$U^{(i)}(\mathbf{r}) = \exp\left[ik(\mathbf{s}_{0\perp} \cdot \boldsymbol{\rho}) + s_{0z}z\right], \tag{10}$$

where $\mathbf{r} = (\boldsymbol{\rho}, z)$, $\mathbf{s}_0 = (\mathbf{s}_{0\perp}, s_{0z})$ and

$$s_{0z} = \begin{cases} \sqrt{1 - s_{0\perp}^2}, & \text{when } |\mathbf{s}_{0\perp}| \leq 1 \\ i\sqrt{s_{0\perp}^2 - 1}, & \text{when } |\mathbf{s}_{0\perp}| > 1. \end{cases} \tag{11}$$

For $|\mathbf{s}_{0\perp}| \leq 1$, the incident field is a homogeneous plane wave propagating in the direction of the unit vector $\mathbf{s}_0$, as seen in the previous section. It has a transverse scale length

$$\lambda_{0\perp} = \frac{2\pi}{k_{0\perp}} = \frac{\lambda}{|\mathbf{s}_{0\perp}|} \tag{12}$$

that is larger than the wavelength (see Fig. 3).

For $|\mathbf{s}_{0\perp}| > 1$, the incident field is an evanescent plane wave with a transverse scale length that is smaller than the wavelength (i.e., $\lambda_{0\perp} < \lambda$). Consequently, evanescent incident fields can probe and *encode* structure on spatial scales smaller

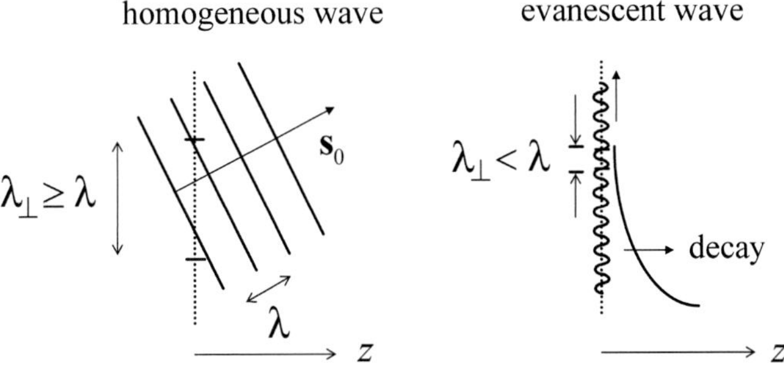

Figure 3 Illustrating the transverse scale length of homogeneous and evanescent waves.

than the illuminating wavelength. Unfortunately, evanescent waves decay exponentially from their point of origin with decay rate

$$\gamma = \frac{2\pi}{\lambda}\sqrt{|s_{0\perp}|^2 - 1},\tag{13}$$

so that a two-fold increase in probe scale length ($|s_{0\perp}| = 2$), for example, requires that the source of evanescent waves be located within a distance of $\lambda/10$ of the scattering structure. This is the reason the resolution of a conventional imaging system is $\lambda/2$; the source and detector are many wavelengths from the scattering medium, so that effectively, only homogeneous waves are present.

Evanescent waves for illumination may be generated by total internal reflection or by diffraction at a subwavelength aperture. The near-field scanning optical microscope (NSOM) is an instrument that uses the second mode of generation to locally confine a probe field at the surface of a sample (see Fig. 4) [1–4].

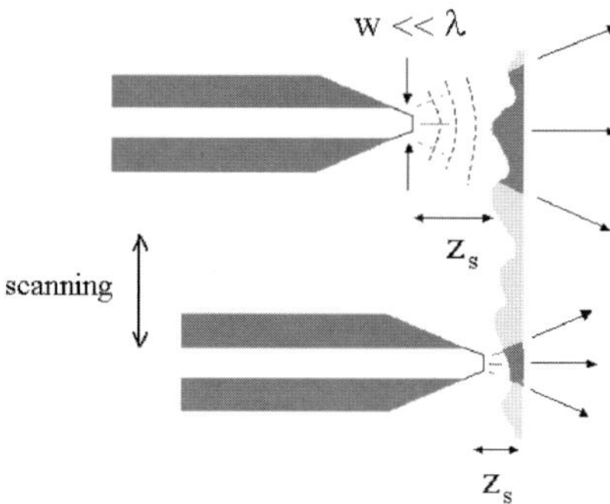

Figure 4 Illustrating the principle of NSOM. Z_s represents the fiber-sample separation.

Scanning the aperture across the sample and recording either the throughput or the reflected field as a function of probe position produces an image with subwavelength resolution. This technique was first proposed by Synge in 1928 [22] and experimentally realized through the use of microwaves by Ash and Nicholls in 1972 [23]. Today, a tapered optical fiber with subwavelength tip cross section w is typically used in place of the aperture for illumination (Fig. 4). When the fiber tip is very close to the surface of the sample, the resolution is on the order of the tip size. As the fiber is removed from the sample, the localization of the incident light (and the corresponding resolution) is reduced due to the loss (decay) of the evanescent waves. Figure 5 illustrates some of the first experimental images taken with an NSOM [1]. One can clearly see the loss of resolution as the fiber tip is incrementally distanced from the sample.

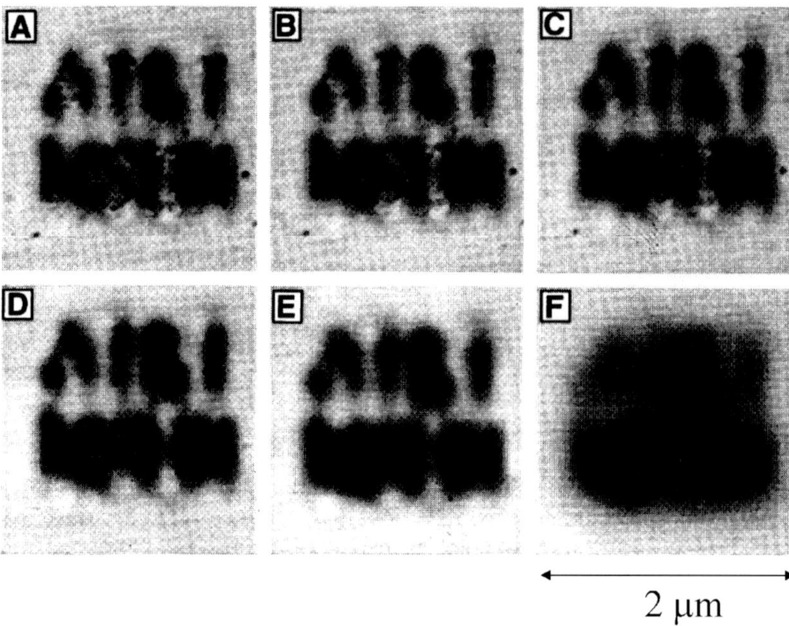

Figure 5 AT&T NSOM images from Ref. [1]. The fiber-sample separations are (a) near contact, (b) 5 nm, (c) 10 nm, (d) 25 nm, (e) 100 nm, and (f) 400 nm. [Reprinted with permission from E. Betzig and J.K. Trautman, "Near-field optics—microscopy, spectroscopy, and surface modification beyond the diffraction limit," *Science* **257**, 189–195 (1992). Copyright (1992) by AAAS.]

NSOM images have striking subwavelength detail, but their interpretation is often problematic, especially for thicker samples. This is due to the fact that an NSOM image is a measure of the field *outside* the sample, and the fundamental relationship (or mapping) between the field and the optical properties of the medium is usually not taken into account. Rather, the mapping is modeled effectively as a (one-view) projection or shadowgram. Furthermore, an NSOM image is typically defocused since the probe field contains a continuous distribution of both homogeneous *and* evanescent waves, and, consequently, its exact composition (and the resulting image) is a function of the fiber-sample separation.

This is illustrated in Fig. 6. We consider a 3D object composed of two stacked planes, one at $z = 0.005\lambda$ and another at $z = 0.405\lambda$ ($z = 0$ being the location of the evanescent wave source). Figures 6(a) and 6(b) illustrate the object structure in the two planes, respectively, and Figs. 6(c) and 6(d) illustrate their individual far zone images.

These are qualitatively similar to the images in Figs. 5(a) and 5(f) obtained from the AT&T NSOM. It is clear that the object structure contained in the plane $z = 0.405\lambda$ cannot be resolved from far zone scattered field measurements. Furthermore, for the composite object, the image (i.e., scattered field) of the un-

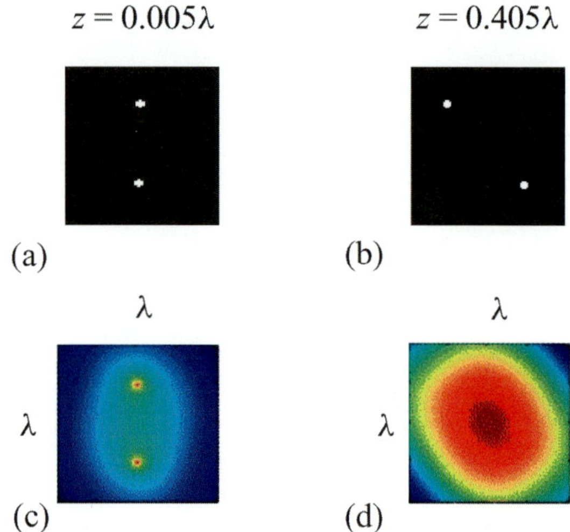

Figure 6 Illustrating NSOM imaging of a thick object.

resolved structure in the plane $z = 0.405\lambda$ will obscure and degrade the image of the structure in the plane $z = 0.005\lambda$.

It is clear from Fig. 6 that the field scattered from the deeper plane is simply defocused. This phenomenon is not particular to the near field, but since a physical lens for the near field is not currently available (due to the irretrievable loss of evanescent waves), a post-processing solution must be sought. That this need seemed to be so long overlooked might be explained by the fact that the data obtained at closest approach so much resembles an actual image. As we shall see in the next section, there exist modalities for which processing of the collected data is absolutely required.

21.4 Three-Dimensional Near-Field Imaging

Total internal reflection microscopy also involves illuminating a sample with evanescent waves [5–7]. In TIRM, an evanescent wave generated by total internal reflection illuminates an object, and the scattered (or radiated) light is collected by a standard microscope objective. Due to the exponential decay of the incident evanescent wave, the interrogation (or excitation) volume is limited in depth to a thickness of $z_p = 1/\gamma$, with more evanescent fields yielding narrower excitation regions [see Eq. (13)]. Consequently, TIRM provides a far-field (diffraction limited) intensity image of a subwavelength region near the exit face of the prism. While it does not provide subwavelength imaging, it does provide subwavelength localization in depth. As such it is extremely useful for surface inspection. TIRM can be taken a step further if images are recorded for a series of distinct evanescent incident fields. In this case, each image is the result of a different (and

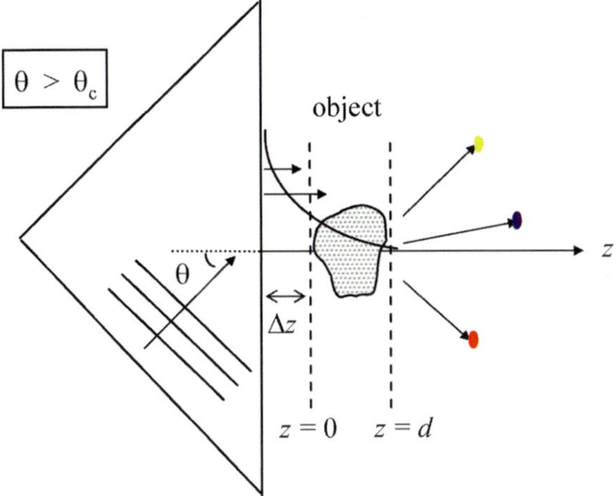

Figure 7 Illustrating the TIRT geometry. θ_c represents the critical angle.

unique) exponential weighting of the susceptibility, and the composite image stack can be inverted (possibly) to provide subwavelength resolution in depth (i.e., the z-dimension). In either the standard or extended TIRM case, the *transverse* resolution is limited to $\lambda/2$ since the excitation volume is not constrained in the transverse dimension and only the amplitude of the scattered field is measured in the far zone.

Total internal reflection tomography [24–27] is a coherent extension of TIRM in which the complex scattered field (both amplitude and phase) in the far zone is recorded for a diverse set of evanescent incident fields. The scattered field data is subsequently synthesized (or inverted) to yield an estimate of the 3D object structure. Central to the inversion is the physical connection between the structure of the susceptibility and the scattered field. This connection is most easily seen in the Fourier domain. Consider an evanescent plane wave with complex wave vector $k\mathbf{s}_0$, generated by total internal reflection, which illuminates a weakly scattering object that is confined to the region $0 \leq z \leq d$ (see Fig. 7).

The amplitude of the evanescent wave at the exit face of the prism ($z = -\Delta z$) is taken to be unity. It can be shown that the scattering amplitude, to first order, is given by [24]

$$a(\mathbf{s}, \mathbf{s}_0) = k^2 (2\pi)^2 \exp\left(-k|\mathbf{s}_{0z}|\Delta z\right) \int dK_z \tilde{\eta}\left[k(\mathbf{s}_\perp - \mathbf{s}_{\perp 0}), K_z\right] I_h\left(K_z; s_z, |s_{0z}|\right),$$

$$(14)$$

where

$$I_h(K_z; s_z, |s_{0z}|) = \frac{\exp\{[k|s_{0z}| + i(ks_z - K_z)]d\}}{k|s_{0z}| + i(ks_z - K_z)}.$$

$$(15)$$

There is no longer a one-to-one correspondence between the scattering amplitude and the 3D Fourier transform of the dielectric susceptibility, as there was in the case of conventional imaging [cf., Eq. (7)]. Specifically, the scattering amplitude $a(\mathbf{s}, \mathbf{s}_0)$ is now proportional to the weighted projection (i.e., the generalized Radon transform) along the K_z axis of all of those Fourier components of the susceptibility that have $\mathbf{K}_\perp = k(\mathbf{s}_\perp - \mathbf{s}_{0\perp})$. The weighting function I_h, which is independent of object thickness, determines the effective number of longitudinal Fourier components that contribute to the scattering amplitude. Its normalized modulus, shown in Fig. 8, is peaked at $K_z = k s_z$ and has a nominal width of $2\sqrt{3}k|s_{0z}|$.

As a result, the scattered field in the direction $\mathbf{s}$ carries information about a single high-frequency transverse Fourier component and many (both low- and high-frequency) longitudinal Fourier components, with the number of longitudinal Fourier components that effectively contribute to the scattered field, increasing with the degree of evanescence $|\mathbf{s}_{0\perp}| = \sqrt{1 + |s_{0z}|^2}$ of the incident field.

By contrast, for a 2D object with susceptibility $\eta(\mathbf{r}) = \beta(\boldsymbol{\rho})\delta(z)$, the scattering amplitude takes the form

$$a(\mathbf{s}, \mathbf{s}_0) = k^2(2\pi)^3 \exp\left(-k|\mathbf{s}_{0z}|\Delta z\right)\tilde{\beta}\left[k(\mathbf{s}_\perp - \mathbf{s}_{0\perp})\right], \tag{16}$$

where

$$\tilde{\beta}(\boldsymbol{\xi}) = \frac{1}{(2\pi)^2} \int_{z=0} d^2\rho\,\beta(\boldsymbol{\rho})\exp(-i\boldsymbol{\xi}\cdot\boldsymbol{\rho}), \tag{17}$$

is the 2D Fourier transform of $\beta(\boldsymbol{\rho})$. In this case, there is a one-to-one mapping between the high-frequency 2D Fourier components of the object and the scattering amplitude. This one-to-one mapping explains the success of NSOM in resolving subwavelength detail in 2D samples.

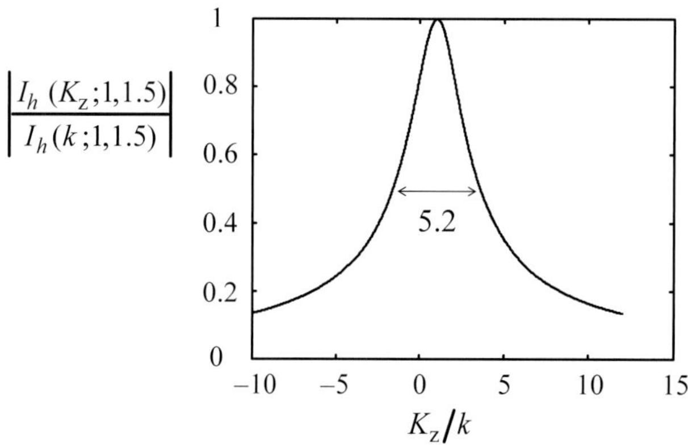

Figure 8 The modulus of the normalized weighting function, plotted for $|\mathbf{s}_{0z}| = 1.5$ and $s_z = 1$.

21.5 Image Reconstruction

As we have seen, for evanescent wave illumination and far-field detection, there is a *many-to-one* mapping between the 3D spatial Fourier components of the dielectric susceptibility and the scattered field. This leads to an inverse problem that is inherently ill-posed (or ill-conditioned in the discrete case) and generally underdetermined [21]. In TIRT, there are two general reconstruction strategies that can be employed for 3D structure determination: Fourier domain sampling and singular value decomposition. In either case, the inversion procedure must involve some type of regularization to deal with the problem of limited and noisy scattered field data. We will now discuss these two reconstruction strategies.

21.5.1 Fourier domain sampling

Fourier domain sampling is a procedure by which the Fourier components of the susceptibility are determined by discrete inversion of the generalized Radon transform represented by Eq. (14) [24,25]. This sampling is typically achieved by one of two measurement schemes. The first scheme involves sampling the Fourier transform of the susceptibility $\tilde{\eta}(\mathbf{K})$ (for a given object orientation) by independent variation of $\mathbf{s}_\perp$ and $\mathbf{s}_{0\perp}$, such that $|\mathbf{s}_\perp|$ remains fixed and $|\mathbf{s}_{0\perp}|$ assumes the discrete values $|\mathbf{s}_{0\perp}|^{(i)} = |\mathbf{s}_{0\perp}|^{(i-1)} + 2|\mathbf{s}_\perp|$, $(i = 1, 2, \ldots, N)$. A contiguous set of annular projection data is obtained, with each annular projection having the same width $2k|\mathbf{s}_\perp|$ but a different weighting $I_h^{(i)}$. This is illustrated in Fig. 9.

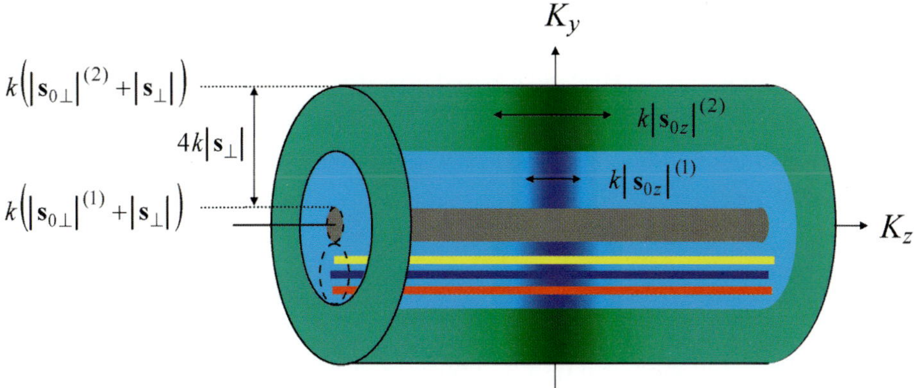

Figure 9 Illustrating two contiguous annular projection regions in the Fourier space of the dielectric susceptibility. The weighting of the projection data in each of the regions is clearly indicated. In addition, several typical projections in the first region are shown, which correspond to the color-coded measurement directions illustrated in Fig. 7.

It follows that the inner and outer radii of the composite annulus are $K_{\min} = k\lfloor|\mathbf{s}_{0\perp}|^{(1)} - |\mathbf{s}_\perp|\rfloor$ and $K_{\max} = k\lfloor|\mathbf{s}_{0\perp}|^{(N)} + |\mathbf{s}_\perp|\rfloor$, respectively. If the measurement procedure is then repeated for all possible object orientations, a complete multiview set of annular projection data is obtained that can be inverted for the 3D Fourier transform of the susceptibility. As an example, for $|\mathbf{s}_{0\perp}|^{(1)} = 1.2$, $|\mathbf{s}_{0\perp}|^{(2)} = 2$, and $|\mathbf{s}_\perp| = 0.4$, we obtain two annular projections that cover the spatial frequency range $0.8 \le K \le 2.4$. Alternatively, we could achieve the same Fourier coverage by choosing $|\mathbf{s}_{0\perp}| = 1.6$ and $|\mathbf{s}_\perp| = 0.8$. Ultimately, the choice depends upon measurement noise and the availability of a given measurement.

Since $|\mathbf{s}_\perp| \le 1$ and $|\mathbf{s}_{0\perp}|^{(i)} > 1$, there will be a spherical region of radius $K_{\min}$, centered about the origin, for which no Fourier information is available. This is not a problem in TIRT, since $K_{\min} \le 2k$, and the low-frequency Fourier components can be determined from conventional scattering data [14]. The upper limit $K_{\max}$, which ultimately determines the spatial resolution, is a function of the distance between the exit face of the prism and the object, the thickness of the object, and the measurement noise. For TIRT, the theoretical maximum is $K_{\max} = k(n + 1)$, so that the achievable spatial resolution is limited to $\lambda/(n + 1)$. For a prism with index of refraction $n = 2.4$, this yields a resolution limit of $\lambda/3.4$.

Alternatively, one can attempt a reconstruction for a fixed object orientation (i.e., a single view). By this scheme, we fix $\boldsymbol{\xi} = \mathbf{s}_{0\perp} - \mathbf{s}_\perp$ and vary $\mathbf{s}_{0\perp}$ and $\mathbf{s}_\perp$ independently. We obtain a linear system of equations with the same 3D Fourier components sampled by weighting functions of varying widths, which can be inverted. We then choose another $\boldsymbol{\xi}$ and repeat the process. The stability of the inversion depends upon the diversity that one can achieve in the variation of $\mathbf{s}_{0\perp}$ and $\mathbf{s}_\perp$.

21.5.2 Singular-value decomposition

The second reconstruction strategy is based upon the singular value decomposition of the linearized scattering kernel [26–28]. It follows from Sect. 21.2 that, for a weakly scattering medium, the scattering amplitude (i.e., the data function) has the general linear form

$$a(\mathbf{s}_\perp, \mathbf{s}_{0\perp}) = \int K(\mathbf{s}_\perp, \mathbf{s}_{0\perp}; \mathbf{r})\eta(\mathbf{r})d^3r \qquad (18)$$

or, in operator notation, $a = K\eta$. The Fourier space analysis leads us to conclude that there exists a great deal of redundancy in the accessible data space. It is clear that a low-pass version of the susceptibility (with subwavelength detail) may be constructed from any number of subsets of the data. To obtain the best possible reconstruction, it is therefore desirable to take some linear combination of the available data to find a best solution. It is sensible to take as the best solution the one that minimizes the squared discrepancy between the forward modal (i.e., operator) acting on the solution and the actual data. That is, a solution that makes

$$\|K\eta^+ - a\|^2 \tag{19}$$

a minimum. In the event that such a solution is not unique, the solution of minimum norm is chosen. The operator K^+ that connects the data to the solution is known as the pseudo-inverse:

$$\eta^+ = K^+ a. \tag{20}$$

To construct the operator K^+ explicitly, the singular value decomposition (SVD) may be employed [21]. This approach offers several advantages, namely that the structure of the linear transformation is readily apparent, the effective degrees of freedom may be observed, and many regularization methods may be implemented by modification of the spectrum that is obtained. The SVD of the kernel K is given by [28]

$$K(\mathbf{s}_\perp, \mathbf{s}_{0\perp}; \mathbf{r}) = \sum_n \sigma_n f_n^*(\mathbf{r}) g_n(\mathbf{s}_\perp, \mathbf{s}_{0\perp}), \tag{21}$$

where σ_n is the singular value associated with the singular functions f_n and g_n. The $\{f_n\}$ and $\{g_n\}$ are orthonormal bases for the object and image Hilbert spaces, respectively, and are eigenfunctions with eigenvalues σ_n^2 of the positive self-adjoint operators $K^+ K$ and KK^+:

$$K^+ K f_n = \sigma_n^2 f_n \tag{22a}$$

$$KK^+ g_n = \sigma_n^2 g_n. \tag{22b}$$

In addition, the f_n and g_n are related by

$$K f_n = \sigma_n g_n \tag{23a}$$

$$K^+ g_n = \sigma_n f_n. \tag{23b}$$

For a band-limited (i.e., physical) operator, the singular values are a monotonically decreasing function of the index n, as shown in Fig. 10 [21].

It is interesting to note that the singular value decomposition is the functional equivalent of eigenfunction decomposition, except that the orthonormal basis functions f_n and g_n for the object space and image space, respectively, are different.

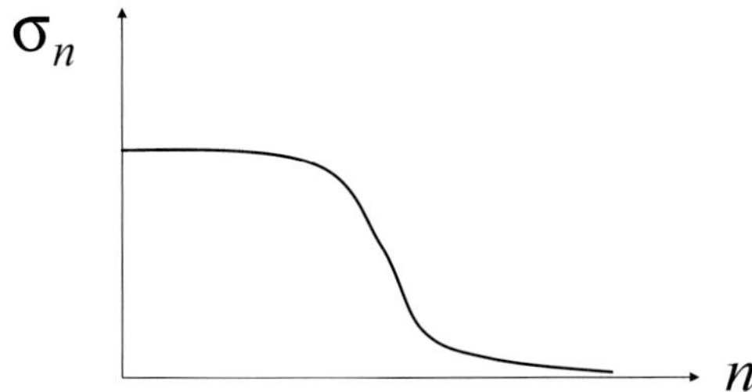

Figure 10 Illustrating the behavior of the singular values for a band-limited operator.

The kernel of the pseudo-inverse is readily obtained by the expression

$$K^{+}(\mathbf{r};\mathbf{s}_{\perp},\mathbf{s}_{0\perp}) = \sum_{n} \frac{1}{\sigma_{n}} f_{n}(\mathbf{r}) g_{n}^{*}(\mathbf{s}_{\perp},\mathbf{s}_{0\perp}), \tag{24}$$

where the sum is carried out over the n such that $\sigma_{n} \neq 0$. In the event that a true inverse exists, this pseudo-inverse reduces identically to it. When a true inverse does not exist, the pseudo-inverse yields the solution that minimizes the error on the orthogonal complement of the null space of K.

In practice, small eigenvalues σ_{n} in Eq. (24) lead to unstable reconstructions [21]. In order to regularize the results and effectively deal with noisy data, the pseudo-inverse may be modified in a number of ways. By the Tikhonov method, the inverse singular values $1/\sigma_{n}$ are replaced with $\sigma_{n}/(\sigma_{n}^{2} + \beta^{2})$, β being a tunable parameter. The modification of the spectrum is equivalent to solving a modified least-squared error problem, where in addition to minimizing the error, the quantity $\| \beta \eta^{+} \|^{2}$ is minimized simultaneously. When the spacing of the singular values is large compared to β, this method produces results very similar to a simple cutoff in the spectrum, i.e., a truncation in the sum over singular values to eliminate terms for which $\sigma_{n} \leq \sigma_{c}$, where σ_{c} represents the cutoff singular value. Regularization effectively imposes a band limit on the reconstructions and so connects the noise level to the resolution.

21.6 Numerical Simulation

To illustrate the utility and power of TIRT, let us consider the following two-dimensional computer simulation. We take two point scatterers of diameter $d = \lambda/50$ to be separated along the x-axis by $\lambda/4$ and along the z-axis by $\lambda/10$ (see Fig. 11).

The distance along the z-axis between the prism face and the first scatterer is taken to be $\lambda/4$. The index of refraction of the prism is $n = 2.4$. We take 21 equally spaced angles of illumination in the backward half-space and 21 equally spaced

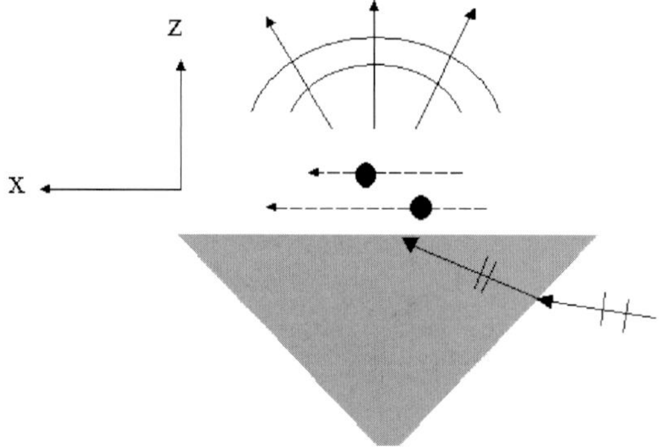

Figure 11 Illustrating the object used for the numerical simulation.

angles of detection in the forward half-space. This corresponds to the following ranges for the incident and scattered transverse wave vectors ($\mathbf{k}_{0\perp} = k\mathbf{s}_{0\perp}$ and $\mathbf{k}_{\perp} = k\mathbf{s}_{\perp}$):

$$0 \leq |\mathbf{k}_{0\perp}| \leq 2.4k$$
$$0 \leq |\mathbf{k}_{\perp}| \leq k. \tag{25}$$

In Figs. 12(a) and 12(b), the transverse information content of the scattered field is illustrated for this case[†] and the case of homogeneous illumination ($n = 1$). Figure 12(c) represents the region of overlap between the two cases. We see that the use of evanescent incident waves increases the information content of the scattered field by roughly 70%, with a corresponding increase in transverse resolution. But the real utility of the TIRT modality is the ability to resolve subwavelength features as a function of depth. This is clearly demonstrated in the following figures. Figure 13(a) illustrates the object reconstruction with no added noise (only machine error noise) and no regularization. The field of view of the figure is $\lambda/2$ by $\lambda/2$. Figure 13(b) illustrates the object reconstruction with 40 dB of additive noise that has been regularized. One can clearly resolve the two spheres. For purposes of comparison, Fig. 13(c) illustrates the object reconstruction for a prism index of refraction of $n = 1$. This corresponds to the case that only homogeneous waves are used for illumination, and we see that it is difficult, even in this noise-free case to resolve the spheres.

[†] In Figs. 12 and 15, we actually show the two-and-a-half-dimensional case (i.e., the case where the scattered field is measured in the entire forward half-space) for clarity. For the 2D case, the bands would be lines along the x-axis.

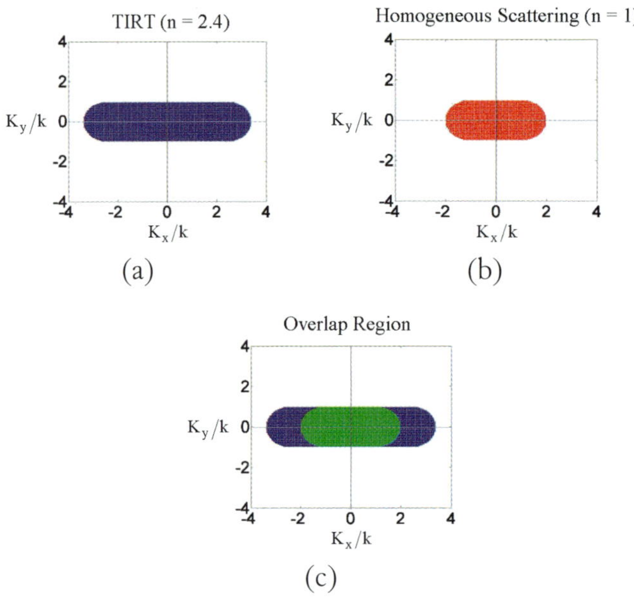

Figure 12 Illustrating the transverse information content of the scattered field for the numerical simulation.

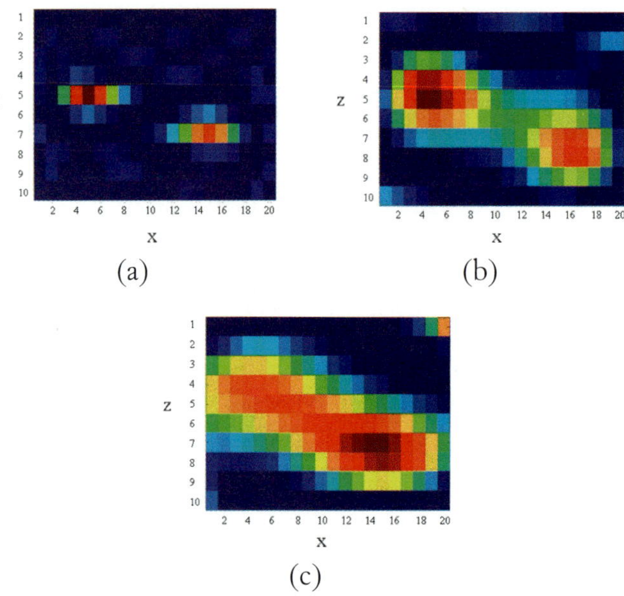

Figure 13 Numerical simulation.

21.7 Experimental Configuration

We are currently constructing a TIRT microscope at the NASA Glenn Research Center for use in biofluids research. Initially, the microscope will be two-and-a-half-dimensional, with a 3D version to follow. The configuration of the microscope is shown in Fig. 14.

The output from a frequency-doubled Nd: YAG laser is split into a sample beam (1%) and a reference beam (99%). The sample beam is scanned by a rotating mirror through an angular range $35° \leq \theta_s \leq 65°$ into a cylindrical glass prism of index of refraction $n = 1.9$. The critical angle of the prism is $\theta_c = 31.8°$, so the scanned beam is totally internally reflected and evanescent waves are generated within the transverse wave vector range $1.09k \leq |\mathbf{k}_{0\perp}| \leq 1.72k$. The scattered light is collected by a high-quality objective with a numerical aperture (NA) of 0.9 (collection half-angle of 64.2°). This corresponds to a scattered field transverse wave-vector range of $0 \leq |\mathbf{k}_{\perp}| \leq 0.9k$. The incident and scattered wave-vector ranges are much more restricted in this case than in the numerical simulation due to practical limitations of the scanning and collection optics, respectively. Finally, the scattered light is heterodyned at a high-resolution CCD camera with the beam from the sample arm, which has been phase modulated and expanded. Consequently, there is a direct mapping between the CCD output and the complex scattering amplitude over the aforementioned range. For each evanescent incident field, four CCD images corresponding to four different reference beam phase shifts $(0, \pi/2, \pi, 3\pi/2)$ are taken to allow unambiguous determination of the phase of the complex scattering amplitude.

For comparison with the numerical simulation, Fig. 15 illustrates the information content of the scattered field for the prototype TIRT microscope. We see that the TIRT modality extends and complements the structural information that one would obtain with homogeneous incident fields alone.

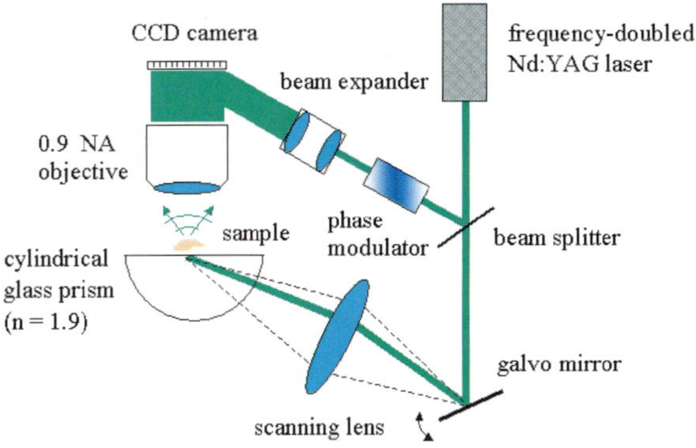

Figure 14 Illustrating the configuration of the prototype TIRT microscope.

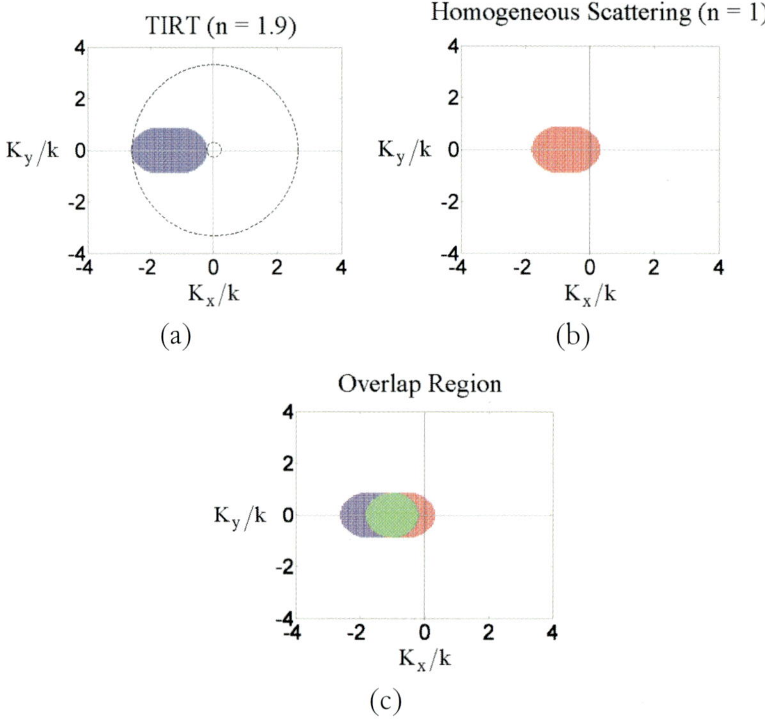

Figure 15 Illustrating the transverse information content of the scattered field for the prototype TIRT microscope.

21.8 Conclusions

We have discussed the role of evanescent waves in achieving the super-resolution of near-field imaging modalities. The essential mechanism, the interaction of super-oscillatory evanescent fields with the subwavelength structure of the sample, may be invoked by means other than the usual probe-induced field localization. In fact, a single evanescent plane wave may be generated at the surface of a prism and used as a source of illumination, as is routinely done in TIRM. A careful analysis of the scattered field reveals that the Ewald sphere of reflection is shifted by the incident wave vector as is well understood in diffraction tomography. However, because the magnitude of the transverse part of the evanescent wave vector is larger than the free-space wavenumber, the region of the Fourier space of the object that is accessible now includes points outside the usual Ewald limiting sphere. For this reason, when data are collected for a range of incident fields, the classical resolution limits may be overcome in a computed reconstruction of the sample susceptibility. This approach yields the additional benefit that the reconstruction is inherently tomographic and so a view of the sample structure as a function of depth may be obtained.

Acknowledgments

DGF would like to acknowledge the support of the NASA Microgravity Fluid Physics Program, as well as the contributions of Marius Asipauskas in the construction of the prototype TIRT microscope. PSC would like to acknowledge support from the National Aeronautics and Space Administration under Grant NAG3-2764 and the National Science Foundation under CAREER grant 0239265.

References

1. E. Betzig and J.K. Trautman, "Near-field optics—microscopy, spectroscopy, and surface modification beyond the diffraction limit," *Science* **257**(5067), 189–195 (1992).
2. U. Durig, D.W. Pohl, and F. Rohner, "Near-field optical-scanning microscopy," *Journal of Applied Physics* **59**(10), 3318–3327 (1986).
3. D. Courjon and C. Bainier, "Near-field microscopy and near-field optics," *Reports on Progress in Physics* **57**(10), 989–1028 (1994).
4. C. Girard and A. Dereux, "Near-field optics theories," *Reports on Progress in Physics* **59**(5), 657–699 (1996).
5. P.A. Temple, "Total internal-reflection microscopy—a surface inspection technique," *Applied Optics* **20**(15), 2656–2664 (1981).
6. D. Axelrod, "Total internal reflection fluorescence microscopy in cell biology," *Traffic* **2**(11), 764–774 (2001).
7. K. Stock et al., "Variable-angle total internal reflection fluorescence microscopy (VA-TIRFM): realization and application of a compact illumination device," *Journal of Microscopy-Oxford* **211**, 19–29 (2003).
8. K. Sarayeddine et al., "Scanning tunneling optical microscopy—a new tool for high-resolution nondestructive testing," *Institute of Physics Conference Series* **98**, 29–32 (1990).
9. J.M. Guerra, "Photon tunneling microscopy," *Applied Optics* **29**(26), 3741–3752 (1990).
10. D. Courjon, K. Sarayeddine, and M. Spajer, "Scanning tunneling optical microscopy," *Optics Communications* **71**(1–2), 23–28 (1989).
11. M.W. Kowarz, "Homogeneous and evanescent contributions in scalar near-field diffraction," *Applied Optics* **34**(17), 3055–3063 (1995).
12. P. Blattner, H.P. Herzig, and R. Dandliker, "Scanning near-field optical microscopy: transfer function and resolution limit," *Optics Communications* **155**(4–6), 245–250 (1998).
13. M. Born and E. Wolf, *Principles of Optics*, 7th ed., Cambridge University Press, Cambridge (1999).

14. E. Wolf, "Three-dimensional structure determination of semi-transparent objects from holographic data," *Optics Communications* **1**, 153–156 (1969).

15. E. Abbe, *Archiv. f. Mikroscopische Anat.* **9**, 413 (1873).

16. L. Rayleigh, *Philosophical Magazine* **8**, 261 (1879).

17. R.W. James, *The Optical Principles of the Diffraction of X-rays*, G. Bell & Sons, London (1948).

18. B.A. Fuks, *Introduction to the Theory of Analytic Functions of Several Complex Variables*, American Mathematics Society, Providence (1963).

19. G.A. Viano, *Journal of Mathematical Physics* **17**, 1160–1165 (1976).

20. T. Habashy and E. Wolf, "Reconstruction of scattering potentials from incomplete data," *Journal of Modern Optics* **41**(9), 1679–1685 (1994).

21. M. Bertero and P. Boccacci, *An Introduction to Inverse Problems in Imaging*, Institute of Physics, Bristol (1998).

22. E. Synge, "A suggested method for extending microscope resolution into the ultra-microscopic region," *Philosophical Magazine* **6**, 356–362 (1928).

23. E. Ash and G. Nicholls, "Super-resolution aperture scanning microscope," *Nature* **237**, 510–512 (1972).

24. D.G. Fischer, "The information content of weakly scattered fields: implications for near-field imaging of three-dimensional structures," *Journal of Modern Optics* **47**(8), 1359–1374 (2000).

25. D.G. Fischer, "Subwavelength depth resolution in near-field microscopy," *Optics Letters* **25**(20), 1529–1531 (2000).

26. P.S. Carney and J.C. Schotland, "Three-dimensional total internal reflection microscopy," *Optics Letters* **26**(14), 1072–1074 (2001).

27. P.S. Carney and J.C. Schotland, "Theory of total-internal-reflection tomography," *Journal of the Optical Society of America A—Optics Image Science and Vision* **20**(3), 542–547 (2003).

28. P.S. Carney and J.C. Schotland, "Near-field tomography," in *Inside Out: Inverse Problems and Applications*, G. Uhlmann and S. Levy, Eds., Cambridge University Press, Cambridge (2003).

A Personal Note

Looking at the participants of the conference and the authors herein, one is struck by the fact that, beyond his specific scientific achievements and scholarly writings, Prof. Wolf has amassed an amazing community of students, former students, collaborators, and researchers who build upon the foundation that he has laid. This community shares an intellectual thirst, a dedication to scholarship, trust, respect, and a somewhat unique sense of camaraderie. We would like to express our appreciation to Prof. Wolf for our time as his students, our ongoing collaboration, and for introducing us to this community in its entirety. He has been and remains for us the prototype mentor, to be imitated but never truly copied in our own labs.

David G. Fischer attended the Johns Hopkins University, receiving B.S. and M.S. degrees in Electrical Engineering. He pursued further graduate work at the University of Rochester, receiving a Ph.D. in Optics under the guidance of Professor Emil Wolf. His thesis involved methods of inverse scattering for random media. He is currently an optical scientist at the NASA Glenn Research Center in Cleveland, OH and codirector of the GRC Biophotonics Laboratory.

P. Scott Carney attended the University of Illinois at Urbana-Champaign from 1990 to 1994, obtaining a B.S. in Engineering Physics. He began his graduate studies in physics at the University of Rochester in the fall of 1994 and joined the research group of Prof. Emil Wolf in the fall of 1995. He obtained his Ph.D.

in 1999 and became a postdoctoral researcher at Washington University in Saint Louis with Prof. John Schotland. Prof. Carney joined the faculty of the Department of Electrical and Computer Engineering at the University of Illinois at Urbana-Champaign in the fall of 2001, where he is today.

❧Chapter 22☙

NANO-OPTICS: ATOMS IN THE NEAR FIELD

Vladilen S. Letokhov

22.1 Introduction

The classical scientific "best seller" by M. Born and E. Wolf [1] already contained some elements of nano-optics (the near-field Mie scattering), though this domain of science has only recently started developing in connection with the vigorous development of nanotechnology. The optical near field differs substantially from the optical far field usually used in optical systems, measuring much more than the wavelength of light. The optical near field arises near structures of subwavelength size or near boundaries. The word "near" always means a subwavelength distance. In other words, the near field is as if "tied" to these subwavelength structures or boundaries; i.e., it does not propagate and contributes nothing to the far field formed at distances much longer than the wavelength of light [2]. The optical near field localized within a subwavelength region forms the basis of nano-optics [3–5], a part of nanotechnology.

The nanolocalization of an optical field makes its intensity highly nonuniform in space. The strong spatial inhomogeneity of the optical near field, first, makes it possible to control the spectral characteristics of atoms placed in it (atoms near nanostructures); and second, gives rise to a gradient force that enables one to control the motion of the atoms [6] and forms the basis of atom optics [7–9]. Finally, the optical near field localized near metal nanostructures can interact in a resonance fashion with plasmons whose frequency is close to that of the optical field, the plasmons themselves being of a subwavelength size. This interaction highly enhances the intensity of the localized field and forms the basis of plasmon nano-optics [10] and entirely new effects, such as the extraordinary optical transmission through subwavelength hole arrays (the Ebbesen effect) [11].

In this chapter, we consider but a single avenue of inquiry in nano-optics that is associated with both the control of the spectral properties of atoms near nanostructures and the motion of atoms in the optical near field.

22.2 Atoms in the Vacuum Near Field of a Nanosphere

The presence of a nanostructure distorts the vacuum field that is usually represented as a 3D expansion of plane waves. As a result, the standard formulas for the spontaneous decay rate of atoms in free space become inapplicable to atoms in the immediate vicinity of nanostructures. The author and others have briefly reviewed the many works that consider this problem [12]. It can be illustrated by the vivid example of the drastic modification of the quadrupole transitions of an atom near a nanosphere [13].

It is well known that the rates of quadrupole transitions in the optical region are lower by a factor of $(a_0/\lambda)^2 \propto 10^{-8}$ to 10^{-6} than those of their dipole counterparts, where a_0 is the Bohr radius and λ is the radiation wavelength. In the presence of meso- and nanostructures, the characteristics of dipole transitions undergo substantial changes [14–16]. The probability of quadrupole transitions can rise materially in the vicinity of a nanosphere.

To explain the physical aspect of the problem, let us consider the amplitude of the decay of an excited atomic state into an unexcited (metastable) state, accompanied by the emission of a photon. In that case, the transition matrix element V_{fi} has the form

$$V_{\mathrm{fi}} \propto \int \psi_{\mathrm{f}}^*(\mathbf{r}) \nabla \psi_{\mathrm{i}} \mathbf{A}(\mathbf{r}, \omega) d^3\mathbf{r}, \tag{1}$$

where $\mathbf{A}$ is the vector potential of the emitted photon taking into account the presence of material bodies, and ψ_{f} and ψ_{i} are the wave functions of the final and initial atomic states, respectively.

As in the case of free space, the wave functions of the atom change more rapidly than the wave function of the photon, which makes it possible to expand the latter in a power series of coordinates in the neighborhood of the atom. With the dipole emission being forbidden, the first term of this series goes to zero, and the magnitude of the matrix element is determined by the characteristic photon wave function gradient in the vicinity of the atom:

$$V_{\mathrm{fi}} \propto \int \psi_{\mathrm{f}}^*(\mathbf{r}) \frac{\partial}{\partial r_{\mathrm{i}}} \psi_{\mathrm{i}}(\mathbf{r}) r_{\mathrm{j}} \frac{\partial}{\partial r_{\mathrm{oj}}} \mathrm{A}(\mathbf{r}_0, \omega) d^3\mathbf{r}. \tag{2}$$

A principal distinction of the case under consideration from the case of free space is that the scale of the photon wave function gradient $\mathbf{A}$ is generally governed

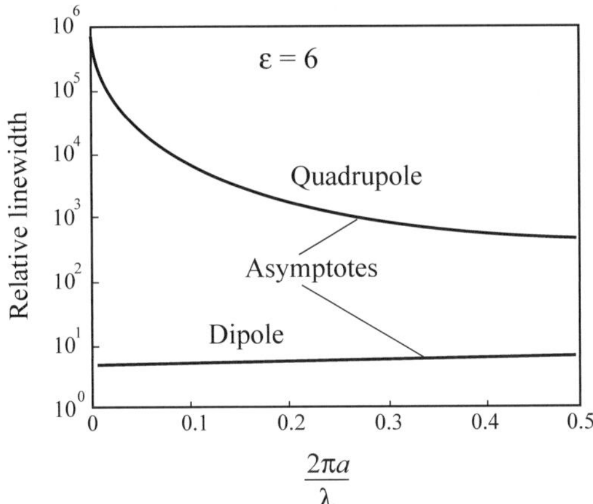

Figure 1 Relative change radiative linewidth (rate of spontaneous emission) for dipole and quadrupole transitions of atom located in close proximity to the surface of dielectric nanosphere (diamond) of radius a, where λ is the wavelength of emission.

not only by the radiation wavelength, but also by the characteristic dimensions of the problem. Moreover, in the case of an atom located near a material body with a small radius of curvature, a, the wave function gradient will mainly depend on this radius and not on the radiation wavelength in free space. As a result, it might be expected that the quadrupole emission probability will rise $(\lambda/a)^2$ times in comparison with that in the case of free space. A still greater increase in the probability should be expected for multipole transitions of higher order. Note the fact that where characteristic geometric parameters of the problem are close to the size of the atomic orbit, the radiation intensity may approach the intensity of dipole transitions.

Figure 1 presents calculation results for the increase of the radiative decay rate of a quadrupole transition of an atom near a dielectric nanosphere with a dielectric constant of $\varepsilon = 6$ (diamond) as a function its radius a. As can be seen, the effect for the quadrupole transition is very substantial.

The nanocurvature of the vacuum near field also has a perceptible effect on nonradiative transitions, for example, the resonance energy transfer between two atomic dipoles separated by the surface of a dielectric nanosphere [17].

22.3 Atom Nano-optics: Photon Dots and Photon Holes

Atom optics based on laser light fields [7] suffers from a number of restrictions of both principal and technological character that are due to the spatially "nonlocalized" character of the laser fields. This makes the elements of atom optics nonlocalized as well. Hence the defects of these elements: the aberrations of atomic lenses, low diffraction efficiency of atomic waves, restrictions on the contrast of interference fringes in atomic interferometers, and so on.

It is obvious from general physical considerations that the use of spatially local-ized potentials is preferable for the construction of the elements of atom optics. To date there are only two types of laser fields known to be spatially localized enough, namely, (a) the evanescent lightwave arising upon the total internal reflection of light (1D localization), and (b) the light field produced upon diffraction by struc-tures with a characteristic size less than the wavelength of light. The best known example for this type of localization is that occurring upon diffraction by an aper-ture small in comparison with the wavelength of light in an ideally conducting screen (Bethe hole), when a local 3D field maximum is formed in front of it, whose size is largely determined by the size of the small aperture [18–20].

A considerable shortcoming of a light field localized near a single aperture is that it is inexorably linked with the field of the attendant standing light wave; this is undesirable for atomic focusing, for atoms moving in this region may suffer spontaneous decay processes that affect the focusing quality. Balykin, Klimov, and Letokhov have proposed new types of spatially localized laser light fields with a characteristic size in the nanometer range that are free from the above shortcom-ing [21].

Figure 2 presents schematic diagrams for the production of a spatially local-ized light *nanofield*. Two plane conducting screens separated by a distance of d of the order of or less than the optical wavelength λ form a plane 2D waveguide for the laser light introduced into it from the side. As is well known [22], for a waveguide consisting of two ideally conducting parallel planes, there exist so-lutions of the Maxwell equation that permit the propagation of radiation though

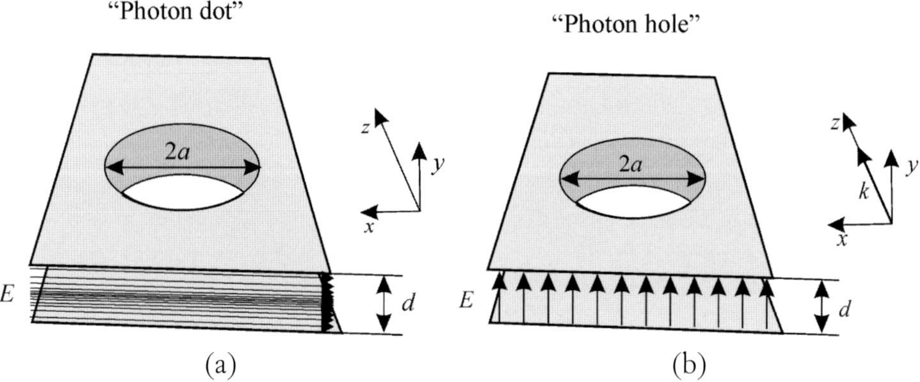

Figure 2 Geometry of forming of photon dot (a) and photon hole (b) by means of two coaxial nanoholes in plane conductive thin waveguide. [Reprinted from V.I. Balykin, V.V. Klimov, and V.S. Letokhov, "Atom nano-optics on the basis of 'photon dots' and 'pho-ton holes'," *JETP Lett.* **78**, 8–12 (2003). Copyright (2003) by IAPC.]

it, no matter how small its thickness d, even if it is much less than the radiation wavelength. Inside the waveguide, this solution coincides with a plane wave whose electric field strength vector is normal to the waveguide planes. Actually, such a system is a two-conductor transmission line and provides for a 2D nanometer-size localization of light. Let two small coaxial holes with a radius of $a \ll \lambda$ be made in the conducting screens [Fig. 2(b)]. If the diameters of the holes are substantially smaller than the wavelength of the radiation introduced, the radiation will practically not escape through these holes, but will be strongly modified near them. Actually, in the vicinity of the holes there takes place the reduction of the field in a region with a characteristic spatial size of the order of the hole diameter, i.e., much smaller than λ, where λ is the radiation wavelength. The volume of this region is $V \sim a^2 d \ll \lambda^3$. Such a field modification can naturally be called a "photon hole."

Another method for localizing light fields within nanometer-size regions is shown in Fig. 2(a). This method is a generalization of the localization near an aperture [20], but it is free from the shortcoming associated with the presence of the standing wavefield. Consider once more two ideally conducting planes with holes, but separated now by a distance of $d = \lambda/2$, i.e., $kd = \pi$. Figure 3 shows the field intensity distribution in the vicinity of the holes in the plane waveguide and inside the waveguide in the case where $d = \lambda/2$, $a = \lambda/4$. As one can see

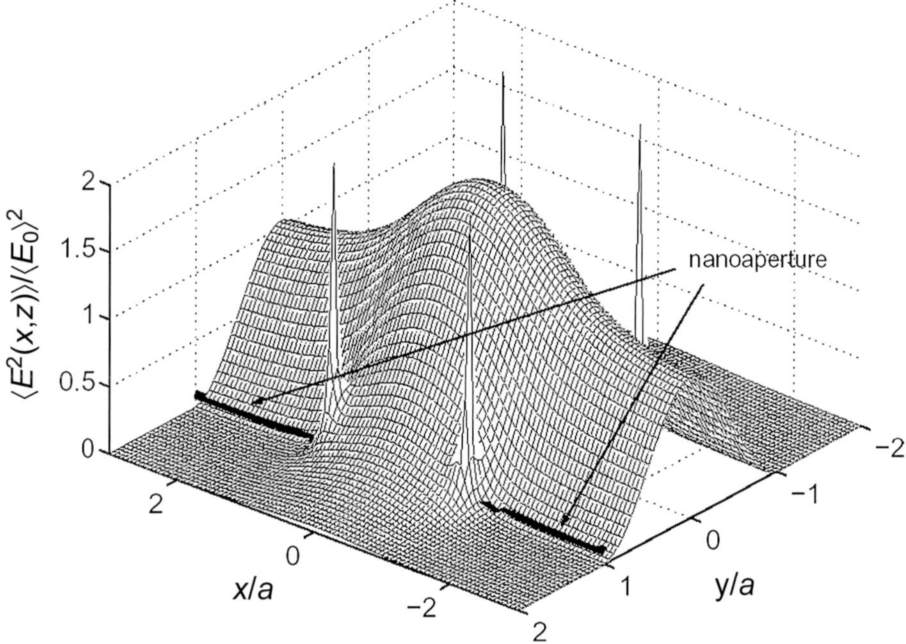

Figure 3 Intensity dependence of light for photon dot for $a/d = 0.5$. [Reprinted from V.I. Balykin, V.V. Klimov, and V.S. Letokhov, "Atom nano-optics on the basis of 'photon dots' and 'photon holes'," *JETP Lett.* **78**, 8–12 (2003). Copyright (2003) by IAPC.]

from the figure, the field drops rapidly enough outside the waveguide in the direction normal to the waveguide planes and has a maximum in the center of the waveguide; i.e., a "photon dot" is formed. The characteristic volume of such a dot is $V = \lambda/2a^2 \ll \lambda^3$. The sharp intensity peaks near the aperture edges are due to the assumption of the infinite conductivity of the waveguide walls. In waveguides with a finite wall conductivity, the peaks will be not so prominent. The magnitude of the maximum (reckoned relative to the case where the holes are absent) at $z = 0$ is twice that in the case of a single hole. This circumstance is due to the structural interference of the fields scattered by the holes and makes it possible to use weaker fields compared to those in the case of a single hole. Spatially localized light fields in the form of photon dots and photon holes can be used for both the focusing and localization of atoms.

It is possible to make a great number of hole pairs (arrays) and the corresponding number of localized fields (0-dimensional dots—photon holes and photon dots). Such an array makes it possible to simultaneously control many atomic beams. The use of such arrays will in turn make it possible to produce periodic arrays of localized atoms—atomic arrays [23]—with a period independent of the wavelength of light. Such periodic arrays may have properties similar to those of the planar photonic crystals [24], but as distinct from the latter, they may combine both arrays of photon dots and arrays of localized atoms. On the whole, the approach suggested in Balykin et al. [21], along with Klimov and Letokhov [19,20], forms the concept of "atom nano-optics"—atom optics based on optical nanofields.

22.4 Atom Manipulation in the Near Field

The concept of controlling atomic motion in a standing lightwave by means of the dipole (gradient) force [25] contains elements of a number of effects of atom optics, such as the reflection, focusing, and channeling (guiding) of atoms (Fig. 4). All these effects, starting with the reflection of atoms from the intensity gradient of an evanescent lightwave field (suggested in Cook and Hill [26] and first tested in Balykin et al. [27]) were investigated at many laboratories. Being unable to consider all of them in this brief paper, we refer the reader to the first book on atom optics [7] and the review [28] devoted to the motion of atoms in the optical near field.

Figure 4 schematically illustrates several methods of controlling atomic motion in an evanescent field that were either suggested or experimentally implemented at my laboratory at the Institute of Spectroscopy of the Russian Academy of Sciences: (a) reflection of atoms from the gradient of an intense light field (suggested in Ref. [26], first experimented in Ref. [27]); (b) guiding of atoms in a hollow optical waveguide (suggested in Ref. [29], experimented in Ref. [30]); (c) trapping of

Atom Optics with Evanescent Waves

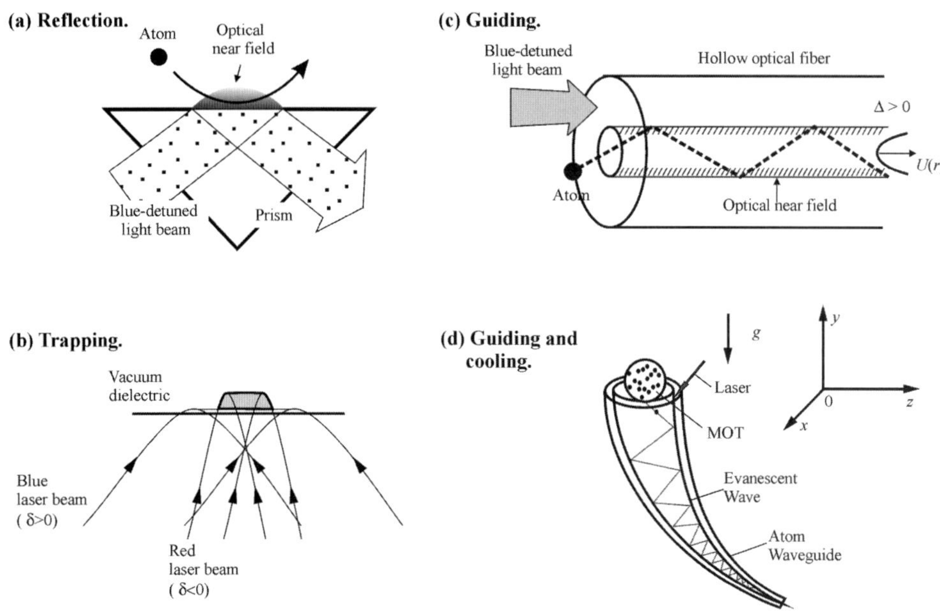

Figure 4 Basic effects of atomic nano-optics with evanescent waves: (a) reflection of atoms; (b) guiding of atoms in hollow fiber; (c) trapping of atoms; (d) reflection cooling and guiding.

atoms (suggested in Ref. [31]), and (d) trapping and cooling of atoms in a hornback waveguide (experimented in Ref. [32]). All these effects form the basis of atom optics.

Figure 5 illustrates the main effects of atom optics that can be observed in the near zone of a subwavelength-size hole in a screen (Bethe hole): (a) focusing of slow atoms [33], (b) trapping of atoms [20], (c) sorting of atoms according to their species [34], and (d) sorting of atoms according to their velocity (Maxwell demon) [35]. Though the near-field atom-optical experiments have only just been started, they hold much promise, especially in regard to the use of photon dots and photon holes considered above in Sect. 22.3.

22.5 Atom in the Near Field and Plasmons

The main problem in nano-optics is the localization of light inside a volume of space (matter) only a few nanometers in size by various techniques (Figs. 4, 5). Surface plasmons that can enhance electromagnetic energy into a tiny volume should play an important role here. We have already mentioned in the introduction the effect of the giant optical transmission of subwavelength, nanometer-scale aperture

Atom Optics in the Near Field

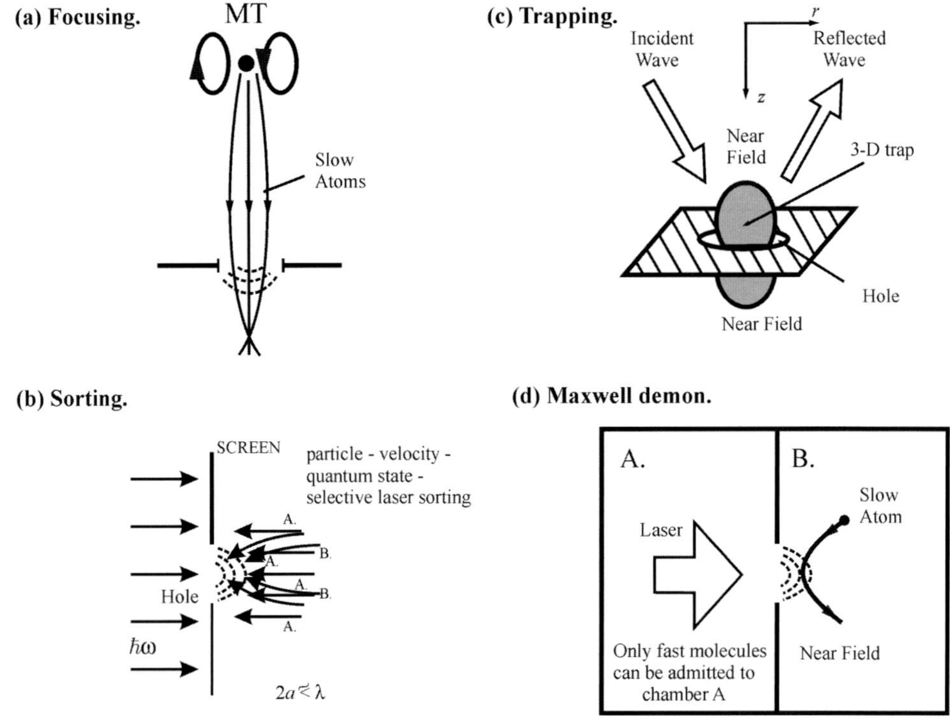

Figure 5 Basic effects of atomic nanooptics in near field of Bethe hole: (a) focusing; (b) trapping; (c) sorting.

in a metal film. This effect arises precisely owing to the plasmons of the metal film, when the incident light is in resonance with surface plasmons [36]. One can foresee the great role of plasmon effects in nano-optics: first, the vibration frequency of plasmons lies in the optical region and can be varied; second, the size of plasmons is in the submicron region; and third, plasmons possess mobility.

Klimov, Ducloy, and Letokhov recently considered a model of an apertureless scanning microscope with a prolate nanospheroid as a tip and excited molecule (dipole) as an object (Fig. 6) [37,38]. A prolate nanospheroid possesses resonance properties owing to the plasmons excited therein. The process of excitation of the object molecule and the process of emission of light by it are separated both in time and frequency. This means that the molecule excitation process is off resonance with the nanoscope needle, whereas the emission band of the molecule falls within the resonance range of the nanospheroid as a tip (Fig. 6).

When the molecular dipole is oriented normal to the surface of the microscopic stage and the nanotip scans the surface past the molecule at various heights h from it (from 0 to 60 nm), the emission rate of the excited molecule is observed to in-

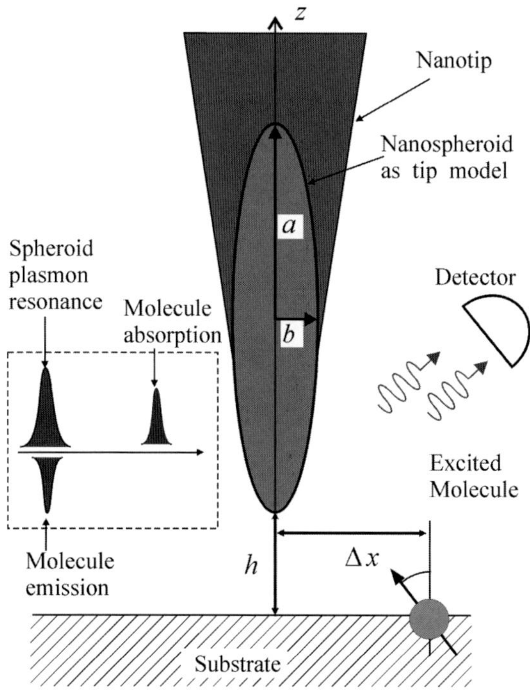

Figure 6 Geometry of an apertureless scanning microscope with an excited molecule as an object and location of frequencies of external pumping laser light for excitation of molecule on surface and plasmon resonance in the nanospheroid tip. [Reprinted from V.V. Klimov, M. Ducloy, and V.S. Letokhov, "Spontaneous emission of an atom placed near a prolate nanospheroid," *Eur. Phys. J.* D20, 133–148 (2002). Copyright (2002) by EDP Science.]

crease in a resonance fashion in a very small region (a few nanometers across) equal to the radius of the nanotip [Fig. 7(a)]. When the molecular dipole is oriented parallel to the surface, the spontaneous emission rate of the excited molecule is observed to be sharply suppressed [Fig. 7(b)]. The size of the region where the radiation intensity is reduced also amounts to a few nanometers; i.e., it coincides with the radius of curvature of the tip. Thus, in the given geometry of the apertureless near field microscope the *spatial resolution* is governed by the *radius of curvature* of the nanotip and hence can be hundreds of times better than the optical wavelength λ.

22.6 Applications

Near-field nano-optics is a new and fruitful domain of nanoscience. Numerous effects arising here make it possible to use the effect of nonpropagating light fields on both the internal degrees of freedom of atoms or molecules (alteration of radia-

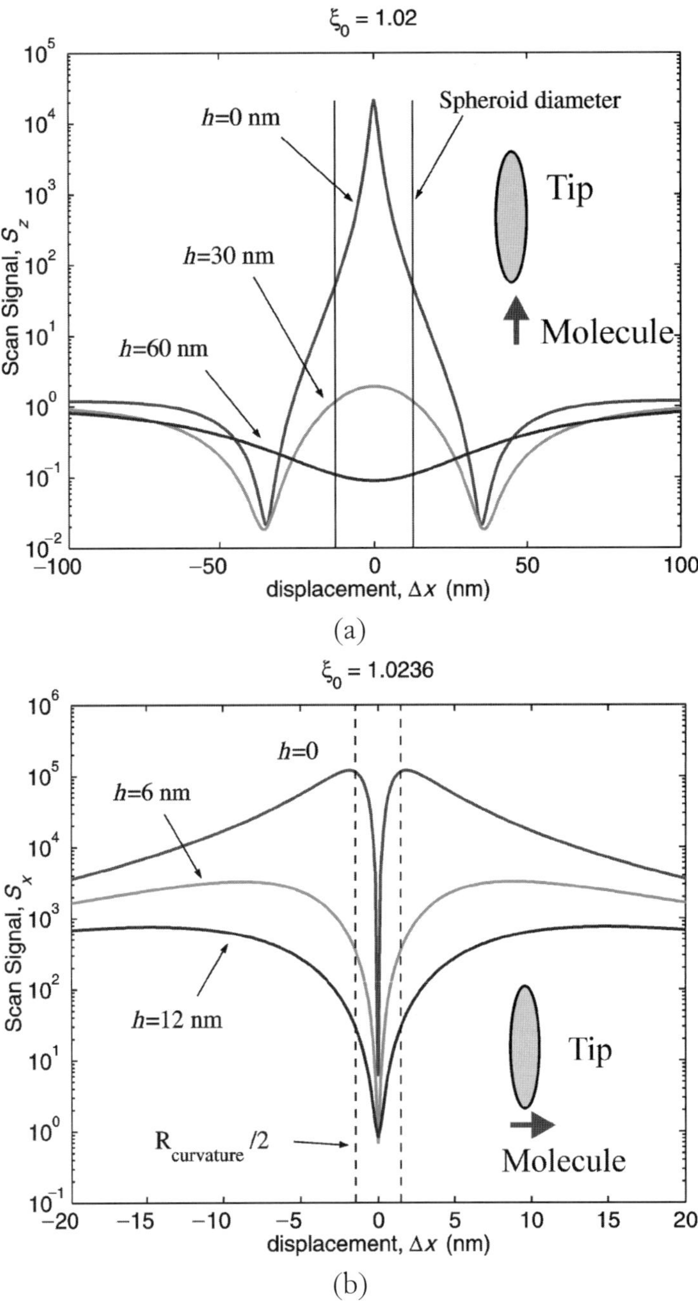

Figure 7 Radiated power (relative radiative losses) as a function of the dipole displacement Δx to a silver nanospheroid (Ag: $\varepsilon = -15.37 + 10.231$, $\lambda = 632.8$ nm, $f = 60$ nm is the half distance between focii of prolate nanospheroid) for various minimal heights h. Case of small detuning from the plasmon resonance ($\varepsilon = 1.02$): (a) dipole is oriented vertically along the z-axis; (b) dipole is oriented horizontally along the x-axis. [Reprinted from V.V. Klimov, M. Ducloy, and V.S. Letokhov, "A model of an apertureless scanning microscope with prolate nanospheroid as a tip and an excited molecule as an object," *Chem. Phys. Lett.* **358**, 192–194 (2002). Copyright (2002) by Elsevier.]

tive transition rates) and the translational degrees of freedom of atoms (reflection, guiding, trapping, cooling, etc.). One can, therefore, foresee various applications, at least in the following fields: (1) spectroscopy with a nanometer-high spatial resolution, (2) nanofabrication technology, and (3) high-density optical storage of information. All this makes nanooptics one of the important areas of nanoscience and nanotechnology.

Acknowledgments

The author is very thankful to Prof. V. I. Balykin and Prof. V. V. Klimov for their cooperation and stimulating discussions.

References

1. M. Born and E. Wolf, *Principles of Optics*, 7th ed., Cambridge University Press, Cambridge (1999).
2. E. Wolf and J.T. Foley, "Do evanescent waves contribute to the far field," *Optics Lett.* **23**, 16 (1998).
3. D.W. Pohl and D. Courjon, Eds., "Near field optics," *Proc. of ARW*, Kluwer Acad. Publ., Dordrecht (1993).
4. M.A. Paesler and P.J. Moyer, "Near-Field Optics," John Wiley & Sons, New York (1996).
5. S. Kawaza, M. Ohtsu, and M. Irrie, Eds., *Nano-Optics*, Springer (2002).
6. V.G. Minogin and V.S. Letokhov, *Laser Light Pressure on Atoms*, Gordon and Breach Science Publ. (1987).
7. V.I. Balykin and V.S. Letokhov, *Atom Optics with Laser Light*, Harwood Acad. Publ. (1995).
8. P. Meystre, *Atom Optics*, Springer (2001).
9. M. Ohtsu, Ed., *Near-Field Nano/Atoms Optics and Tecnhology*, Springer (1998).
10. S. Kawata, Ed., *Near-Field Optics and Surface Plasmon Polaritons*, Springer (2001).
11. T.W. Ebbesen, H.J. Lezec, H.F. Ghaemi, T. Thio, and P.A. Wolff, "Extraordinary optical transmission through sub-wavelength hole arrays," *Nature* **391**, 667–669 (1998).
12. V.V. Klimov, M. Ducloy, and V.S. Letokhov, "Spontaneous emission of atoms in the vicinity of nanobodies," *Quantum Electronics* **31**(7), 569–586 (2001).
13. V.V. Klimov and V.S. Letokhov, "Quadrupole radiation of an atom in the vicinity of dielectric microsphere," *Phys. Rev.* **A54**(3), 4408–4423 (1996).
14. J. Gerstein and A. Nitzan, "Spectroscopical properties of molecules interacting with small dielectric particles," *J. Chem. Phys.* **75**(3), 1139–1152 (1981).

15. H. Chew, "Transitions rates of atoms near spherical surfaces," *Phys. Rev.* **87**(2), 1355–1366 (1987).

16. V.V. Klimov, M. Ducloy, and V.S. Letokhov, "Radiative frequency shift and linewidth of an atom dipole in vicinity of a dielectric microsphere," *J. Mod. Optics* **43**, 2251–2267 (1996).

17. V.V. Klimov and V.S. Letokhov, "Resonance interaction between two atomic dipoles separated by the surface of a dielectric nanosphere," *Phys. Rev.* **A58**, 3235–3247 (1998).

18. H.A. Bethe, "Theory of diffraction by small holes," *Phys. Rev.* **66**, 163–182 (1944).

19. V.V. Klimov and V.S. Letokhov, "A simple theory of the near field in diffraction by a round aperture," *Optics Comm.* **106**, 151–154 (1994).

20. V.V. Klimov and V.S. Letokhov, "New atom trap configurations in the near field of laser radiation," *Optics Comm.* **121**, 130–136 (1995).

21. V.I. Balykin, V.V. Klimov, and V.S. Letokhov, "Atom nano-optics on the basis of 'photon dots' and 'photon holes'," *JETP Lett.* **78**, 8–12 (2003).

22. L.A. Vainstein, *Electromagnetic Waves* (Russian), Radio and Comm., Moscow (1988).

23. V.S. Letokhov, "Electromagnetic trapping of cold atoms: an overview," in *Trapped Particles and Fundamental Physics*, S.N. Atutov, R. Calabrese, and L. Moi, Eds., Kluwer Acad. Publ., Dordrecht (2002).

24. E. Yablonovich, "Photonic band-gap structures," *J. Opt. Soc. Am.* **B10**, 283 (1993).

25. V.S. Letokhov, "Doppler line narrowing in a standing light wave," *JETP Lett.* **7**, 272–274 (1968).

26. R.J. Cook and R.K. Hill, "An electromagnetic mirror for neutral atoms," *Optics Comm.* **43**, 258–260 (1982).

27. V.I. Balykin, V.S. Letokhov, Yu. B. Ovchinnikov, and A.I. Sidorov, "Quantum-state-selective mirror reflection of atoms by laser light," *Phys. Rev. Lett.* **60**, 2137–2140 (1988); Erratum **61**, 902 (1988).

28. V.V. Klimov and V.S. Letokhov, "Atom optics in laser near field," *Laser Physics* **6**, 475–500 (1996).

29. M.A. Ol'shanny, Yu.B. Ovchinnikov, and V.S. Letokhov, "Laser guiding of atoms in a hollow optical fiber," *Optics Comm.* **98**, 77–79 (1993).

30. M.J. Renn, D. Montgomery, O. Vdovin, D.Z. Anderson, C.E. Wieman, and E.A. Cornell, "Laser-guided atoms in hollow-core optical fibers," *Phys. Rev. Lett.* **75**, 3253–3256 (1995).

31. Yu.B. Ovchinnikov, D.V. Larushin, V.I. Balykin, and V.S. Letokhov, "Cooling of atoms during reflection from the evanescent light wave," *Pis'ma Zh. Eksp. Teor. Fiz.* **62**, 102–107 (1995).

32. V.I. Balykin, D.V. Larushin, M.V. Subbotin, and V.S. Letokhov, "Increase of atomic phase density in hollow laser waveguide," *JETP Lett.* **63**, 802–807 (1996).

33. V.I. Balykin, V.V. Klimov, and V.S. Letokhov, "Laser near-field lens for atoms," *JETP Lett.* **59**, 235–238 (1994); *J. Phys. II France* **4**, 1981–1997 (1994).

34. V.V. Klimov and V.S. Letokhov, "Selective sorting of neutral atoms and molecules by the gradient dipole force in the near field of laser radiation," *Optics Comm.* **110**, 87–93 (1994).

35. V.S. Letokhov, "Laser Maxwell's Demon," *Contemporary Physics* **36**(4), 235–243 (1995).

36. T. Thiro, K.M. Pellerin, R.A. Linke, H.J. Lezec, and T.W. Ebbesen, "Enhanced light transmission through a single subwavelength aperture," *Optics Lett.* **26**, 1972–1974 (2001).

37. V.V. Klimov, M. Ducloy, and V.S. Letokhov, "A model of an apertureless scanning microscope with prolate nanospheroid as a tip and an excited molecule as an object," *Chem. Phys. Lett.* **358**, 192–194 (2002).

38. V.V. Klimov, M. Ducloy, and V.S. Letokhov, "Spontaneous emission of an atom placed near a prolate nanospheroid," *Eur. Phys. J.* **D20**, 133–148 (2002).

Vladilen S. Letokhov was born on November 10, 1939, Irkutsk, Siberia, USSR. Education: 1957–1963: Moscow Physical-Technical Institute, Dolgoprudnyi, Moscow Region, USSR; 1963–1966: P.N. Lebedev Physical Institute, Moscow, graduation under Prof. N.G. Basov. Professional Experience: P.N. Lebedev Physical Institute, Moscow: Researcher (1966–1970); Ph.D. (1969); Degree of Science (1970); MIT Cambridge, Visiting Professor (1970); Institute of Spectroscopy of USSR Academy of Sciences, Troitsk, Moscow Region, Russia: Associate Director for Research (1971–1989), and Head of Laser Spectroscopy Dept. (1970-present); Professor of Quantum Optics Moscow Physical-Technology Institute, Dolgoprudnyi, (1972-present); UCLA, Los-Angeles, Blacet Lecturer in Physical Chemistry (1989); Israel Academy of Sciences, James Franck Lecturer (1989); Bayreth University, Emil Warburg Lecturer, Germany (1990); Iowa University, Ida Beam Lecturer, USA (1990); Cleveland Clinic Foundation, Intern, visiting Professor, USA, (1991), Berkeley University, Regent Professor, CA, USA (1993); University Paris-Nord, Part-time Professor, France (1993–1996); University of Arizona, Part-time Professor, USA (1996–1997); École Normale Supérieure, Paris, Part-time Condorcet Chair (1998); Lund University, T. Erlander Professor (2000); Lund University, Lund Observatory, Visiting Professor (2001–2004). Awards:

Lenin State Prize (1978); Honorary Jubilee Intern. Medal of 600 Years Anniversary of Heidelberg Univ. (1986); Docteur Honoris Causa, Univ. Paris-Nord (1995); Quantum Electronics Prize of European Physical Society (1998); Rojdestvenskii Prize of Russian Academy of Sciences (2001); State Prize of Russian Federation in science and technology (2002). Coeditor and Member of Editorial Board: numerous international journals. Associations: Fellow of Optical Society of America, USA (1977); External Member of Max Planck Society, FRG (1989); Member of European Acad. of Arts and Sciences (1996); Fellow of World Innovation Foundation (2000); Member of European Academia (2002); Member of Leibniz Society, FRG (2003). Publications: more than 800 scientific articles and 12 monographs in the field of laser physics, spectroscopy, chemistry and biomedicine, including 15 recent papers on atomic astrophysics.

❧Chapter 23❧

Coherence Issues in Flatland

Adolf W. Lohmann, Avi Pe'er, and Asher A. Friesem

E. A. Abbott wrote his famous story 120 years ago [1]. He describes the life of the 2D creatures living in Flatland. We, as 3D creatures, are able to inspect everything in Flatland without leaving a trace behind. We also may influence the laws of physics in Flatland, for example by tilting it, which would change the gravity in Flatland.

We have shown how to convert the Flatland world from pure science fiction to experimental reality, at least the optical aspects of reality [2–4]. Our experimental setup is shown in Fig. 1. A 3D point source and a 3D lens illuminate Flatland ($y = 0$) with a tilted plane wave. The Flatland physicist will observe a wavelength, which differs from the 3D wavelength λ by a directional factor:

$$\Lambda = \frac{\lambda}{\cos\alpha}. \tag{1}$$

The tilt angle α, and hence the Flatland wavelength, can be varied easily by the 3D physicist, who shifts the 3D point source along the y-axis and thereby varies α.

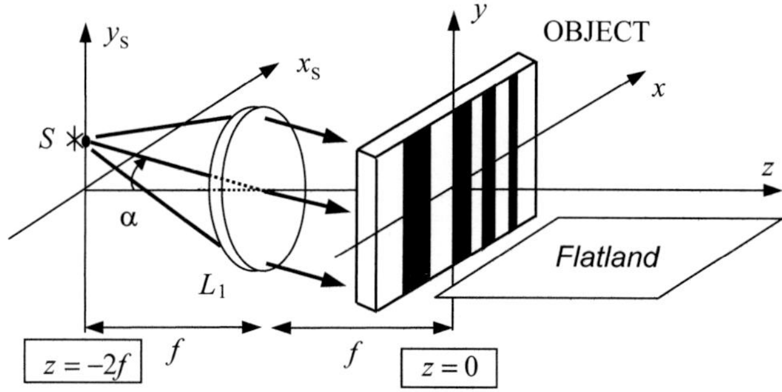

Figure 1 Generic setup for optics in Flatland. A tilted plane wave illuminates the object $U_0(x)$. The wave field $U(x,z)$ may be observed at a plane $y =$ constant, at $z \geq 0$.

491

The theory is based on the 3D wave equation:

$$\Delta_3 V(x,y,z) + k^2 V(x,y,z) = 0, \tag{2}$$

$$\Delta_3 = \frac{\partial^2}{\partial x^2} + \frac{\partial^2}{\partial y^2} + \frac{\partial^2}{\partial z^2}. \tag{3}$$

The tilted plane wave in the region $z < 0$ is

$$V(x,y,z) = \exp\left[ik(y\sin\alpha + z\cos\alpha)\right]. \tag{4}$$

The object $U_0(x)$ at $z = 0$ generates immediately behind it

$$V(x,y,z=0) = U_0(x)\exp\left[iky\sin\alpha\right]. \tag{5}$$

The y-dependence will remain the same for $z > 0$ because there is no piece of hardware with a y-dependent structure. Hence, the 3D wavefield is separable:

$$V(x,y,z) = U(x,z)\exp\left[iky\sin\alpha\right]. \tag{6}$$

Insertion into the wave equation yields

$$\exp(iky\sin\alpha)\left[\Delta_2 U(x,z) + k^2(1 - \sin^2\alpha)U(x,z)\right] = 0, \tag{7}$$

that leads us to the 2D wave equation

$$\Delta_2 U(x,z) + k^2\cos^2\alpha\, U(x,z) = 0, \tag{8}$$

$$\Delta_2 = \frac{\partial^2}{\partial x^2} + \frac{\partial^2}{\partial z^2},$$

$$k\cos\alpha = \frac{2\pi}{\lambda}\cos\alpha = \frac{2\pi}{\Lambda}. \tag{9}$$

The wavelength in Flatland is apparently:

$$\Lambda = \frac{\lambda}{\cos\alpha}. \tag{10}$$

So much about the theory, which has been presented in more detail before [2]. We did show that it is really Λ, not λ, that determines the outcome of historical interference experiments such as Young's double-slit experiment [3]. Grating diffraction served as a tool for measuring the wavelength. It yielded Λ, not λ.

We used monochromatic laser light as before, when manipulating the 2D coherence, as observed in Flatland. To that end we blurred the effective size of the point source (Fig. 1). A vertical stretching in y-direction meant a blur of the angle α and hence also a blur of the 2D wavelength $\Lambda = \lambda/\cos\alpha$. Hence, in this manner we manipulated the spectral aspect of coherence. The spatial aspects of partial 2D coherence can be manipulated by blurring the lateral x-location of the point source. This blurring procedure may be implemented, for example, by moving the point source laterally while recording the outcome of an experiment. Instead of moving the source laterally, it is more convenient to move the lens laterally.

In a more general experiment one might also modulate the radiance of the moving point source. That corresponds to a partial coherence experiment with a temporally incoherent light source, which has a spatial structure. In other words, this would be the implementation of the van Cittert-Zernike theorem.

That theorem describes a special case of the Wolf equations [5]. They are the generalization of the Helmholtz wave equation from strictly coherent to partially coherent. The Helmholtz equation is

$$\Delta V(x,y,z) + k^2 V(x,y,z) = 0, \tag{11}$$

abbreviated as

$$H[V(x,y,z)] = 0,$$
$$H = \Delta + k^2. \tag{12}$$

The definition of the coherence function is in our notation:

$$\Gamma(x_1,y_1,z_1;x_2,y_2,z_2) = \langle V(x_1,y_1,z_1)V^*(x_2,y_2,z_2)\rangle. \tag{13}$$

The operators H_1 and H_2 commute with the averaging brackets. Hence, it is

$$H_1[\Gamma] = (\Delta_1 + k^2)\Gamma = 0, \quad H_2[\Gamma] = (\Delta_2 + k^2)\Gamma = 0, \tag{14}$$

where

$$\Delta_1 = \frac{\partial^2}{\partial x_1^2} + \frac{\partial^2}{\partial y_1^2} + \frac{\partial^2}{\partial z_1^2},$$
$$\Delta_2 = \frac{\partial^2}{\partial x_2^2} + \frac{\partial^2}{\partial y_2^2} + \frac{\partial^2}{\partial z_2^2}. \tag{15}$$

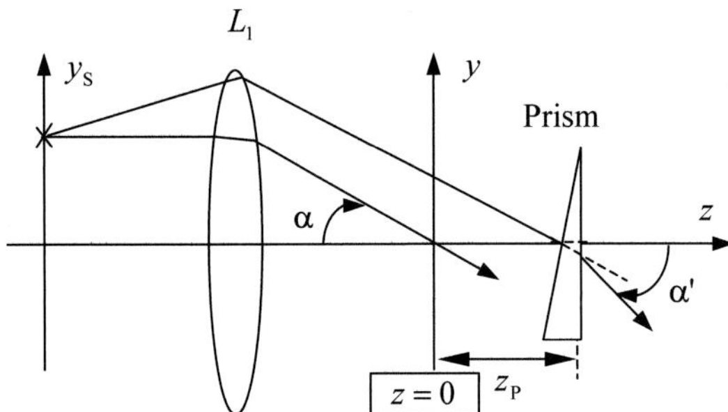

Figure 2 In-flight change of the wavelength, due to deflection by a prism.

These are the Wolf equations, written for the 3D world. But it is obvious from the previous discussions that Wolf's equations also hold in Flatland.

The 3D physicist who provides Flatland with light may use a He-Ne laser for convenience. The 2D wavelength will be larger because of the cosine term in Eq. 11. The tilt angle at the object plane ($z = 0$) depends on the geometry of the setup (Fig. 1). The 3D physicist is capable of changing Γ in flight, for example by inserting a prism at $z = z_p$ (Fig. 2). The new wavelength beyond $z = z_p$ may be larger or smaller, depending on the direction of the prism wedge (in y-direction). With this amazing capability, the 2D physicist can observe a Hubble effect somewhere far to the right (z: large). Two stars can be implemented as two slits at $z = 0$:

$$U_0(x) = A_1\delta(x - x_1) + A_2\delta(x - x_2). \tag{16}$$

Their wavelengths Λ_1 and Λ_2 will be different if two different prisms are placed upon the two slits. If the cosmology in Flatland is based on a big bang hypothesis, and if the Flatland physicist knows about the Doppler effect, the physicist will declare the star with the larger Λ to be farther away than the other star. We, with our superior 3D intelligence, of course know the true reality. Are there perhaps hidden somewhere some super-smart ND creatures ($N > 3$)?

Before closing we would like to express our respect to the two pioneers of coherence theory: Emil Wolf and Leonard Mandel. In addition to many specific contributions, they coordinated and fostered the field of coherence theory by starting the series of Rochester coherence conferences in 1961 and by writing a conclusive monograph [6]. In Fig. 3(a) you see a computer-generated Fourier hologram, whose macroscopic brightness distribution shows Emil. The holographic information is encoded as micro positions of many dots (128 × 128). The optical reconstruction shows Leonard as Fig. 3(b). For the sake of balance we made a similar experiment, but with switching the roles of Leonard Mandel and Emil Wolf [Figs. 3(c) and (d)].

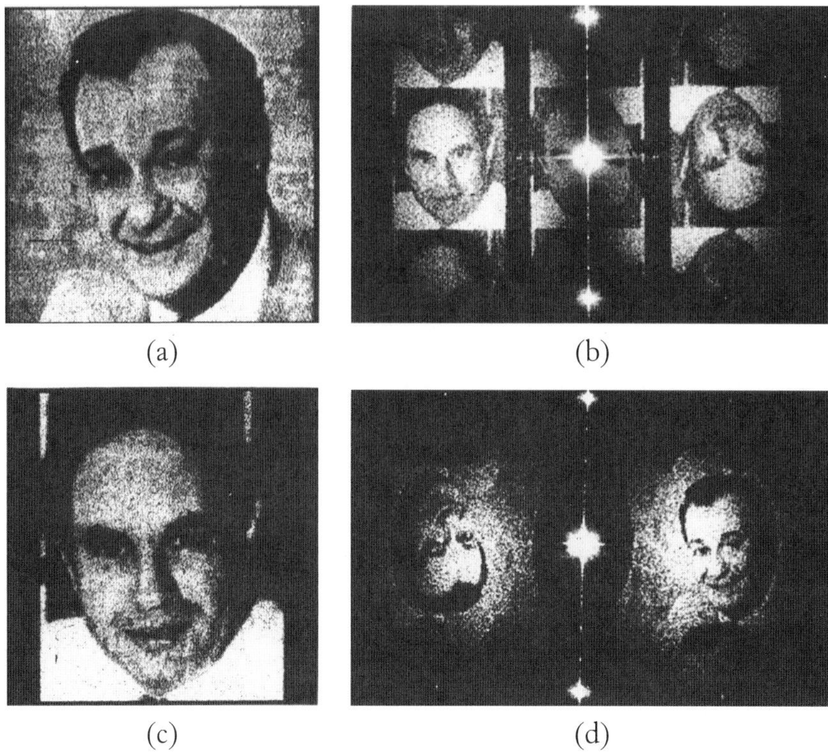

(a)   (b)
(c)   (d)

Figure 3 The Wolf-Mandel entanglement. Figure 3(a) shows a computer-generated Fourier hologram, where the macroscopic intensity at the hologram plane matches the face of Emil Wolf; the optical reconstruction shown in (b) matches the face of Leonard Mandel. Figure 3(c) and (d) are similar, but with the roles of Wolf and Mandel interchanged for the sake of balance.

References

1. E.A. Abbott, *Flatland, a Romance of Many Dimensions*, 6th ed., Dover Publications Inc. (1952).
2. A.W. Lohmann, A. Pe'er, D. Wang, and A.A. Friesem, "Flatland optics I: Fundamentals," *J. Opt. Soc. Am. A* **17**, 1755–1762 (2000).
3. A.W. Lohmann, A. Pe'er, D. Wang, and A.A. Friesem, "Flatland optics II: Basic experiments," *J. Opt. Soc. Am. A* **18**, 1056–1061 (2001).
4. A.W. Lohmann, A. Pe'er, D. Wang, and A.A. Friesem, "Flatland optics III: Achromatic diffraction," *J. Opt. Soc. Am. A* **18**, 2095–2097 (2001).
5. E. Wolf, "A macroscopic theory of interference and diffraction of light from finite sources. II. Fields with a spectral range of arbitrary width," *Proc. Roy. Soc. London* **230**, 246–265 (1955).
6. L. Mandel and E. Wolf, "Optical Coherence and Quantum Optics," Cambridge University Press, Cambridge (1995).

Adolf W. Lohmann was born (1926) and raised in Northern Germany, close to Hamburg, where he studied Physics. As assistant professor he stayed at Braunschweig and at Stockholm. His main interest was then and still is "Optical Information Processing" (the title of his book). While at IBM in San Jose, California, he developed computer generated holograms (1961–1967). After six more years in California at the University of San Diego he returned to his home country, to the University at Erlangen. Following his official retirement in 1992 he spent about half of his time at various research facilities in USA, in Mexico and mainly in Israel. He was President of ICO (International Commision for Optics) in 1978–1981. He received the Max Born Medal of the Optical Society of America.

Avi Pe'er received his B.Sc. degree in physics and computer science from Tel-Aviv University, Israel (1996) and his M.Sc. degree in Physics from the Weizmann Institute of Science, Rehovot, Israel (1999). He is currently working on his Ph.D. thesis in Physics at the Weizmann Institute. The M.Sc. thesis was in the field of optical processing with totally incoherent light, and the Ph.D. thesis is in the field of nonlinear optics, focused on the special characteristics of broadband down converted light, and its applications. Mr. Pe'er is a student member of the Optical Society of America (OSA).

Asher A. Friesem received B.Sc. and Ph.D. degrees from the University of Michigan in 1958 and 1968, respectively. From 1958 to 1963 he was employed by Bell Aero Systems Company and Bendix Research Laboratories. From 1963 to 1969, at the University of Michigan's Institute's of Science and Technology, he conducted investigations in coherent optics, mainly in the areas of optical data processing and holography. From 1969 to 1973 he was principal research engineer in the Electro-Optics Center of Harris, Inc., performing research in the areas of optical memories and displays. In 1973 he joined the staff of the Weizmann Institute of Science, Israel, and was appointed Professor of Optical Sciences in 1977. He subsequently served as Department Head, Chairman of the Scientific Council, Chairman of the Professorial Council and Member of the Board of Governors of the Weizmann Institute of Science. In recent years his research activities have concentrated on new holographic concepts and applications, optical image processing, electro-optic devices, and new laser resonator configurations. He has served on numerous programs and advisory committees of national and international conferences. Among other posts, he currently serves as a Vice President of the International Commission of Optics (ICO) and is also Chairman of the Israel Laser and Electro-Optics Society. He is a fellow of OSA and of IEEE, a member of SPIE, and a member of Sigma Xi. Over the years he has been a Visiting Professor in Germany, Switzerland, France, and the U.S.A., has authored and coauthored 200 scientific papers, coeditor of three scientific volumes, and holds more than 25 international patents.

INDEX